MODERN STRUCTURAL DESIGN

A project primer for complex forms

ANDREW WATTS

Birkhäuser Newtecnic

CONTENTS

INTRODUCTION

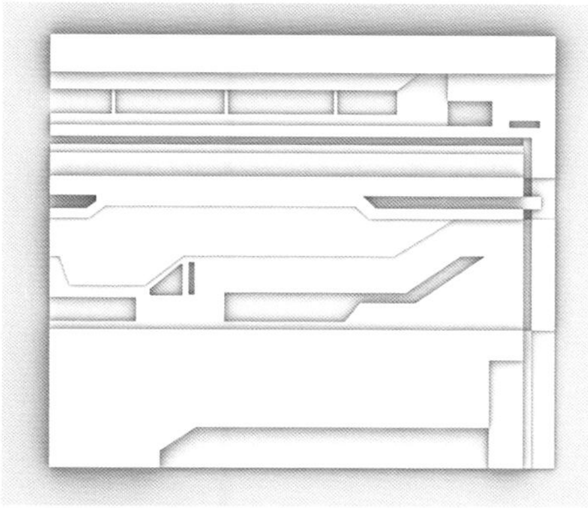
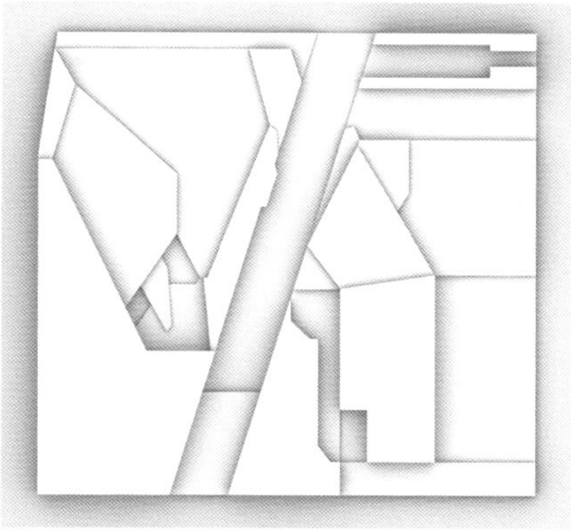

Examples of spatial and volumetric compositions used in projects, shown in low-relief. The compositions illustrate the use of overlaid volumes and spaces.

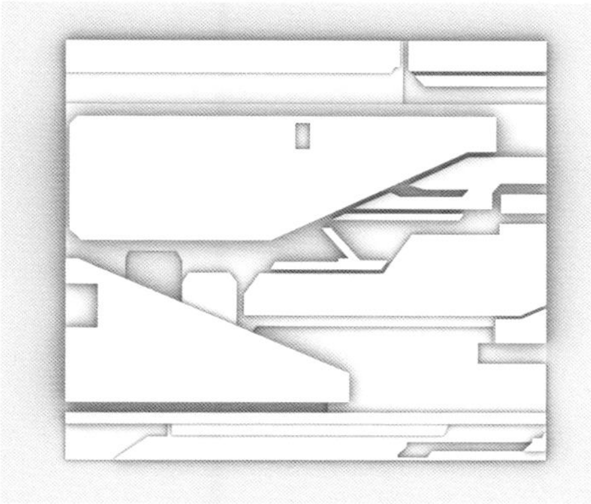
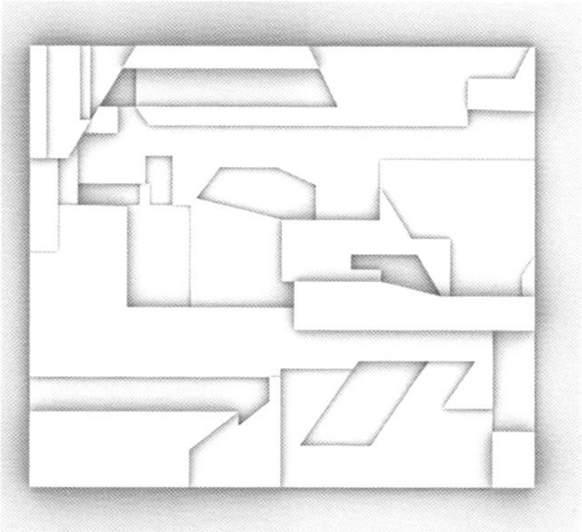
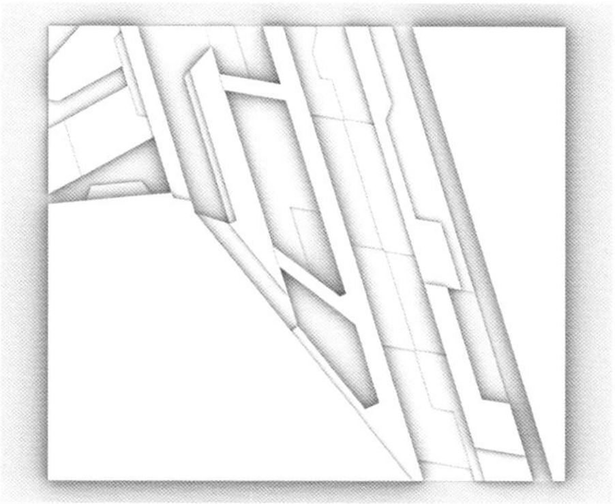

The example compositions shown here are independent of the strategies for structure, environment and spatial arrangements described for each project.

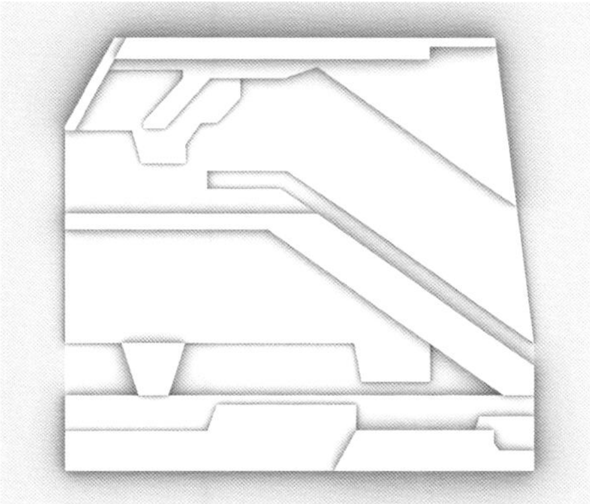
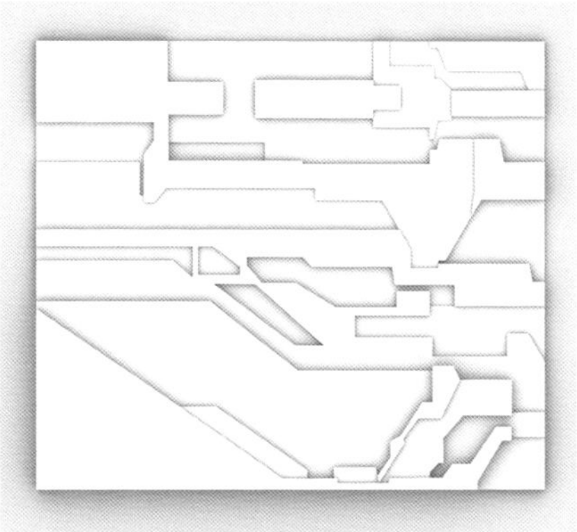
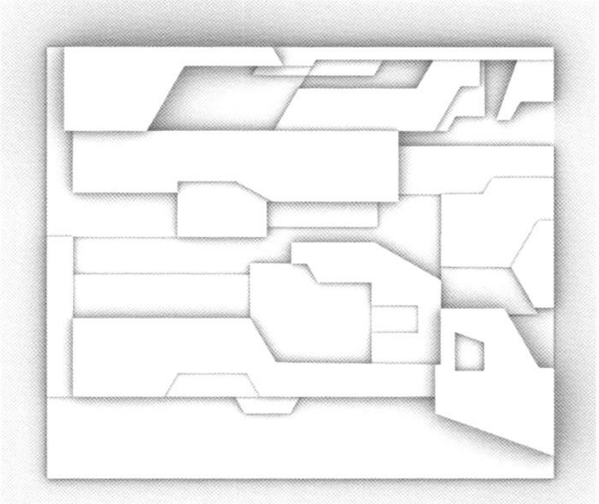

The compositions are independent of scale; being used across the full range of design tasks, from masterplanning to facade design and internal volumes.

Themes explored in this book

This book sets out 18 building design primers which explore the following themes:

- Linked structural forms for activity space and support space.
- Support of passive and active systems within facade zones.
- Floor plate structures which suit natural cross ventilation.
- Unity of exterior form and internal space to reduce secondary structure.
- External space between building forms, linked by bridges, in place of enclosed space.

The introductory texts to each project set out a specific theme which has been applied to the project. The themes are all generated around the principles of structures for naturally ventilated activity space, structures for support space which are either integrated into the facade zone or set adjacent to the activity zone, and how they might be made more efficient in order to reduce energy consumption during the construction and operation of buildings.

The designs make use of **linked structural forms for activity space and support space** as a result of the two spatial types being very different from one another. Activity spaces are designed to be as open as possible; support spaces make use of enclosing walls and floors. The structure supporting services, support spaces for staircases, lifts and services equipment are often separate from the structures used for the main spaces or for the activity spaces in the building.

Structures enable **the support of passive and active systems within facade zones** for natural ventilation, daylight control and solar control. Support for environmental controls are set within facade zones.

Structures are optimised for **floor plate structures which suit natural cross ventilation**. Floor plates occupy a width which suits the free passage of air from one side of the building to the other. Their structural design responds to the need to form very different enclosures in response to the environmental zones which are designed to be naturally ventilated, daylit and protected from solar gain. **A unity of exterior form and internal space reduces the need for secondary**

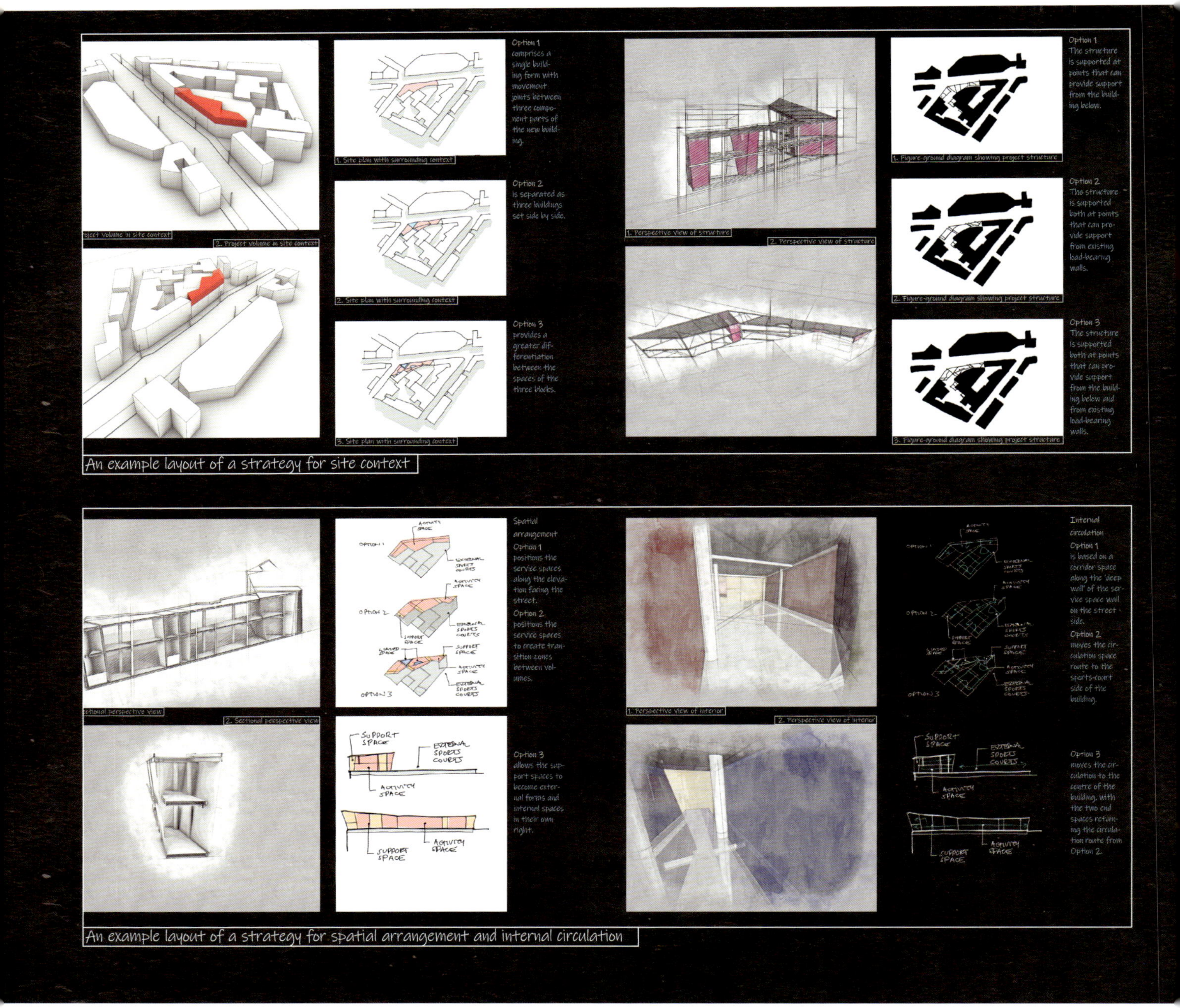

structure. Internal spaces are expressed externally as the same 3D forms in order to reduce the need for secondary support between the outer envelope and the inner space. **The external space between building forms is linked by bridges in place of enclosed space**. Bridges connect a larger number of smaller blocks as an alternative to a smaller number of large-sized blocks. This approach creates open spaces between blocks. The structural design of this spatial arrangement provides clear open spaces, with service space or support space being outside the activity spaces.

Strategies for building design at an early stage of a project

This primer for building design projects, like its companion volume *Modern Environmental Design*, sets out design strategies used in example buildings which have been designed by Newtecnic. The designs are theoretical versions, or generalised versions, of projects designed by Newtecnic. **Strategies have been used in the early stages of the designs shown in this book: site context, spatial arrangement, internal circulation, structure, environment and typical bays.** Each project demonstrates buildings of complex form which can be constructed from commercially available systems. The text for each strategy aims to illustrate themes from associated strategies for the same project; strategies are not 'siloed'. The general principles of these systems are set out in the *Modern Construction Handbook* (Birkhäuser, 2019) and *Modern Construction Envelopes* (Birkhäuser, 2019). These two books show how a technical understanding of systems can contribute to the creation of internal spaces and external forms, together with an environmental design which reduces, as much as possible, the energy consumption of buildings and well as the energy used in their construction. A reduced energy consumption includes an ability to recycle and reuse components and assemblies used in the construction of the building; this approach can result in building forms which look different from their predecessors as a result of the opportunities presented by construction technologies. This design approach does not begin from the point of view of providing technical

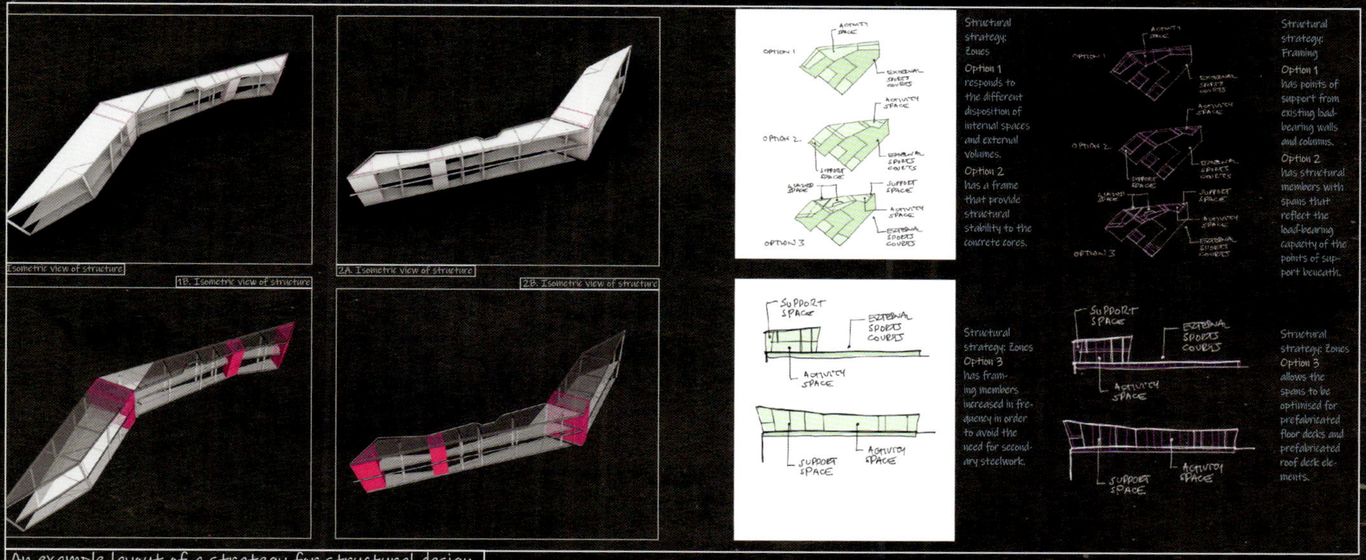

An example layout of a strategy for structural design

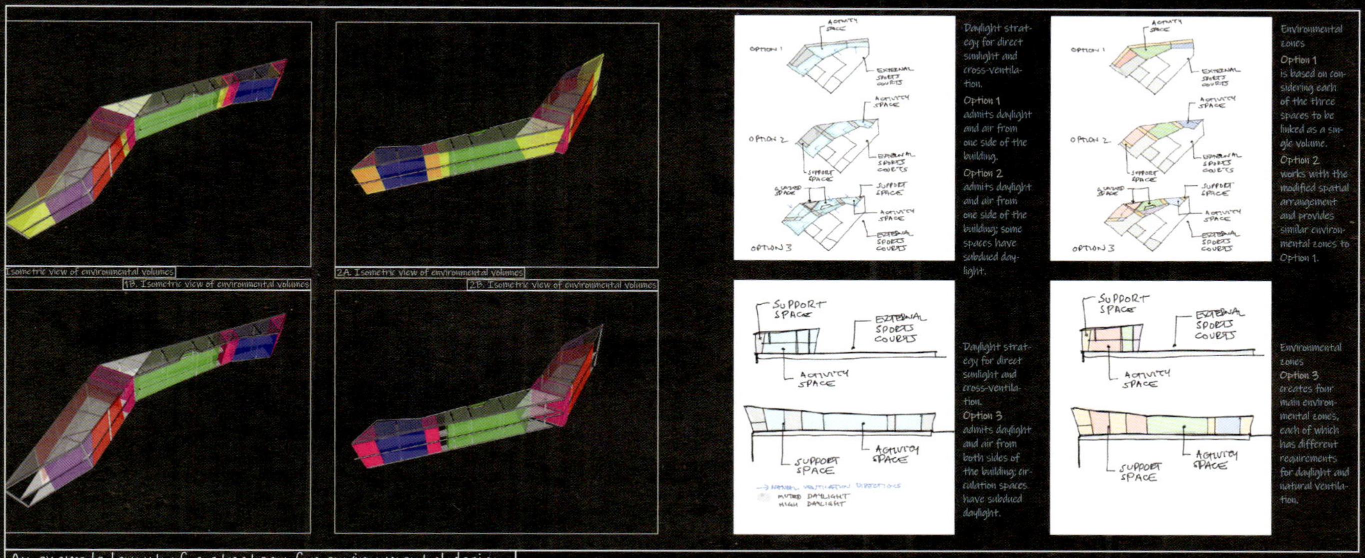

An example layout of a strategy for environmental design

solutions that enable novel building forms to be realised, but instead uses technologies to create space through structure, environment, external envelopes and internal envelopes of spaces. A key aspect of this design approach is to integrate internal wall finishes with external envelopes; deploying construction technologies used for structural design and for environmental design.

The design strategies shown in this book focus on buildability, the energy consumption of the building when in use, and how well the building fits within the surrounding cityscape or landscape. The project examples illustrate outputs at an **early stage** of a building design: how the structural strategy might respond to the other strategies of spatial arrangement, internal circulation, internal environment and typical bay design which are specific to the project. Each building design is explored through three options for each of the key aspects of **early stage** building design. An objective of each strategy is to optimise the materials used in the construction of a building and to optimise the energy required to operate the building when in use.

Design strategies

The structural strategy integrates issues from the other strategies to create an early-stage design which informs the spatial and environmental design. The structural design can partly determine the approach taken to environmental and envelope design. The strategies for each of the projects draw general principles from each aspect of the building design and do not set out the specific conditions of any project; nor does the text elaborate on the specific layouts or construction systems chosen. Instead, general principles and wider issues within each design draw together information of interest to the reader.

Qualifying comments

The information set out in this book is intended to be a source of general information, not to be used for specific application to any project. This book does not necessarily endorse or justify the strategies set out, since techniques and priorities in building design are in a continual state of change and development.

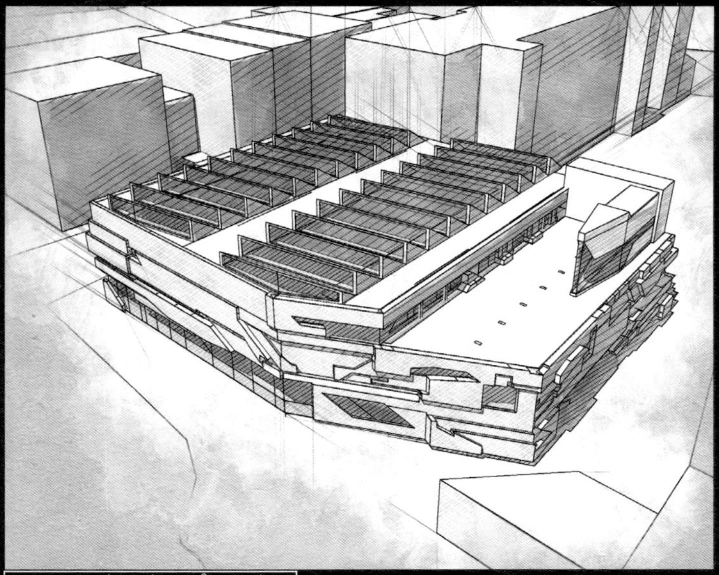

1. Perspective view of exterior

2. Perspective view of exterior

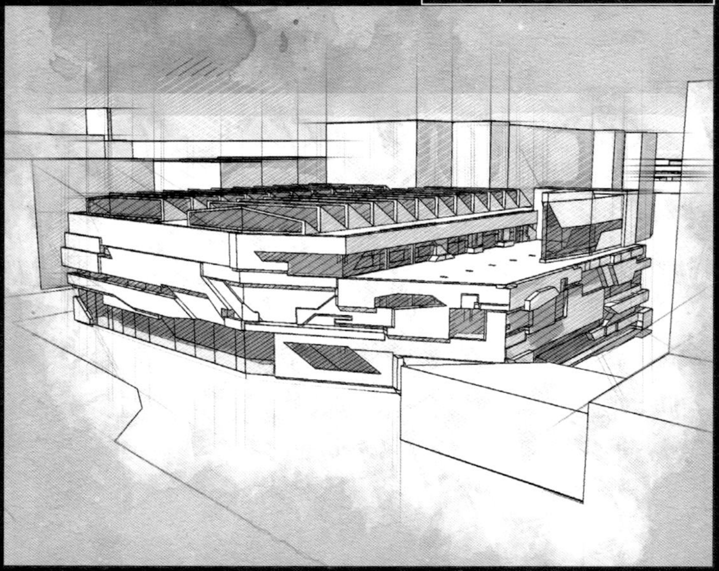

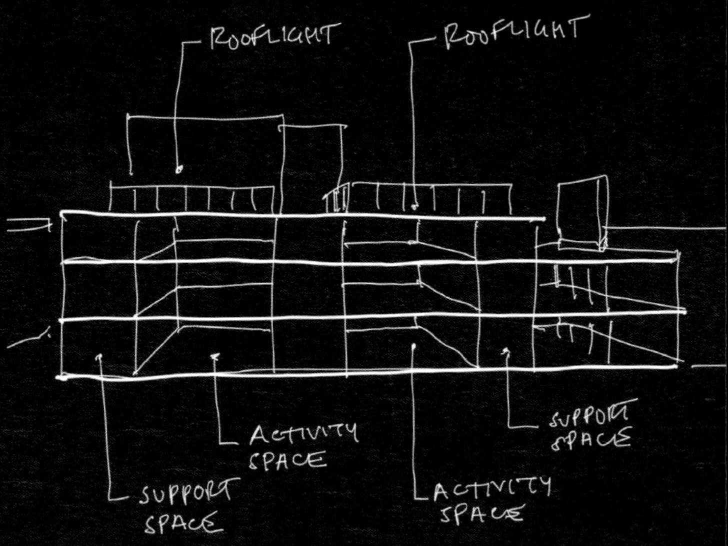

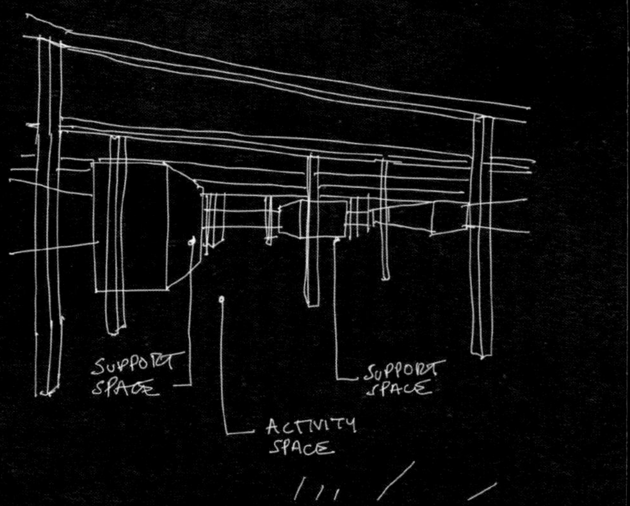

Mass customisation in the fabrication of structure

An aim of this project is to reveal as much of the structural design as possible in all scales of the building's fabric and assemblies: from the small scale of doors, to walls, to interior spaces, to the larger-scale volumes that comprise the overall form of the building. This project creates more than a building design as an architectural expression of a 'map of functions'; more than a design based on a diagram of the functions carried out in the building. Instead, the architectural expression of the building design is one of 'differentiation'; revealing the components and assemblies that comprise the construction. Elements are visually differentiated by their progressive stepping up of scale; from the scale of a door, to the wall that encloses it, to the space that encloses the walls, to the external forms that comprise the building.

Key elements of the building design relate to each other visibly as a result of their juxtaposition; a door and its frame set into an interior wall with its own edges and recesses, for example. Elements of the design are juxtaposed rather than blended together in order to allow each assembly, such as a door or an area of external wall, to include more functions than are typically the case when the door is integrated into the language and technology used for the adjacent external wall. Some projects benefit from the use of a single system of facade: for example, to enclose a building; other projects benefit from the approach described here of juxtaposing doors, interior walls, and so on in the interests of allowing each assembly to be optimised for its own performance, irrespective of the performance of adjacent assemblies. This latter approach is used on this project and has the benefit of optimising the energy used in the construction of the building by reducing the quantities of material used in its fabrication; allowing some buildings to be constructed using technologies that are local to the project. This thinking brings the benefit of encouraging local companies to participate in their realisation. The alternative approach is to bring in large construction firms from outside the sphere of influence of the project. The approach described here is achieved through the use of mass customisation, which provides a high degree of differentiation between

3. Perspective view of exterior

4. Perspective view of exterior

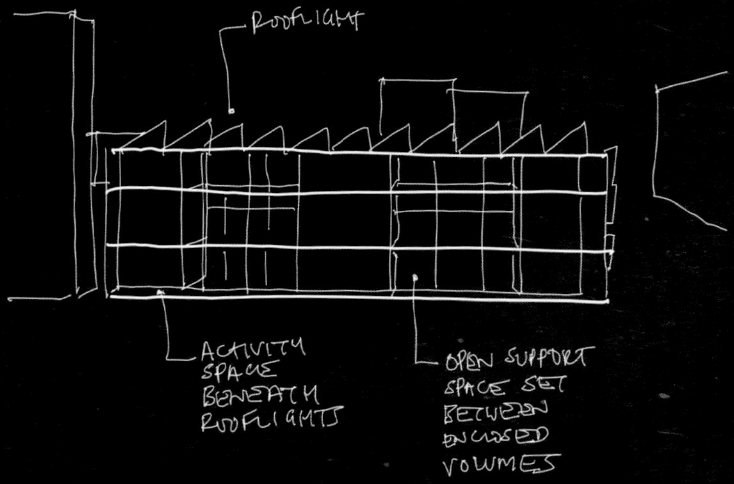

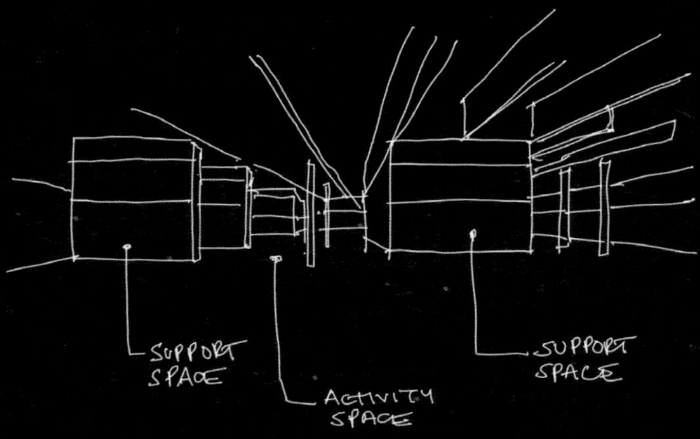

elements that comprise the structural design. The approach of mass customisation used in this design was based on using readily available steel sections and pressed steel profiles; combining them to create a combined primary and secondary structure. This method provides the advantage of integrating points of support for facades and roofs as well as for framing to support interior walls and ceilings within the building. The design uses the customary juxtaposition of assemblies provided by different suppliers, but integrates them by anticipating what savings can be made, both in time and cost, by combining their different requirements. This is effectively a project designed as a kit of parts from different suppliers which is brought together as the result of a structural assembly in which the fixing points are already provided. This approach reduces the need for coordination during the construction stage; bringing coordination back into the design stage where possible. The assembly technology specific to this project uses a suite of connectors for pin joints and moment connectors that allow them to be fabricated off site for rapid installation on site. Components are made off site, with a minimum of connections being formed on site as a result of the fixings being optimised in their fabrication with that requirement in mind. Connections make use of steel shelf angles, for example, to ensure that beams and columns can be bolted in place in a way which reduces the depth of floor buildup; bringing the benefit of a smaller area of facades being needed as a result of the reduced floor-to-floor height. Structural components were optimised for their fabrication off site, making it easier to assemble components on site.

These components were further optimised for fabrication by reducing the range of material sizes that coordinated with a limited range of connector types and sizes. A key approach was to provide connections in the steel structure by integrating support brackets, pre-drilled holes and welds in the fabrication shop and not on site. The main benefit of this approach was to reduce costs, construction time and material waste by focusing on the design of a structure that integrated anticipated installation requirements of adjacent systems including facades, roofs, interior walls and environmental control systems.

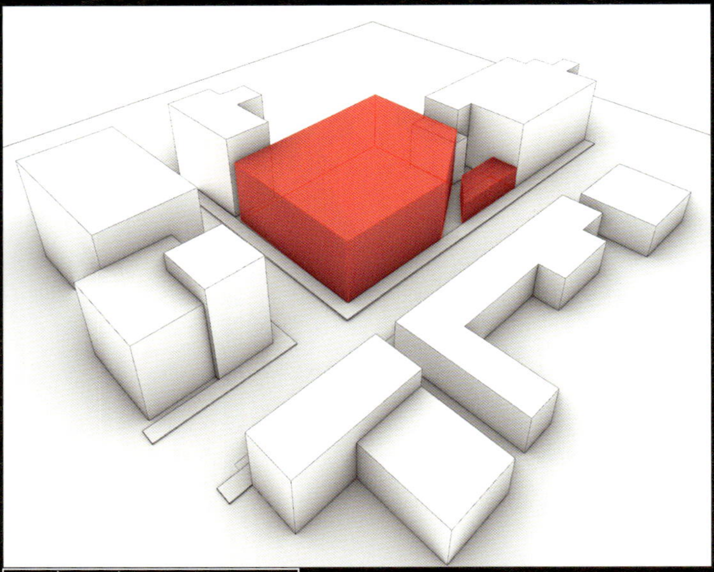

1. Project volume in site context

2. Project volume in site context

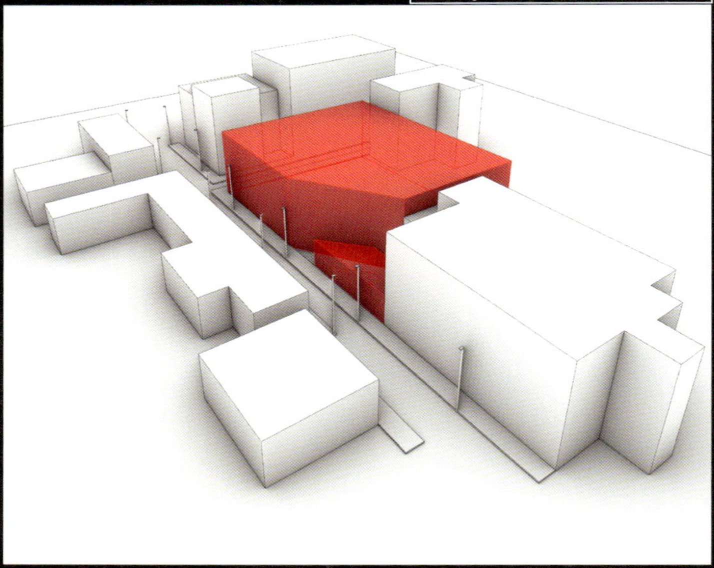

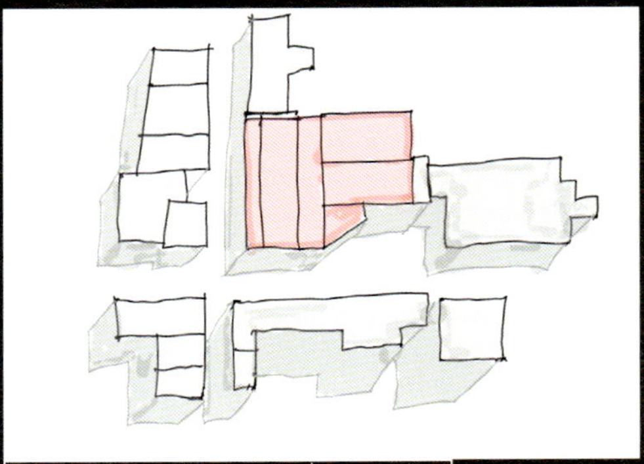

1. Site plan with surrounding context

Option 1 creates a rectangular space which can be considered to be an 'outdoor room'.

2. Site plan with surrounding context

Option 2 adds a small building volume on the street and partly encloses the external space, which now becomes a courtyard.

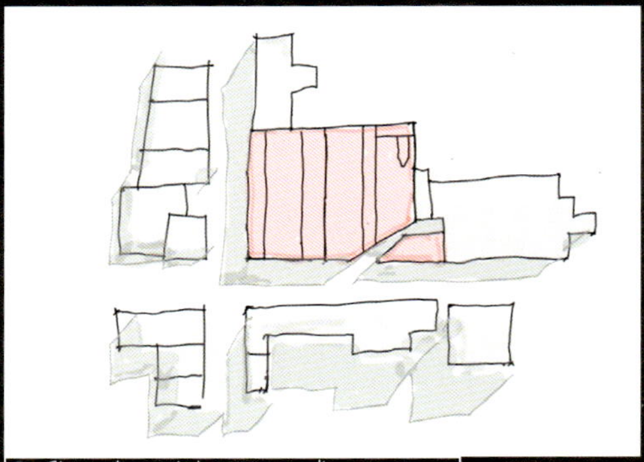

3. Site plan with surrounding context

Option 3 forms an angled entrance slot between the two building volumes that face onto the street - enclosing the courtyard.

Site Context

This project is for an e-sports gaming venue in an urban location. The climate is generally warm and dry throughout the year but the interior requires control of strong daylight and the effects of the sun. The new building is positioned on the site in a way that allows old and new to be both linked and aligned with the existing streetscape rather than be set apart from the existing urban fabric. The new building has additional elements that sit on top of the immediately adjacent existing buildings, which are owned by the same organisation. For all three options shown here, the building is arranged in order to create the entrance space between an existing building and the new building. This approach allows external spaces to be created between old and new urban fabric; linking spaces which are sheltered from the full effects of the sun, located at both street level and roof terrace level.

A clear spatial relationship between the 'activity' spaces, or main spaces, and the 'support' spaces, or service spaces, was achieved through a high degree of prefabrication of building components and assemblies. Prefabrication has the benefit of reducing the environmental impact of a large-scale set of forms and spaces by reducing the time required for construction. It was considered essential that neighbouring businesses and residents should not experience a significant reduction of amenity and utility in their daily lives around the construction site. A high degree of prefabrication was planned for structural components, for the external envelope (facades and roofs) as well as parts of the support spaces, including forms set onto the top of existing structures. The resulting approach is a mixture of prefabricated building assemblies delivered by truck and semi-prefabricated structural and interior based components. Prefabricated modules of mechanical and electrical services are used where possible. The relatively small scale of the building lent itself to a high degree of prefabrication; an approach usually justified in terms of cost for larger buildings only.

Option 1 arranges the building as rectangular forms which are positioned side by side. The largest box-like volume is set out to align with the streets which enclose the site. One end of this form abuts a neigh-

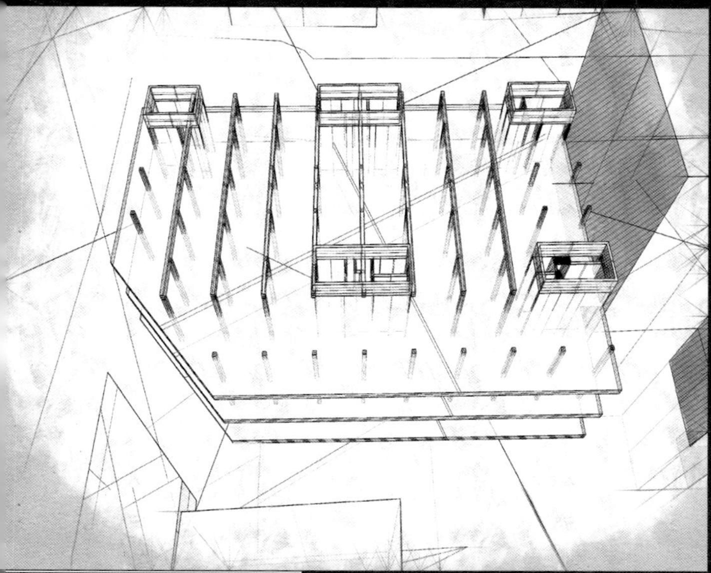

1. Perspective view of structure

2. Perspective view of structure

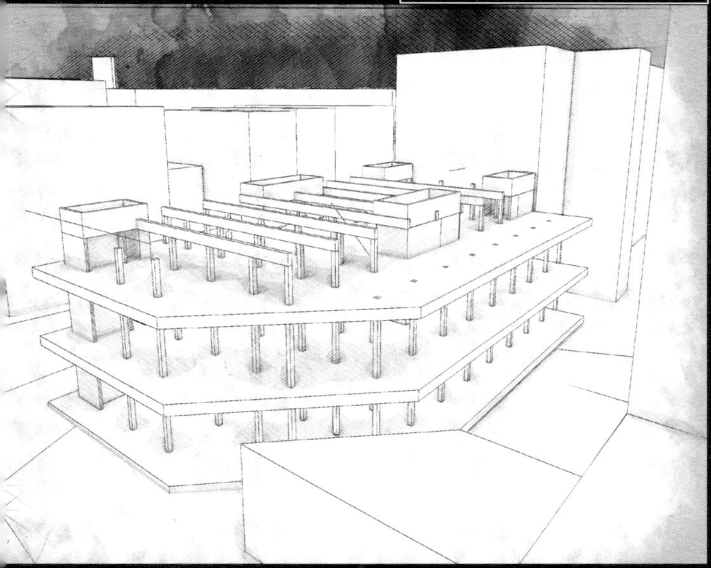

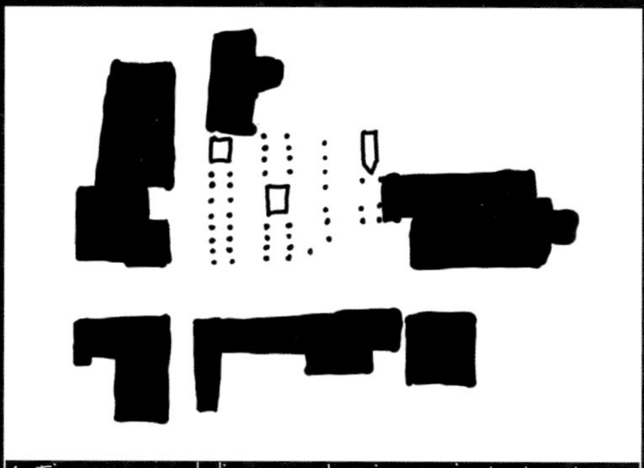

1. Figure-ground diagram showing project structure

Option 1
The structure continues the rectilinear pattern of the street. The column grid is visible through the glazed openings in hours of darkness.

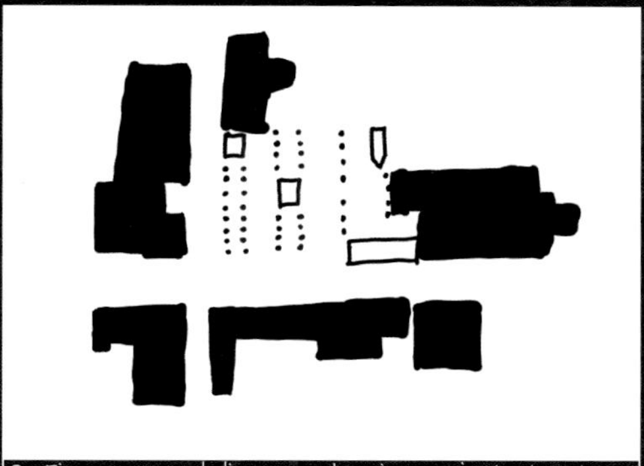

2. Figure-ground diagram showing project structure

Option 2
The column grid of the small block added to the structure can be seen as being part of the adjacent building.

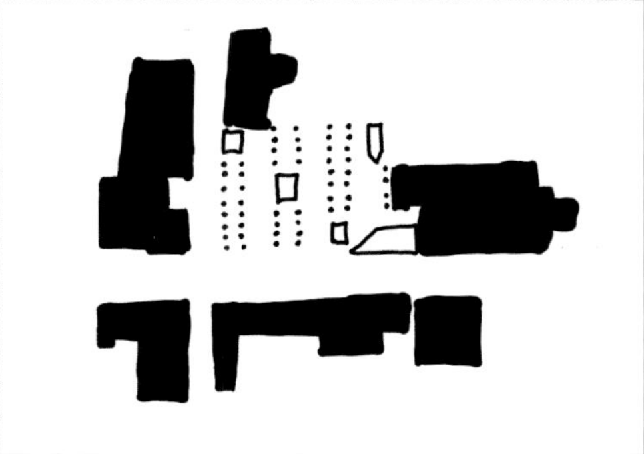

3. Figure-ground diagram showing project structure

Option 3
The new addition is seen as a separate structure and form from the new, larger building.

bouring building; the other end has a facade that recreates a street frontage that has been absent in recent years.

A smaller rectilinear form is set alongside the larger one containing a circulation space, expressed as an independent volume in order to ensure that the overall massing of the e-sports centre does not resemble a single form that would be out of scale with neighbouring buildings. This part of the city is well established and does not suit interventions of buildings which are larger than those already constructed. This approach also allows the structure to be assembled on site in closely linked phases, which has the benefit of minimising any disruption to neighbouring businesses during its construction.

A third volume forms a link building between the circulation space and the existing building on one edge of the site. The arrangement of spaces in Option 1 leaves a rectangular space which can be considered to be an 'outdoor room'. This external space did not form an easy relationship with the streetscape, leaving a gap between storefronts rather than a four-sided outdoor room. Option 2 addresses this issue by adding a small building volume on the street and partly enclosing the external space, which now becomes a courtyard. One of the corners of the large rectangular volume had a chamfer introduced to one corner that created an entrance from the street within its own triangular-shaped external space. This solution also resulted in an awkward external space which was neither public nor private; neither a part of the street nor a part of the e-sports centre.

This contradiction is resolved in Option 3 by forming an angled entrance slot between the two building volumes that face onto the street - enclosing the courtyard. This approach provided a set of separate structures which can be assembled in a sequence to suit the owner. The separate structures benefit from the smaller scale elements required structurally. As a result of the use of small-scale structural forms, formed as a juxtaposed group, an additional volume was added to the roof of the main building. This volume contributes to the urban forms of neighbouring buildings which comprise a mix of different scale; this rooftop enclosure adds to the local set of small-scale buildings.

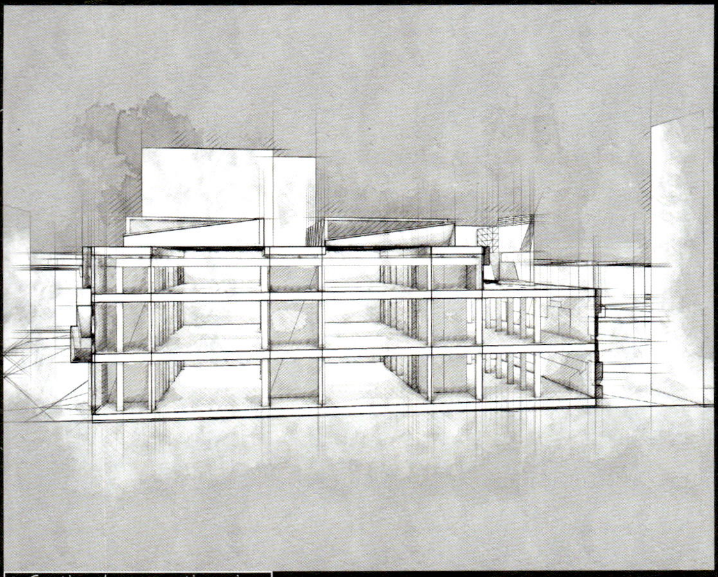

1. Sectional perspective view

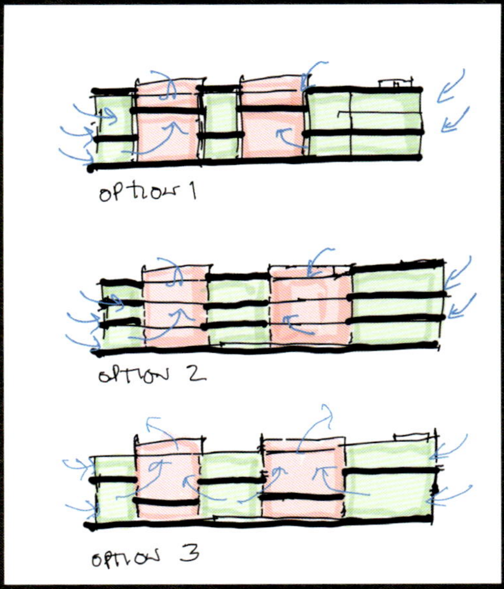

Spatial arrangement
Option 1 reveals the volumes externally as a set of juxtaposed facades.
Option 2 uses a regular grid of columns and beams to create continuous floor plates.

2. Sectional perspective view

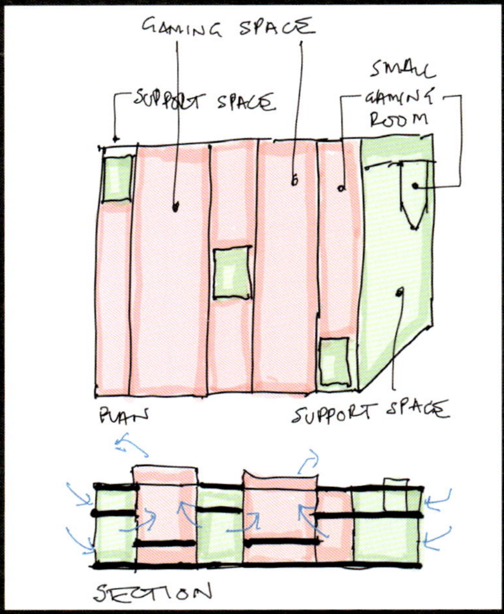

Option 3 has floor plates supported by rows of columns and cores which form spines down the long centre of the building.

Spatial arrangement

The spatial arrangement for all options is a response to changing daylight throughout the day, sunlight through the year and prevailing wind directions in different seasons. The three options shown here are for different ways of introducing daylight and natural ventilation within this spatial arrangement. All three options have the same floor area allocated for each function, but the space is subdivided and distributed differently in each case.

Option 1 shows a set of spaces formed as separate volumes. The volumes are revealed externally as a set of juxtaposed facades; volumes are revealed internally by internal walls that divide spaces to form a set of large-size rooms. Each room gives no indication that the overall form of the building is subdivided with lightweight partitions. This solution results in an irregular grid of beams but a regular grid of columns. Circulation through spaces is one of a transition from room to room, with staircases and lifts/elevators being set by the organisational strategy. Each activity space is ventilated naturally through the facade; every such space has an external wall. Each space is mechanically ventilated separately as spaces are used intermittently. Daylight is admitted through the facades on the two lower levels and through the rooflights at the top level. Option 2 uses a regular grid of columns and beams to create continuous floor plates onto which partitions can be arranged, largely free of structural constraints.

This provides a regular set of columns on the facades; an additional secondary structure is required to support the facades, which do not have an arrangement which matches that of the regular structure. Spaces are lit generally from the facades in response to the need for flexibility in the use of space. The building is also ventilated naturally through opening vents on all facades. Each activity zone has its own mechanical ventilation provision, with support spaces having air supplied from the two main spaces.

Option 3 is arranged with two zones for gaming on each of three floors, separated by cores which form spines down the long centre of the building. There are two additional spine cores along the outer edges of

. Perspective view of interior

2. Perspective view of interior

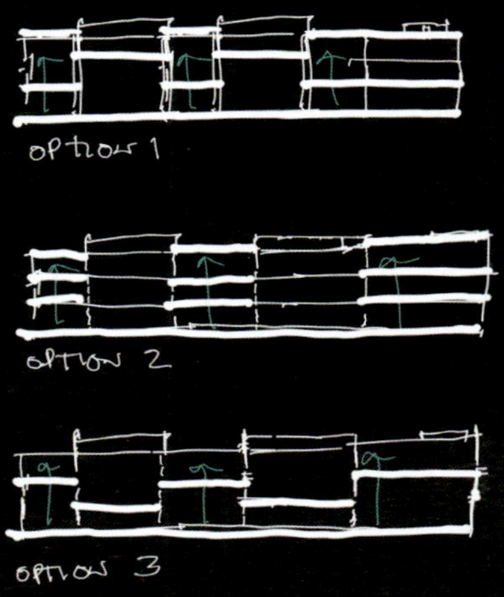

Internal circulation

Option 1 arranges circulation from a single entrance at one corner of building.

Option 2 arranges circulation from entrances at each end of the building.

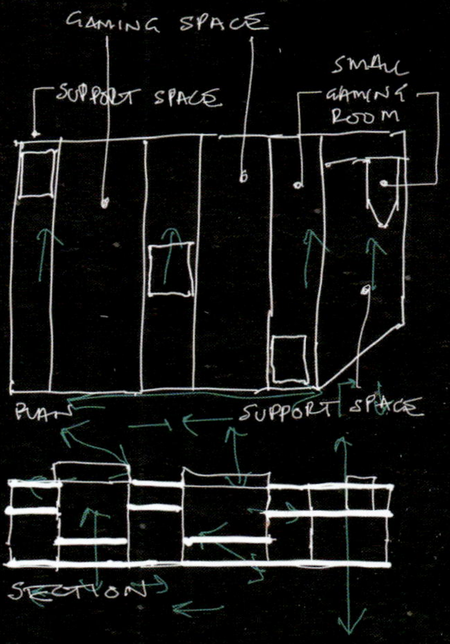

Option 3 arranges circulation from a single entrance at the centre of the building.

the building, on the long sides of the building; the two shorter ends are fully glazed but solar shaded to allow in different amounts of daylight to suit the changing activity within. There are two additional cores along the outer edges of the building, on the long sides of the building; the two shorter ends are fully glazed but solar shading allows different amounts of daylight to suit the changing activity within.

There are two gaming zones on each of the three floors to suit current expectations of the popularity of playing a specific game in a given-format of furniture. Zone 1 accommodates rows of tables in a chess competition-style arrangement. Zone 2 has tables arranged as islands. Zone 3, on the top floor, is arranged in the form of an amphitheatre, while Zone 4 has larger rooms where the gaming tables are arranged in a circle.

Internal circulation

Primary access up and down the building for all options is provided by stairs located at both ends of a central circulation spine. The stairs rise from ground floor to a lobby in each of the upper floors. The third floor accommodates an amphitheatre space which is positioned in order to provide a sense of arrival at a 'terminus space' after moving past the sequence of spaces below. The spine cores set on the long sides of the building are arranged as a sequence of open spaces; gamers pass through the central axis of the building; a glazed corridor that provides interaction with other gamers; a requirement of this gaming community. In terms of sequence of movement, upon entering the building and passing the reception area, gamers are in a larger open area and traverse this area through openings in the central support spine of the building to spaces behind.

The building is divided into activity areas and support areas; the activity areas being those primarily for gaming, and support areas being primarily staircases, restrooms and a cafe. The emphasis is on creating as many links as possible between people; gamers are encouraged to traverse the glazed corridor link spaces, allowing interaction between games and individuals where and when needed.

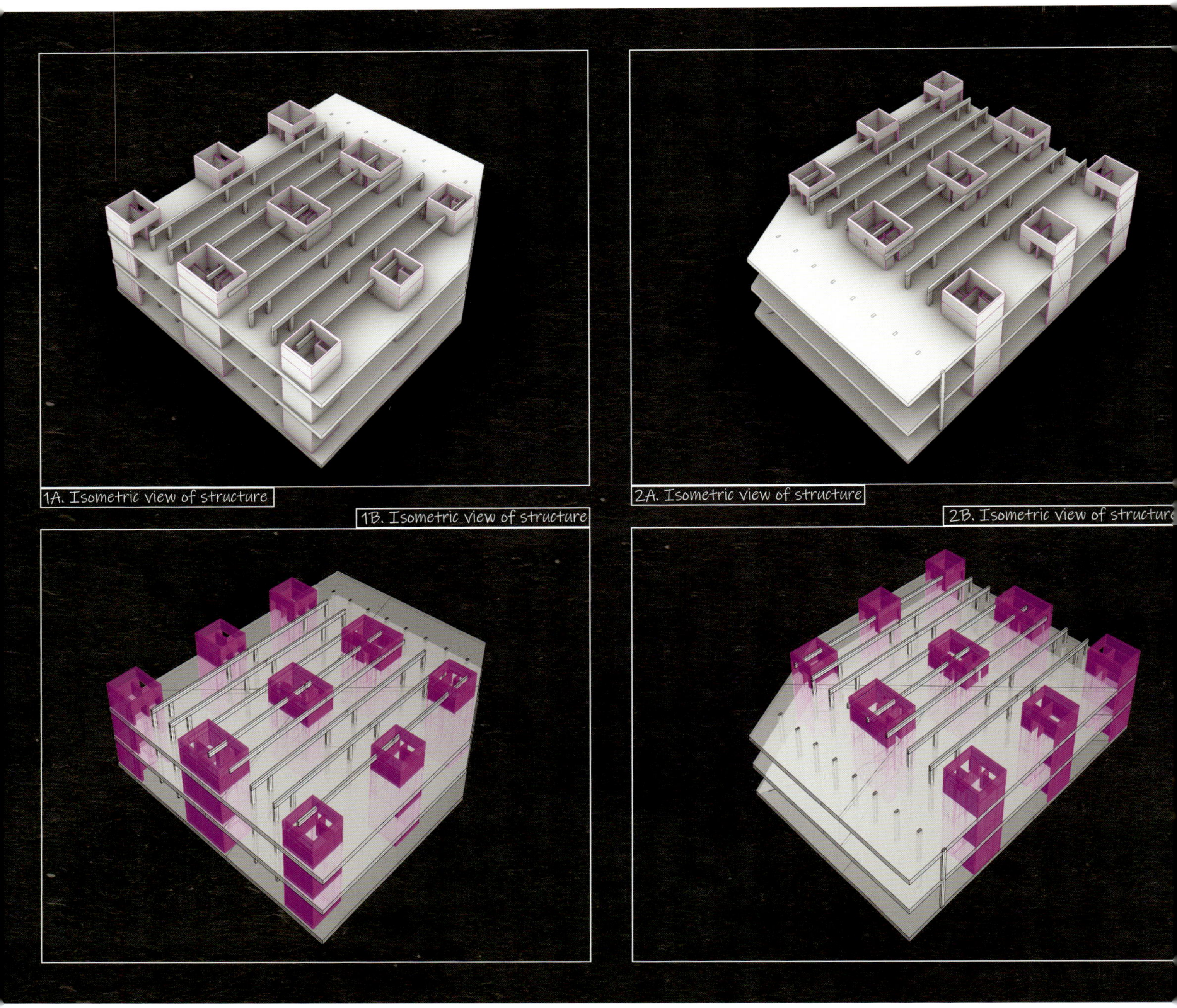
1A. Isometric view of structure
1B. Isometric view of structure
2A. Isometric view of structure
2B. Isometric view of structur

Structural strategy

The structural concept is to add new floor spaces and new circulation spaces alongside the existing structures to avoid the need for demolition of any of the existing building, which would create an environmental hazard. The new structures are set adjacent to existing structures; demolition of existing structures is avoided. New structures are designed to be as light in weight as possible in order to reduce the loads on existing structures, as well as avoid the need for strengthening existing structures.

The new structures are as independent of existing structures as possible in order to have more freedom to be optimised to suit their own requirements of long spans, to create internal space which is as clear as possible of columns and load-bearing walls. The building is conceived to have a traditional workshop-style atmosphere; a steel frame structure was selected to suit that aim visually and economically. The structure of all options comprises hot-rolled steel columns and beams, with pressed steel joists to form floor decks; a timber-look floor finish is used for all levels including external areas. Acoustic insulation from floor to floor was not considered to be a key performance requirement, as ambient noise is considered to be part of the experience of visiting the venue.

Facades are non-load-bearing cladding panels fixed to the edges of floor beams where they interface with primary structure. Internal walls of the building are lined in dark panelling to provide the required atmospheres of muted daylight and soft electrical light during hours of darkness. The dark walls of the interior contrasts with the facades as seen from the outside; facades have their depth modelled in order to create the required visual identity.

The structural strategy was to provide a series of juxtaposed small-scale steel-framed forms, set together in a way that makes their interaction and their interplay a key visual aspect of the building's design. The structural design juxtaposes a series of frames of mixed form and construction that results from the interpretation of the spatial arrangement being used for a set of relatively small spaces of different charac-

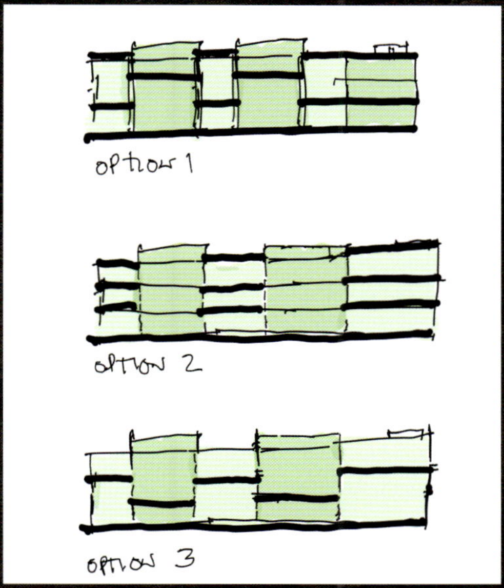

Structural strategy: Zones

Option 1 has a stepped section created by floor plates supported adjacent to cores.

Option 2 has a set of openings between floor plates. The column layout is as Option 1.

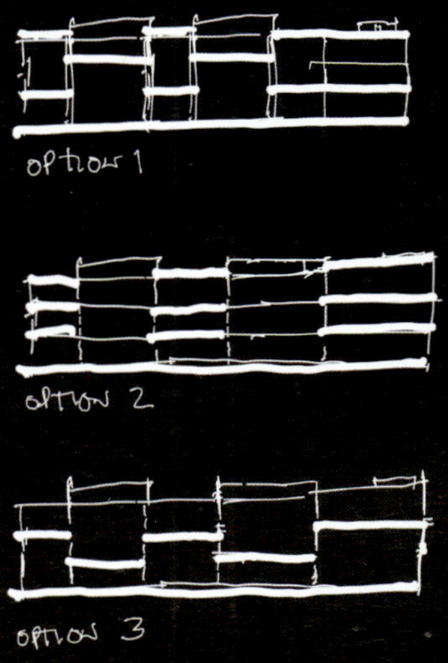

Structural strategy: Framing

Option 1 is a concrete frame with concrete core walls.

Option 2 is largely a steel frame.

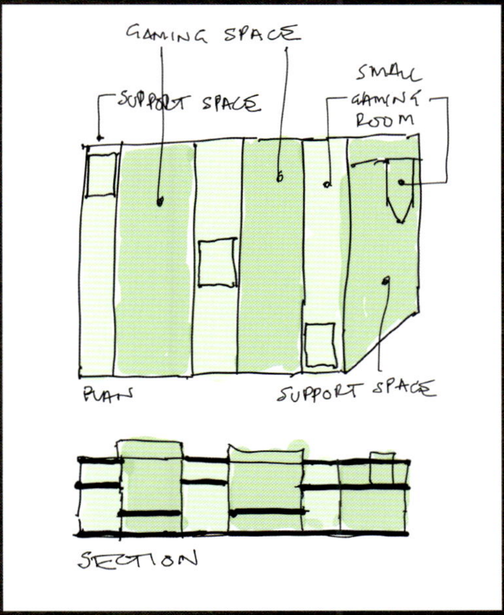

Structural strategy: Zones

Option 3 has a stepped section with fewer floor plates than Options 1 or 2.

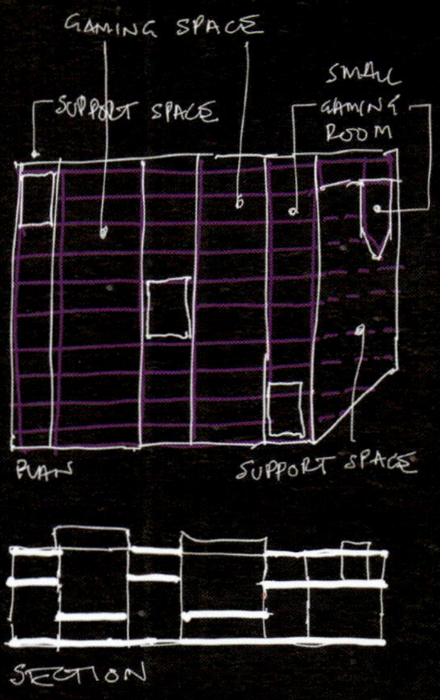

Structural strategy: Zones

Option 3 is a mixed solution of reinforced concrete and steel.

ter that are set together to form a single venue. The spatial arrangement is derived from the program/brief developed with the client as the early stage of the design progressed. The primary structural frame comprises hot rolled steel sections forming a frame with a grid of 6.0 metres by 6.0 metres across the open gaming areas.The total width of the building is and length of the building is not a multiple of these dimensions; the cores are of a size that allows this grid to fit within the site.
The steelwork is exposed to view and is fire protected with an intumescent coating.
The cores, which are braced with steel sections, form part of the structure. The ground floor deck is formed as a reinforced concrete slab with a timber-look floor finish. The roof is formed as a series of vertically set trusses which are linked with inclined steel sections to create a 'sawtooth' configuration. The vertical trusses are clad in glass, while the inclined steelwork is clad in an opaque roof of standing seam metal with thermally insulated panels set beneath them. The trusses are conceived as a series of light-gauge steel members, forming a series of Vierendeel trusses that span across each of the 6.0-metre wide gaming areas from facade to central spine wall. A Vierendeel truss configuration is used in order to coordinate with the rectangular glazing panels. The vertical face of each truss is glazed with sealed units; natural ventilation for winter and mid-season is provided by 'chimney' units fixed through the opaque inclined panels.
The service cores that form a spine down the centre of the building are prefabricated in response to the densely occupied site context. Modules are storey-height, 2.5 metres wide and 6.0 metres long, allowing them to be transported by road on a trailer. Additional finishes applied to the modules bring the width of core elements to 3.0 metres.
Floor decks and roof trusses are made in modules of the same size in both length and width. The structure is assembled as a mixture of site-based assembly and installation, and a prefabricated kit of parts. Consequently, the time-based advantages of quickly installed modules will not be lost as a result of adjacent site-based tasks, as the courtyard space can be used as a lay-down space for partial assembly.

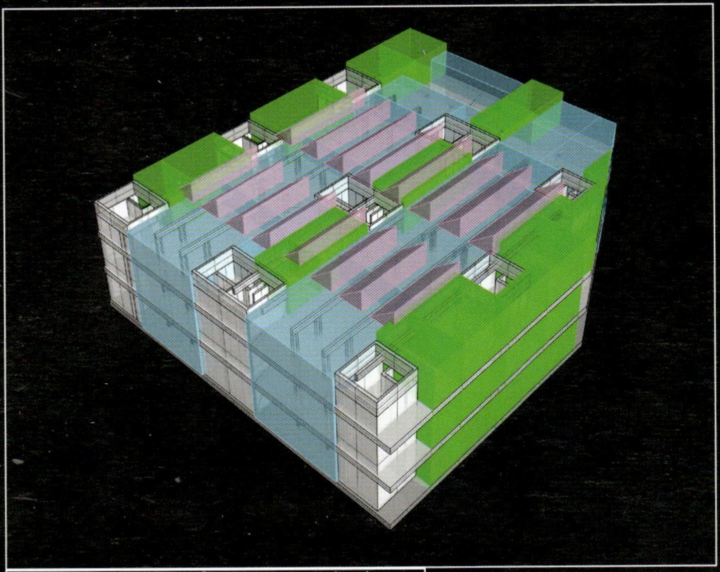

1A. Isometric view of environmental volumes

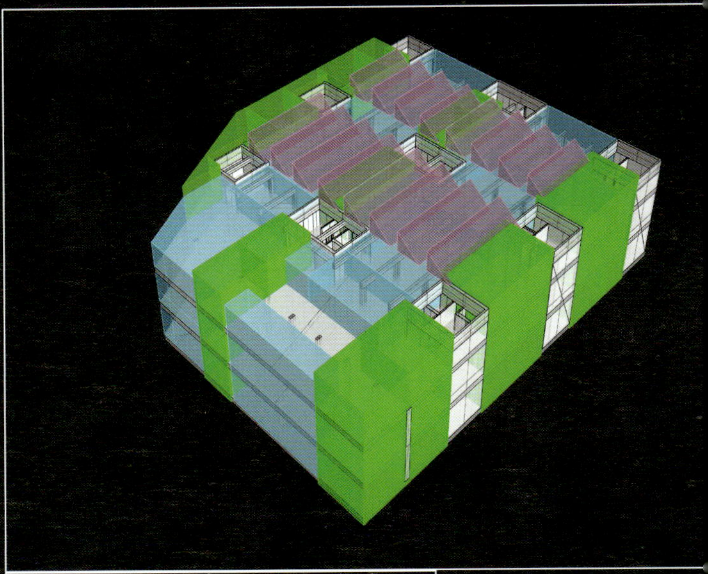

2A. Isometric view of environmental volumes

1B. Isometric view of environmental volumes

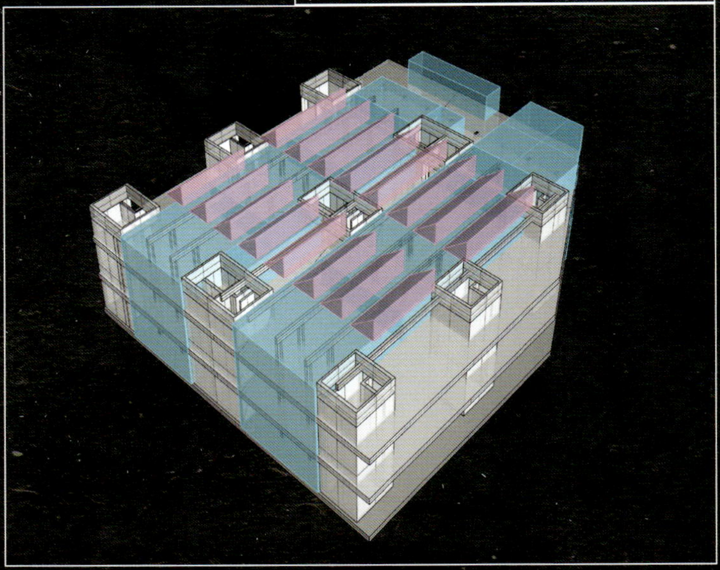

2B. Isometric view of environmental volume

Environmental strategy

One requirement of the program/brief was not to rely on air conditioning as the only means of providing fresh air to the building, as the environmental cost was considered to be too high; this climate has traditionally been preferred for its ambient outside conditions. This approach has the benefit of bringing a specific atmosphere to a range of entertainment-based spaces for use both during the daytime and at night-time as a result of its limited heating, cooling and humidity control. The use of natural ventilation, rather than mechanical ventilation and air conditioning systems, is made possible by the passive solar shading of the facades and by night-time ventilation from the operable rooflights. The sawtooth arrangement of the roof provides north light to admit natural daylight, reducing the solar gain and glare associated with this climate at different times of the year. The inclined outer plane of the sawtooth roof allows the possibility of photovoltaic panels to be installed on the outer face of the roof in order to provide power for electrical lighting and the natural ventilation controls, including fans, during winter and the mid-season. The inner spine of the building is supplied by cooled air ejected from the circulation spaces through the two main gaming areas on each floor, and the auditorium and smaller gaming areas on the upper floor. Environmental control assemblies and electrical lighting are also assembled in the workshop and are fixed to the modules during their construction.

Option 1 places the service equipment in cores containing stairs and restrooms. Activity spaces are considered to form a single environmental zone on each floor level, as a high degree of flexibility of use is required. Volumes are clustered together to suit the requirements of mechanical ventilation for large, open floor plates on all floor levels. A limiting factor in this design is the lack of natural ventilation resulting from the arrangement of service spaces.

Option 2 develops a strategy for natural ventilation by integrating, environmentally, the preference for smaller, linked internal spaces of building form both in the spatial arrangement and the massing of forms. A services 'spine' is introduced into the rectangular volume set between

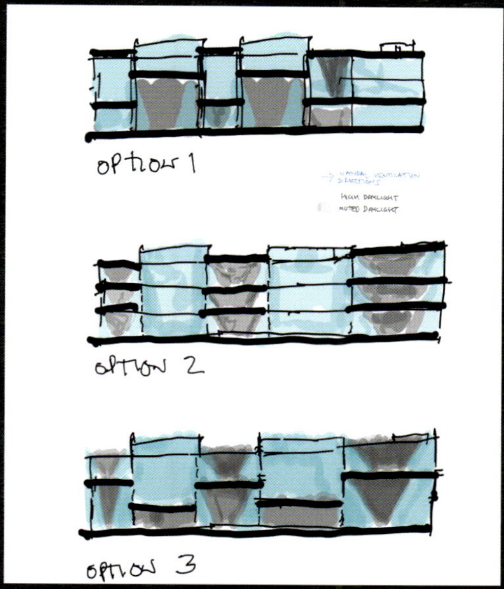

Daylight strategy for direct sunlight and cross ventilation.
Option 1 admits daylight and air to lower floors with a stepped section.
Option 2 admits daylight and air to lower floors with voids between floor plates.

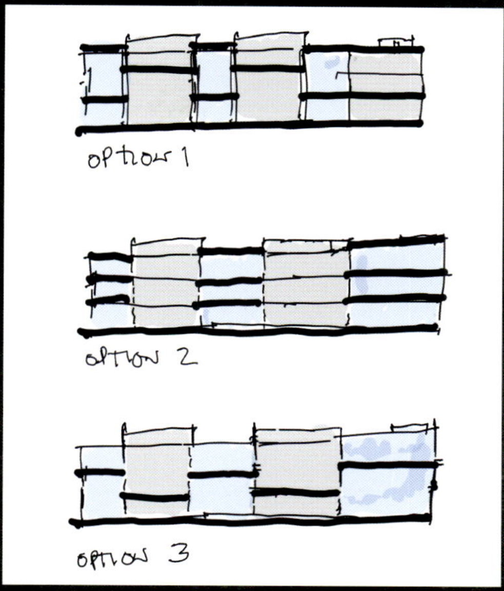

Environmental zones
Option 1 places the service equipment in cores containing stairs and restrooms. Activity spaces are considered to form a single environmental zone on each floor level.
Option 2 develops a strategy for natural ventilation by integrating, environmentally, the preference for smaller, linked internal spaces.

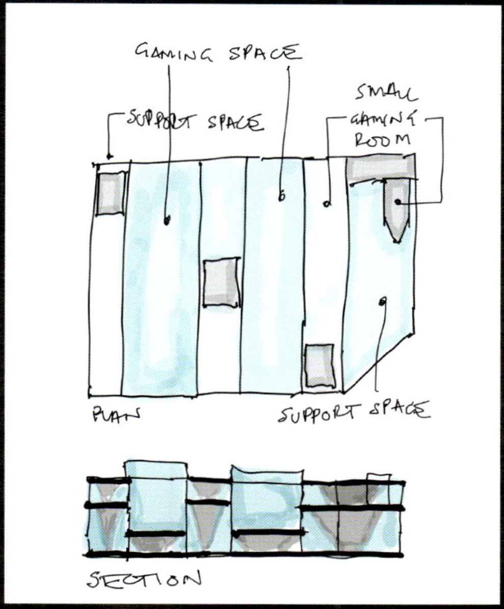

Daylight strategy for direct sunlight and cross-ventilation.
Option 3 admits daylight and air to lower floors with a stepped section with fewer floors than Option 1.

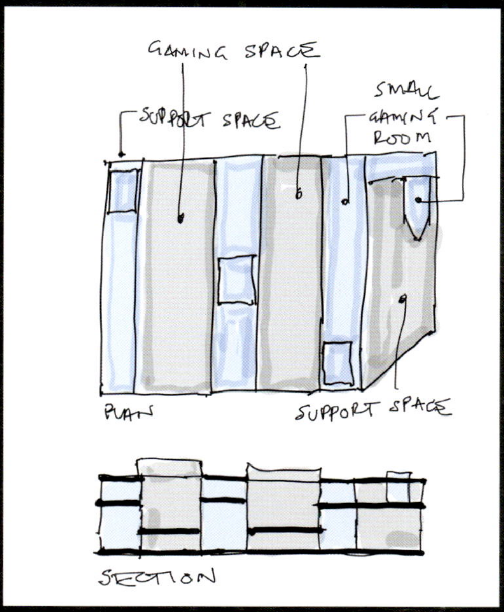

Environmental zones
Option 3 has an increased differentiation between internal volumes and an outside room.

the large rectangular volume and the link block. This spine is set lower than the roof level of the large rectangular block, allowing cross ventilation at that upper level. The top of the link block is set at a height lower than that of the spine, allowing all three constituent spatial volumes to be cross ventilated. The link block is ventilated with opening windows on the courtyard on one side of its external wall, to the opposite external wall at the rear of the site.

The introduction of the courtyard allows the link block to have natural ventilation and adequate daylight without experiencing significant levels of traffic noise from the street. The building form facing the street is ventilated naturally on the courtyard side, but can also be mechanically ventilated from the street side, to suit the time of day and season of the year.

The courtyard in Option 2 provides an 'outdoor room' experience that is open to the street in one corner; it is well shaded from direct sunlight for most of the day, in both the mid-season and in winter. Fabric canopies are deployed in the courtyard during summer months to allow this external space to be used to greater advantage. The approach of increased differentiation between internal volumes and an outside room is further developed in Option 3, where a pointed-shaped volume is added to the spatial composition. The additional small volume enclosing the courtyard – added to the design in Options 2 and 3 – is supplied by air-handling ducts that are located at high level between the two sets of forms, above the entrance. Although this small volume is ventilated on one side only, it is arranged over two floors in order to function independently of the other parts of the project. The space set at ground level has doors opening out onto the courtyard; the opposite facade facing the street has ventilation slots for a mixed mode of mechanical and natural ventilation. The upper floor is a significantly taller space, allowing natural ventilation at high level, where warm air generated within the building is drawn up and exhausted from the interior. The introduction of the angled ground-level entrance slot provides greater environmental control within the courtyard space, which benefits all parts of the building which it serves.

Typical bay adjacent to facade

Facades are made from metal and glass panels in response to the high level of differentiation required across the elevations. Each elevation is divided into a set of juxtaposed elements that reflect the functions going on behind; each facade is a billboard for the activity it encloses.

Each of the support spaces immediately adjacent to the facades is expressed as a set of glazed and opaque boxes which are stacked one above the other to reflect the functions within, rather than being given a single visual expression through their height as is more commonly the case. Areas of facade are subdivided as a set of interlocking volumes that form the external wall zone.

The facade zones are treated similarly on all three facades, with the fourth chamfered elevation being glazed with passive solar shading provided by glass louvres. The facade integrates environmental control ducts, with inlets and outlets arranged adjacent to the main gaming areas. A 'shop window' is set near the entrance to provide a specific point of identity for the gaming inside – where the outer glazed walls on two sides form the shop window – and for letting people know what is going on inside. Conversely, players inside the building can feel that they are on show: part of a set of performances being given at the same time, one next to the other, throughout the height of the building. As part of this idea of visibility of activity from inside to outside, a glazed 'bubble', or oriel window, provides a public-facing gaming space on one diagonal side of the facades; a space which forms a breakout space. In this space, gamers are playing in full view of the public; the passing public are their spectators and momentary audience.

The facades have a mixture of solar shading panels of different design as a result of the need to control levels of daylight and sunlight for specific activities; this building is not a uniformly darkened space. Parts of the facade have horizontally proportioned windows set at a high level on every floor, which are set back from the outer surface of the walls in order to provide passive solar shading. This geometry of facade panels allows the glass to be clear; one of the priorities of the brief/program was to have a mixture of additional glazed solar shading layers in some

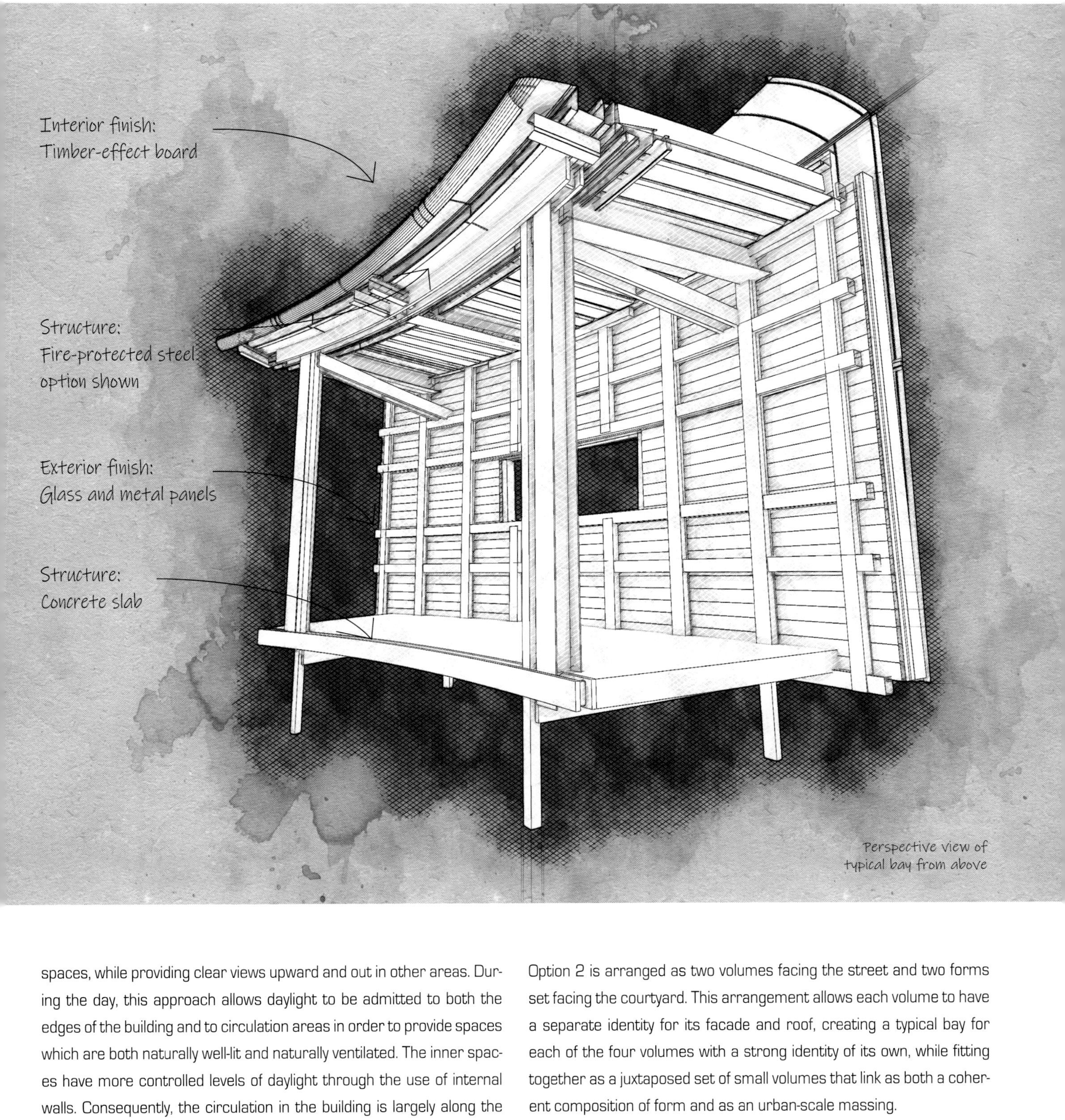

Perspective view of typical bay from above

spaces, while providing clear views upward and out in other areas. During the day, this approach allows daylight to be admitted to both the edges of the building and to circulation areas in order to provide spaces which are both naturally well-lit and naturally ventilated. The inner spaces have more controlled levels of daylight through the use of internal walls. Consequently, the circulation in the building is largely along the central spine. Breakout spaces within the building provide views out through the building facades, resulting in a mixture of glazed facades in small areas and solar-protected facades elsewhere.

Option 1 has the smallest area of facade and roof, providing an economical solution to both the volume enclosed and the surface area of external envelope. Each of the three forms that comprise this option require different facade systems in order to give each form its own volumetric identity within the massing. The use of a single facade language was discarded at an early stage to avoid a homogenous appearance to this building, which instead requires the articulation of a set of linked forms, each with a specific identity, rather than as continuous facades.

Option 2 is arranged as two volumes facing the street and two forms set facing the courtyard. This arrangement allows each volume to have a separate identity for its facade and roof, creating a typical bay for each of the four volumes with a strong identity of its own, while fitting together as a juxtaposed set of small volumes that link as both a coherent composition of form and as an urban-scale massing.

Option 3 has an arrangement of subdivided spaces within each volume that comprise the overall building form. Each volume has a functionally balanced distribution of internal spaces; the structure contributes to this solution by providing a regular structural grid in both the support spaces and activity spaces. The subdivision of volumes leads to some internal spaces requiring mechanical ventilation, while those adjacent to facades benefit from natural ventilation. The addition of the pointed rooftop form allows each set of interior spaces and external volumes to have specific visual identities that fit with the urban scale of neighbouring buildings, both in their massing and in the scale of assemblies and geometries used for the facades and roofs.

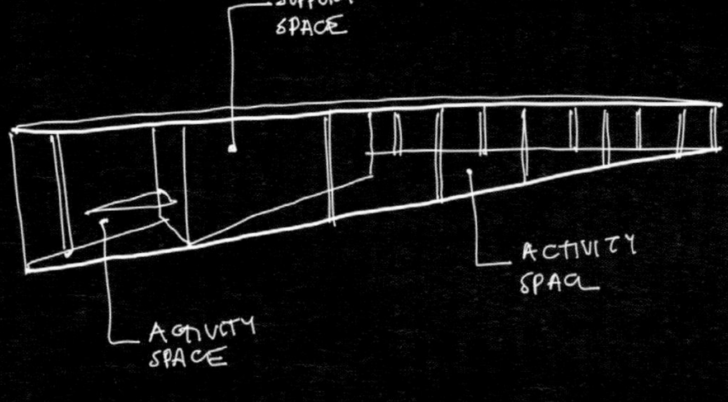

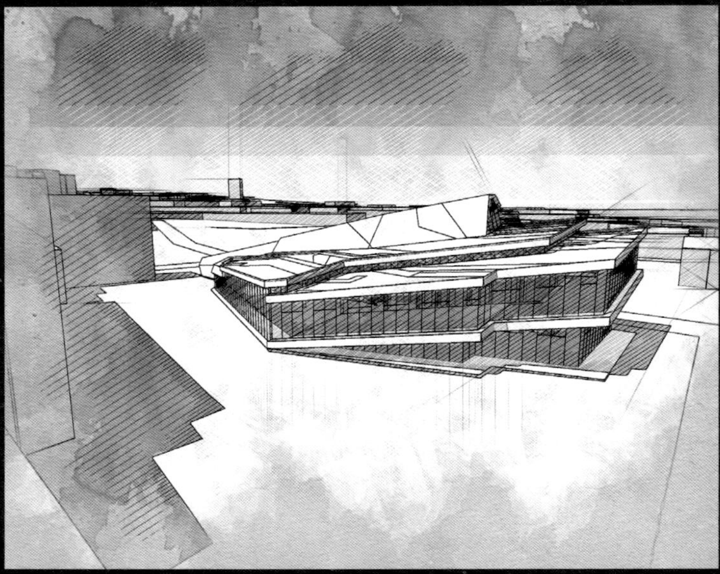

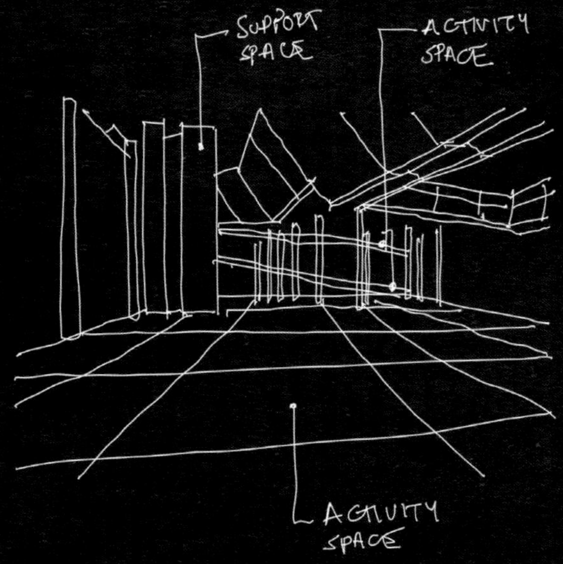

Integration of facade and structure in a small-scale building

An essential design concept for the structural design of this building was to use the facade-supporting frame to replace columns at the perimeter of the building. The steel frame that supports the facade panels also supports the edges of the floor deck; structural columns are set 3.0 metres away from the facade, within the interior spaces of the building. The steel frame at the perimeter of the floor decks combines column support for floor decks with support for facades. The primary structure is formed as a braced frame using hot rolled steel members. Floors are constructed with light gauge steel joists and a timber-look deck that integrates the complete structure into a single framed entity. The resulting structural design is a structural steel 'cage' that combines structural elements that support floor decks and facades as a single entity, and provides an alternative to using a primary structural frame with a facade, with its own support, set adjacent to the primary structure. In some areas, parts of the facade are required to span vertically through double-height spaces without being restrained by the primary structure. The aim here was to avoid having a separate system of steel frames supporting the glazing in the double-height spaces, and curtain wall panels on the other parts of the facades. The steel frame supports the double-height spaces around the complete perimeter of the building, using that same structural concept to support floor decks in places where they meet the facades. This structural approach suits the relatively small size of the building; this is an alternative approach to that of taking construction technology aimed at larger-sized buildings and applying it to a building of this comparatively smaller size.

Internal columns are set out on a rectangular grid of 6.0 metres x 6.0 metres with perimeter columns, set at 3.0 metres, aligned with this grid. This approach provides a sense of visual lightness within the building as a result of this structural approach, but still benefits from a regular structural frame within the building, with columns arranged to suit the required size of structural bays and internal space-defining elements. There is no requirement for floor decks to span long distances

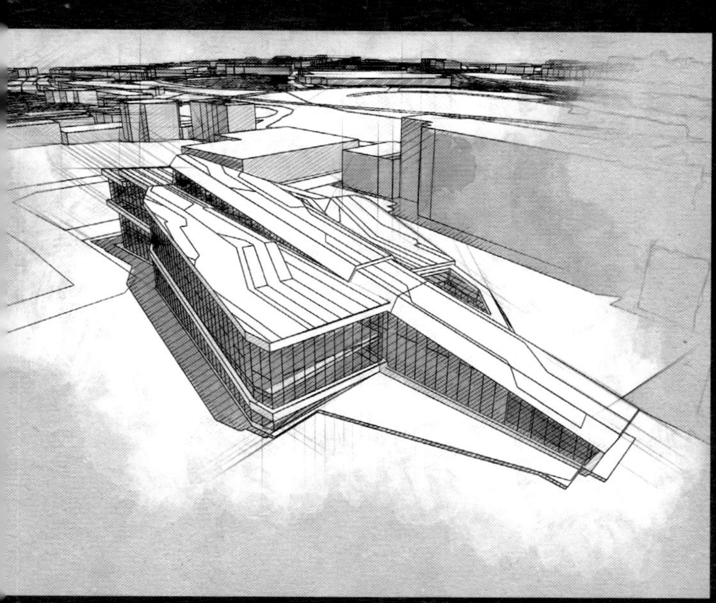

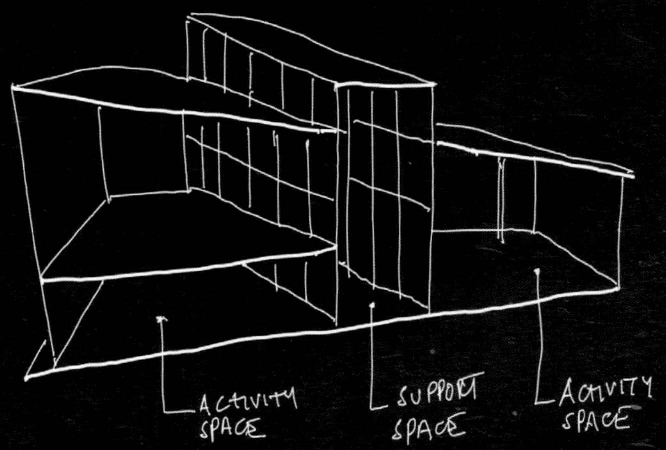

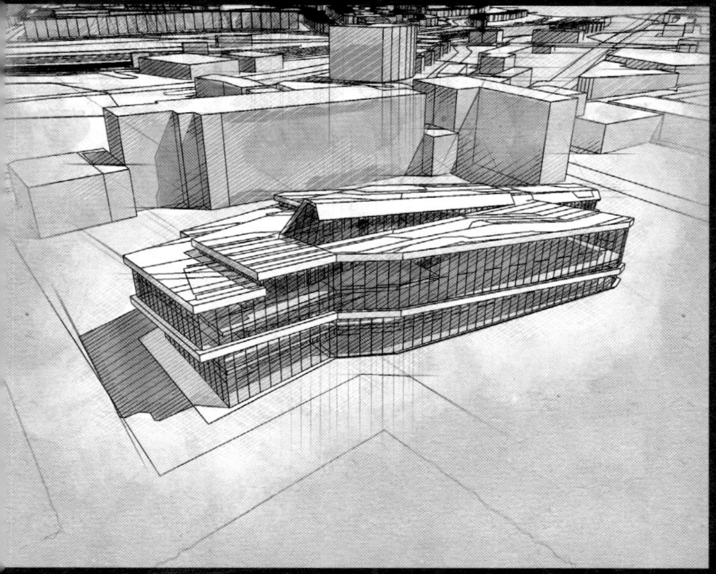

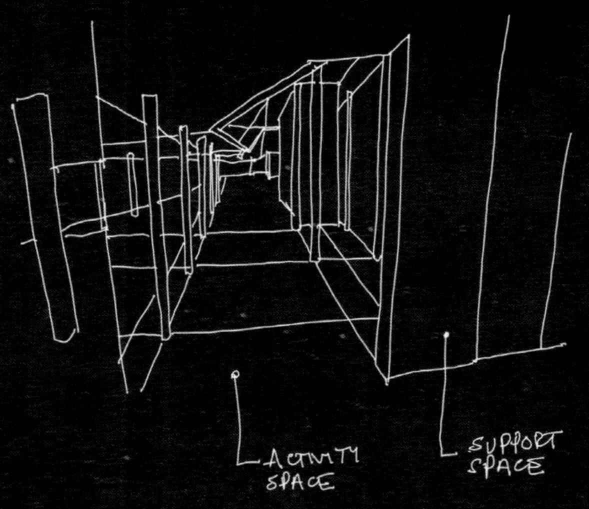

between the facade and the inner core; large areas of open space are not required. Instead, smaller internal spaces are defined as bays by the column set at the midway between the outer wall and the inner core. The inner cores are also made as steel frames using a similar arrangement of members, but here the steel columns are set into the depth of the opaque walls enclosing the cores. An arrangement of steel frames is set between reinforced concrete columns because it was considered visually important to provide a contrast between the art works on display and the clear spaces created within the building; a neutral backdrop to the works on display. GRC panels (glass-fibre reinforced concrete) are fixed to the outer face of the steel frame and the outer face of the support spaces in the centre of the building.

The approach of having a rectilinear grid set into a nonrectilinear-shaped building provides a visual harmony of framing at the perimeter of the building without setting out the facade on a completely different grid. The column arrangement results in the glazed facade panels being of different width in elevation, but of visually constant width when viewed perpendicular to the facades. This approach of a regular structural grid, and regular panel width, results in a sense of visual calm and order within the internal spaces of the building. This method still allows the overall form of the building to be non-rectilinear in response to the planning of the internal spaces as well as the geometric relationship of the building form to its surrounding external spaces.

The structural design provides a way of achieving the requirements of responding geometrically to surrounding buildings and external spaces as well as creating a visually calm backdrop for works of art exhibited within the gallery. This approach avoided the use of an irregular structural grid whose arrangement would have competed with the exhibits. The content and setting out of exhibits is expected to change frequently in response to the need to accommodate travelling exhibitions of paintings, sculptures and installations. The design aims to meet the need to house installations which often require a large floor area. This greater level of interaction between sculpture, installation and the visitor are accommodated in this structural design while remaining economical.

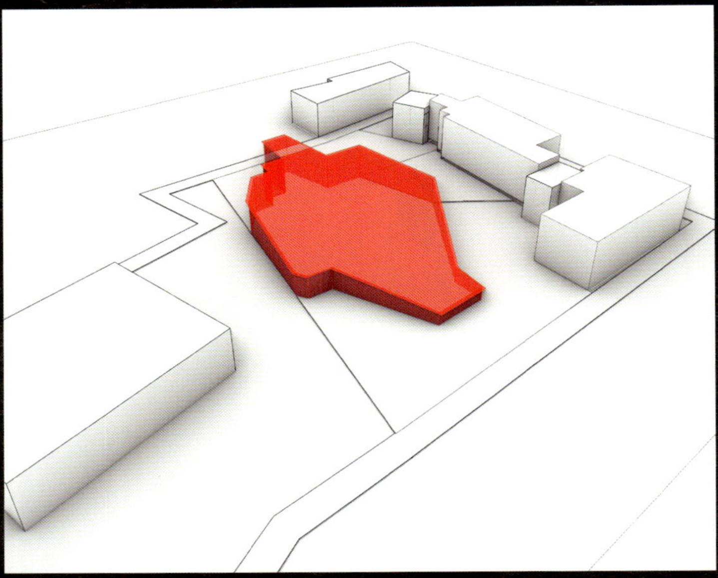

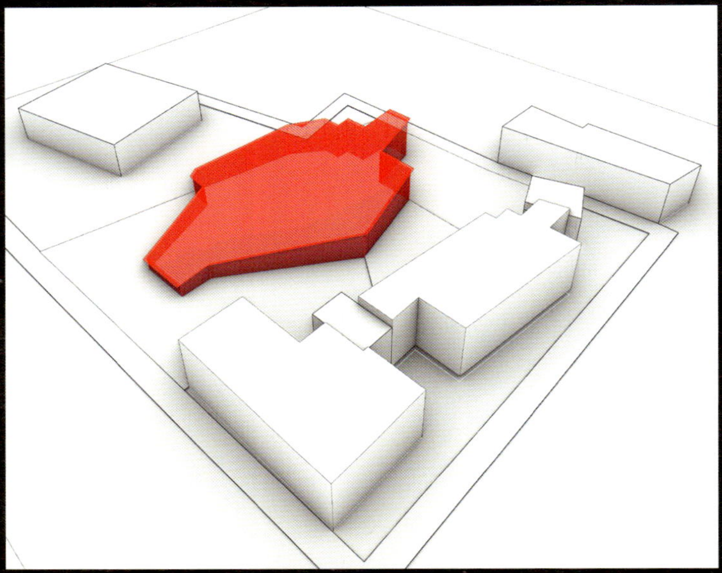

Option 1
creates a set of surrounding public spaces, bringing those external spaces into the sphere of influence of the building.

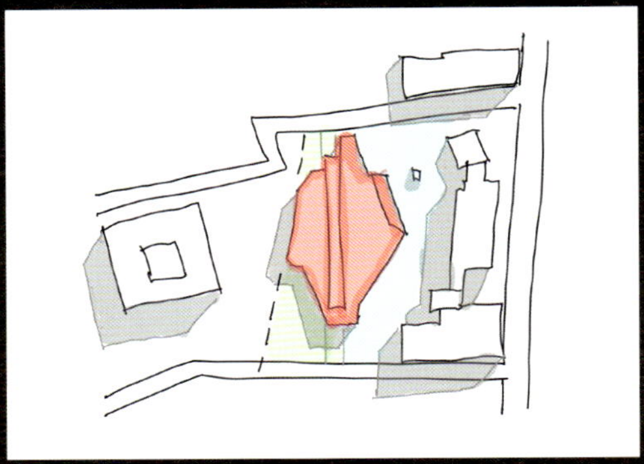

Option 2
uses 'lines of influence' from the facades which extend out as low walls of seating and planting.

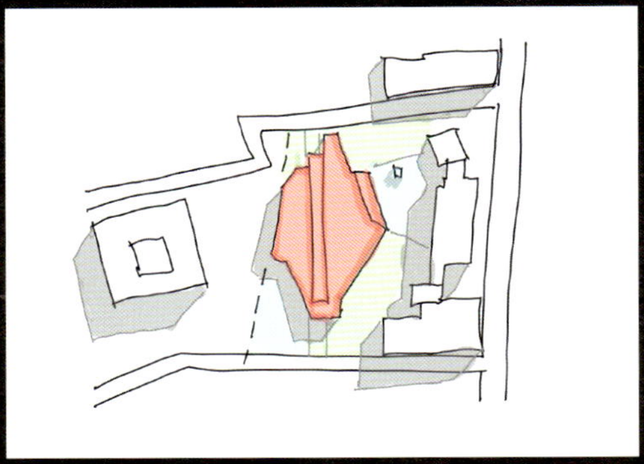

Option 3
uses fewer lines of influence, and does not use the external low-level walls as a place-making device.

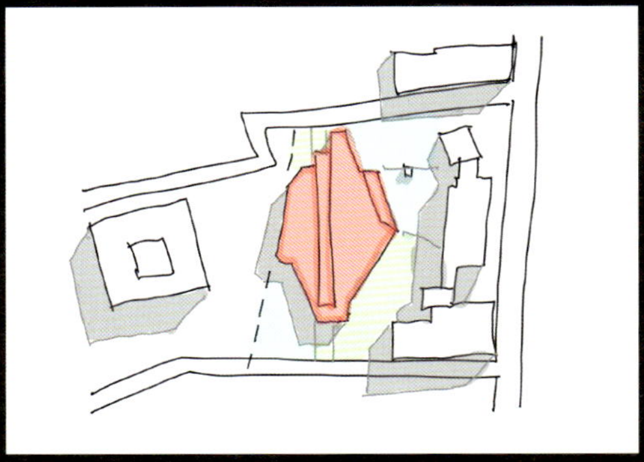

Site context

This project is an art gallery in a business district, located in a hot humid climate near the equator, where environmental conditions throughout the year are relatively constant. These climatic conditions allows the environmental design to be a single approach which is deployed throughout the year and consequently does not require a facade that responds to changes in environmental requirements; environmental performance is for relatively constant conditions. The building is essentially side-lit from glazed walls, with small roof lights to provide natural daylight in areas that are required for exhibition purposes.

Option 1 is based on taking the arrangement of the building as determined by the requirements of the brief/program and the use of both internal and external spaces. The form of the building is such that it creates a set of surrounding public spaces, bringing those external spaces into the sphere of influence of the building. The column grid is extended out into these spaces and is expressed as a decorative ground finish of the external paved areas. A large-scale sculpture is set at the midway between one end of the building and the adjacent buildings in order to frame a public space. Option 2 uses 'lines of influence' from the facades which extend out as low walls of seating and planting. These are low-level walls for seating that extend out from the fabric of the building, such that the external spaces become 'outdoor rooms'. Large-scale sculptures remain in the same position as in Option 1. The low-level walls partially enclose distinguishable external spaces which are defined both by these walls and by adjacent existing buildings. Option 3 uses fewer lines of influence, and does not use the external low-level walls as a place-making device. External spaces are paved and blend into adjacent external spaces such that the space in front of the building – and to the northeast of the building – are defined as public spaces, with the spaces around them left untouched.

Option 3 is the preferred spatial arrangement. In terms of access, Option 1 has pedestrian access from the two ends of the site, providing a visual continuity through the building, with the ability to have two entrances and exits along the single service spine of the building. The

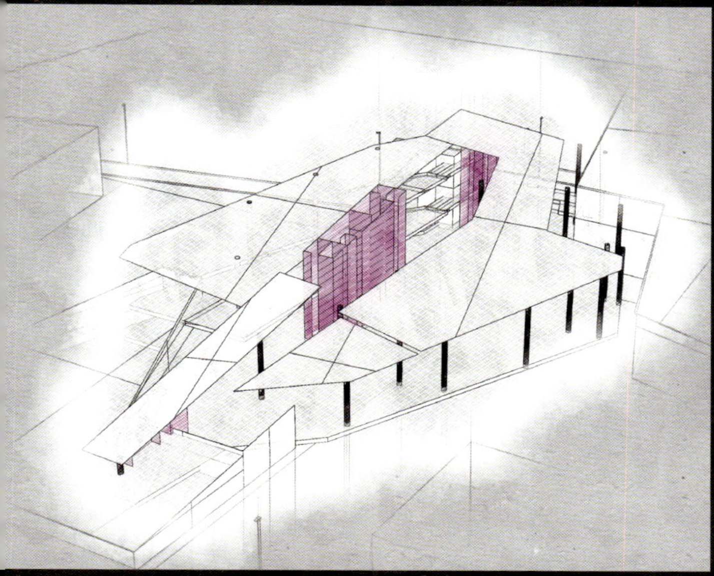

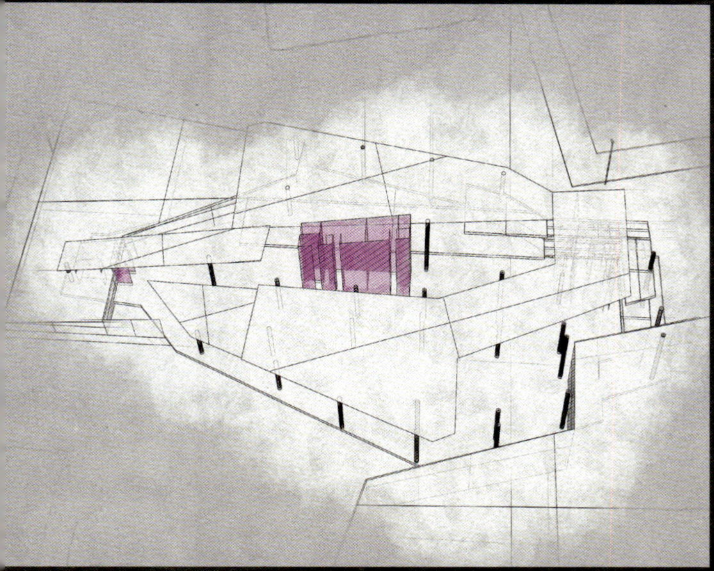

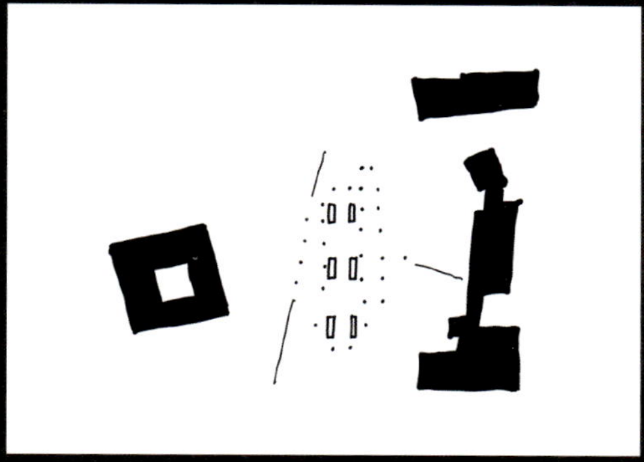

Option 1
Dark green spaces are highly tempered but are reduced to accommodation volumes only.

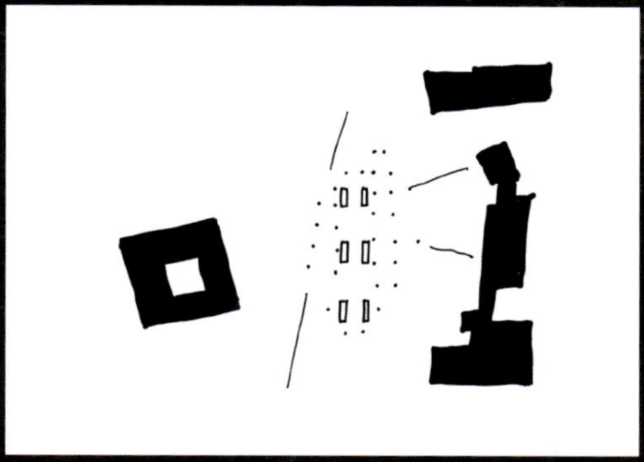

Option 2
Dark blue spaces are less tempered than dark green spaces but are more tempered than light blue circulation space.

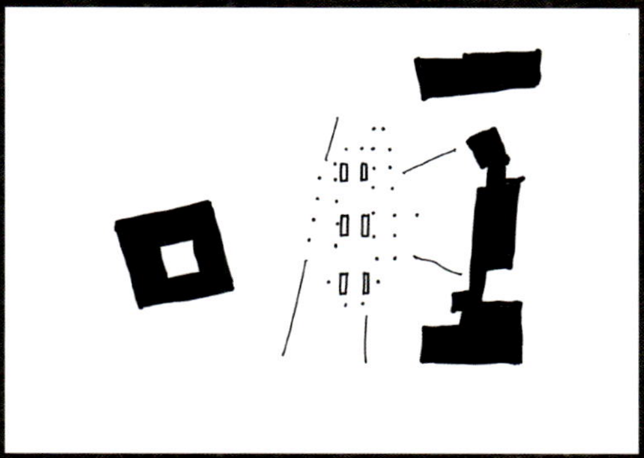

Option 3
Light blue spaces are the least tempered spaces, these spaces interface directly with external space.

two spaces adjacent to the main space, as per Option 1, become an external space for large-scale sculptures and installations, to encourage public involvement. The more private and enclosed nature of the gallery in Option 2 provides the same access as Option 1, though the routes are less clearly defined, allowing for access to potentially be provided along the lines of influence. External spaces can become alternative points of entry into the gallery as access points along the edges and corners of the building, rather than in the more classic entrance at the projecting ends of the service spine of the support spaces.

This approach of alternative entrances at the corners of the facades is strengthened by the use of sliding doors to create entrances for specific exhibition spaces, making the entry into the building less about classic access routes and more about flexibility of entry into the gallery. Option 3, the preferred option, creates external spaces to the southwest and northeast side of the facades, where the complete facade becomes part of the entrance experience, such that all spaces can be used internally and externally and can be linked; the main walls become a part of the entrance experience. This creates the approach of walls becoming entrances, encouraging the external spaces to be linked directly to those spaces, such as exhibitions being accommodated in both small spaces and residual spaces. The building presents itself as a series of internally and externally located performance stages for presenting large-scale artwork, installations and performance art. There is a clearly defined extension of internal spaces which become external rooms; extensions to the small residual spaces in the main building. The adoption of Option 3 allows the external spaces around the building to become almost as important as the spaces within the building; a set of 'outdoor rooms' as extensions of the internal spaces. This approach of visual continuity of space from inside to outside is achieved with the facades, designed more as 'secure partitions' than external walls. The glass forming the outer wall reflects daylight during the day, creating a sense of enclosure from the external spaces, but during hours of darkness the internal and external forms are unified visually; the glass is not seen as a division between inside and outside.

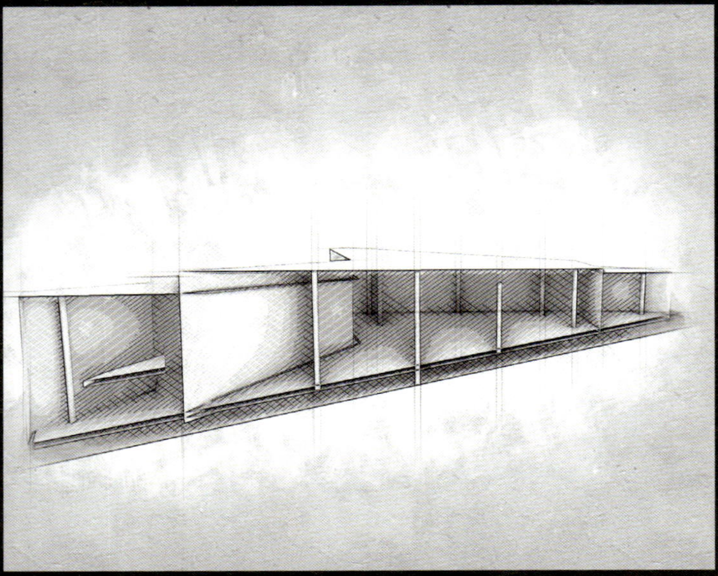

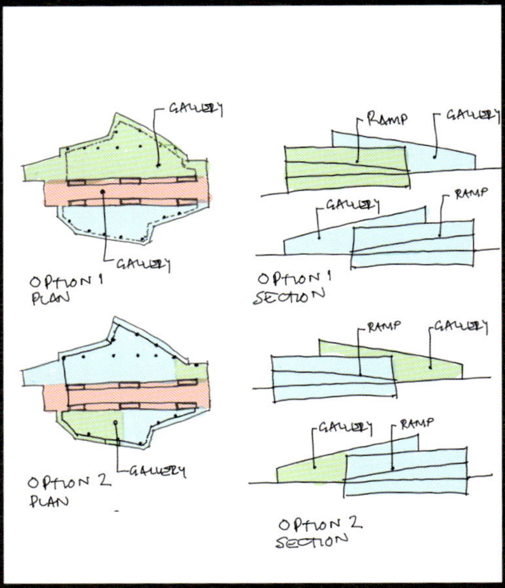

Spatial arrangement
Option 1 uses a column grid which is not parallel with the cores.
Option 2 sets the cores parallel with the column grid.

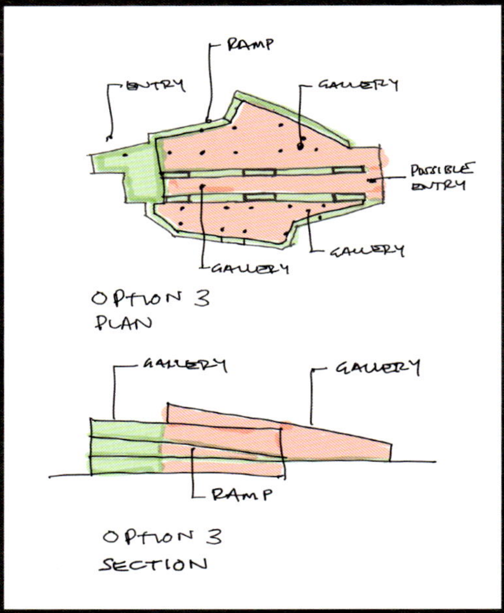

Option 3 adds ramps around the perimeter of the facades as a route around the gallery.

Spatial arrangement

Option 1 uses inclined floor decks for the exhibition spaces enclosing the gallery. This approach provides a clear open floor space where the circulation is accommodated inside the circulation spine down the centre of the building; the central spine space has level floor surfaces. Movement of visitors down the building is set at a gentle slope from the far end of the building back towards the entrance. A link takes visitors back into the core, leading them down to street level.

Alternatively, the floor decks in the inclined areas drop right down to the entrance level. The space underneath the slab reduces in height down to ground floor level; the space beneath is occupied by a small-area service zone which scoops down below ground level in order to make headroom in the service space sufficiently high for the installation of service equipment.

Option 2 uses the double-height spaces to form the central circulation space. This use of space allows the central volume of the building to become a primary central exhibition space: one that can be seen, and accessed, directly from the entrances at each end of the building. This central gallery can also be glimpsed from the smaller exhibition spaces which are positioned adjacent to this central exhibition space; staircases to access these adjacent spaces are inserted into the side wings that extend away from the central space.

Option 3 uses the spatial arrangement of Option 1, where the central circulation spine area is set into the centre of the building, but with the surrounding exhibition spaces having flat decks instead of inclined decks.

Access ramps are positioned at the edges of the building next to the facade, such that upon entering these exhibition spaces the visitor can walk down the ramp slope, down to the ground floor area. Circulation space in the centre is also used for smaller exhibition spaces. Side spaces surrounding the central circulation spine are level rather than inclined; circulation takes place on the facade to provide both views outside and a buffer zone between the outside space and the enclosed exhibition spaces of the building's wings.

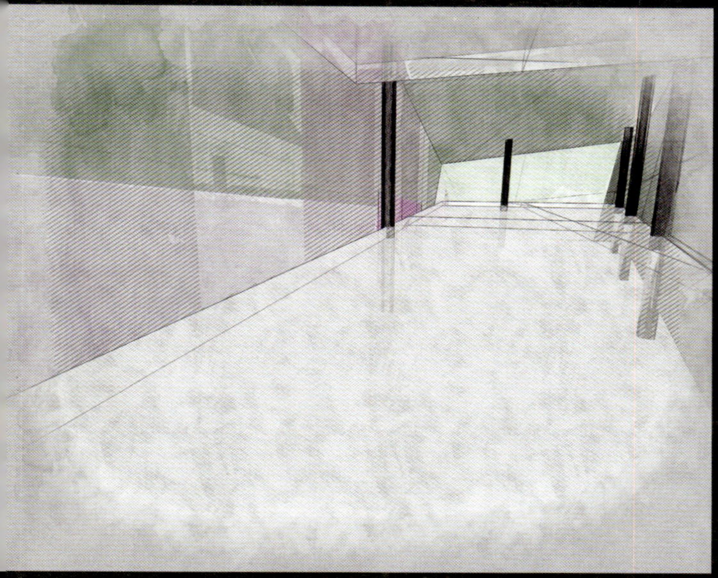

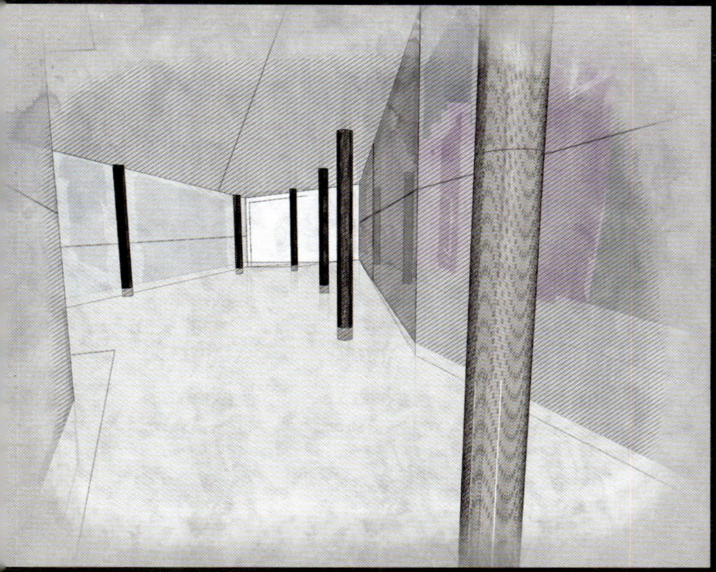

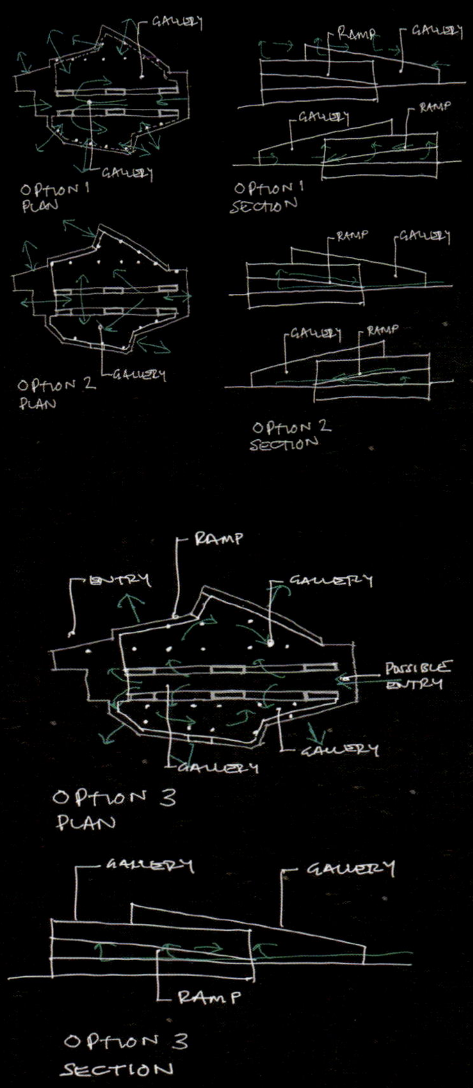

Internal circulation

Option 1 arranges circulation from a single entrance at one end of building.

Option 2 arranges circulation from entrances at each end of the building.

Option 3 arranges circulation from openings in facades.

Internal circulation

This is a building with possible circulation arrangements set out as three options as a result of different uses of the building. These options can vary from using only parts to using all the exhibition spaces for a single show. Option 1 is based on a single occupancy model for the building in which any exhibition occupies the full extent of the building, with the possibility of having the ground floor of one of the wings as a separate permanent exhibition, or alternatively as a separate temporary exhibition.

The circulation concept of Option 1 is one of a single flow of visitors through the building; through the cores and along the gallery to the far end of the building, then turning left or right. At this point the visitor can follow gently inclined slabs back down to the ground floor of one wing; the experience of the second wing is shaped by returning to the first floor and through a different part of the circulation spine and then following the inclined ramp down. Option 2 assumes that the central space is a double-height space, and there are cores arranged along its length. The main circulation occurs along that double-height central space, which is that which is open, and has staircase cores and washrooms and service spaces attached to the sides of the central spine core. This makes movement through the building less fluid than it might otherwise be, but the focus becomes one of moving around the central space without realising what is taking place in the other central space. Stairs and a lift/elevator at the far end of the building direct visitors down the inclined sloping floors back to the entrance.

Option 3 overcomes the circulation issue of Option 1 by the visitor being asked to constantly move downwards along an inclined floor deck. Visitor circulation occurs around the edges of the building such that, upon entering the building, the ramps up around the side are the primary circulation method: rising to the back of the first-floor level at the back of the building, where a promenade returns to the entrance. Views are set up, both on the left- and right-hand sides, through into the side spaces, allowing the visitor to move in and out of those spaces on a meandering route.

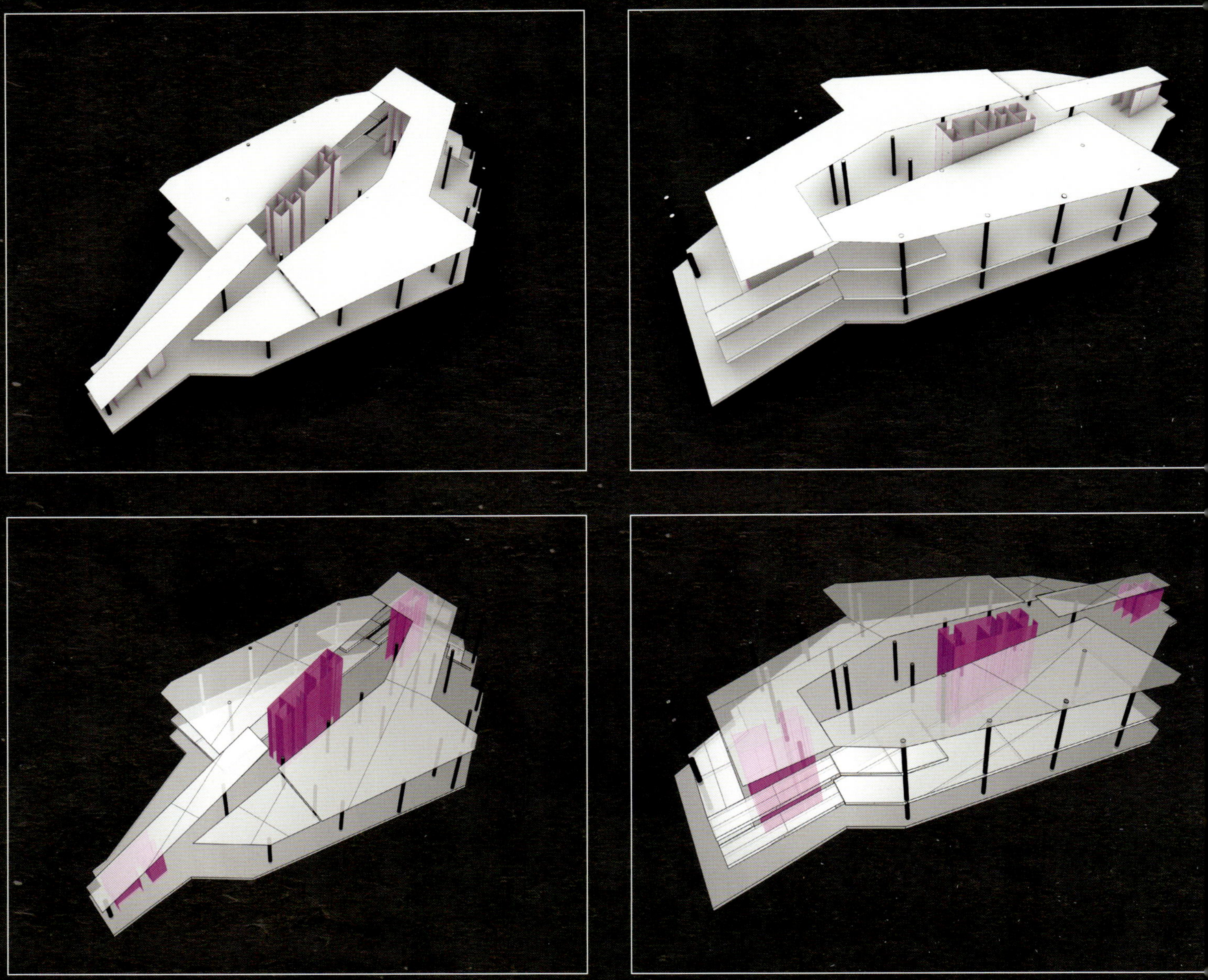

Structural strategy

The roof structure is a significant part of the structural design. The roof is formed as a folded plate with openings for rooflights cut out of the folded deck. The steel-framed roof is supported on reinforced concrete columns at 6.0 metre centres onto which the visually contrasting folded plate roof is set. This roof design allows large spaces to be created internally with a column arrangement that is independent of the form of the roof above, creating as much emphasis on the exhibition of sculptures and paintings as on the space itself.

The concept of the structural design is to distribute loads as evenly as possible across all members in order to avoid structural components having different depths when forming beams or trusses between columns. This concept was required to function alongside a need to ensure that the vertical framing members supporting the facade form an integral part of the external wall enclosure, from both a structural and visual point of view. The steel columns supporting floor slabs along the spine cores are reduced in size but increase in frequency in order to be integrated within the depth of core walls. This approach allows the columns to have a double role of supporting the edges of floors and roof while also providing primary support for internal walls that enclose the central circulation and exhibition space. This approach is in contrast with an alternative strategy, which was not followed, of having a large-scale structural frame with reinforced concrete cores.

The structure, facade and internal cores are integrated from a single set of materials, allowing the structural assemblies to have a visually closer fit in integrating the elements of column, roof, spine wall and floor structure. All elements of this fire-protected structure are exposed where possible, avoiding the need for suspended ceilings beneath the roof structure and beneath floor slabs. This approach is one of significantly reducing the need for internal finishes, providing a greater ability for the structure to be to be adjusted or adapted more easily at a

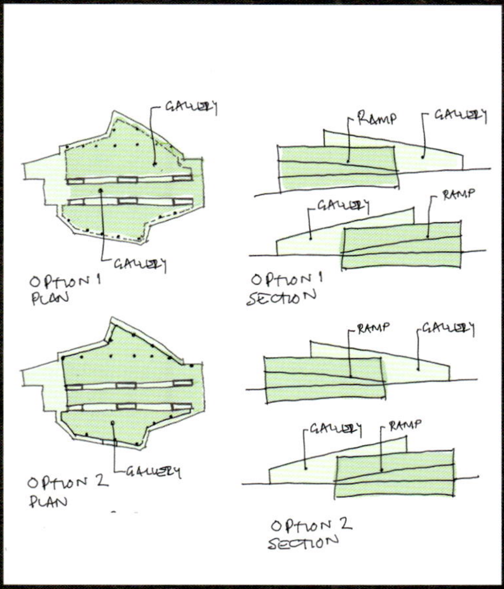

Structural strategy: Zones
Option 1 arranges columns in a regular grid which extends to the edges of the floor slab.
Option 2 introduces a perimeter structural frame in structural steel.

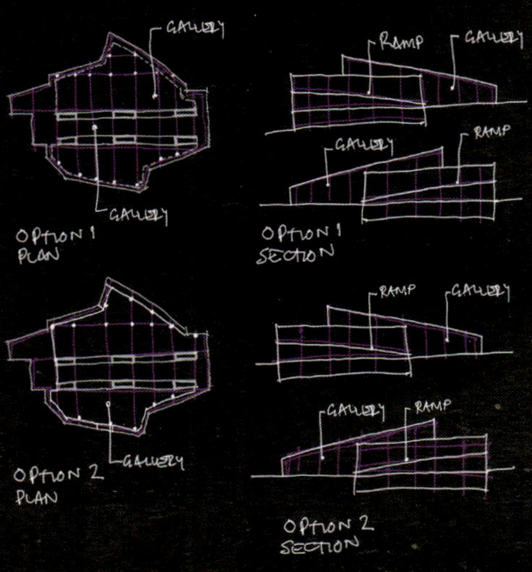

Structural strategy: Framing
Option 1 encloses the core spaces and dividing space into sets of spaces.
Option 2 supports the edges of the roof and floor slabs with primary structure as well as providing primary points of support for the glazed facade panels.

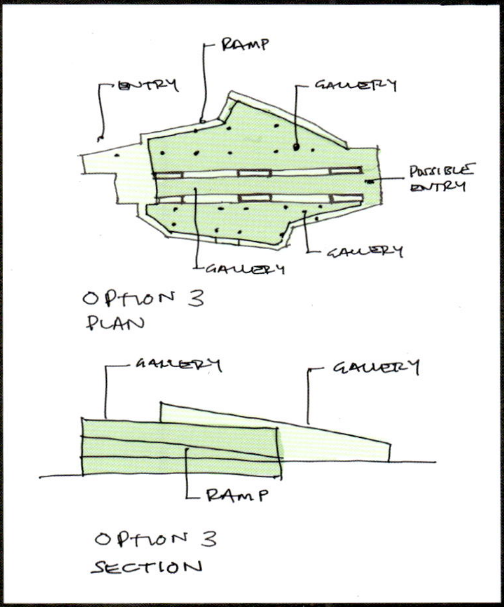

Structural strategy: Zones
Option 3 uses structural steel framing for floors, roof, facade supports and support for the central cores.

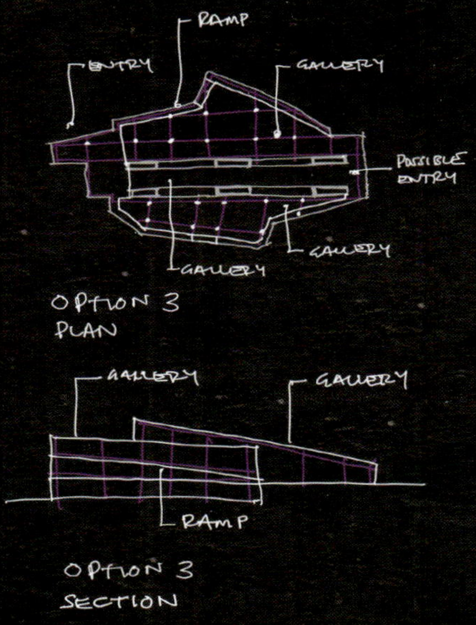

Structural strategy: Zones
Option 3 integrates facade and primary structure by using structural steel framing for floors, roof, facade.

later stage of the life of this gallery. The structure of the central space comprises a steel frame that encloses cores in each side of this space, providing a 6.0-metre-wide space that extends the length of the building. Enclosing core areas and support spaces occupy zones 3.0 metres wide on each side of the central gallery zone. The steel-framed cores that enclose staircases are strengthened with bracing to avoid the need for reinforced concrete walls. Shafts enclosing lifts/elevators are also supported with steel frames and are enclosed in glass partitions on two sides. The central core provides a continuous structural 'spine' down the centre of the building, with the outer facades avoiding the need for cross bracing: providing the building with its visual emphasis of framing that supports the cores and facades which contrast visually with floor plates.

In Option 1, columns are arranged in a regular grid which extends to the edges of the floor slab. The structure was envisaged to comprise circular reinforced concrete columns and floor slabs. The facades were independent of this primary structure, both geometrically and as a metal-framed system of glazed panels. This approach was developed in Option 2 by introducing a perimeter structural frame in structural steel that would support the edges of the roof and floor slabs as well as providing primary points of support for the glazed facade panels.

Option 3 developed this theme of the integration of facade and primary structure by using structural steel framing for floors, roof, facade supports and support for the central cores. This approach ensured an ease of integration of framing of different scale, from the smaller size arrangement of structural members used in the facades and central cores to those used for the primary columns and floor plates. This approach results in visible framing to the cores that contributes to the pared-down look of the structure in contrast to the visual richness of the artwork exhibited.

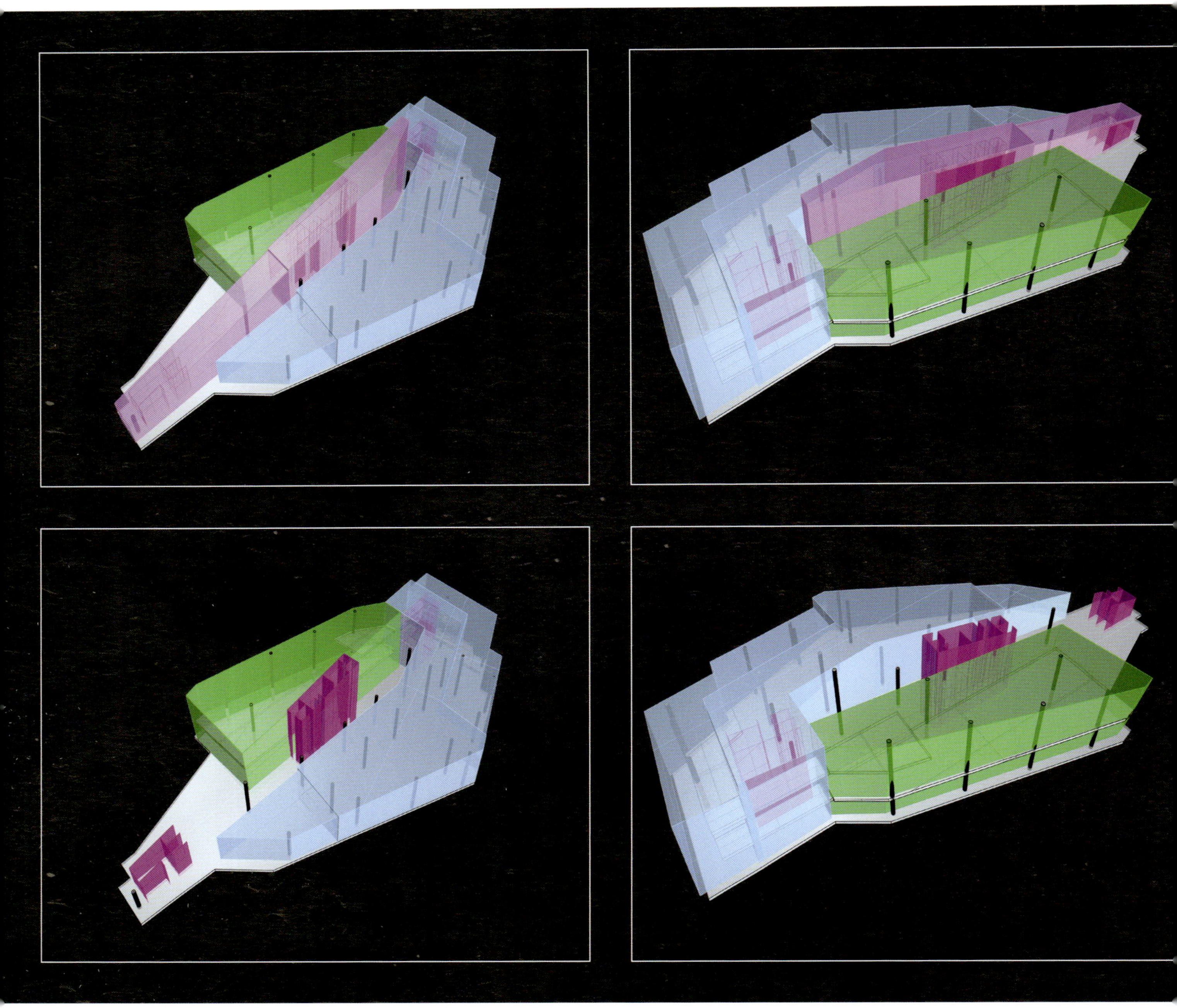

Environmental strategy

A primary requirement of the art gallery is to have a series of spaces which are lit from the side, and which have very controlled top-lit areas capable of being adapted to admit daylight from both the sides and from the top. This requirement is in response to the gallery having drawings, paintings and sculptures exhibited within the gallery, and associated large-scale sculptures being exhibited outside the gallery in the adjacent public spaces.

Directed daylight is required for the exhibition of sculpture pieces within the gallery, as many of the pieces were created with a view to them being public art set in an outdoor location. Light fittings fixed on the facades project light onto sculptures beneath. A key requirement of the exhibition of sculptures is to reproduce, as closely as possible, the daylight conditions outside without the use of electrical light during daylight hours. Some sculptures are exhibited within the building rather than externally in order to protect works from the effects of the external climate.

Environmentally, the building is formed in three zones for daylighting, which correspond to those zones for mechanical ventilation. The central spine is the support space with additional small exhibition spaces within it. Exhibition space is set on both sides of the central core, comprising a lower-level exhibition area, with columns sloping up for 2.0 to 6.0 metres down to a 2.0 to 4.0-metre height at their periphery. On the other side the roof steps up to 8.0 metres and steps down to a 5.0-metre perimeter at the edges which creates a variety of lighting conditions.

Option 1 has three environmental zones, based around the orientation of the sun, changing levels of daylight and shadow patterns on internal walls at different times of year. The central zone uses a circulation spine as an area where the 'waste' heat or cooling energy from both wings located at the sides of the building are used to ventilate the central space. One wing has a second environmental zone. The third wing, because of its different orientation and its smaller volume and narrower shape, has a third environmental zone. Each zone is separated

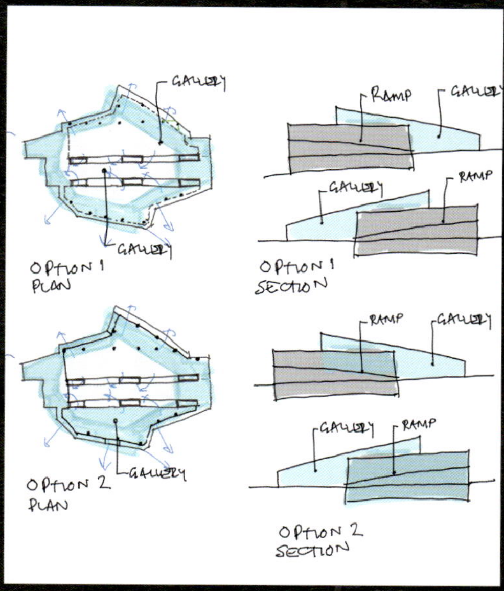

Daylight strategy for direct sunlight and cross ventilation.

Option 1 arranges the building as two blocks separated by a service zone.

Option 2 widens the central circulation space in order to add the main entrance floor.

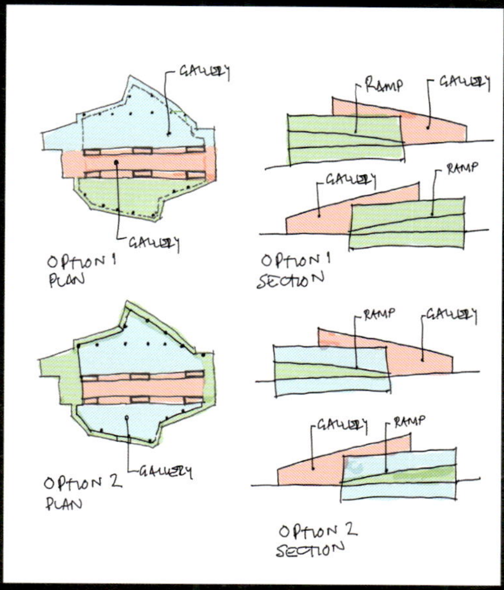

Environmental zones

Option 1 has three environmental zones; one zone per wing and a zone for the central space.

Option 2 accommodates the wings into a single zone and introduces a single zone for the perimeter.

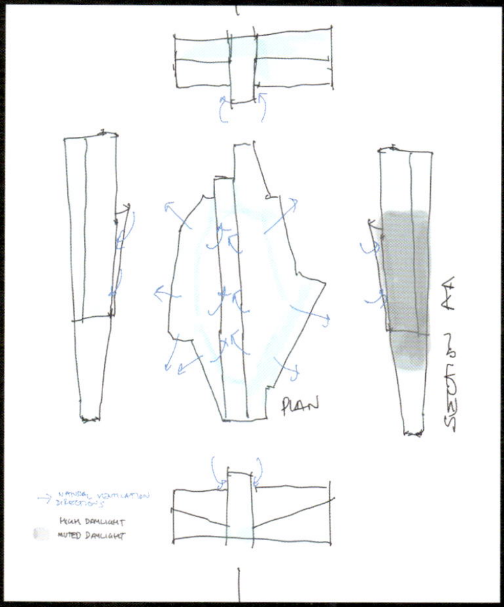

Daylight strategy for direct sunlight and cross ventilation.

Option 3 arranges the building as two larger blocks with communal spaces on the ground floor and accommodation above.

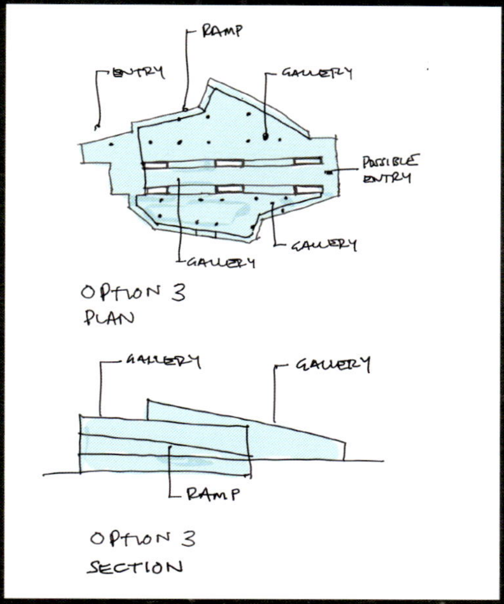

Environmental zones

Option 3 accommodates all spaces, including support spaces, as a single environmental zone.

by the central core. This core is used as a spine to distribute ducted air. As a result of the use of the central space as a double-height open volume, this space becomes the primary gallery to be ventilated, with its temperature and humidity closely controlled for exhibits. The larger spaces surrounding the spine have air supplied as air recycled from that central space.

The central space is top lit in part, having varying requirements for heating and cooling during the day and across the year. The spaces immediately on either side of the central core are linked by the ramps, such that two spaces, the upper and lower gallery spaces, are open and linked together. These spaces are treated as a single environmental zone; air is moved mechanically from space to space to suit the changing needs of each gallery as a result of the changing exhibitions, and their individual requirements, as well as changes in solar gain at different times of day.

A key feature of this office building is the external mounting of the mechanical ventilation installation to provide flexibility of use while retaining clear floor plates with cores that do not need to accommodate environmental controls. The air-handling units are small scale rather than the customary centralised units that are typically mounted at the top of the building. Instead, air-handling units are mounted within the roof zone. This approach provides a flexibility of use across the different exhibition spaces.

The environmental design of the building divides the volume into environmental zones, even though many spaces are linked. Zones are enclosed by partition screens which are deployed to suit the specific requirements of an exhibition.

The environmental zones can be arranged to follow either the servicing of the zones stacked vertically, or they can be arranged horizontally. Each zone potentially has an independent environmental control system that can respond to the different exhibition requirements in each zone, so that the building is divided environmentally either into three vertical strips, or into three superimposed floor zones.

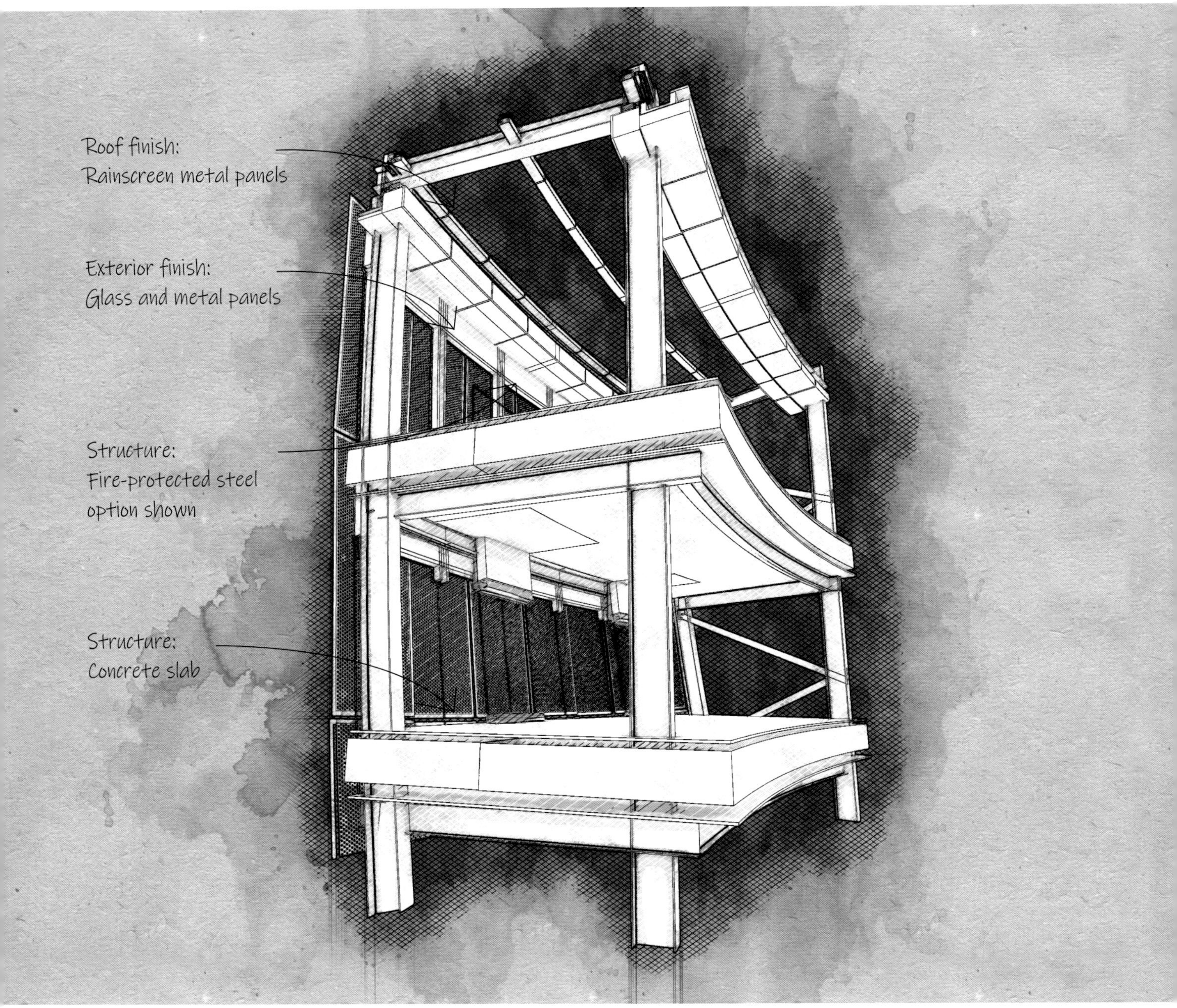

Typical bay adjacent to facade

The building is used both during the day and during hours of darkness; the need for a primary spatial experience of natural light was not a key driver in the brief/program. As a result, the side elevations have some transparency to allow passers-by to see what is going on inside the building, and draw visitors who would not normally be drawn to an art gallery to see the art works within. The envelope of the building has walls in floor-to-ceiling glass, which are framed and stiffened by the primary structure behind each glass joint.

The roof is clad in a rainscreen system; metal rainscreen panels are set on top of a thermally insulated roof. The end of the gallery that faces the landscape has a series of folded planes forming ramps and staircases leading to the upper mezzanine level at one end of the building. The mezzanine provides a facade-focussed exhibition space for paintings that can be seen in relation to the sculptures below.

The mezzanine aims to exhibit the sketches and maquettes of the sculptures made in preparation for the final pieces, which can be seen in the main gallery spaces below. The central spine core provides structural support at the centre of the roof and stability to the roof planes which extend out to the glazed walls, where they are supported on steel posts at 1.5-metre centres.

The roof structure comprises a series of triangulated trusses with openings that allow light transmission, based on the 1.5 metre width of glass that was used. The underside of the roof has translucent ceiling panels behind which are mounted the environmental control system and the electrical lighting arrays, accessed from within the space itself. The use of the ceiling, as well as a continuous walkway to allow access to all parts for the lighting system, avoided the need for the setting up of exhibitions and lighting arrangements independently of exhibitions going on below. The building is primarily mechanically ventilated. This is common with art galleries in order to protect the works of art exhibited. The outer glazed walls are arranged as double-glazed units which span between the structural members at 1.5 metres. This loose-fit glazing system allows the glass to be fixed directly to the primary structure set

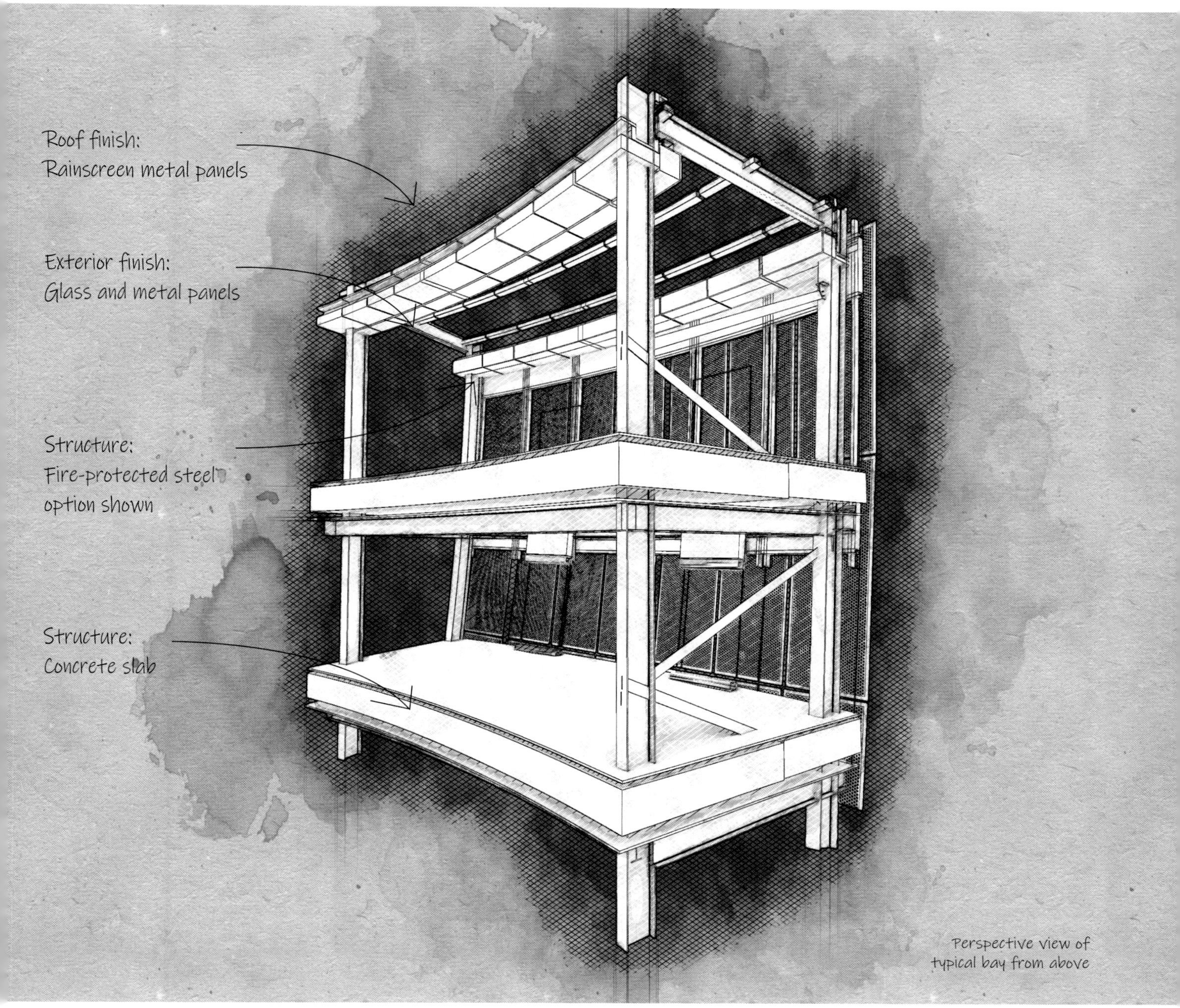

Perspective view of typical bay from above

at 1.5-metre centres, avoiding the need for secondary element glazing. The glazing system was developed in order to ensure a single contractor could construct the primary structure, the secondary structure, roof elements and glazing.

The exhibition spaces are arranged on a 2.4-metre by 2.4-metre grid, defined by a grid of cable fixings, used to support art works. Cable fixings are set within the depth of both the floor and the ceiling. Moveable wall panels can be positioned above floor level and below ceiling level as required. Within a typical bay of the gallery, art works are arranged along the central core, with additional screens mounted 6.0 metres away from the central core. Paintings are mounted on sliding screens which are able to be moved both parallel and perpendicular to the structural grid of the building, with screens capable of being raised or lowered to suit exhibits. As the screens do not extend from floor to ceiling, they can slide on cables attached to the floor deck, which has fixing hooks set into the floor beneath covers which are set flush with the floor finish; similar concealed hook-fixing units are located within the ceiling. These moveable partitions slide up and down in response to the needs of exhibiting larger-sized paintings, which require a flat vertical surface onto which they are mounted. The central core is used for the support space of entrance facilities, cloakrooms, restrooms and a small café; an outside seating area is set in the space beyond the central core of the building; the circulation spine leads to the 'outdoor rooms', where sculptures are exhibited in their preferred external environment.

These outdoor rooms do not have permanent pavilion forms, which would have marked their edges or corners, located within them. Instead, the facades have canopies which are attached to the facades at different times of year. These canopies are in the form of temporary tent-like forms and roof-only enclosures which project out from the facades rather than being separate from the gallery building. This provides the opportunity for temporary floor decks to be deployed outside the building, adjacent to the facades. It is envisaged that the floor decks will be of the same design as the interior floor finish and will become an extension of the typical bay immediately behind the facade.

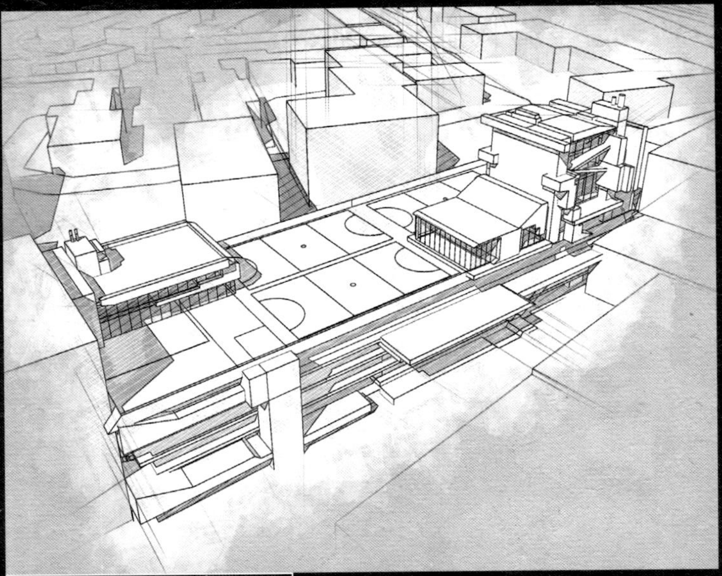

1. Perspective view of exterior

2. Perspective view of exterior

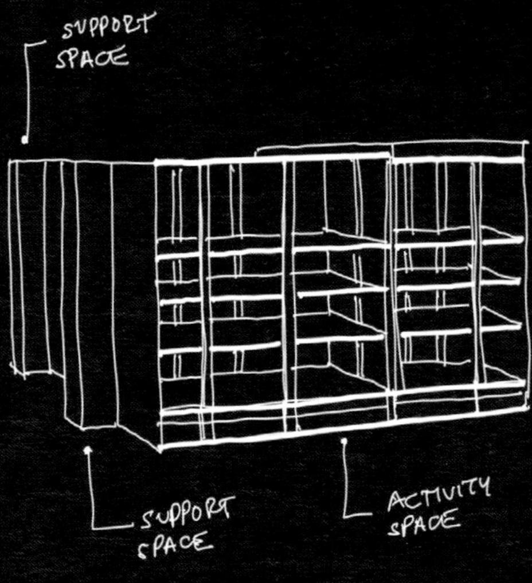

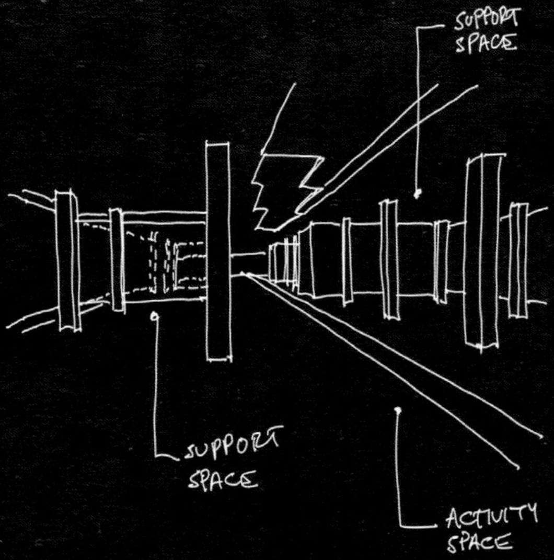

Exoskeletons and woven structures

Two approaches were explored for the structural design of this project; both have been used in different locations in this project. The two approaches are as follows:

1. An 'exoskeleton', where part of the primary structure is externally-located.
2. A 'woven' structure, where structural members are part of single structural form.

Exoskeleton

The exoskeleton allows internal spaces to have fewer internal columns and walls which are considered to be primary structure. This approach allows the internal face of external walls to follow a continuous line which is uninterrupted by columns adjacent to the facades, or at slab edges. Non-load-bearing partitions are supported on the primary structure. This method provides a 'loose fit' within floor plates, where internal partitions can be positioned, moved and rebuilt adjacent to the external walls to suit the need to change the spatial relationships at short notice. The exoskeleton can also be formed from load-bearing walls as well as columns. These walls are usually formed in reinforced concrete, and have the opportunity to be supporting structures for cores that accommodate a mixture of 'support' spaces comprising stairs, lifts/elevators, washrooms and service spaces which are not expected to change configuration during the lifespan of the building. In this configuration, external structural walls are used to enclose fixed support spaces, and floor plates which are column-free at their edges are used for 'activity' spaces such as teaching rooms and internal circulation.

Woven structure

In the woven structure, structural members – including structural members supporting facades – are of different scales of assembly but are considered to be part of a single structural form. Structural members are 'woven' together; the scale of the structural arrangement of members will vary widely to suit the very different needs of facade

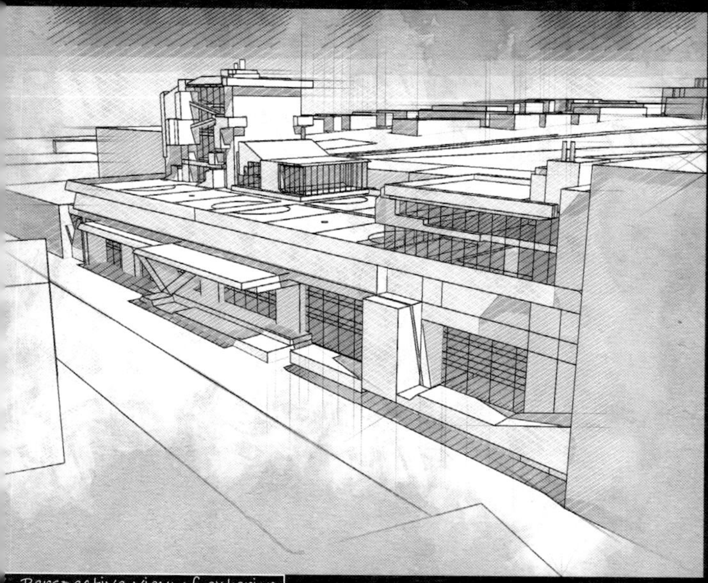

. Perspective view of exterior

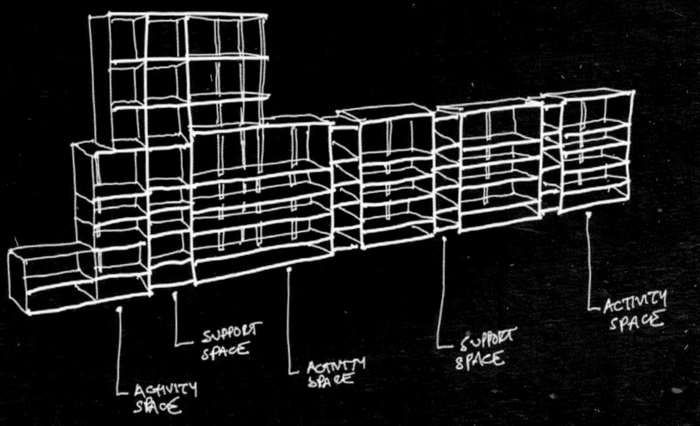

4. Perspective view of exterior

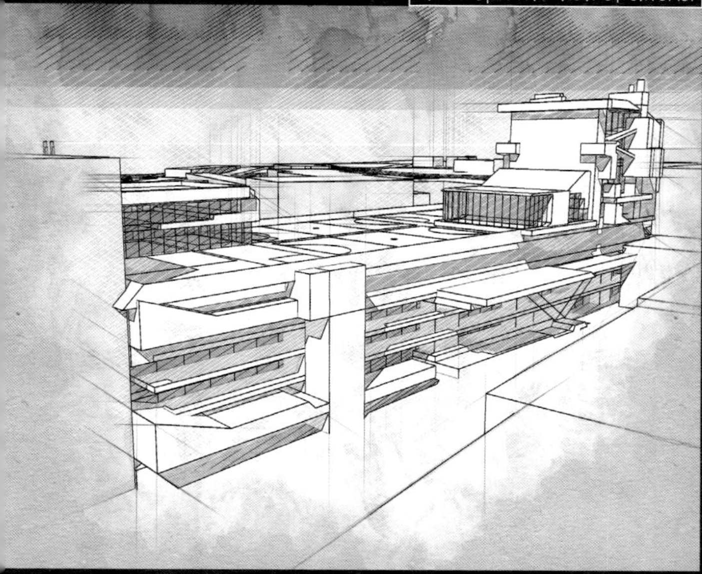

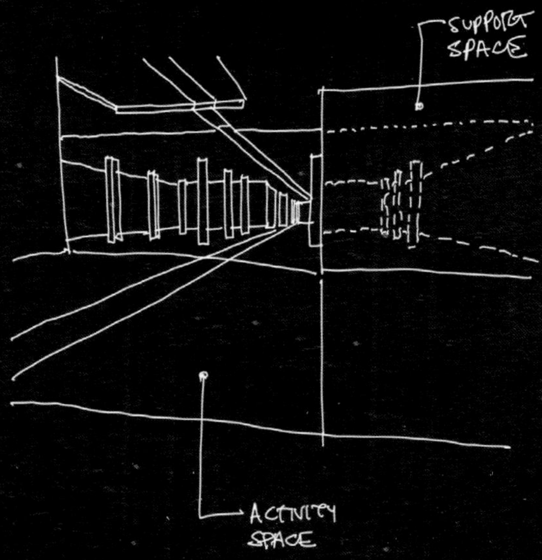

support, floor support and staircase support. With this approach, the structures forming the spatial arrangement of volumes are used to create the complete form of the building, without the need for additional structure to support internal partitions; all structural members contribute to the primary structure supporting the building. Wall panels and finishes, both internal and external, are fixed directly to this structure. This approach has less flexibility for changing internal walls during the lifespan of the building.

Use of both structural configurations in this project

For this building, an exoskeleton is used in the lower parts of the building in order to create a structure of external walls and volumes which provide both a primary structure and an enclosure within that primary structure. Internal columns are reduced or removed as the result of the load-bearing, box-like forms. Within this exoskeleton is a woven structure of columns, beams and slabs which creates the frame for the internal spaces that are visually differentiated in order to define walls and floors of internal spaces. The woven structure makes use of a regular braced frame of beams, columns and slabs but creates a high level of differentiation in the arrangement of structural members in order to create spaces which are individual and non-modular, while remaining economical in cost. A degree of modularity of setting out is created by the exoskeleton that does not need to follow the irregular grid of the woven structure. This approach is an alternative to using staircase and lift enclosures to provide elements of structure. This approach provides a set of visually 'nested' structures that offer flexibility by breaking down the structure of a large-scale building into a set of smaller structures. Two small-scale structural forms are set at roof level; they are structurally continuous with the volumes within the podium below. This mixture of exoskeleton and woven internal structure allows the quantities of material used in the construction of the structure to be minimised by using different structural concepts to accommodate a variety of spatial enclosures, with different expectations of being column-free or part of a set of support spaces.

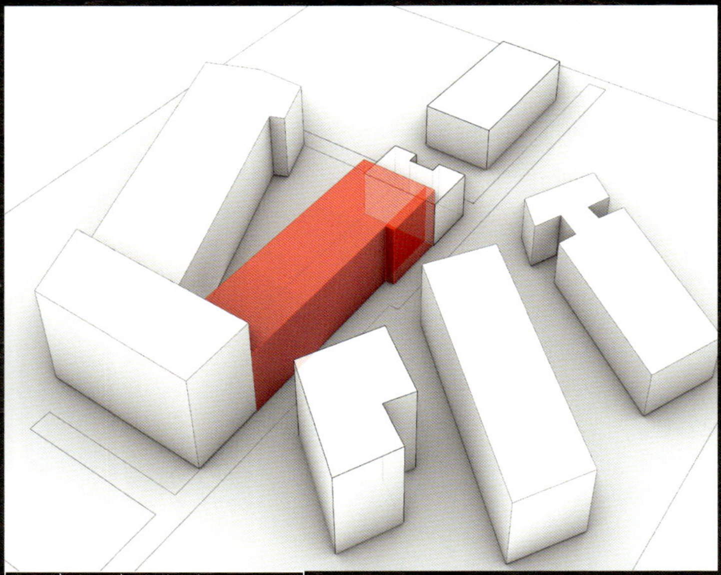

1. Project volume in site context

2. Project volume in site context

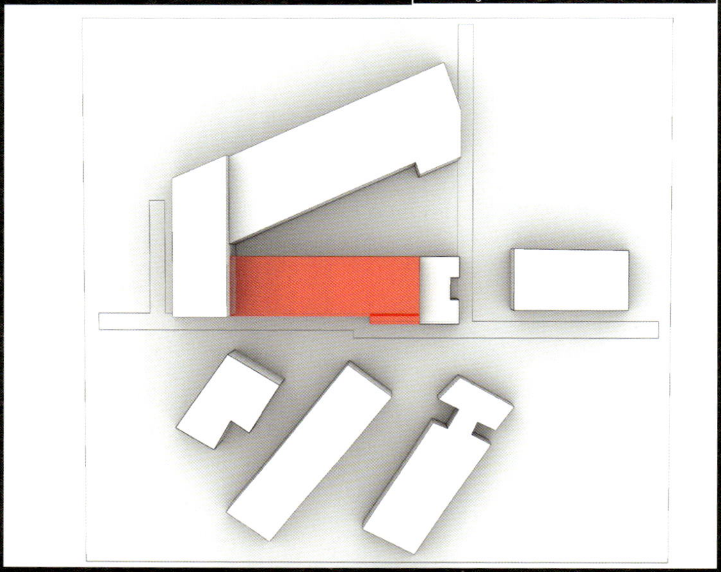

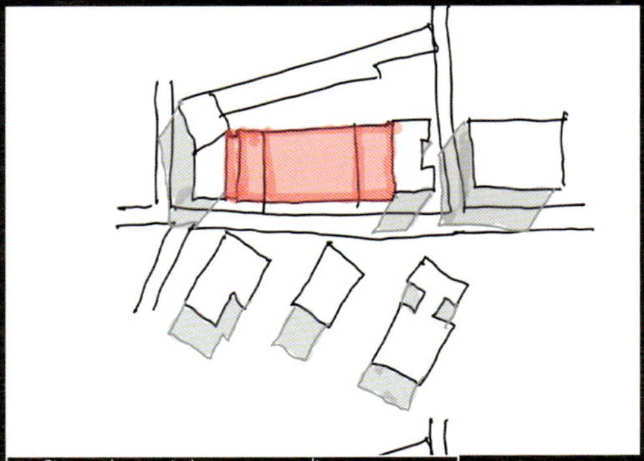

1. Site plan with surrounding context

Option 1 uses the site context of setting taller forms at the ends of the long volume, abutting taller adjacent buildings.

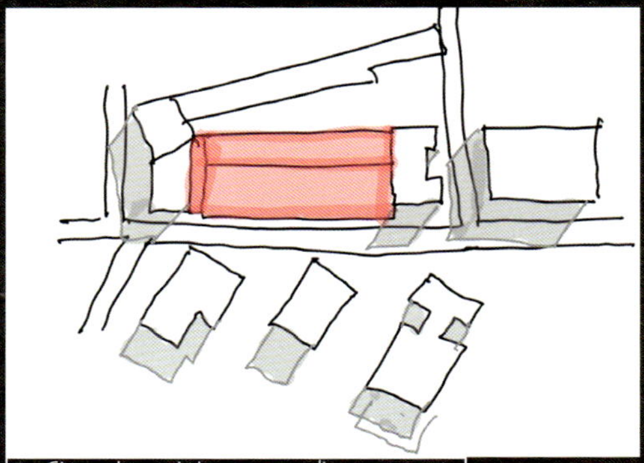

2. Site plan with surrounding context

Option 2 puts a thin strip of raised building along the back of the site.

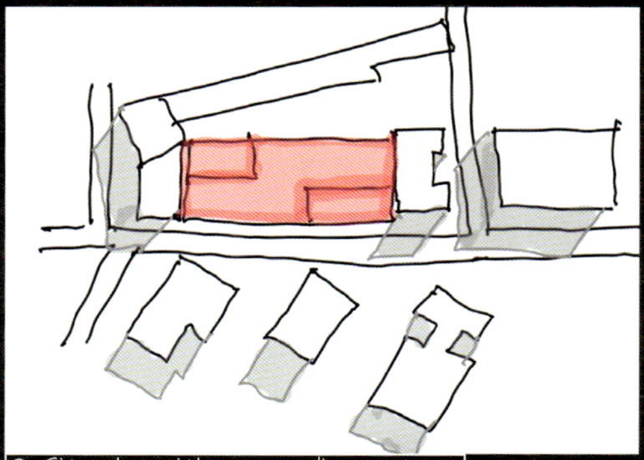

3. Site plan with surrounding context

Option 3 has separate building forms for the upper volumes, optimised for both circulation space and for access to internal spaces.

Site context

This university teaching building for sports science is set over a total of ten floors. The volumes comprise a podium, a series of rooftop sports facilities, and two small teaching blocks set at roof level. The upper part of the building, set above roof level, is a sports science centre with outside sports pitches as part of the teaching facilities. The structure, with its characteristic inclined members, contains stairs for key building-user circulation routes through the teaching spaces. Circulation occurs around the edges of the podium block and around the edges of the two blocks set at roof level. The rooftop buildings are divided into a series of tall spaces for teaching and gymnasia, combined with the long, low spaces within the podium.

This project addresses the issue of combining buildings of different scale into a single composition, without the constituent blocks becoming separate buildings, by superimposing dissimilar volumes and interlocking them geometrically. An alternative approach to the brief/program would have been to create a set of separate buildings arranged adjacent to one another, or a single rectilinear building block. Both solutions would have resulted in floor plans with deep floor plates and with little natural daylight, as the roof plane is needed to accommodate sports pitches. The design described here is a set of volumes which are very different in scale and function, allowing them to be identified individually as a set of juxtaposed small-scale buildings rather than as a set of spatially limiting superimposed floor slabs of repeated size. The advantage of the approach taken is the ability to create a cluster of small-scale building forms, each at the scale of the surrounding cityscape. Each form has its own identity.

Option 1 uses the site context to put to the taller buildings at the ends of the long volume, abutting taller adjacent buildings. This arrangement is needed in order to provide the large-scale accommodation in the lower part of the building, while smaller but well-lit teaching spaces are provided above. The massing of the site results in taller spaces being required. The disadvantage of this option is that large areas of internal circulation are required in each of the two blocks at each end.

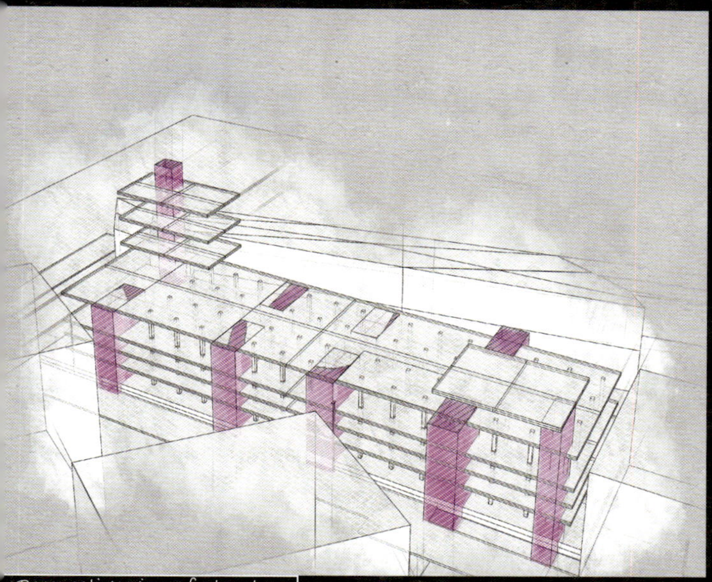

. Perspective view of structure

2. Perspective view of structure

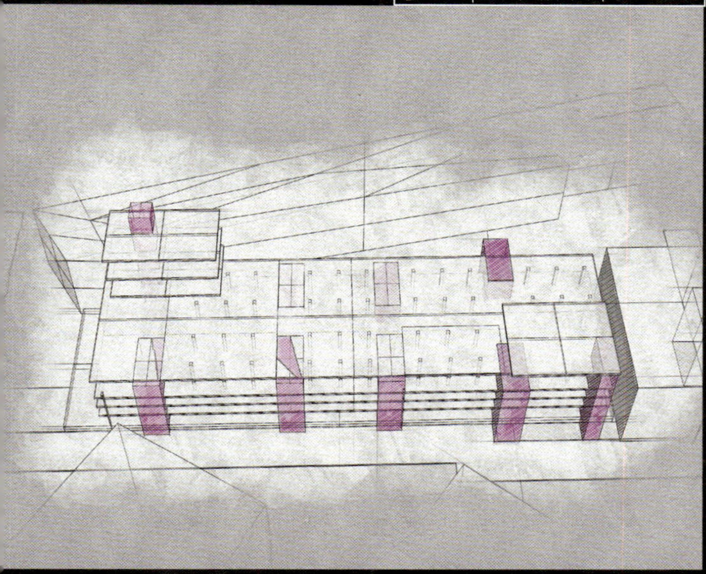

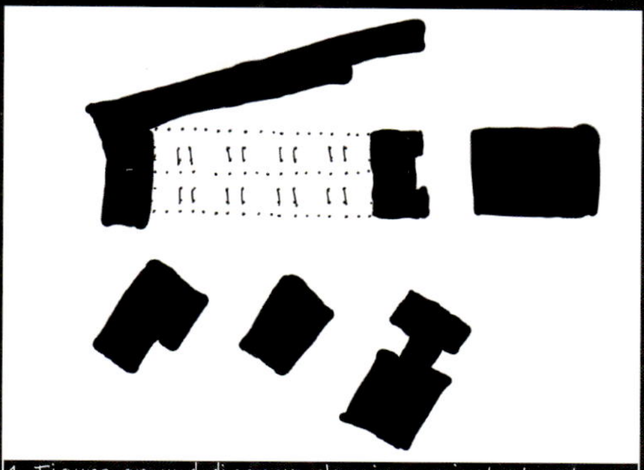

1. Figure-ground diagram showing project structure

Option 1
The structure fills the site with a grid of columns, assisted by cores in a rectilinear arrangement.

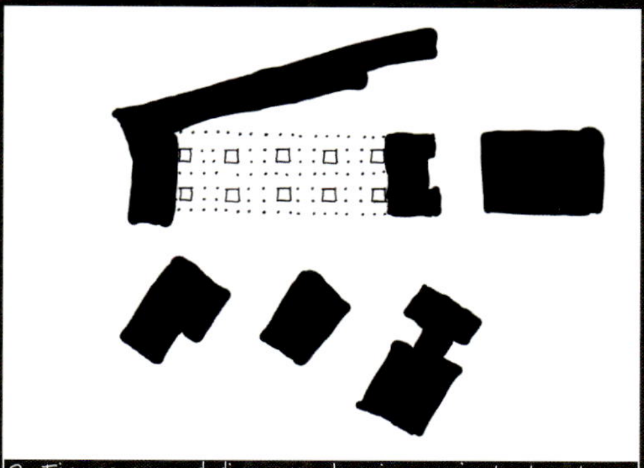

2. Figure-ground diagram showing project structure

Option 2
The cores are arranged in a non-rectilinear arrangement.

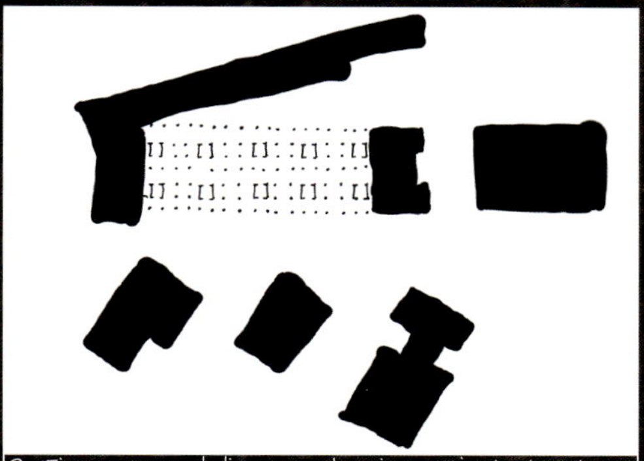

3. Figure-ground diagram showing project structure

Option 3
The cores are arranged in order to provide stability to the structures that continue up from the roof of the podium.

Option 2 puts a thin strip of raised building along the back of the site. This keeps this high-volume form away from the street where it would be visually dominating. In terms of urban design, the taller space was set at the back of the site in order to maintain the line of the existing streetscape, but alternatively the long thin band of raised upper volume could have been placed along the street without negatively impacting the open massing around it.

Option 3 has separate building forms for the upper volumes, optimised for both circulation space and for access to the different spaces required, from the hall to individual teaching spaces. The arrangement of forms, when seen at the scale of the surrounding urban environment, is such that each volume fits against the existing taller building next to it; continuing the scale and form of that existing building. The other volume is set away from the street or upper level fits, with the upper level which has the taller volume positioned alongside.

This arrangement fits comfortably into the existing urban forms, and is sympathetic to those forms in that it continues the volumetric arrangement of surrounding buildings. For all options, the existing buildings 'bookend' the building at the long ends; these existing buildings are owned by the university. As a result, the new building has direct links into the existing buildings on several levels, unifying the two separate buildings and bringing them into a single set of spaces which are bridged.

The functions of each part of the building are currently independent; the changes in the design allow for the existing fabric to be integrated rather than demolished and replaced with a single new building. As a result, the new building is smaller than would otherwise have been needed if the existing buildings were demolished. The new design aims to integrate the new with the old, giving continuity with the existing streetscape in terms of overall massing while creating a habitable roof in the form of courts for team sports at roof level. Many of the surrounding buildings have flat roofs which are not used; this project aims to bring an animation and use to roofscapes rather than limiting their use to weather protection and enclosure.

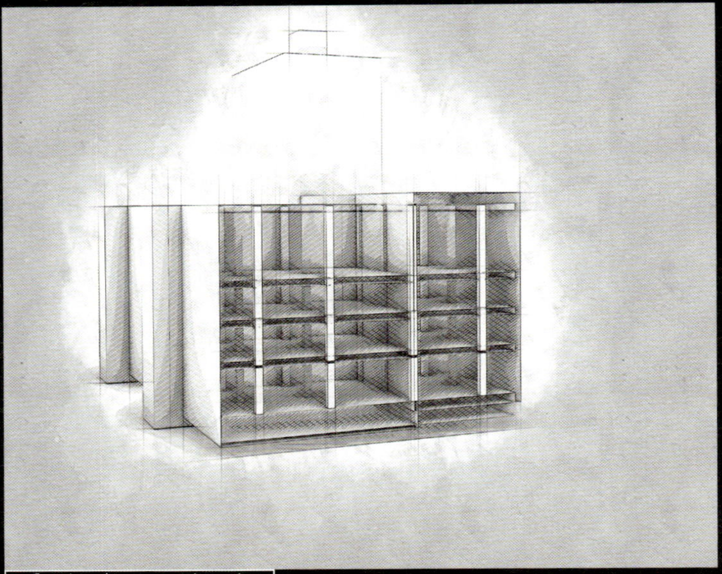

1. Sectional perspective view

2. Sectional perspective view

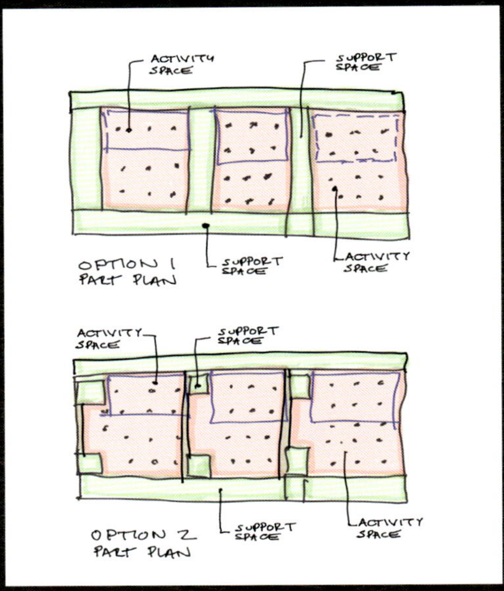

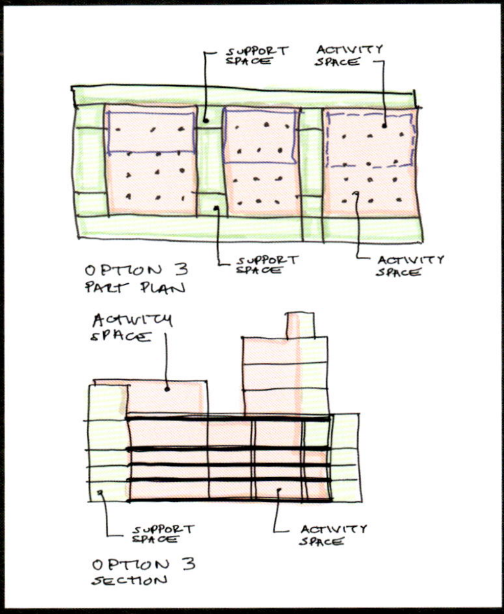

Spatial arrangement

Option 1 uses the 'bookended' volumes extending above the podium building, and provides access along the front of the building on the street side.

Option 2 sets the rooftop mini-buildings at the courtyard side of the site, away from the road.

Option 3 provides access to the school either by way of the street or by way of the courtyard.

Spatial arrangement

Spatial arrangement options are based on superimposed but dissimilar volumes, each option having its own spatial requirement.

The spatial arrangement in Option 1 uses the 'bookended' volumes extending above the podium building, and provides access along the front of the building on the street side, as well as a secondary circulation space on the opposite side facing the courtyard. This option creates open space within the podium of the building with access on all four sides to teaching spaces at upper levels. Spaces within the podium are side-lit and are of similar height. A disadvantage of this option is that primary circulation routes extend the full length of the building on both sides of the podium, with corridors linking these primary routes.

Option 2 sets the rooftop mini-buildings at the courtyard side of the site, away from the road; alternatively, this arrangement could be mirrored so that the roof-level blocks are set on the street side of the site. The current arrangement of Option 2 allows access to be situated mainly on the street side rather than the courtyard side. The position of the projecting rooftop buildings suggests that the courtyard side would be the preferred side for primary access, allowing vertical circulation immediately into the taller parts of the building. Circulation across the width of the building is provided by three corridors, providing access to a large gallery and mixing space set on the street side of the school, which could be closed off to public access. The disadvantage of this option is that it concentrates the circulation of students on one side of the building, whereas for a large-scale university school building of this type, it is preferred to even out the distribution of people's movement, so that there are no congested areas.

Option 3 provides a more balanced distribution of movement around the site, combined with flexibility of access, by having access to the school either by way of the street or by way of the courtyard. Either access route provides the required functionality with the two buildings projecting above the podium, set at two opposing diagonal corners of the site, which creates a series of spaces that interlock the two mini-towers with the podium beneath.

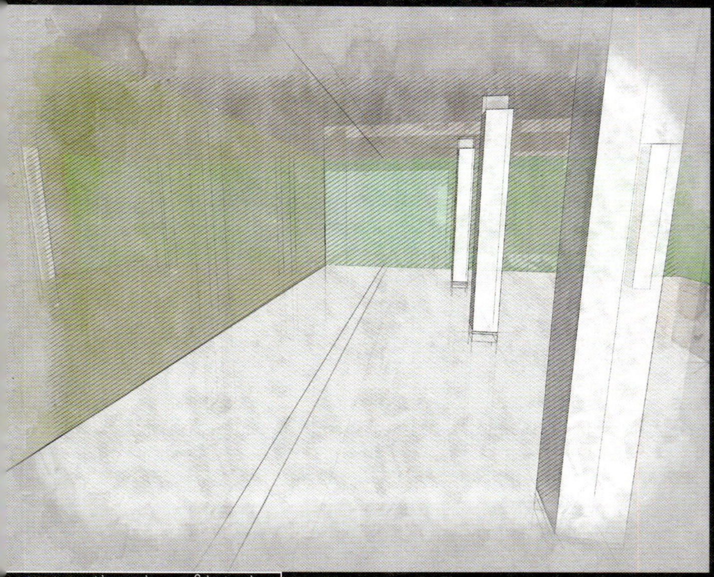

. Perspective view of interior

2. Perspective view of interior

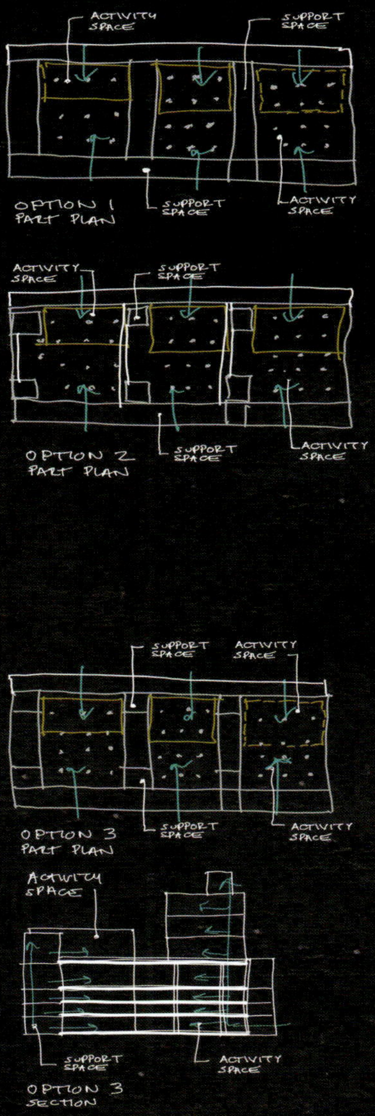

Internal circulation

Option 1 has primary entrances on the street side of the building, and a secondary entrance on the courtyard side.

Option 2 traverses the width of the building allowing for smaller spaces to be accessed within the depth of the site, and suit access to the blocks that project above roof level.

Option 3 has two central corridors as primary circulation routes across the building.

Internal circulation

For all three options, circulation routes through the building are set along the edges, where there are deep spine cores, partly glazed on both sides. Staircases allow students to pass freely from floor to floor; circulation areas also accommodate support spaces for teaching spaces as well as the large sports hall adjacent to the entrance. The vertical circulation zones provide access upwards to both the rooftop sports area and to the rooftop buildings. The internal circulation of all three options is based on being able to cross the width of the building down three internal circulation zones, as shown in Option 3. The circulation of Option 1 assumes that there are primary entrances on the street side of the building, and a secondary entrance on the courtyard side. This entrance arrangement can be swapped to suit changing access arrangements for all students. The large number of students using the building makes it necessary to have a long entrance space on the street side of the building to provide semi-public access for staff and visitors. Secondary access and maintenance access can be made on the courtyard side. The long, thin circulation zones on both sides of the building provide access to the full width of the front of the building without the need for a space that links the two zones as part of the domain with public access. This arrangement creates a clear separation between the semi-public and secure areas, teaching rooms and lecture spaces. Option 1 also provides access into the two bookend volumes projecting above the podium roof. The circulation zones of Option 2 across the width of the building allow for smaller spaces to be accessed within the depth of the plan, and suit access to the blocks that project above roof level.

Option 3 has circulation corridors which are wider than in Option 2 in order for the two central corridors to become primary circulation routes across the university department. The two long facades have circulation areas which are reduced when compared with Option 2. The arrangement of the circulation in Option 3 is the same for both sides of the building as a result of providing corridors which extend the full width of the floor plans.

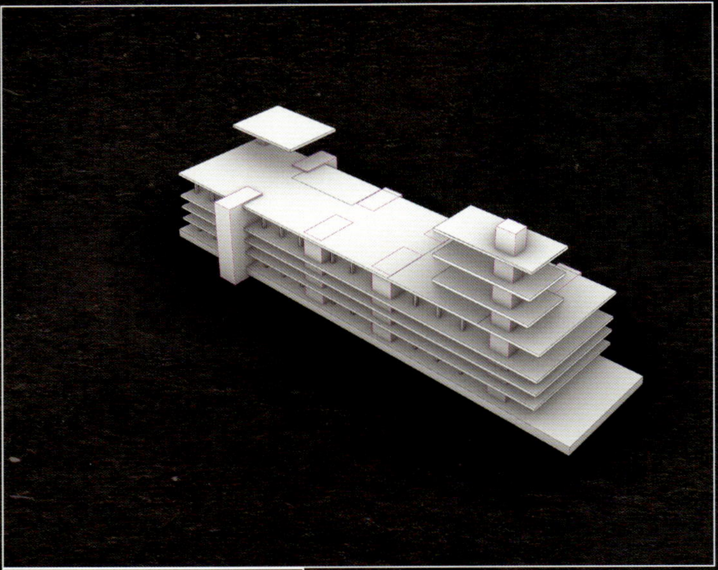

1A. Isometric view of structure

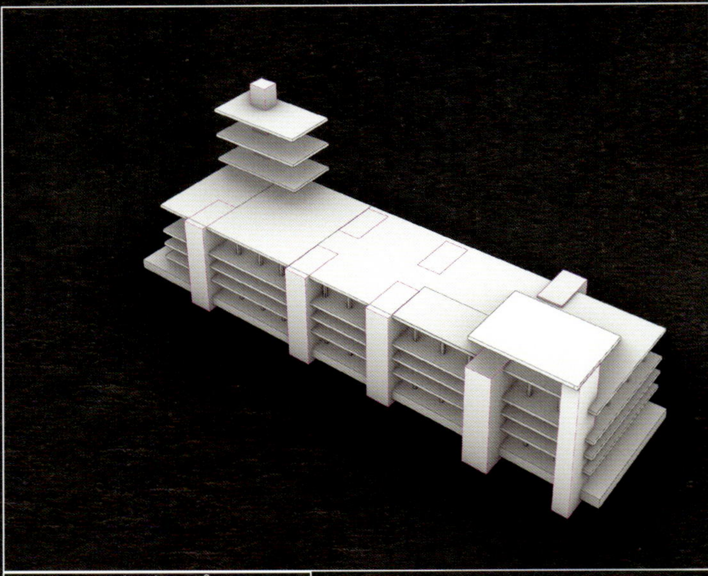

2A. Isometric view of structure

1B. Isometric view of structure

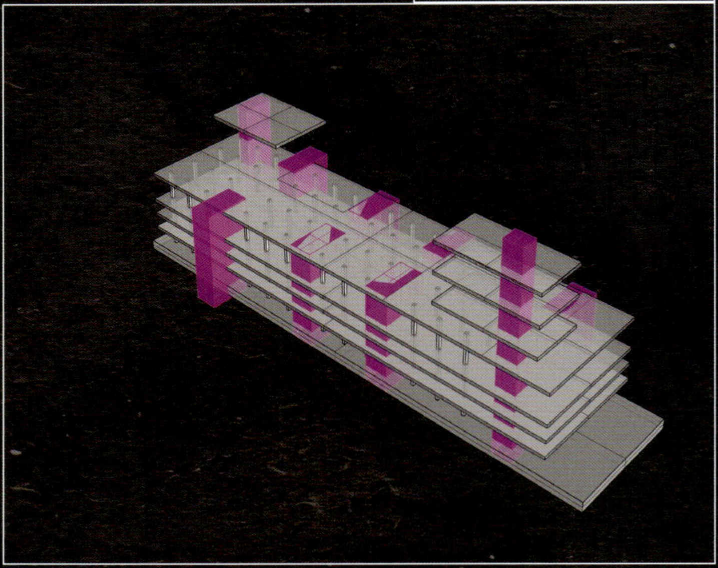

2B. Isometric view of structure

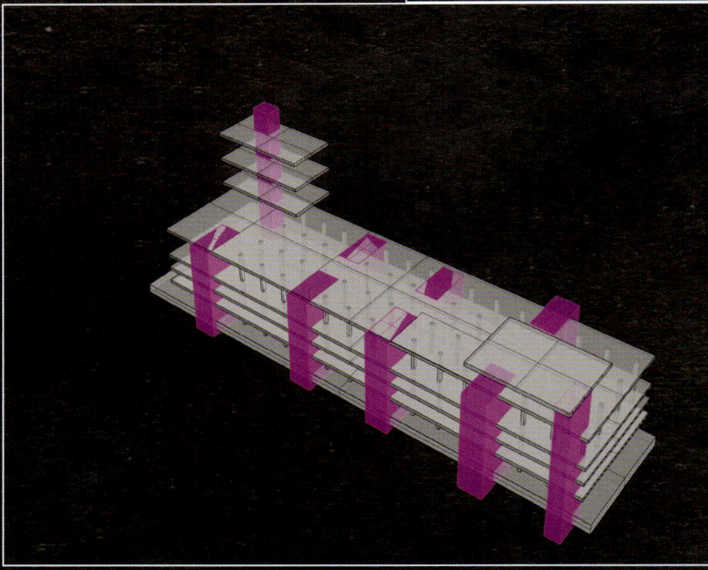

Structural strategy

The structural design is based on sets of reinforced concrete frames, walls and slabs which are of different scale, and which use exoskeletons at the perimeter of each building form and woven structures within each exoskeleton. The 'scale' of each set of reinforced concrete frames changes from regular structural grids of columns, beams and floor slabs to portal frames and long span trusses; allowing a single construction technology of post-tensioned reinforced concrete to be used, with its benefits of economy. The single construction technology used for all reinforced concrete elements includes the brackets and connections that join concrete components; a more economical solution than attempting to connect the structural members directly. The large structural frames and box-like forms comprise post-tensioned beams, floor units and long span trusses spanning spaces of different size; combining those with smaller-scale sections used as columns and beams at the corners of each building form. This solution requires each building volume to have a movement joint set between adjacent volumes which have different structural 'scales'; that is to say fundamentally different spans forming the structural grid of each 'mini-building'. The introduction of movement joints addresses the different characteristics of structural movements experienced between these mini-buildings. The movement joints provide opportunities to introduce thermal separation between each of the structures; each mini-building is thermally independent of its neighbour from the point of view of the environmental strategy. Double-wall construction is used at movement joints in order to provide an opportunity for acoustic separation as well as to provide a fire break for compartmentation between the parts of the overall building.

The structural concept for Option 1 is to have two braced frames as separate structures extending from ground floor level to the top of each of the mini-towers set at podium roof level, surrounded by the podium structure. These mini-towers are further differentiated into smaller-scale forms that create the more complex enclosures and spaces required within the building. The two mini-tower structures stand inde-

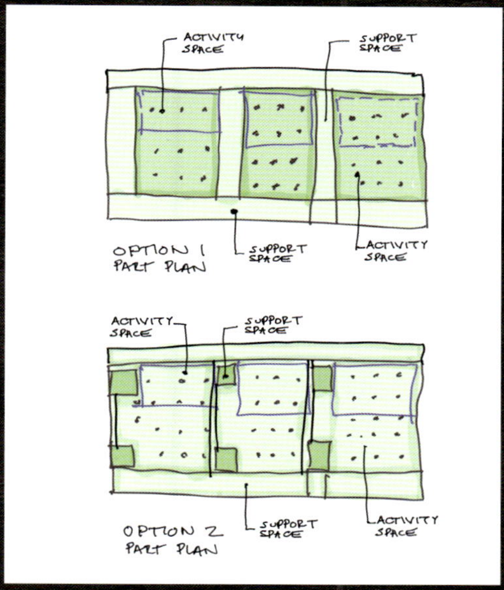

Structural strategy: Zones
Option 1 has two braced frames as separate structures extending from ground floor level to the top of each of the mini-towers.
Option 2 provides greater possibilities for the future subdivision of space by making the structural grid smaller, rather than larger.

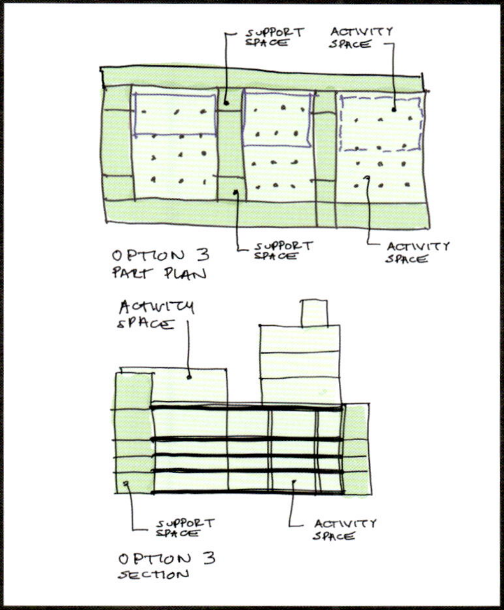

Structural strategy: Zones
Option 3 arranges the volumes of the two projecting mini-towers as separate structures.

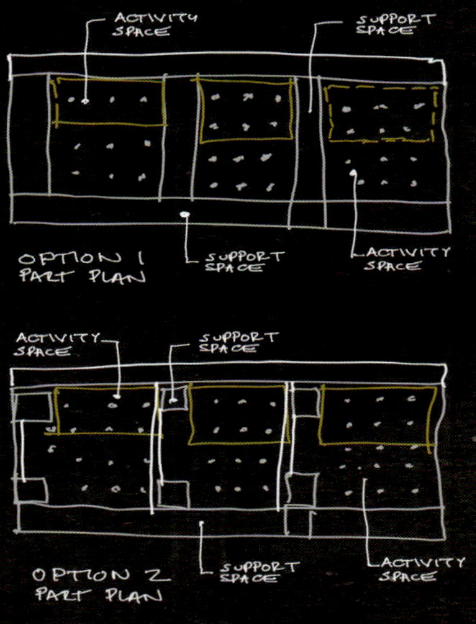

Structural strategy: Framing
Option 1 has mini-towers which are differentiated into smaller-scale forms that create the more complex enclosures and spaces required within the building.
Option 2 provides a smaller grid for greater points of support for adding new accommodation at roof level.

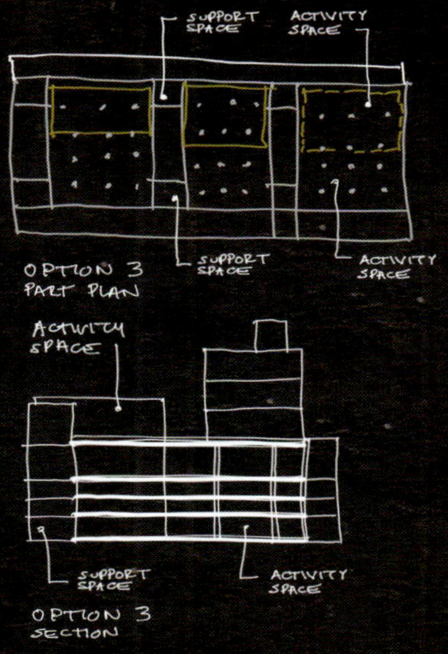

Structural strategy: Zones
Option 3 subdivides the podium into three linked structures.

pendently of the central podium building, which is set adjacent to them but which has its independent set of structures. This approach allows each structure within the overall design to be optimised for reduction of material quantity while performing with structural efficiency; the structural needs of the towers are different from those of the longer-spanning spaces in the podium building, particularly at roof level.

Option 1 comprises a long thin mini-building forming a bar-shaped form at roof level that projects above the podium roof level, forming a separate structure from the podium building beneath. The podium has large-scale spanning trusses and beams that form most of the spaces within; the bar-shaped structure encloses much smaller-scale spaces. Teaching spaces in the mini-towers have no columns in any classroom space; this was set as a standard requirement in the brief/program.

Option 2 accommodates the entrance spaces at ground floor which projects behind the tower part of the podium level. This option aims to provide greater possibilities for the future subdivision of space by making the structural grid smaller, rather than larger. A smaller grid provides greater points of support for adding new accommodation at roof level, where most changes and developments are expected to take place. The denser grid of columns provides more possibilities for a range of new floor plates at the roof podium level.

Option 3 arranges the volumes of the two projecting mini-towers as separate structures. The podium is subdivided into structures; one immediately adjacent to each of the mini-towers at each end of the site, a second structure set between the mini-towers, and a third podium structure in the middle which extends the full length of the site. The three podium structures suit the arrangement of circulation routes around the building. This structural approach allows each part of the building to have a structure which is optimised; not only for the quantity of material used, but also to allow adjacent spaces within the building to have different structural forms. The approach also allows the use of a single structural system throughout the building without being constrained by the choice of structural forms; each structure is optimised for the functions it serves.

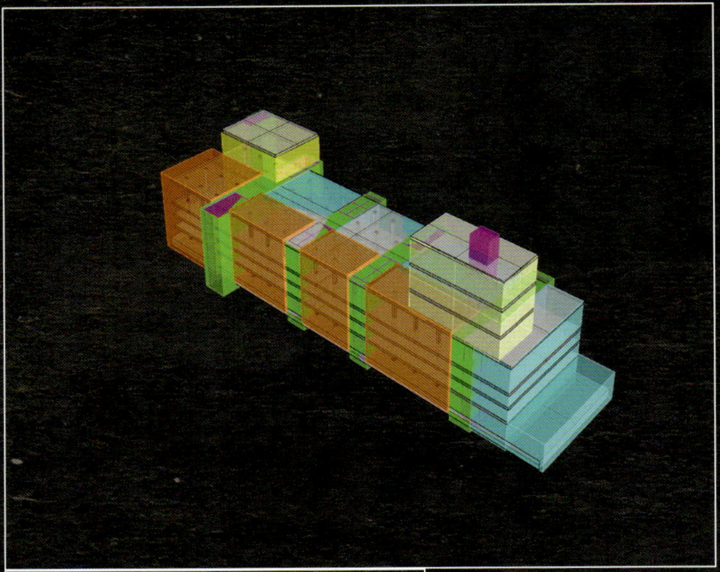

1A. Isometric view of environmental volumes

1B. Isometric view of environmental volumes

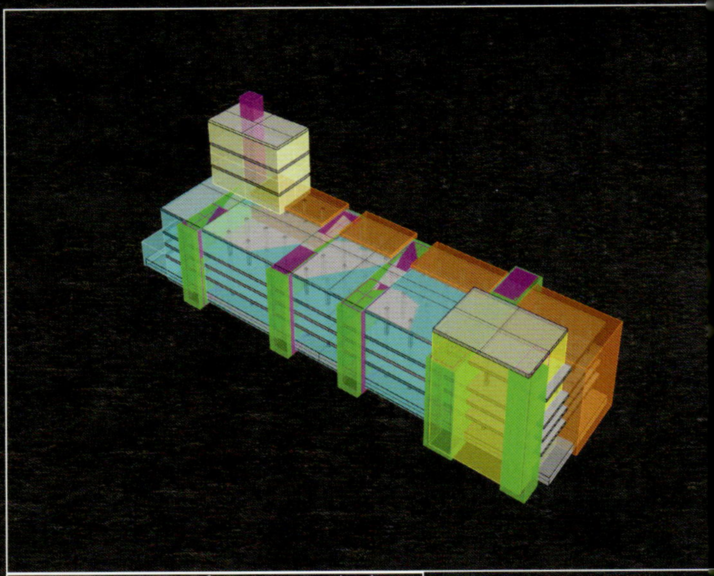

2A. Isometric view of environmental volumes

2B. Isometric view of environmental volumes

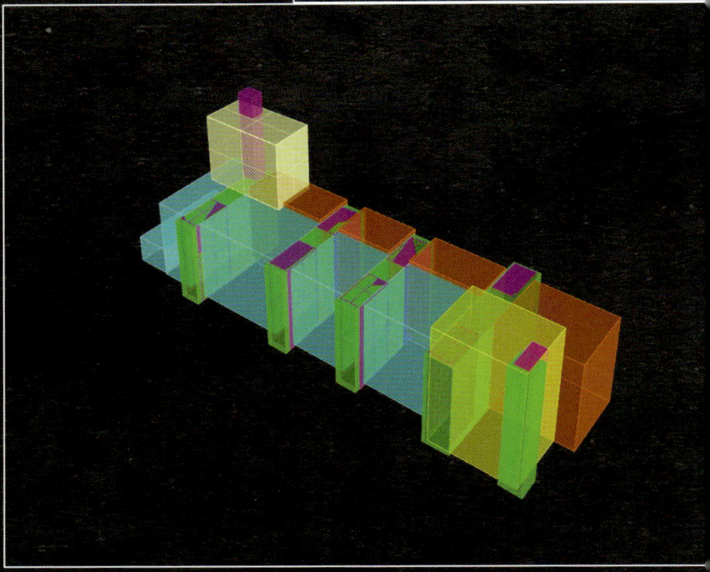

Environmental strategy

The separation of the constituent parts of the building into mini-buildings, with their different scales of construction and structural system, allows each mini-building to form its own specific environmental zone; each zone corresponds to a grouped set of spaces with similar environmental performance requirements. The environmental strategy for all zones centres around maximising the provision for natural ventilation where possible and a thermal performance of the building fabric to suit heating and cooling systems that work well with natural ventilation. Some of the mini-buildings have lower levels of thermal insulation than their neighbour, but higher thermal mass; others have an opposite configuration depending on their frequency of use. Larger spaces, such as sports halls, have more regular hours of use than smaller spaces; they are continuously heated or cooled during the day and are naturally ventilated, to some extent, at night. Smaller-sized spaces do not benefit as much from natural ventilation at night as a result of their configuration; they rely more on mechanical ventilation than the larger-scale sports halls. As result, the structural strategy – which provides exposed surfaces that contribute thermal mass – and the environmental strategy, which aims to minimise energy consumption within the building, are linked accordingly.

The mini-buildings that are set above podium roof level are self-contained with regard to their functions, frequency of use and environmental performance. As a result, the mini-buildings above roof level form their own environmental zones; they are typically more highly thermally insulated than spaces in the podium below. When configured as mini-towers they have passive solar shading provided by the geometry of the projecting beams and walls that are integrated into both the primary structure and the external envelope. Spaces in the podium below have electrically-operated solar shading devices as a result of the changes of use within those spaces during the day. Shading devices in the podium are deployed typically from season to season rather than from hour to hour, so these devices have fixed positions for summer, winter, and mid-season. The environmental strategy for Option 1 is one of

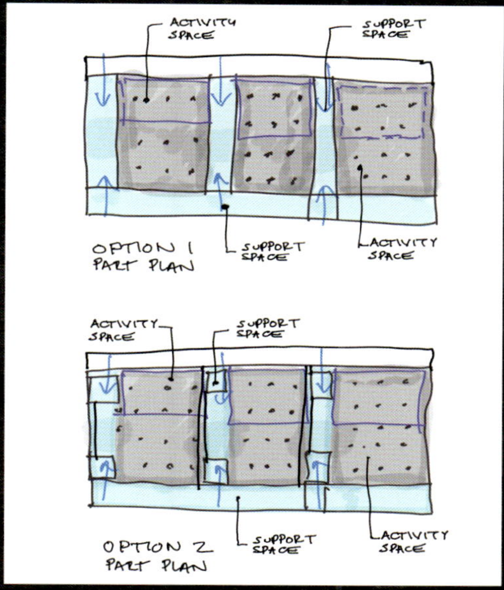

Daylight strategy for direct sunlight and cross ventilation.

Option 1 admits daylight and air into circulation spaces. Activity spaces have controlled daylight and sunlight.

Option 2 admits daylight and air into circulation spaces and cores.

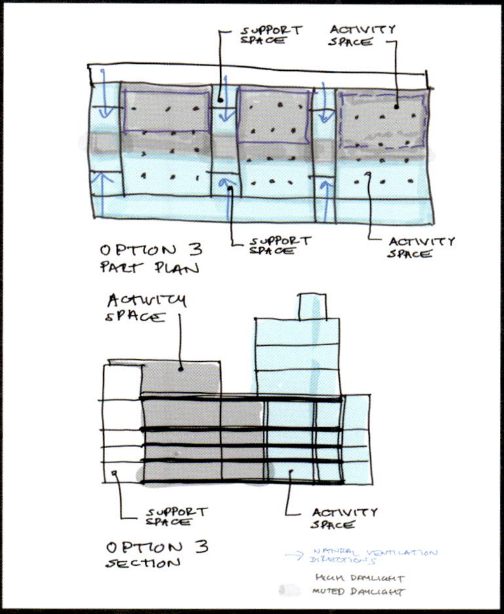

Daylight strategy for direct sunlight and cross ventilation.

Option 3 admits daylight and air into some activity spaces as well as cores.

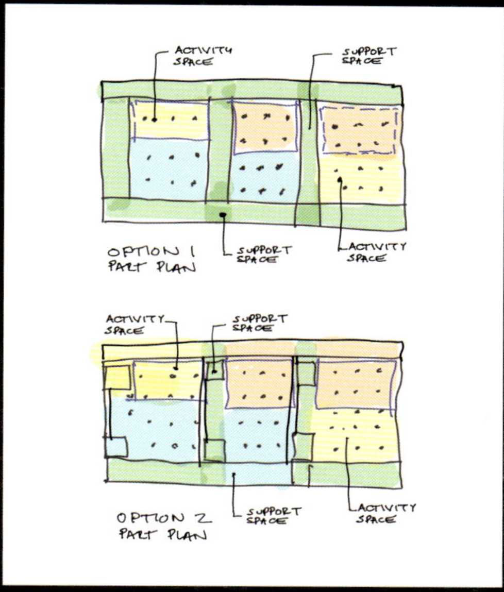

Environmental zones

Option 1 has zones where each environmental zone is separated physically from its neighbour.

Option 2 uses the vertical circulation zones of stairs and lifts in the podium as a single environmental zone, with the podium spaces comprising a second environmental zone.

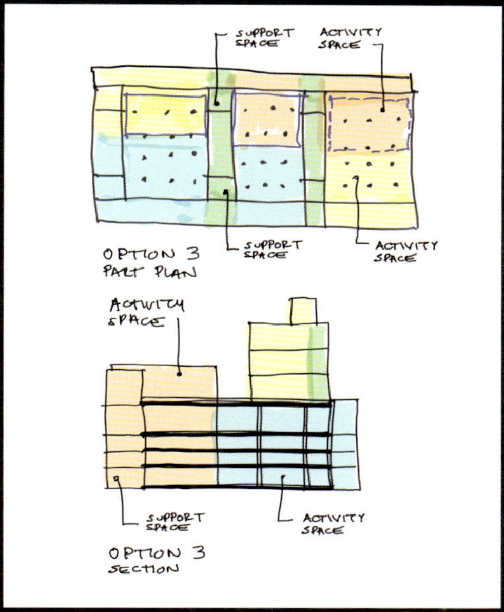

Environmental zones

Option 3 has shared environmental zones at the lower levels.

zones where each environmental zone is separated physically from its neighbour. The building is divided into three volumes for the podium and for the mini-towers. Circulation zones in Option 1 are treated as an independent environmental zone within the podium and the mini-towers above; each part of the building having its own circulation zone which is set adjacent to the corresponding environmental zones within each mini-building. The advantage of this approach is to allow circulation areas to become environmental 'buffers' or separators between adjacent spaces.

Option 2 uses the vertical circulation zones of stairs and lifts in the podium as a single environmental zone, with the podium spaces comprising a second environmental zone. The two mini-towers have separate environmental zones, as in Option 1, with the ability to sub-divide environmental zones further in order to reduce energy consumption within specific spaces.

Option 3 has shared environmental zones at the lower levels, where the primary spaces in the podium supply air to adjacent circulation spaces as a result of sports halls having extra capacity in their supply of air during parts of the day; air can be redirected. Spaces in the podium share and re-distribute fresh air with adjacent rooms and halls during the day as a result of this spare capacity. Local adjustments of ventilation between spaces in the podium reduce overall energy consumption across different environmental zones. An advantage is that the circulation spaces can benefit from the heating or cooling energy that is not required in the main spaces of the podium. In order to provide a single set of circulation zones within the building, different environmental zones can contribute fresh air to the circulation zones at different times of day. Option 3 is the preferred option as it provides the most amount of flexibility for future development by using the routes for internal circulation set out in the corresponding spatial arrangement and associated structural design.

In all three options, the environmental zones and their distribution are independent of the specific spatial arrangements within each environmental zone.

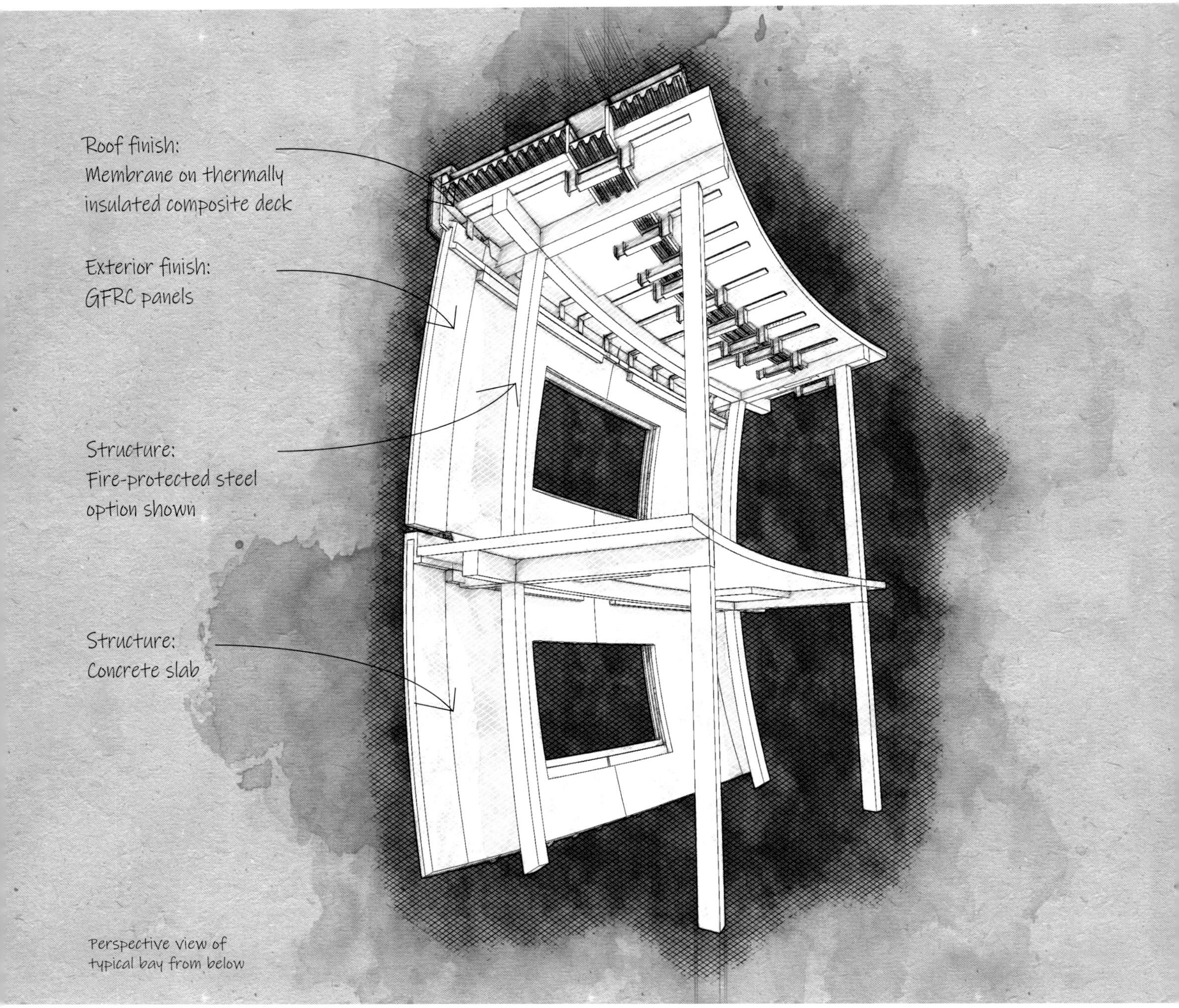

Perspective view of
typical bay from below

Typical bay adjacent to facade

The building has larger-scale opaque and glazed facade elements within the podium, and similar smaller-scale elements on the mini-buildings set above podium level. Areas of opaque facades comprise an outer face of GFRC (glass-fibre reinforced concrete) panels that also provide passive solar shading in response to local climate conditions, as well as controlling levels of daylight within the building and reducing solar gain. The mini-towers set above podium level each have an energy recirculation system such that the environmental controls for these two blocks are independent of each other and of the podium. Each building has its own roof-mounted environmental equipment; the one that is required for the podium is set at podium roof level, away from the mini-towers. Each element functions as an independent building which brings the advantages of energy efficiency to each element.

The external envelope uses the same language of opaque panels and glazing for all three options for spatial arrangement, circulation, structure and environment. The specific arrangement of panels is changed within each set of options. The system is one of floor-to-ceiling panels with a specific opaque/glazed mix within each panel, supported on the edge of each floor slab restrained at ceiling level. This approach avoids the need for curtain walling, which would require an additional layer of solar shading fixed to it. The floor-to-ceiling glazing system allows a separate solar shading panel to be used, which is fixed directly to the floor slab behind the panel.

This choice of slab-level panel has the benefit of providing additional acoustic separation and fire separation between floors. The slab-level panel provides both an economic method of enclosing part of the building as well as creating a high level of differentiation required between different types of spatial enclosure. Some panels are glazed from floor to ceiling and provide only facade enclosure and solar shading; others enclose technical equipment, such as ducts and small air-handling units. This technical equipment would have been required on the roof, where space is already occupied by sports facilities. This equipment is instead attached to the sides of the building. This approach makes

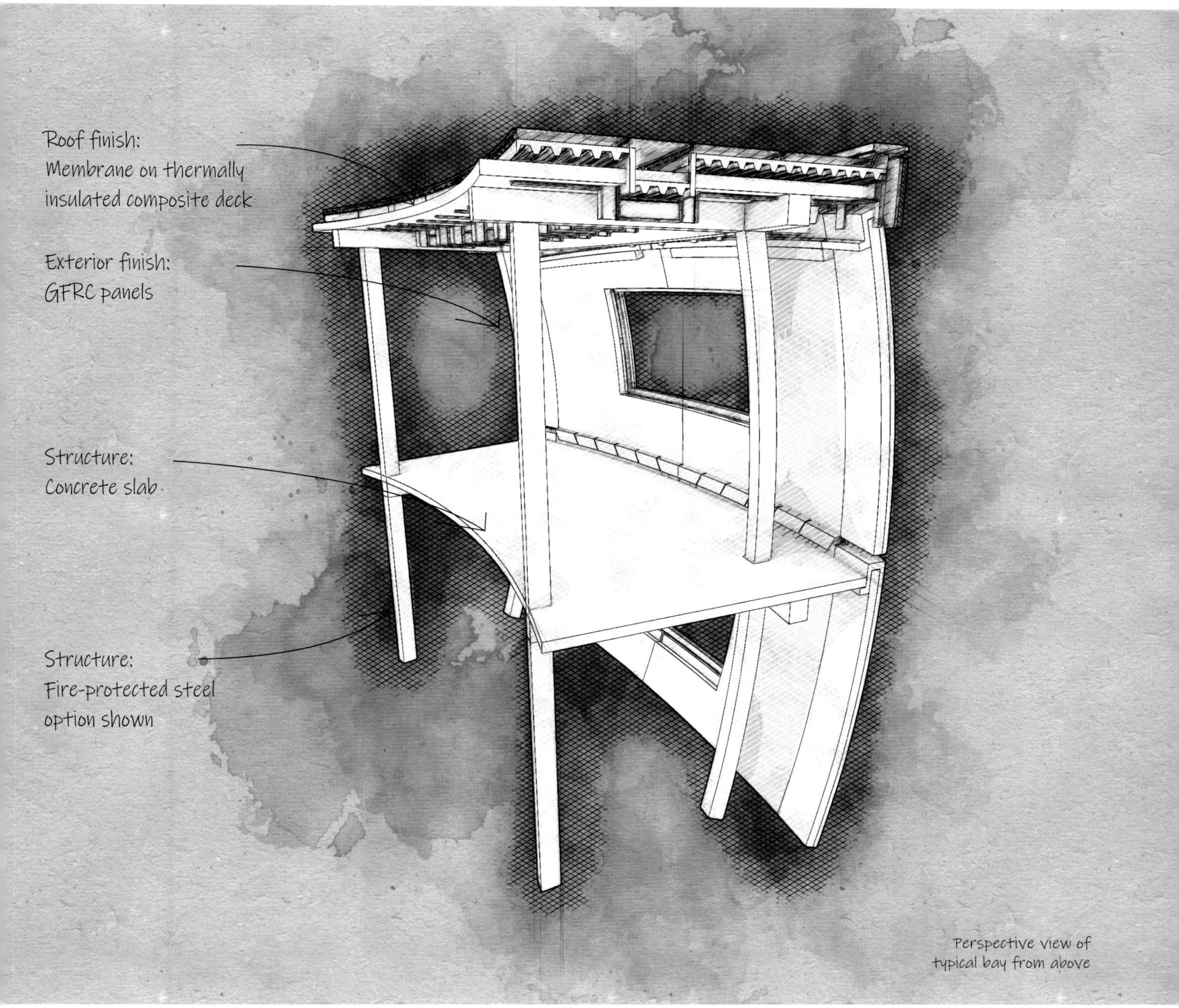

Perspective view of
typical bay from above

the integration of environmental controls straightforward by putting the equipment near the point of use; providing simpler controls for a mixed mode of both mechanical ventilation and natural ventilation. With electrically operated fans for natural ventilation and associated air-handling units for the tempered air, the facade system can provide passive solar shading together with a high degree of ventilation control. This approach also creates a visual unity for the highly differentiated external envelope. The juxtaposition of opaque panels and glazed panels allows each part of the building to be expressed at the smaller scale of the adjacent buildings rather than establishing a single large-scale visual expression.

These small-scale facade elements are no taller than 3.0 metres; combining to provide an overall sense of scale for each of the mini-buildings that comprise the design. The mini-buildings make use of the language of windows and doors taken from the existing buildings which are adjacent to the new building; there is a common identity between the new and the old. The external envelope of the new building is seen as an 'upgrade' of the language of windows and doors used in neighbouring existing buildings.

A typical bay, the junction with the roof, includes a zone that creates a wide parapet to separate the areas for sports courts from the facade facing the street. This edge of the building accommodates an array of horizontally set nets to collect items which might have found their way over the top of adjacent fences. This soft edge to the podium roof also accommodates air-handling ducts and service equipment beneath the soft netting. The netting is up to 6.0 metres wide in places in order to accommodate storage beneath for items of sports equipment, when not in use. The arrangement of parapet netting follows that of the facade framing immediately adjacent; the framing from the external walls appears to fold horizontally and terminates at the edge of the sports courts. The typical bay accommodates a siphonic drainage system that draws water off the roof in the event of heavy rainfall, avoiding the possible deterioration of sports court surfaces. The system works as gravity drainage when not required to perform siphonically.

1. Perspective view of exterior

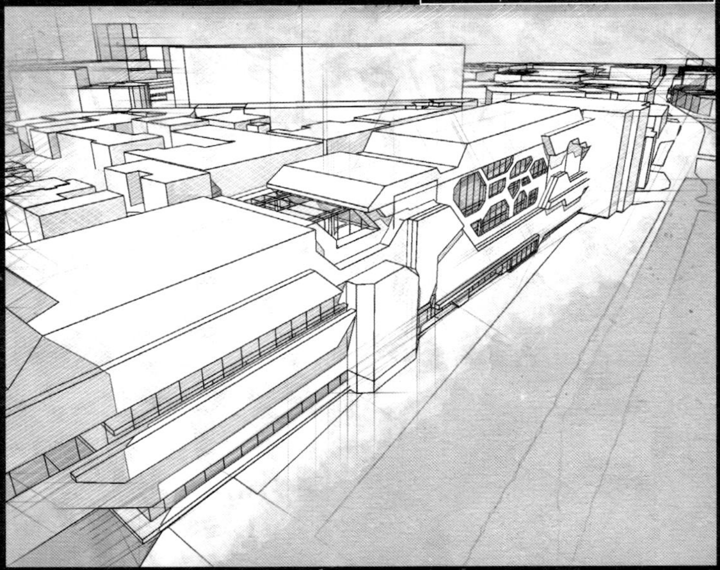

2. Perspective view of exterior

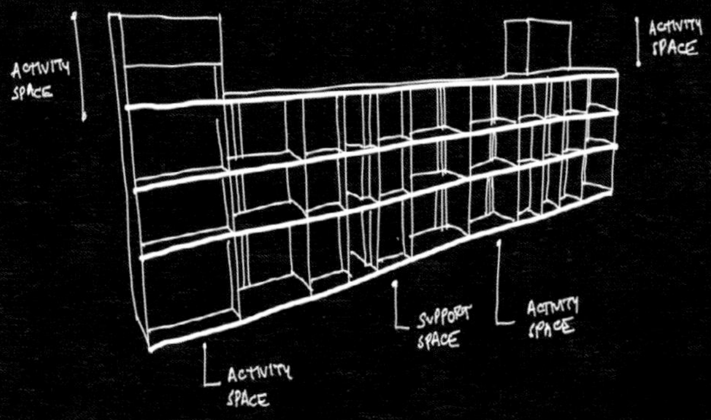

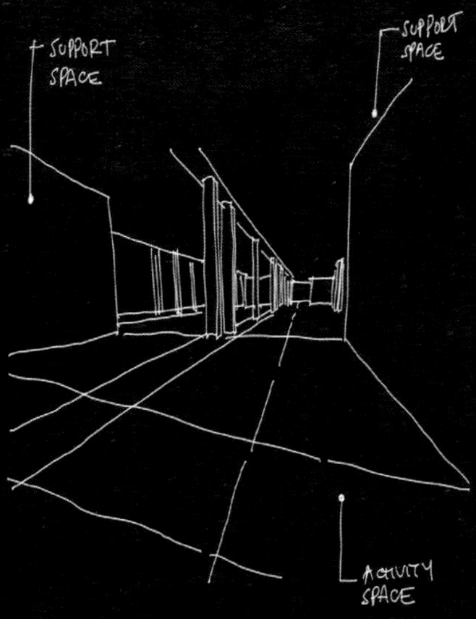

Scaling structure to create differentiated space

This project uses the structural approach of being able to scale down from large-scale to small-scale structures by separating structures with large-scale enclosures from adjacent smaller-scale forms within the same building.

A primary design consideration for the structure of this project is that different-sized volumes, including circulation spaces, are seen as having different structures as forms in their own right. As a result, an 'exoskeleton', or externally positioned structure is created on the two long sides of the facades; the two short sides are formed as party walls which are joined to the long facade structures to form a single exoskeleton box-like structure at the perimeter of the building. Inserted into the exoskeleton box are framed structures set on different structural grids; these elements are joined to the exoskeleton but are separated by a continuous zone of circulation space to separate the external 'facade box' from the internal 'boxed spaces'. This approach allows the outer facade to visually 'map' the circulation around the building for students, while allowing the visual expression within the building to reflect the arrangement of internal spaces; creating a 'building within a building'. Large spaces are formed from sets of smaller spaces; halls and and gyms are formed by a building-up of scale of space-creating components, from door to wall to small room to large room; spaces are created in this hierarchy by using the exoskeleton facade structure as the generator for spatial arrangement.

In this project the exoskeleton extends to the glazed areas of the building providing structure that can accommodate air-handling ducts and service runs across the extent of the facades. In addition, the structural facade form accommodates steps in the floor slabs where they intersect with the external walls. Consequently, the facades do not need to be rectilinear and do not need to follow the norms of curtain walling. Instead, the exoskeleton facade can accommodate glazed windows. This approach avoids the need for a full curtain wall to form an external envelope, with the needs of curtain wall for continuity from floor to floor and an allowance for structural movement, with its associated horizon-

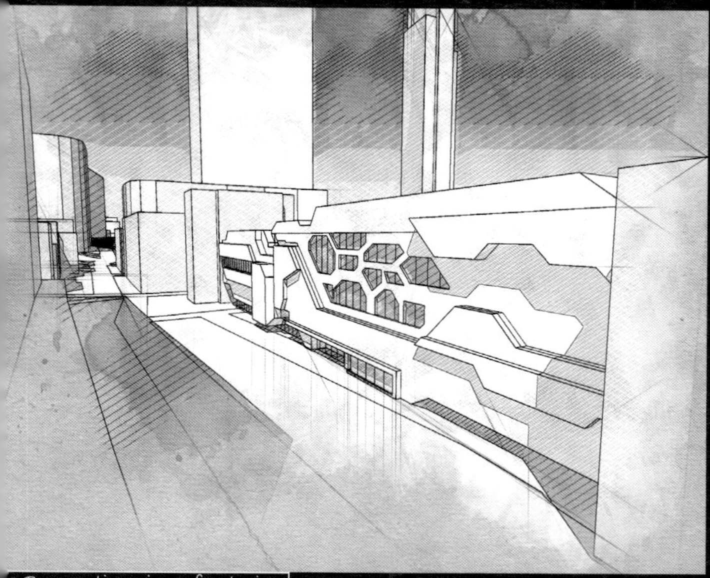

. Perspective view of exterior

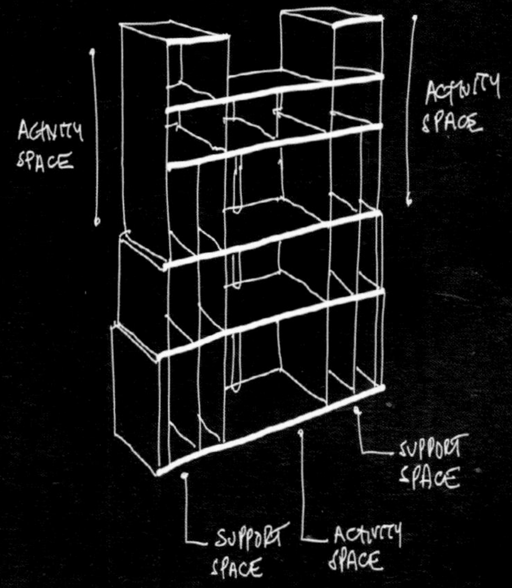

4. Perspective view of exterior

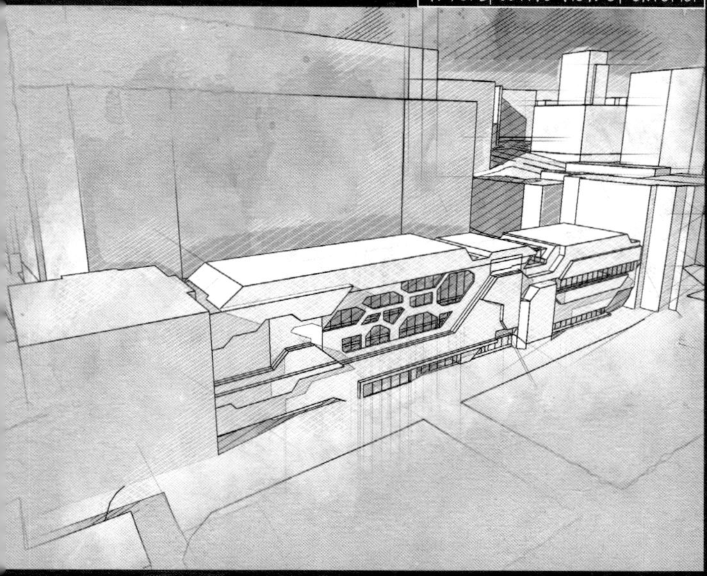

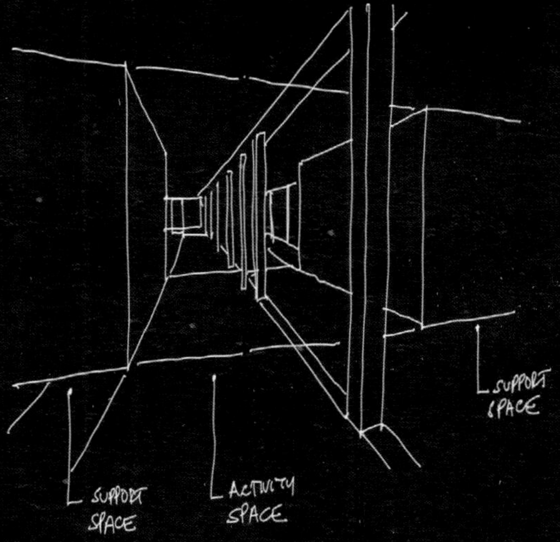

tal joints at each floor level. The approach of using an external structure provides the ability to create differentiation in the facade design in creating windows that address the specific needs of the interior spaces immediately behind. In this project, the spaces enclosed by windows, rather than curtain wall, are classrooms, corridors, sports hall and the school hall. The use of the exoskeleton as a primary structure removes the need for columns behind the external wall, as slab edges provide more flexibility for structural support in assisting the creation of the visual expression of the windows.

Another benefit of the exoskeleton in this project is its configuration as a twin-layer reinforced concrete wall where the concrete is exposed on its outside face. This wall type is formed as a diaphragm of two linked concrete walls with a layer of thermal insulation set between them, allowing the concrete to be structural and be used as an exposed material that avoids the need for additional external layers of thermal insulation and rainscreen panels, as is typically used to enclose reinforced concrete walls. This use of an exposed diaphragm wall permits the expression of exposed concrete on its outside face, providing a visible expression of the formwork used in its creation. The concrete walls are formed using reusable steel formwork moulds, used in preference to the coated plywood sheets that are difficult to reuse. The introduction of exposed reinforced concrete provides a richness of texture of surface finish to be brought to the material, as was more customary in earlier uninsulated applications of exposed reinforced concrete. In this project, the concrete is formed in large areas of smooth white concrete with an architectural finish; formed not as a flat surface but as stepped planes that give a three-dimensional relief to the facades. Joints can be formed with precision at the change in depth of the formwork; casting shadows across the facades. This approach allows for different textures and surface finishes of the exposed concrete walls to be introduced, allowing the scale of the concrete panels to respond to the cityscape in which the building is located. The exoskeleton incorporates slots for rainwater pipes and electrical services which are installed within the facade zones; accessible from inside the building.

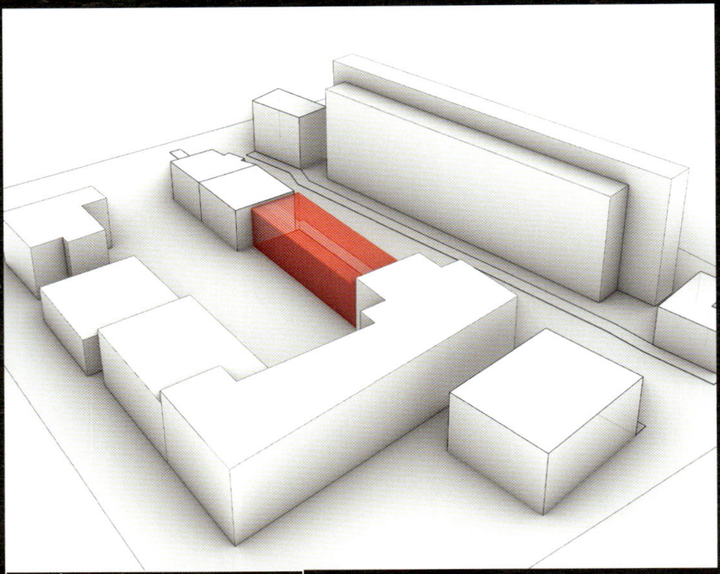

1. Project volume in site context

2. Project volume in site context

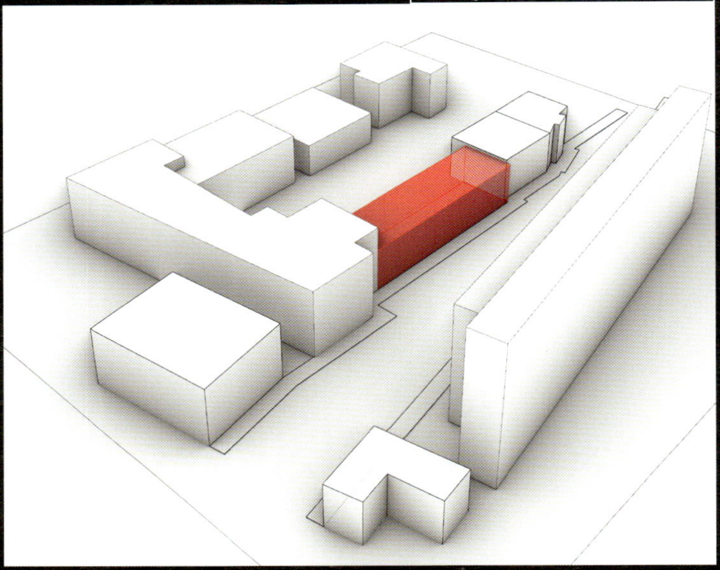

1. Site plan with surrounding context

Option 1 creates a partly sculpted facade that is created on the side of the building adjacent to an existing car park.

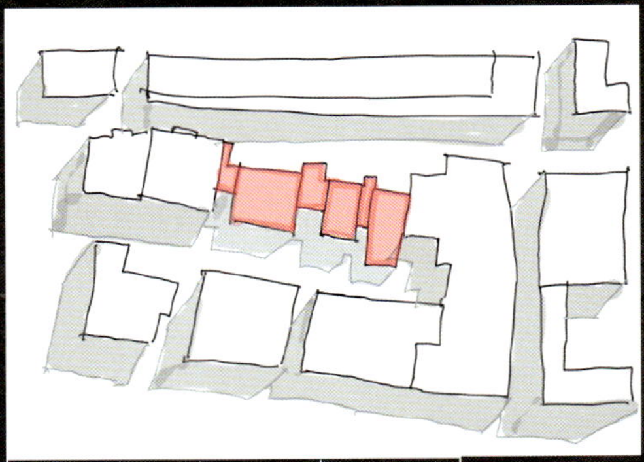

2. Site plan with surrounding context

Option 2 follows the line of the existing buildings that form the streetscape, rather than the existing line of the road.

3. Site plan with surrounding context

Option 3 creates a simple rectilinear form that fits in with the adjacent urban forms of the street and provides an economic 'built volume'.

Site context

This school building is arranged over six floors with small entrance spaces arranged along the length of the facades; stairs rise from these entrances up to the main school lobby and auditorium above. The auditorium is set immediately above entrance spaces. Adjacent teaching rooms are interlinked; set next to one other as a mixture of side-lit and top-lit spaces to provide a high level of variation in the daylighting of each space to suit the different teaching environments required for a specific subject. Arts subjects are largely top-lit; science subjects are side-lit, while some music rooms are entirely enclosed. This school has teaching spaces which are highly visible and highly articulated, as seen on the facades from outside the building, to facilitate a visual 'reading' of the functions within.

Option 1 for the site context aligns the school building with the line of the street; a partly sculpted facade is created on the side of the building adjacent to an existing car park. This approach allows the functions of rooms within to be expressed freely, and a geometry that is consistent with the internal arrangement, as required by the brief/program. In this option, the primary entrance is on the car-park side, away from the street, where there is more room for vehicles to stop and turn.

Option 2 follows the line of the existing buildings that form the streetscape, rather than the existing line of the road; dividing the brief/program into zones of space which are 'extruded' vertically, visually, to create a set of linked spaces of very different size. There is an expression of those smaller-scale compositions in the facades of both the car-park side and the road side. A limitation of Option 2 is that spaces are stacked vertically, one above the other, as the building takes on the form of a vertical 'extrusion' of smaller spaces. Option 1 allows more freedom in the planning of spaces by having a smoother box-like form with a projecting volume to accommodate the larger spaces in the brief/program.

Option 3 uses a simple rectilinear form that fits in with the adjacent urban forms of the street and allows for an economic 'built volume' to be created. The complexity of the brief/program is revealed in the

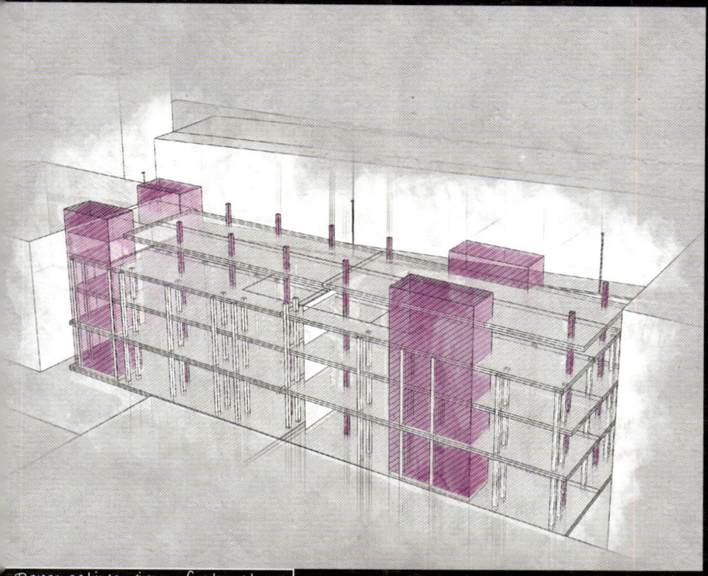

Perspective view of structure

2. Perspective view of structure

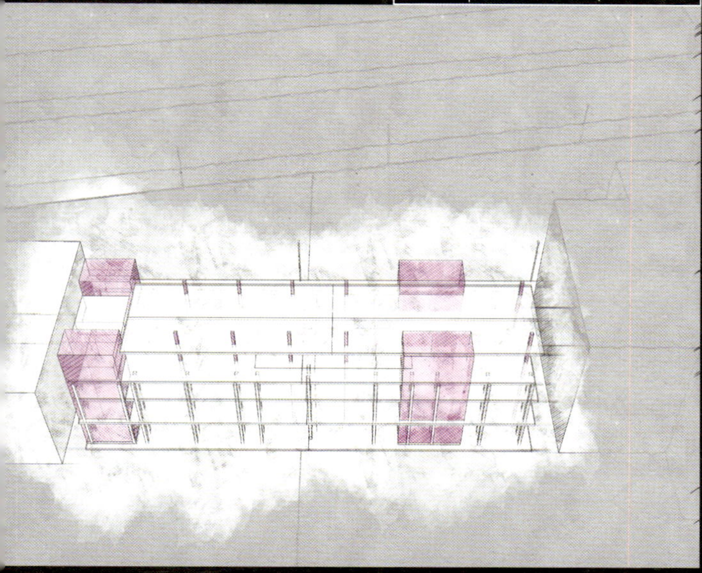

1. Figure-ground diagram showing project structure

Option 1
The structure forms a regular grid with cores that traverse the width of the building.

2. Figure-ground diagram showing project structure

Option 2
The structure forms a visually less regular grid with cores that are positioned to suit the structures above the roof level.

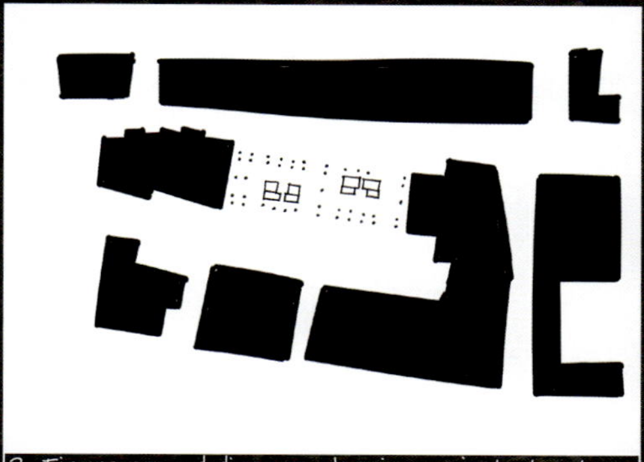

3. Figure-ground diagram showing project structure

Option 3
The structure forms a visually less regular grid with cores that are positioned to suit structures above the roof level.

vertical section through the building rather than in the horizontal plan, such that the forms of the building are stacked vertically above each other. The spatial forms nest as a set of extrusions which are extruded horizontally rather than vertically. This arrangement allows spaces to have levels of daylight and acoustic performance which meet the brief/program in a way in which a rectilinear classroom as a teaching space or performance space cannot provide. The volume also allows the provision of an internal sports hall at roof level, which accommodates internal space for team sports with small numbers of spectators seated.

The site context for Option 3 allows two adjacent existing buildings to be brought into the overall building composition without the need to demolish them. All three options work with the existing buildings that are integrated into the design of the new building. This approach avoids demolition, where possible. Consequently, the structural and environmental strategies consider existing buildings, or parts of the existing buildings, adjacent to the new building with a view to providing a continuity of streetscape and of urban form.

The roofscape follows the profile of the interior spaces beneath, appearing to 'shrink-wrap' the building to follow the forms of the interior spaces. This approach is taken in preference to the introduction of a flat roof that would result in pockets of unused space beneath, or would create distortions in the heights of those spaces. The roof wraps the spaces beneath, ensuring that only useable space within the building is mechanically ventilated with the associated cooling or heating energy required to temper those spaces. The roofscape is weatherproofed with a membrane in a colour that closely matches that of the external walls; forming a continuous wrap around the building which is economical and whose form can be seen from neighbouring buildings which rise above this school building. An advantage of the use of concrete is the ability to create forms which appear to be a set of small-scale buildings which form a group of intersecting volumes; each being a technically-driven form in its purpose as a teaching environment at a scale which corresponds to its immediate urban context.

1. Sectional perspective view

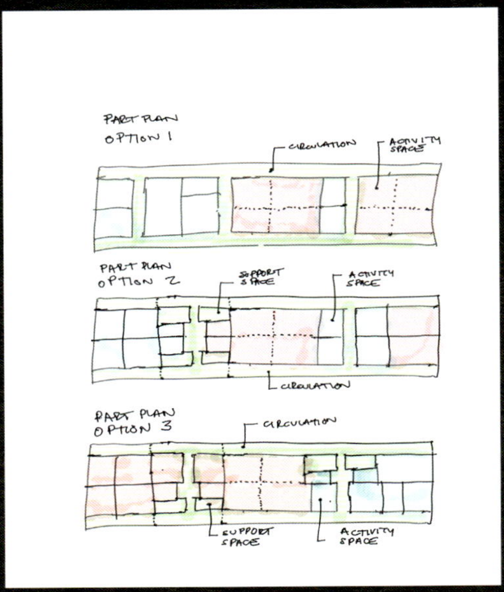

Spatial arrangement

Option 1 has a series of floor arrangements which create double-height spaces and single-height spaces.

Option 2 varies the floor heights in the vertical section to provide a greater variety of spaces.

2. Sectional perspective view

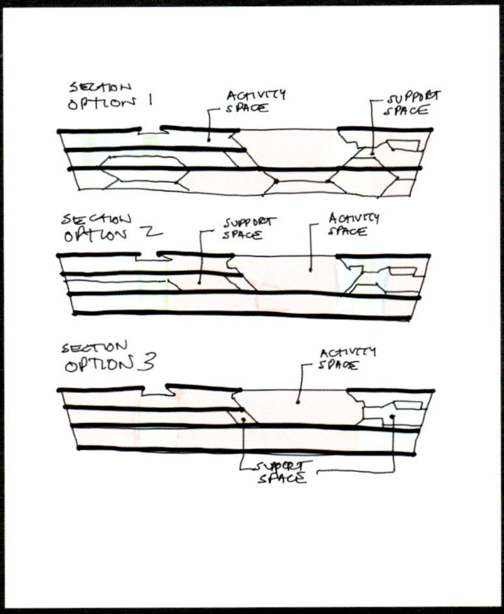

Option 3 provides a straightforward set of floor plates with the circulation zones set at the edges of the building.

Spatial arrangement

The building has a close integration of a large number of functions, with a central space that provides an auditorium and mixing space for students. The spatial concept of the building is to enter directly into circulation paths along the long edges of the building from outside, instead of providing a single entrance area. This approach creates a visual 'reading' of the environment in much the same way that a full-height, large-scale entrance space achieves in a more traditional design. The spatial arrangement of this design uses circulation space to reveal the extent of the building and the activities within, with views down the horizontal plan arrangement of circulation routes rather than up to the edges of circulation spaces immediately above the entrance. The entrance to this school takes the student immediately into the heart of activity in the building; guided around circulation spaces in which the spaces within the building are only revealed en route. Only when the visitor is in the central hall at the upper level can surrounding spaces be appreciated to their full extent. There are two ways of viewing activity in the building from circulation routes: From within the central hall, where activities within the school can be glimpsed from a static position, or by moving around the school building and experiencing the spaces as dynamic views from circulation routes, where different spaces are revealed through movement around the building, both vertically and horizontally. Option 1 has a series of floor arrangements which create double-height spaces and single-height spaces as required. The floor to-ceiling heights are consistent in this option. Option 2 varies the floor heights in the vertical section to provide a greater variety of spaces to suit the needs of the brief/program. A disadvantage of this approach is that there are many different floor levels and floor-to-floor heights that respond to the spatial requirements of the brief/program, but which make movement around the building more difficult than Option 1. Option 3 provides a straightforward set of floor plates with the circulation zones set at the edges of the building; rising and falling along ramps and stairs in order to reach different sets of spaces on various floor levels. The section through the building is consistent from long facade to long facade; the

. Perspective view of interior

2. Perspective view of interior

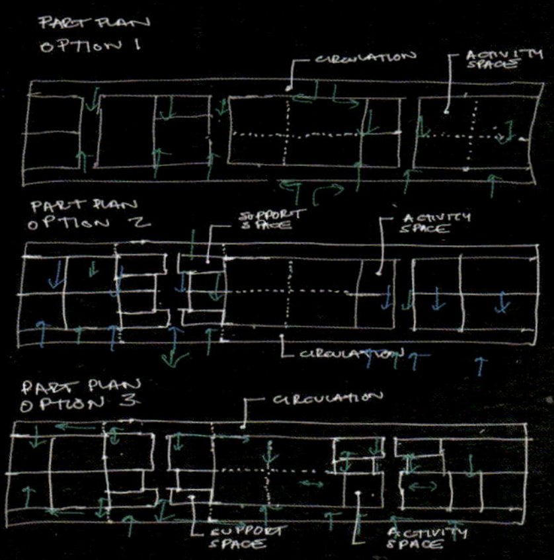

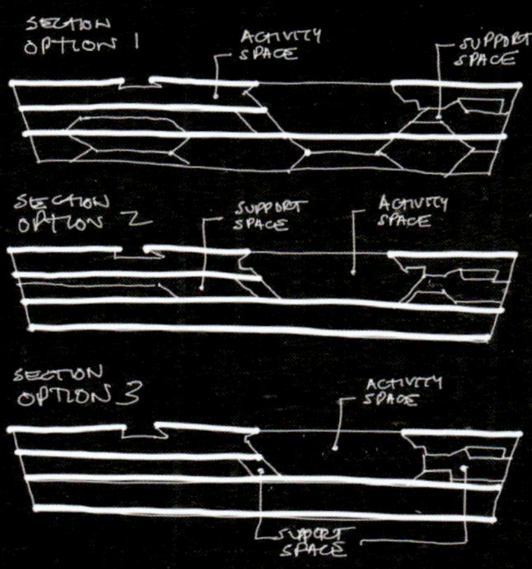

Internal circulation Option 1 arranges circulation along the road facade adjacent to the wall.

Option 2 arranges circulation to follow the indented profile of the facade.

Option 3 arranges circulation along the edges of the building with access directly into spaces which mostly extend from one long side of the building to the other long side.

two sides share the same facade arrangement that reflects the equal arrangement of circulation spaces on both sides of the building. The spatial arrangement device of using the edges of the building allows corridor link spaces to cross the building transversally; again avoiding the need for a central or multi-storey atrium space to provide a visual focus for the design. Instead, there is a network of routes along the edges of the building that allows the full extent of the floor plates to be used, as well as allowing floor plates to be arranged independently of neighbouring spaces.

Internal circulation

The school building has a circulation route on each floor; none are superimposed from floor to floor, but can be seen offset from each other. When the viewer looks down from the circulation space on one floor, the circulation spaces on the floors below can be seen. Floors are not superimposed vertically, but are instead offset in order to provide oblique views around the building that increase the sense of the building as a community. This is very much a 'nest' of smaller and larger volumes fitted together; each building having different performance requirements and a different energy management system.

The internal circulation in all cases is set along the edges of the building. In Option 1 the circulation is along the road facade adjacent to the wall. In Option 2 the circulation follows a similar concept but is indented in order to follow the profile of the facade. In Option 3 there are two straight rectilinear circulation zones, following each of the long walls. Option 3 is the most straightforward for circulation horizontally and vertically within the building, with connecting corridors on different floors. The main circulation for Option 3 is along the edges of the building with access directly into spaces which mostly extend from one long side of the building to the other long side. Some classrooms are divided such that they can be accessed from each side of the building.

The circular routes around the building create an easy reading of functions within the building as is registered on the facades, allowing those in neighbouring buildings to appreciate its clarity of form.

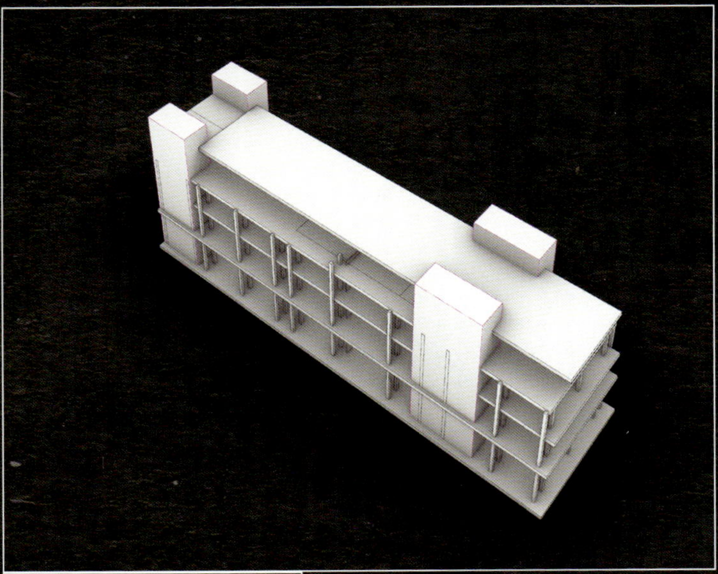

1A. Isometric view of structure

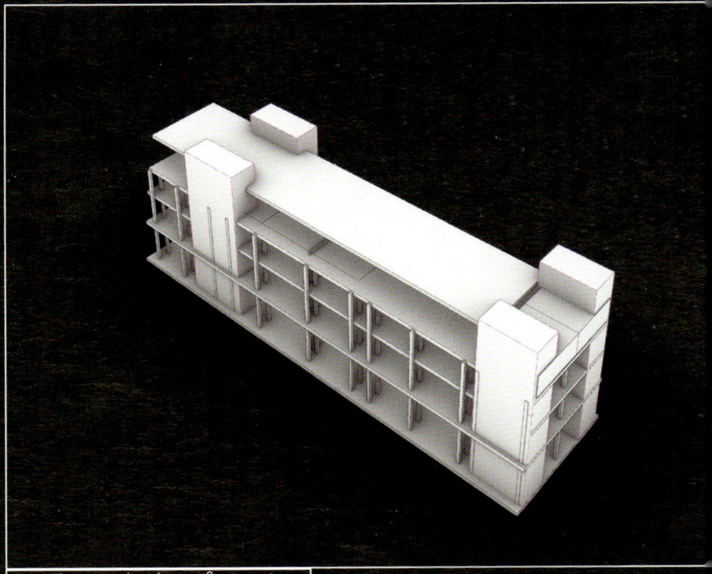

2A. Isometric view of structure

1B. Isometric view of structure

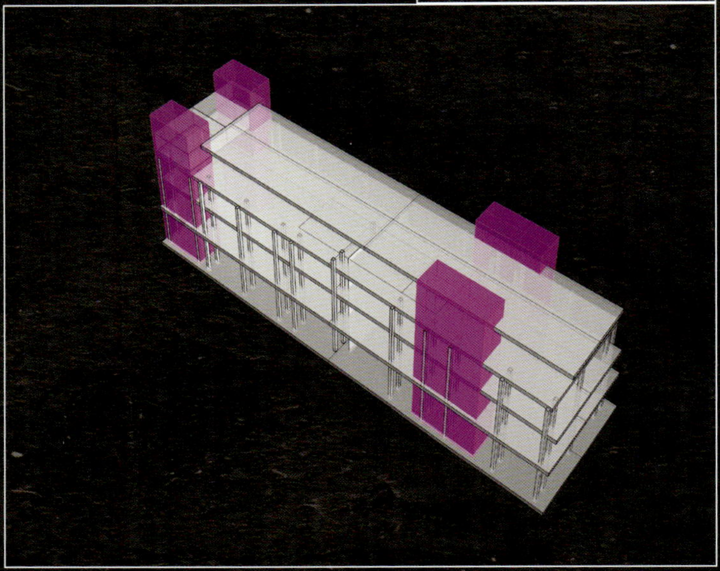

2B. Isometric view of structur

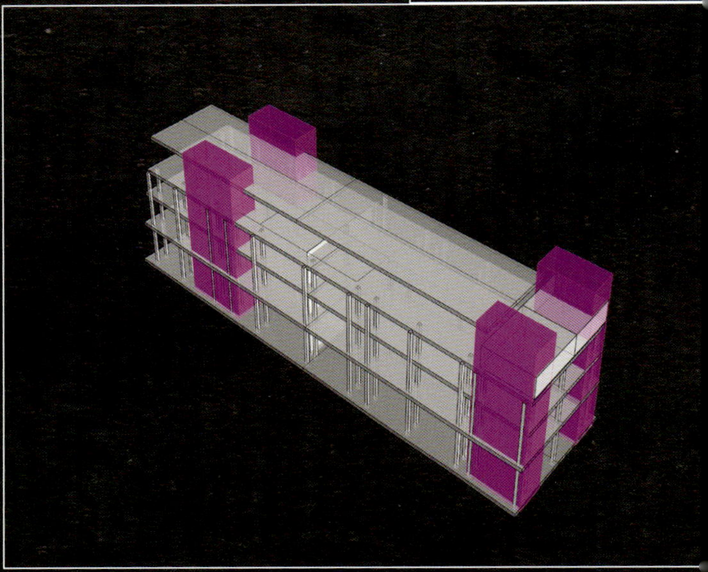

Structural strategy

The structural strategy provides a high level of differentiation of internal spaces without using a single structural type; the aim is not to have a single structure of floor plates with some gaps in them to create double-height spaces, but instead to have different juxtaposed structures which respond to very different functional requirements, but which are all fitted together like a puzzle. Consequently, the structural concept is one of a mixed structure in reinforced concrete and steel to provide the spaces required from an environmental and brief/program point of view. Small-scale spaces such as workshops have straightforward rectilinear spaces, while the requirements for lecture theatres are dictated by the 3D geometry of their layout, which requires that volume to be integrated neatly with adjacent volumes. The raking required in the lecture theatres is achieved without the need for losing usable space in adjacent spaces.

The geometry of the structure is dictated by the requirements of the spaces within, resulting in structural forms that appear to be moulded. Consequently, the building comprises a series of reinforced concrete structures, interwoven with steel-framed structures which weave in and out of the concrete walls and slabs. To avoid complex structural behaviour between concrete and steel components, steel structures and concrete structures are set alongside one another, not superimposed; movement joints are set between those adjacent structures. This approach provides the opportunity for acoustic insulation and thermal insulation between adjacent structures that have very different functions as internal spaces.

There are three structural options. Option 1 is based on the spatial arrangement in site-context Option 1; essentially a series of columns and flat slabs in a rectilinear arrangement. There is a small projecting element, in plan, where the structural grid turns parallel with the edges of the slab projecting out from the building.

Option 2 is a series of rectilinear slabs that are cut into the form of the building to create the profile seen in site context Option 2, but is essentially a rectilinear slab arrangement with columns set out at consistently

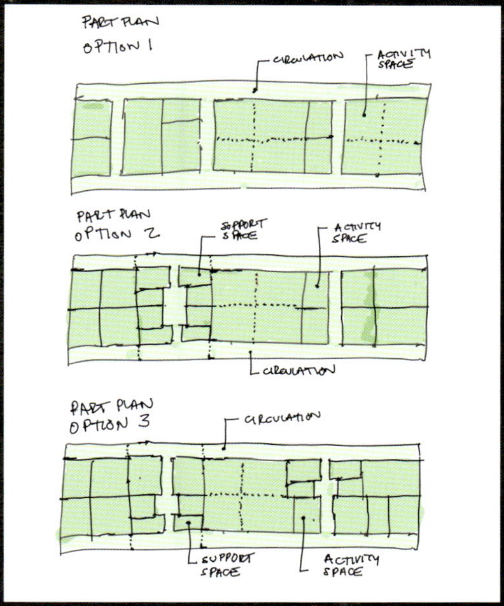

Structural strategy: Zones
Option 1 is based on the spatial arrangement in site-context Option 1.
Option 2 is a series of rectilinear slabs that are cut into the form of the building to create the profile seen in site context Option 2.

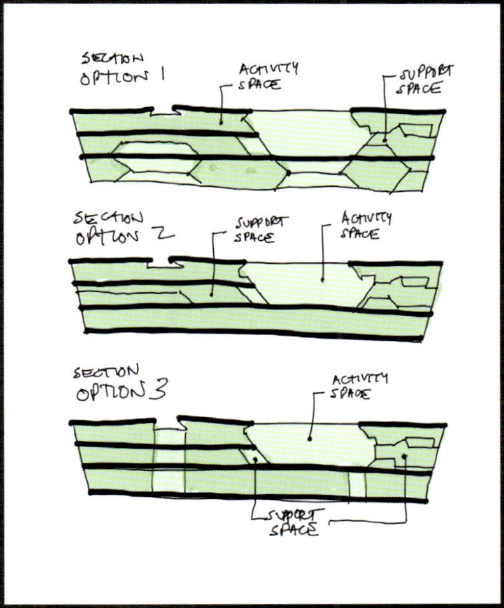

Structural strategy: Zones
Option 3: uses wide-spanning slabs with columns set out in a double row on the edges of the building.

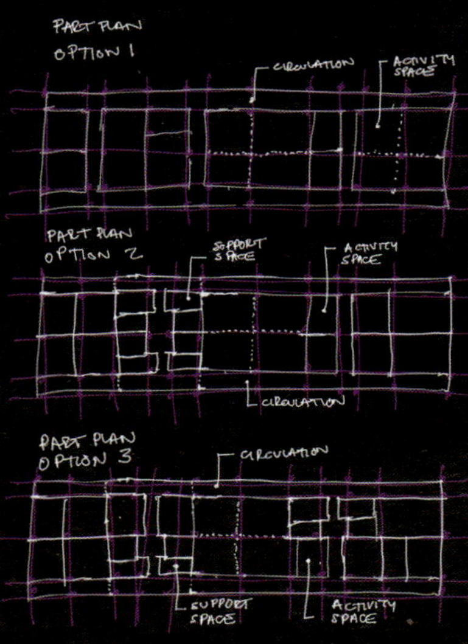

Structural strategy: Framing
Option 1 uses a series of columns and flat slabs in a rectilinear arrangement.
Option 2 uses a rectilinear slab arrangement with columns set out at consistently dimensioned centres.

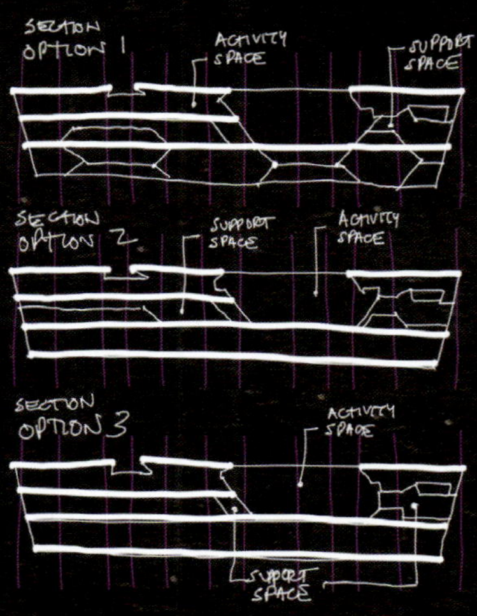

Structural strategy: Zones
Option 3: uses an outer wall comprising a 'moulded' frame that takes up the shape of the facade as a reinforced concrete wall cast in pieces, either on site or as precast concrete components.

dimensioned centres. Option 3 uses a different concept, where the columns are arranged in a rectilinear manner on the inside of the building, but become wide-spanning slabs with columns set out in a double row on the edges of the building. This double row of columns frames the circulation routes visually both horizontally and where vertical movement occurs, with the staircases set within service cores attached externally to the building. The outer wall comprises a 'moulded' frame that takes up the shape of the facade as a reinforced concrete wall cast in pieces, either on site or as precast concrete components. There is a long-span set of prestressed floor slabs across the building. Option 3 follows the layout of spatial arrangement Option 1.

The structural strategy is one of using the side shear walls, as backpacks, to which the environmental control systems are attached, as well as a series of winter gardens on one of the short elevations and a five storey winter garden on the other short side of the building. This provides a series of backpacks that are attached to the primary structure to create opportunities for additional daylight and to allow stepping outside the world of teaching and university environments into an outdoor space above ground level which allows walking around and observing of the surrounding campus as well as the ability to turn around and observe the part of the building in which one is standing, which is a usually a module of the complete building; a part of the complete building in miniature.

The tower comprises five modules stacked onto one another, each module having a different character and spatial arrangement, reflecting the requirements for varied teaching spaces and communal spaces for the students. The driving force behind the design is to be able to provide continuous useable space and a continuous clear relationship between walls, doors, open space rooms and complete buildings, as is the case in historic university buildings.

People are able to move around from space to space, moving through sets of linked spaces that are created from structural forms to create internal spaces and forms which can be seen both externally and from within the building.

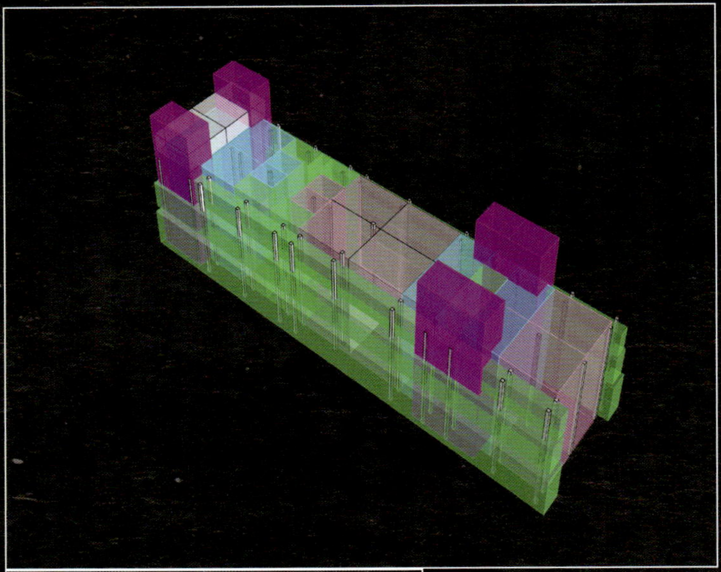

1A. Isometric view of environmental volumes

1B. Isometric view of environmental volumes

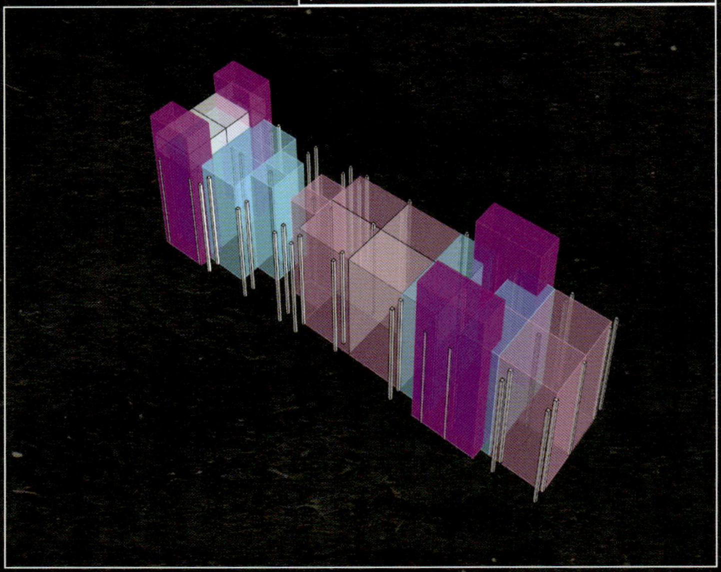

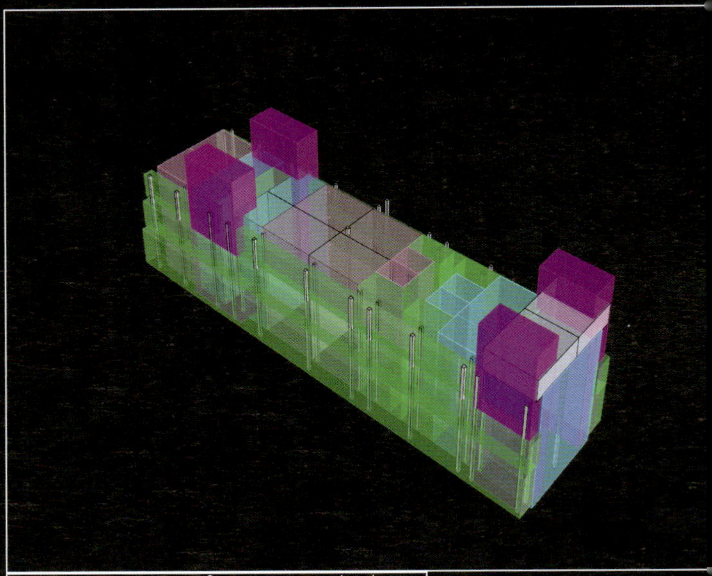

2A. Isometric view of environmental volumes

2B. Isometric view of environmental volume

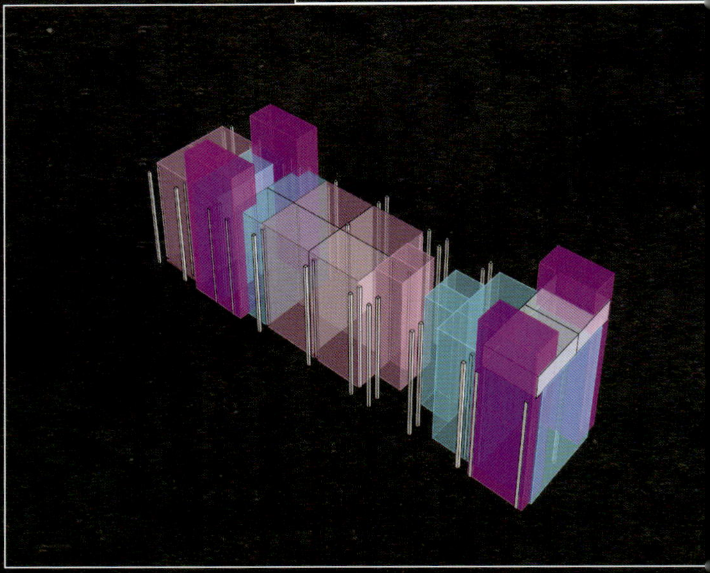

Environmental strategy

The environmental strategy focuses on reducing the energy use for the building at the small scale of individual rooms and subdivisions of large spaces. This approach allows the building user to reconfigure the current set of interlocking spaces to create new combinations of space without needing to change the environmental design. This reconfiguring of volumes is made possible by considering each space within the building as a separate environmental zone. This aim is made possible by creating an energy management system for each space so that the environmental performance of combinations of adjacent volumes is understood by the building user. This approach ensures, for example, that fresh air is redirected – either to or from – spaces which are combined at a later stage. The approach also ensures that, if a large hall is only partly used for example, it is the section of the hall being serviced and not the complete space. The benefit of this strategy is the ability to reduce energy used for environmental control accordingly. This building has no single installation of environmental control equipment; instead each set of spaces is supplied locally from its adjacent roof level in order to distribute services locally, allowing each set of spaces to have its energy consumption optimised for its specific needs and its own patterns of function and occupation. A network of primary air ducts and service distribution is located on both of the long facades in order to provide localised supply and control. This approach introduces the visual language of the framing members which are used both for the primary structure as well as for the distribution of air around the building from ducts.

The environmental strategy for the building is such that, for site-context Option 1, the building has two groups of environmental zones; one for the main activity spaces adjacent to the roadside facade, and a second group of environmental zones for the larger spaces set on the courtyard side. For Option 1, there is a group of environmental zones for smaller classrooms and teaching spaces along the roadside, and a set of zones for the larger volumes comprising the assembly hall and sports halls.

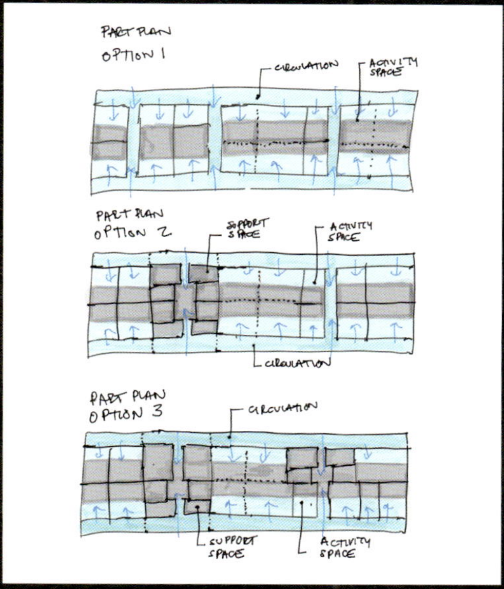

Daylight strategy for direct sunlight and cross ventilation.
Option 1 admits daylight and air along the full extent of the long facades.
Option 2 has subdued daylight in cores and some internal spaces. Air is introduced on the long facades.

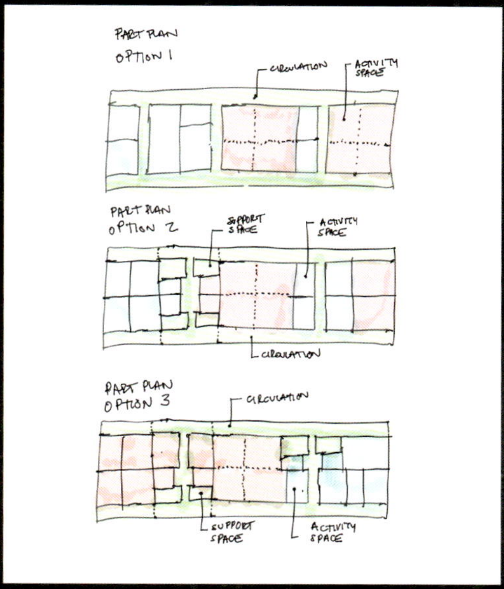

Environmental zones
Option 1 has a zone for the main activity spaces adjacent to the roadside facade, and a second group for the larger spaces set on the courtyard side.
Option 2 uses each 'extruded' block of spaces as an environmental zone in its own right.

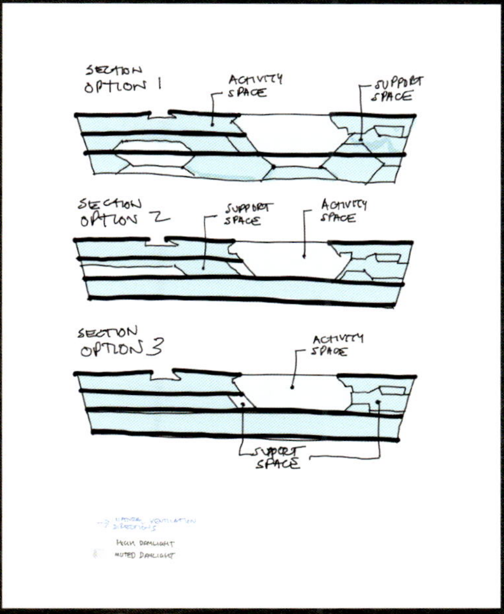

Daylight strategy for direct sunlight and cross ventilation.
Option 3 has subdued daylight in cores and some internal spaces. Air is introduced on the long facades.

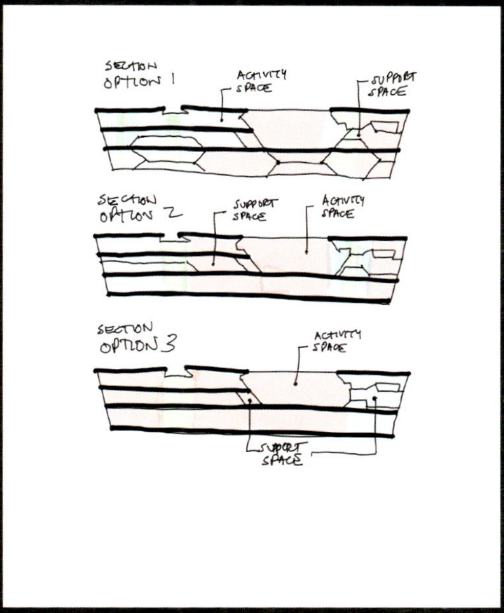

Environmental zones
Option 3 has given each space its own environmental zone, comprising a series of 25 environmental zones across the building expressed as discrete geometric forms on the long facades.

Option 2 uses each 'extruded' block of spaces as an environmental zone in its own right, creating a total of five environmental zones governed by the size of the subdivision of the building form rather than a grouping of the functions themselves. Where functions are grouped, such as in teaching spaces, those grouped environmental zones are subdivided by volume rather than by function.

In Option 3, each space is its own environmental zone, comprising a series of 25 environmental zones across the building expressed as discrete geometric forms on the long facades; each geometric form is 'extruded' horizontally through the building to create the volume of each zone. This is the preferred option as each environmental zone has opening windows or opaque facades that correspond on opposite facades. This allows the concept of fan-assisted cross ventilation to be used between facades, with fresh air being drawn across the width of the building, above some spaces, where it is drawn down for use in the spaces below. The strategy for natural ventilation matches the arrangement of the facades, allowing air to be drawn into the building at different rates and at different times to suit the functioning of spaces behind the external walls. This approach of drawing air across the building through spaces, rather than the air being ducted around the building, is one of reducing the energy consumption for an equivalent all-mechanical system by allowing the air to flow through the spaces rather than through ducts only.

This approach to ventilation also allows closed volumes such as the sports hall and main school hall to be self-contained, with a mixed mode of natural ventilation and mechanical ventilation to suit their frequency of use and the numbers of people using these spaces. The natural cross-ventilation strategy of Option 3 allows this air to be partially ducted away to other areas of the building without the need to fully duct spaces such as washrooms, staircases and service spaces. This approach of moving air around the building at different times of day to suit the use of internal spaces ensures that energy consumption for ventilation within the buildings is reduced as much as possible to that which is required of an efficiently run system.

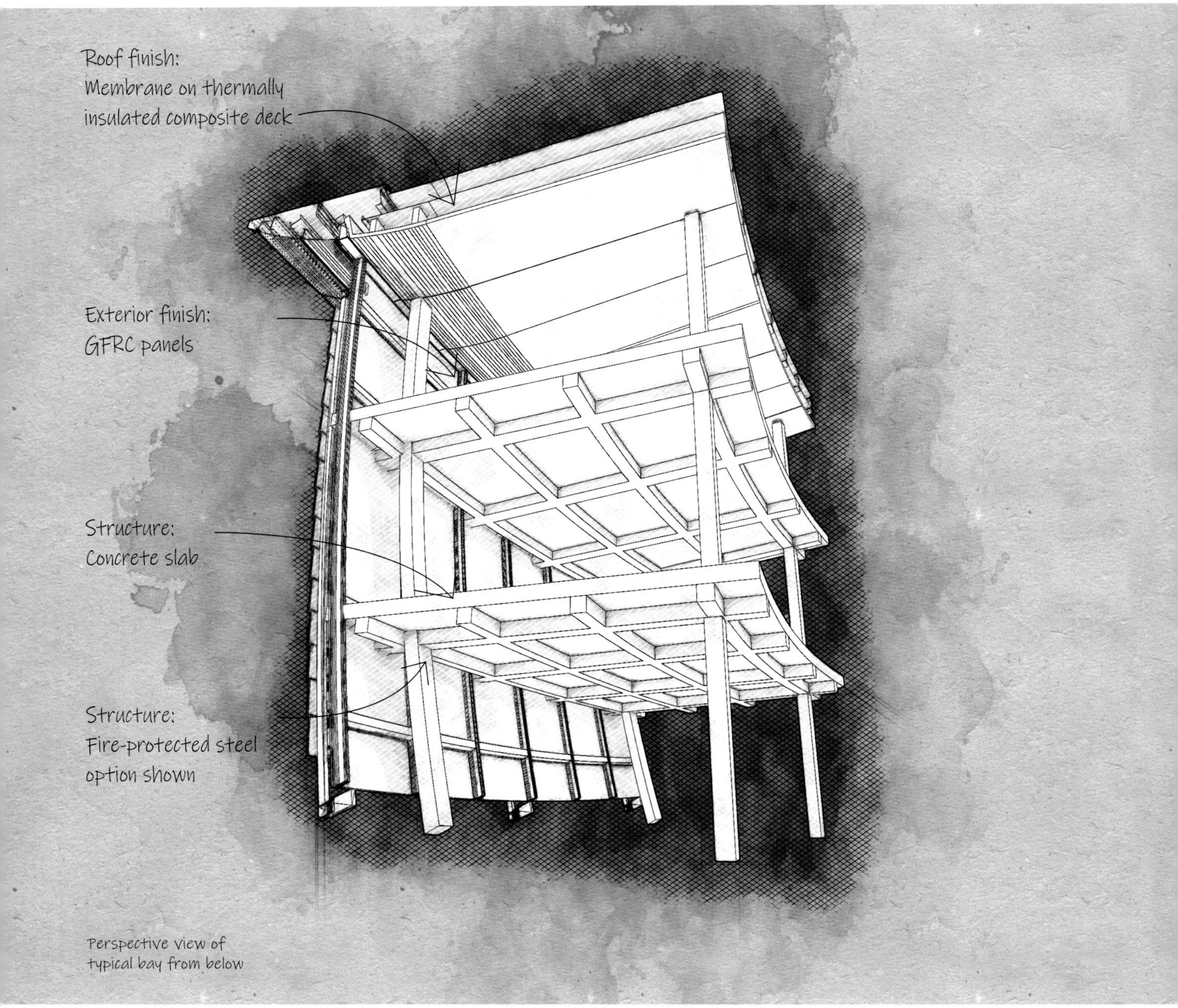

Perspective view of typical bay from below

Typical bay adjacent to facade

The facades of the building are all based on a mixed reinforced concrete and steel frame that provides support for floor slabs set immediately behind the structural facades. Steel is selected over concrete in places where it is easier and faster to construct; ease of connection with floor slabs and stairs, for example. Each of the structural facades on the long elevations supports long-spanning beams which extend the full width of the building. The long facades also support stairs and environmental control equipment which is fixed to the structure of the facades. The structure of the facade comprises columns, horizontal beams and inclined bracing elements that create openings for glazed panels that are set into openings formed in the structure rather than having facade panels which completely cover the reinforced concrete structure. The structure is thermally insulated and clad in GFRC (glass-fibre reinforced concrete) panels. The opaque areas of the external envelope, clad in GFRC panels, provide a smooth and durable finish, together with the high level of thermal insulation required by the design. The arrangement of the facades is one of a mix of opaque panels and glazed panels; glazing is fixed within each of the openings, and is fixed directly to the supporting steel frame in order to avoid both the need for curtain walling and the need for orthogonal arrangements of panels required by many facade systems. This design approach allows lines of opacity and transparency to be followed for floor slabs, stairs and services distribution; whereas environmental zones set adjacent to the facades require equipment to be housed behind panels locally. This approach of having a structural frame that is clad in opaque panels, with glazed panels set into that frame, provides long facades that respond more closely to needs of daylight control, solar control and thermal insulation than could be provided by a more conventional facade system.

The facades integrate a provision for services distribution around the building; services are also integrated within the external envelope; this approach allows the internal spaces to be subdivided and reconfigured as required during the lifespan of the building. The facades provide the primary environmental interface to ensure that as much of the inter-

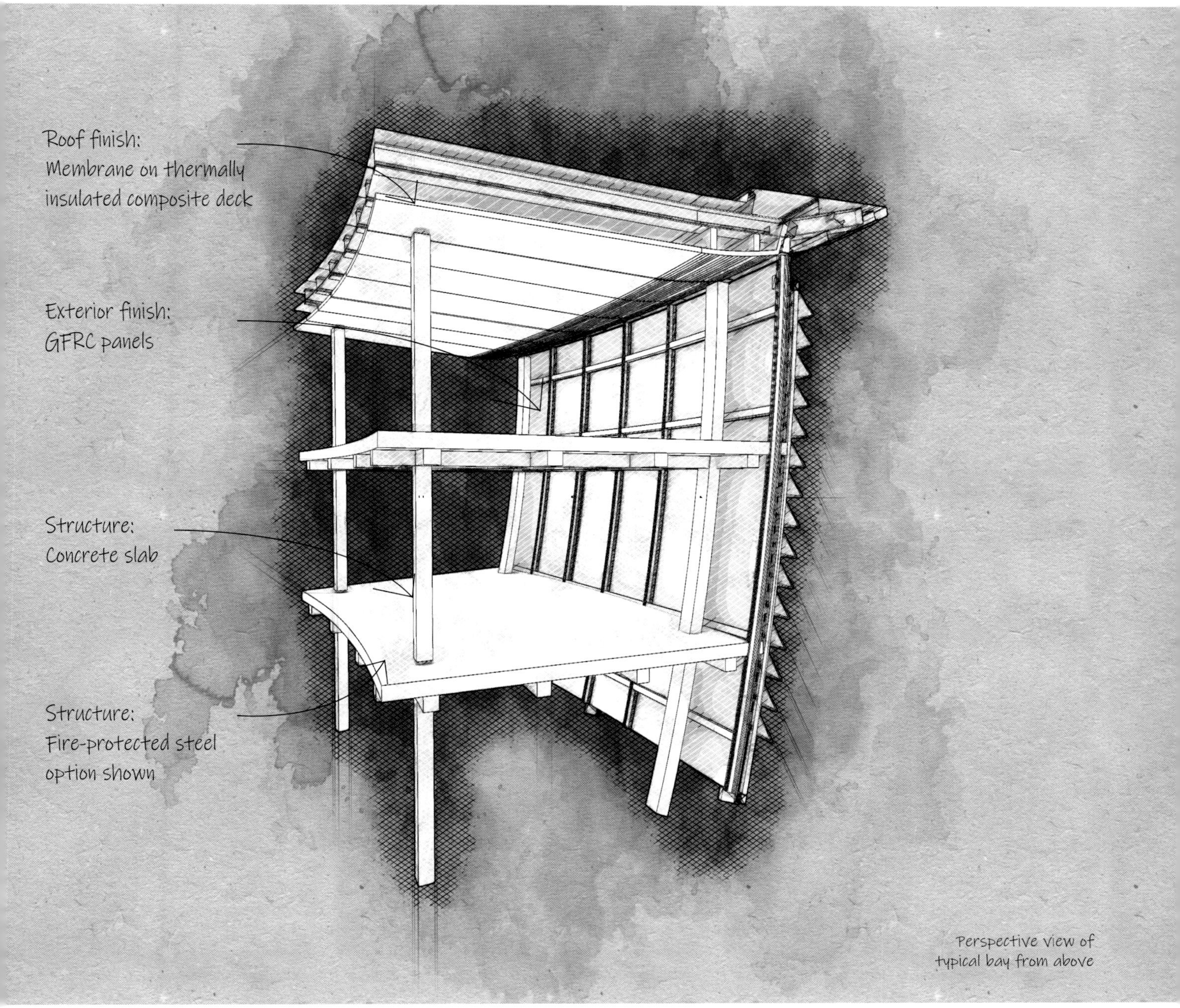

Perspective view of typical bay from above

nal space as possible can be treated as both a 'tabula rasa' and as a 'well-lit box', but the different amounts of solar shading and the thermal insulation can be varied for different times of the year, and for different functions.

Floor-to-ceiling glazing is set either side of the central stack, visible from the outside of the building in order to provide as much transparency and visibility as possible into circulation spaces. The glazing wraps around the long elevations, where it terminates against opacified glazing to reduce the amount of sunlight traversing from outside to inside. Activity spaces are set behind the opacified glazing; this area does not need high levels of daylight. The opacified glass provides consistent levels of daylight within the internal spaces across the long facades of the two towers adjoining it, wrapping around the shorter end of the tower. The inclined roofs are opaque.

The use of opacified glazing with partial light transmission avoids the need for a double skin facade or for external solar shading, while not needing to use the more regular solution of windows and adjacent opaque walls, which would have created a visual hierarchy in the space between those seated next to a window and those seated without a window.

The cross bracing used to stiffen the two long facades is visible through the opacified glass such that the shadow of the cross bracing can be seen both during daytime and more easily during hours of darkness, when the primary structure forms a shadow on the glass, through which viewers experience the building from the outside. The full height glazing has horizontal bracing which also serves as solar shading.

At roof level the reinforced concrete walls, in diaphragm configuration, fold at the edge to form a visual continuity in the external envelope. The diaphragm configuration is not required to continue across the roof; instead a single reinforced concrete slab extends across it, folding its shape to provide the required form of the building. The outside of the slab is thermally insulated and covered with GFRC panels which are waterproofed with a membrane. The roof panels are sealed at the joints between panels; a rainscreen configuration is not required.

1. Perspective view of exterior

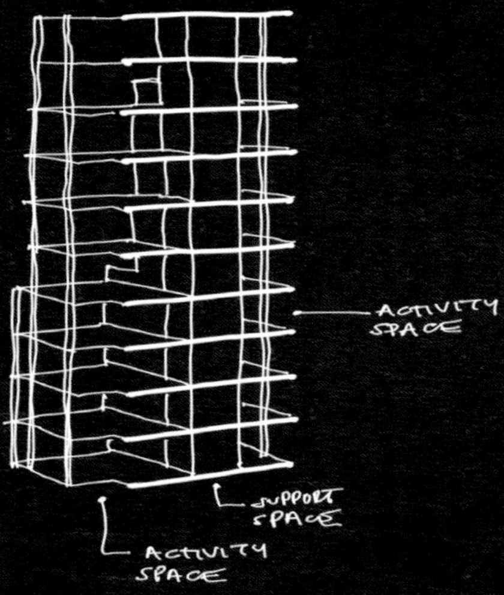

2. Perspective view of exterior

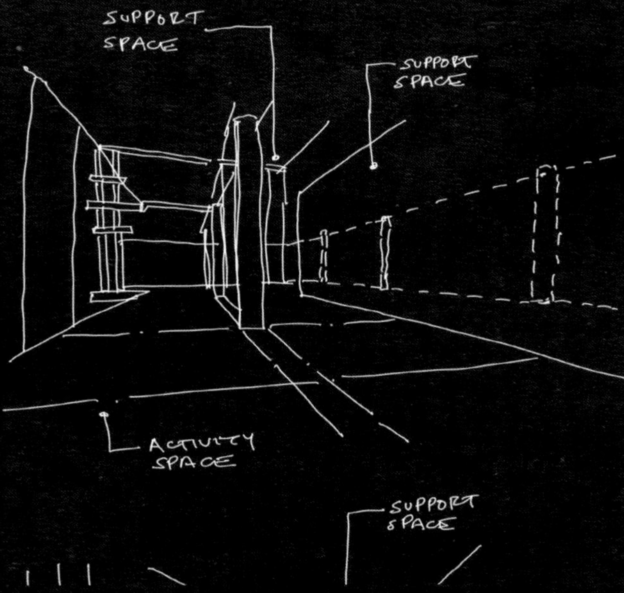

The use of reinforced concrete in facades

This project required the integration of a digitally based research and development facility with design and office space, meeting rooms and prototyping workshops. As a result, parts of the building are required to have opaque panels; other areas require daylight admitted into offices. Opaque panels in the external envelope are formed as reinforced concrete walls that are sufficiently articulated visually to respond to the mixed use of the building, set within an urban environment. The project requires integration in a commercial- and office-based city context and not an edge-of-city environment, where a much larger site would have been required for a low-rise design. The company occupying the building wanted to benefit from the urban infrastructure surrounding the site; the company has another facility nearby. The building makes use of reinforced concrete facades as seen on neighbouring civic buildings built in earlier years. This choice of material led to a high level of articulation of individual elements which reflect the organisation of the spaces behind and reflect the volumes of space created inside the building. This approach allows the building to be highly differentiated in its form, yet retain a visual calmness associated with a regular office block as previously constructed in the surrounding buildings.

This balance of visual calm and differentiation while providing opaque facades for approximately 50% of the external envelope was achieved using permanent formwork where the concrete is cast in place. This method casts concrete by using the outer wall cladding panels as formwork which remain in place after the concrete is cast. In this project, the outer panel is made from GFRC (glass-fibre reinforced concrete); a layer of thermal insulation is positioned in the zone between the formwork and the concrete wall. The structural concrete wall that is cast behind the GFRC permanent formwork panels is not required to follow the geometry or joint lines of the external panels. The inner face of this wall, seen inside the building, can be cast using regular formwork with interior finishes added afterwards, as is commonly the case. The permanent formwork walls are used in areas where the shape of the facades would be expensive to achieve using regular methods of

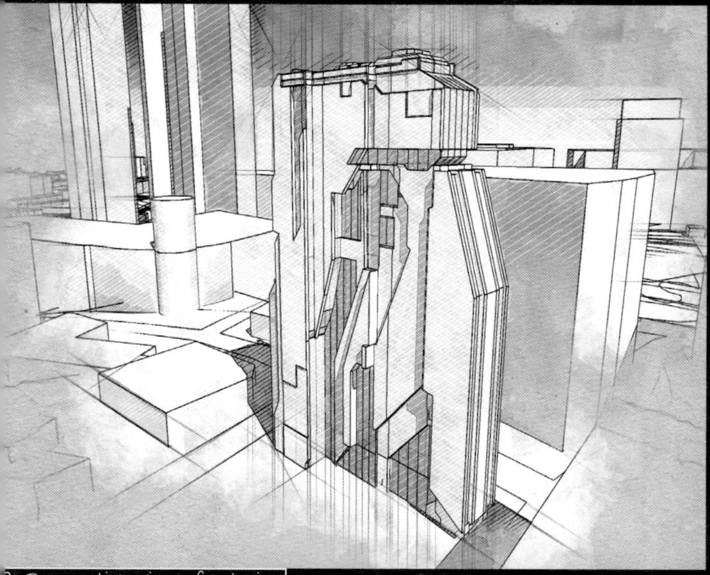

3. Perspective view of exterior

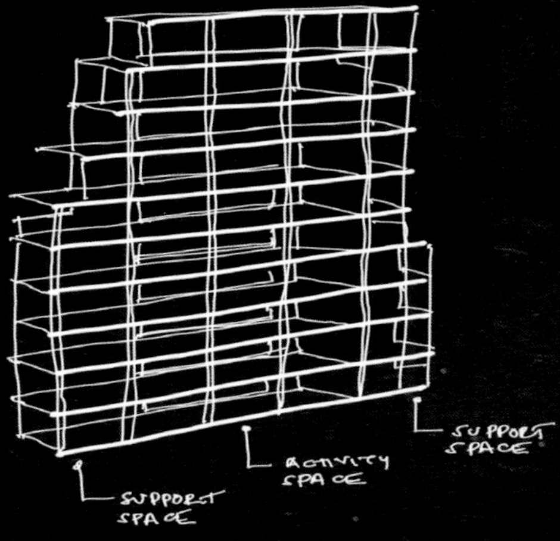

4. Perspective view of exterior

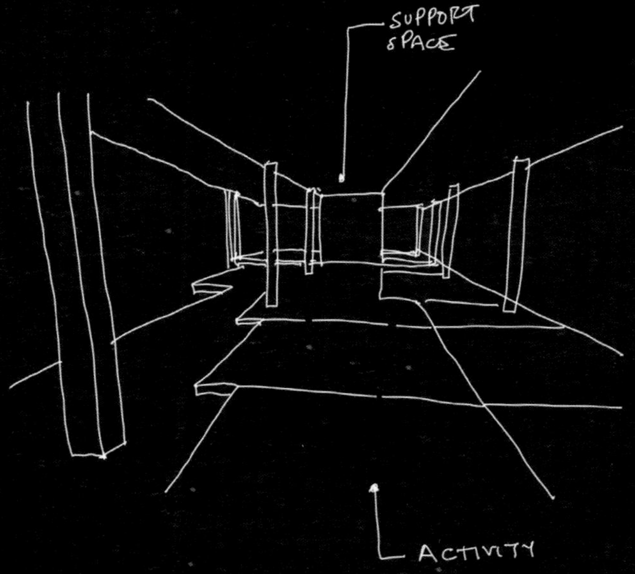

casting concrete. Elsewhere, facades made use of a cladding system of GFRC panels supported by a reinforced concrete frame behind. In some areas, a steel frame was used in a mixed-mode construction with concrete frames. Steel frames were used in areas where the building is glazed; concrete was used as the structural material for columns and walls where the building is faced in GFRC panels. The building was required to accommodate equipment for methods of prototyping based on using digitally controlled machines and robots to make relatively small-scale assemblies as part of the research and development processes. The associated high-volume manufacture of these assemblies is conducted in a factory elsewhere; a single-storey factory building. The processes to be conducted in this building constitute the research-based product development that informs the large-scale manufacturing located elsewhere. This is not a building for large scale manufacturing; this is a facility for developing methods of manufacture and an associated workflow. The research work carried out in the building reviews manufacture and develop the next generation of fabrication methods for new products; the design of the building aims to support that intention. The opaque facades address the need to provide a research-driven prototyping environment where highly glazed facades are not required, but where daylight is required to be admitted from one direction. In this case daylight is admitted from part of the sides of the buildings. The structural concept of a mixed mode of permanent formwork and GFRC cladding allows this building to meet a brief/program for mixed use and mixed activities within a building set within an urban environment, as opposed to an edge-of-city environment. This approach to the structure of the external walls was independent of the strategy taken for the floor slabs, cores and internal columns which use conventional framing and service cores. This separation of the internal structure from the external-edge-supported structure allowed the omission of bays of columns in addition to a regular series of columns in steel or concrete used behind opaque walls. For economy, these framed areas are in-filled internally with a mixture of concrete blockwork and dry wall to suit the internal spaces.

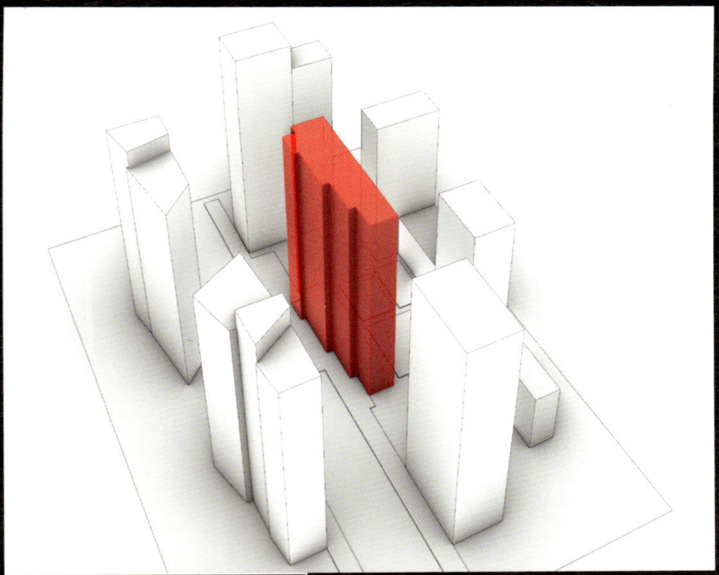

1. Project volume in site context

2. Project volume in site context

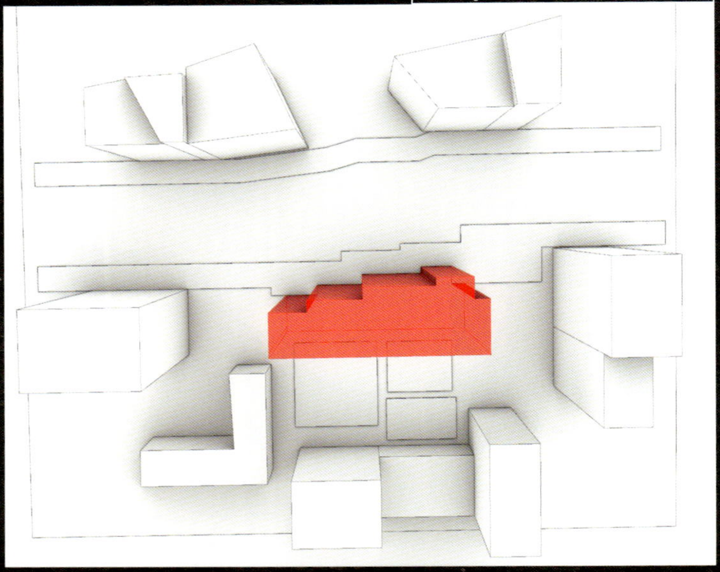

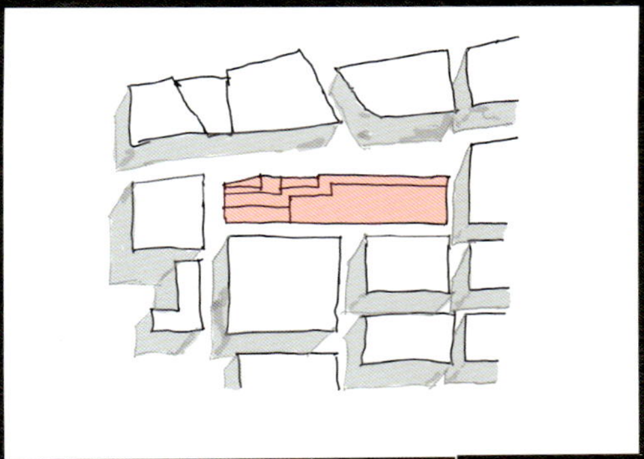

1. Site plan with surrounding context

Option 1 divides spaces for offices, studios and prototyping on the basis of a modular design; all three functions have access to natural daylight and natural ventilation.

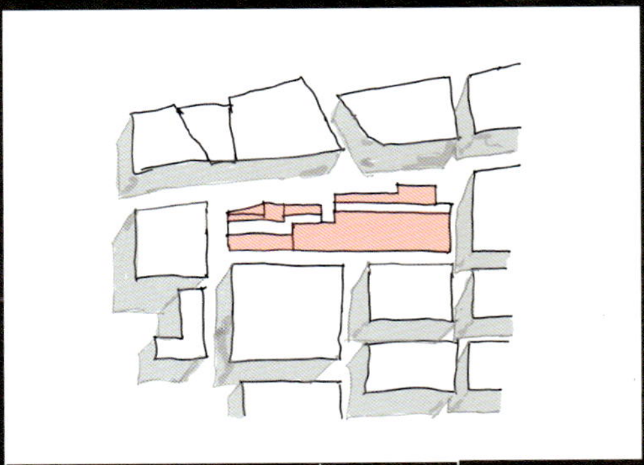

2. Site plan with surrounding context

Option 2 divides the site into two approximately equal parts, where access is provided along two routes down the centre of each of the subdivided blocks.

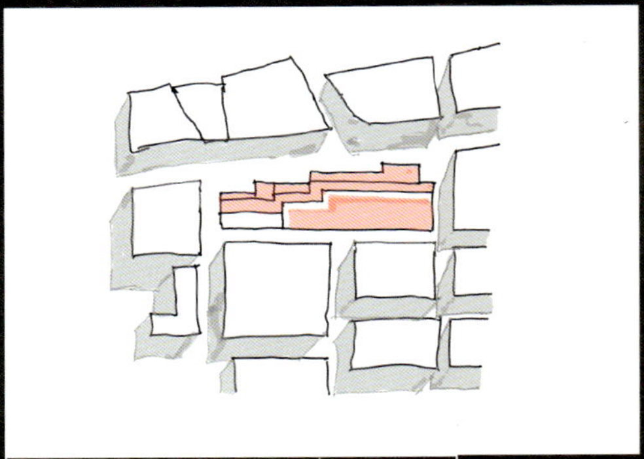

3. Site plan with surrounding context

Option 3 uses a decentralised approach to movement around the site and to the subdivision of space by creating 'indents' into the sides of the building.

Site context

This building accommodates a mix of functions within a single building volume. The design accommodates a vertically arranged set of spaces for a manufacturing-based company; functions of design, engineering, and prototyping are accommodated as a set of nested 'villages', or modules of space, within the building. Spaces are inward looking; glazed internal walls face an internal natural ventilated void; external walls are either highly shaded or opacified, with small bands of glazing crossing the facades. Within the building each set of spaces responds to the brief/program: opaque external walls enclose prototyping space, while office-based and studio-based activities are accommodated in spaces with more controlled daylight conditions. All functions face one another across the open spaces within the building rather than being outward-looking. From the outside, the different functions within the building and their relationship to one another is reflected in the volumetric arrangement of the facades. Option 1 for the site context divides spaces for offices, studios and prototyping on the basis of a modular design; all three functions have access to natural daylight and natural ventilation. This option uses the centre of the building as a service support spine for the vertical and horizontal movement of the occupants. Service spaces accommodate mechanical and electrical services; these support spaces are distributed on both sides of this spine. They can be made larger or smaller to suit the individual requirements by moving the prefabricated internal walls; the size of spaces can be changed to suit the changing needs of prototyping, design and administrative activities.

Option 2 divides the site into two approximately equal parts, where access is provided along two routes down the centre of each of the subdivided blocks. This approach provides daylight to all spaces; larger spaces are for prototyping, being more suited to the centre of the building where larger floor plates are created.

Option 3 uses a decentralised approach to movement around the site and to the subdivision of space by creating 'indents' into the sides of the building to create two types of workspace; smaller spaces which

. Perspective view of structure

2. Perspective view of structure

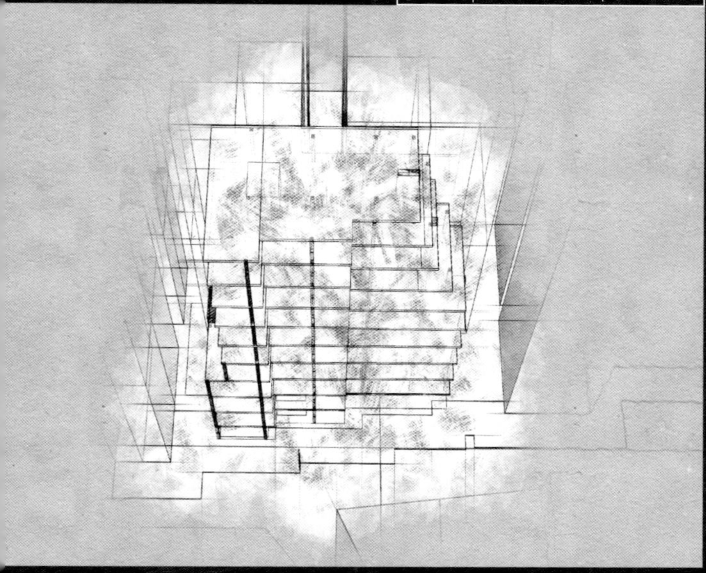

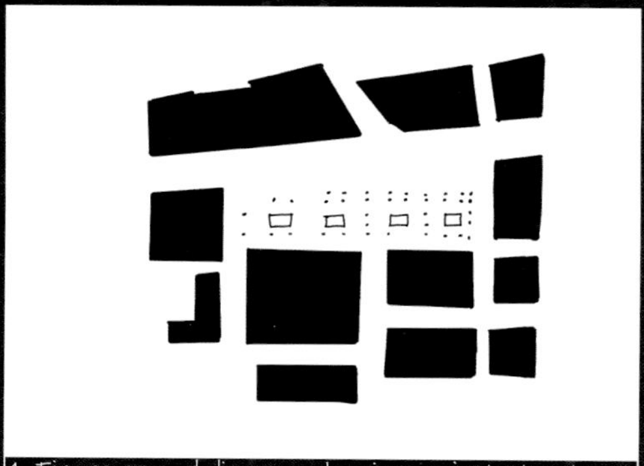

1. Figure-ground diagram showing project structure

Option 1
The reduction of scale of the building form is achieved as a set of visually linked structures.

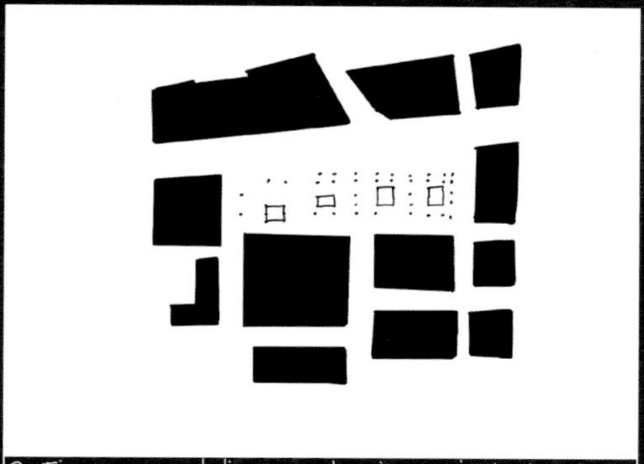

2. Figure-ground diagram showing project structure

Option 2
The visually linked structures extend through different vertical heights in the building.

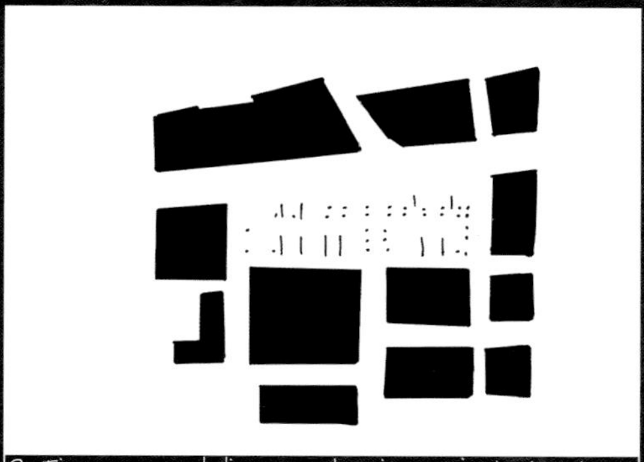

3. Figure-ground diagram showing project structure

Option 3
The visually linked structures are expressed visually through the arrangement of cores.

are set around the indents and larger manufacturing halls. The halls do not have natural daylight in this option as a result of concerns over glare from contrasting light levels caused by shadows. However, prototyping spaces benefit from borrowed light from surrounding glazed volumes within the building; daylight is provided on one long side of the building. Prototyping spaces are subdivided into different-sized spaces and provide a set of 'nests', or groups, where manufacturing in larger spaces and the surrounding office space is linked into three distinct parts. The approach provides close links to natural daylight for office space, while creating the sense of a single community of office spaces that may be partially enclosed or may be completely open plan. Adjacent manufacturing spaces are for small-scale manufacturing but are linked to the office spaces. This link avoids the need for corridors that are dedicated to complete building access from one side of the building to the other.

Because of the nature of this building as being relatively tall in a civic environment, the relationship of the opaque facades to the glazed facades was essential in creating a massing of form that is harmonious with its environment. This relationship is one of interacting with that environment and contributing the building as a piece of urban design to the cityscape. As a result, the building is subdivided into smaller elements, which are not recognisably individual buildings in their own right, but which are forms interconnected with one another to create an overall form of building which is sympathetic to its context. The additional height required as a result of the space planning of the brief/program was achieved by making the building narrower and more slender, allowing the massing to be seen on all four sides. Passers-by and those looking out from neighbouring buildings are able to take in the form of the building at a glance and see that it comprises a set of forms that accommodate a complex, mixed-activity factory-based environment; a place to work and to produce products without the building taking on the characteristics of factory buildings that were until recently been expected to be located on the edge of towns and cities rather than at their heart.

1. Sectional perspective view

2. Sectional perspective view

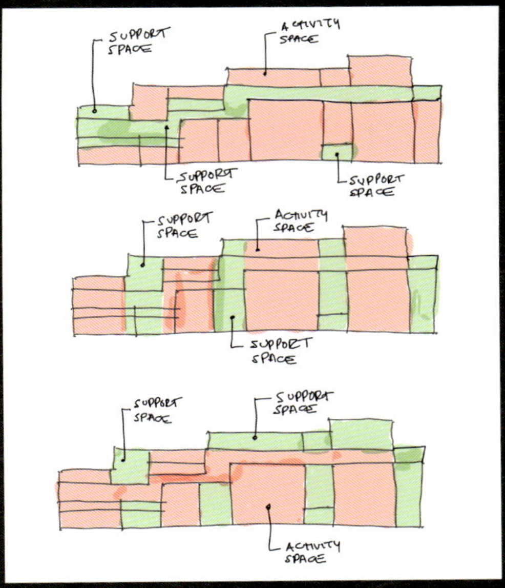

Spatial arrangement Option 1 is based on a central corridor arrangement.

Option 2 has offices that are set along the sides of the building, with prototyping positioned in the middle of the building.

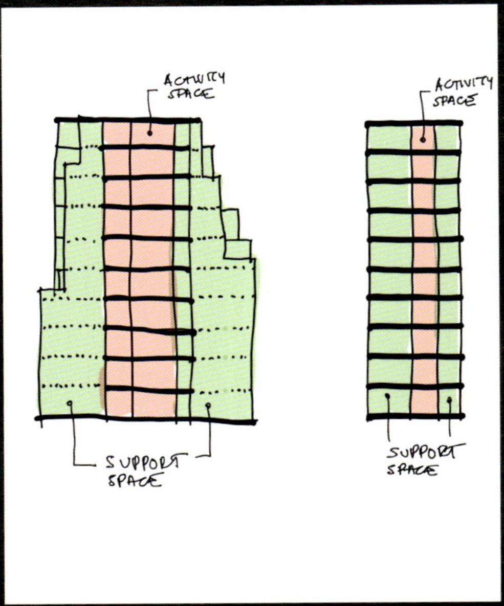

Option 3 creates office spaces that surround the glazed slots, the indents and building.

Spatial arrangement

This small site requires a prototyping building and its associated R&D and administration facilities to be accommodated in a single building on a relatively small site. The building is located at the edge of a manufacturing district and is also adjacent to a residential area. The approach is to take some product development processes which would be expected to be in a low-rise suburban factory environment, and put them into a compact, taller building where functions are set out vertically rather than horizontally. The arrangement of spaces is such that there is a planned minimum movement of materials and assemblies between the start of the design and prototyping processes and the assembled prototypes, which are sent for dispatch on the ground floor. There is no simulation of a production line; assemblies are put together in relatively small spaces; teams undertaking the assembly of related products work in assembly halls on different floors. The facility is vertically stacked rather than being arranged horizontally. The spatial arrangement of a typical floor in Option 1 is based on a central corridor arrangement. This solution provides access from each end of the building with a central spine, providing access into all spaces from this corridor. In Option 2, offices are set along the sides of the building, with prototyping positioned in the middle of the building. Access from administrative office space and design offices to the prototyping spaces is made directly on both sides. In Option 2, the larger spaces in the centre of the subdivided blocks are given over to prototyping, which is provided with less natural daylight than the surrounding office spaces; this arrangement suits prototyping needs. The office spaces effectively surround the prototyping spaces. One of the office spaces is given over to access into the building for visitors with associated meeting rooms.
The spatial arrangement of a typical floor in Option 3 creates office spaces that surround the glazed slots, the indents and building; those indents providing access into the building at ground level and potentially providing circulation spaces on the upper floors. Additional access can be provided at the ground floor. Prototyping spaces are arranged so that they relate to the office space immediately adjacent.

. Perspective view of interior

2. Perspective view of interior

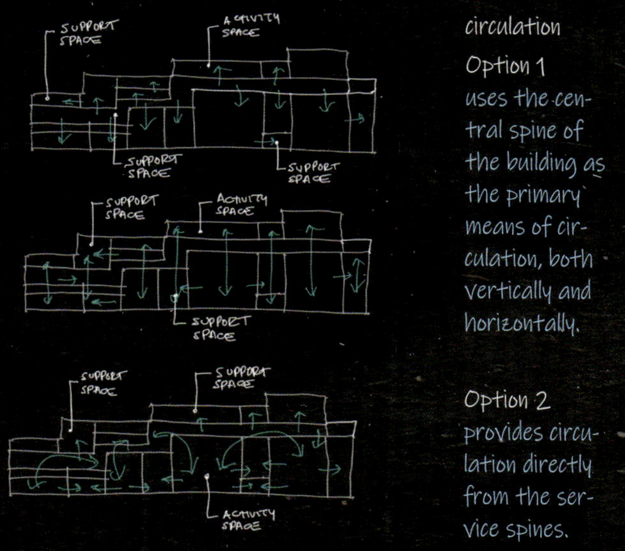

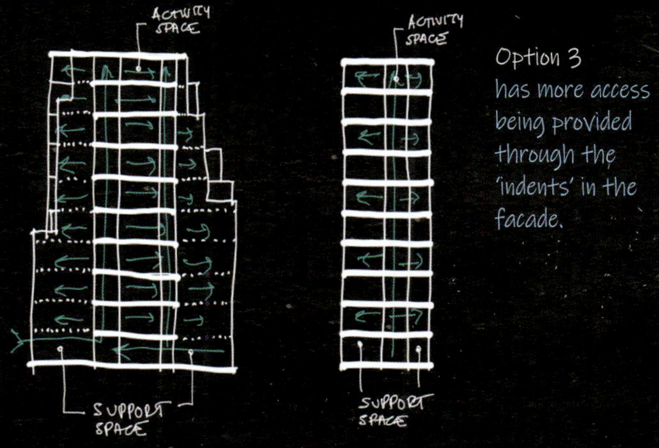

Internal circulation

The internal circulation of Option 1 is based on using the central spine of the building as the primary means of circulation, both vertically and horizontally, with access to spaces immediately adjacent to this spine. One of the spaces, which accommodates visitor entry, allows for meeting room access to be given within that subdivided space, but also provides partial access to the central core space for use of shared service facilities such as restrooms. This entry space creates an easy access route to upper floors. Option 2 provides circulation directly from the service spines. Access for the public and for people visiting from outside the company is by way of spaces that are a subdivision of one of the blocks. The disadvantage of this layout is the use of two separate circulation spines, which reduces overall access through the building. A further circulation route is added to join the two circulation spines that comprise the two halves of Option 2; the advantage of this arrangement is that circulation is reduced for staff working in the building. Option 3 has an internal circulation which is more 'porous' than the other two options, with more access being provided through the 'indents' in the facade to give access both to office space and to the prototyping space beyond. Circulation of occupants within those spaces, together with vehicle access, can be provided for the prototyping spaces on the outer wall, which are set on the opposite side from the indented access. Circulation of occupants across the width of the building is provided to individual spaces that belong to each of the office groups; this arrangement encourages an interaction across departments and between activities within the building. The stepping of the circulation route into the plan creates individual spaces for work rather than simply circulation spaces to walk through.

Internal circulation changes slightly from floor to floor – as the amount of prototyping space increases in relation to adjacent office space as revealed on the facade – but this change is gradual from floor to floor such that the communication remains constant on each from floor plate to floor plate, providing a sense of spatial continuity whilst providing the required proportion of office to prototyping space.

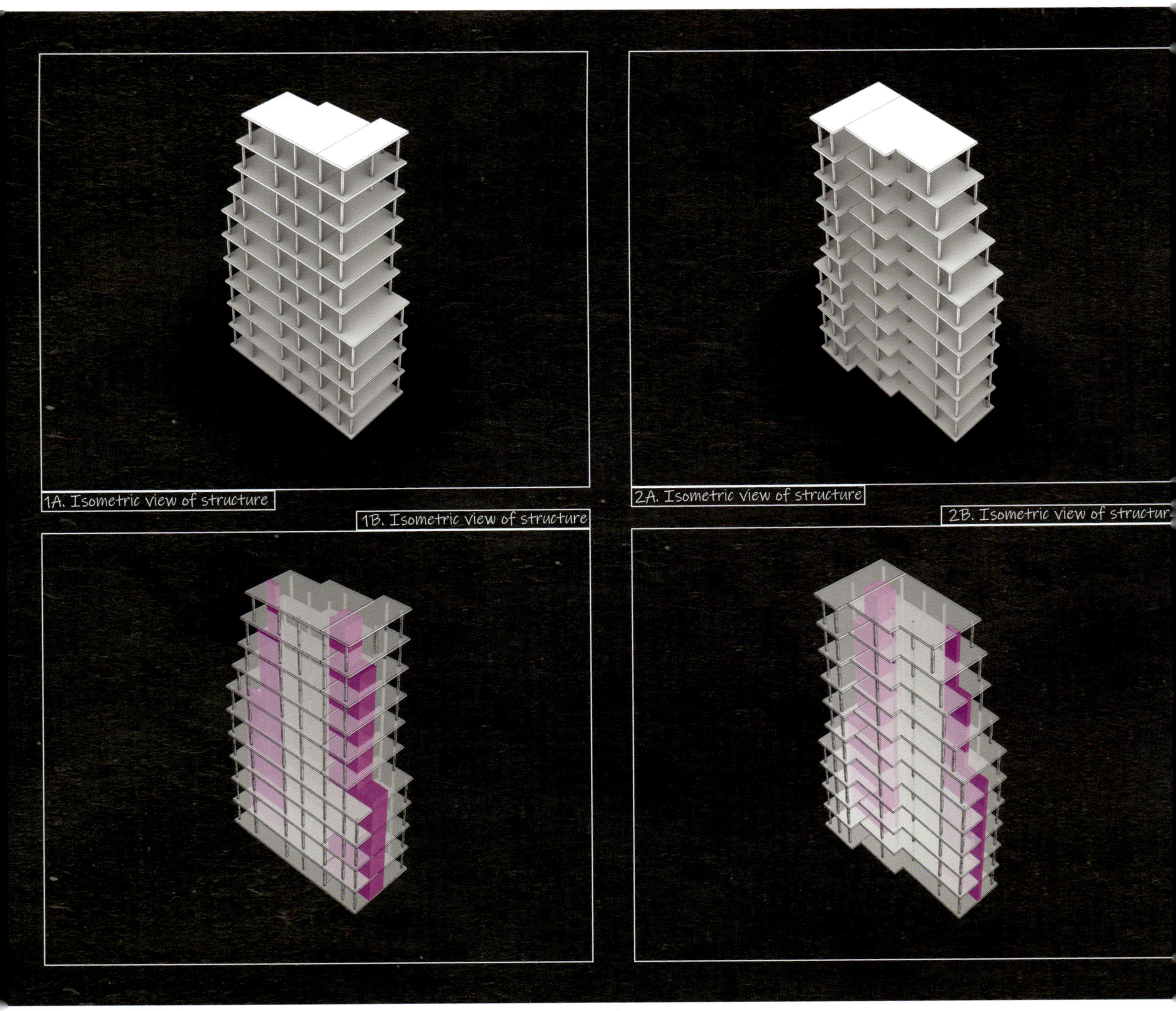

Structural strategy

There is no large open space within the building that would link spaces visually; views through 'gaps' in the spatial arrangement fulfil that function of being able to see all activities from at least one point in the building. The arrangement of volumes within the building is achieved with a structural strategy of using a steel frame module of size 6.0 x 9.0 metres; some spaces are 6.0 metres high.

Prototyping spaces are arranged on each floor plate using the steel framing members set vertically and horizontally, onto which steel composite decks are inserted into the construction; also set out in a modular manner. The structural arrangement comprises a grid of steelwork to provide a high level of flexibility of use in order to respond to anticipated near-future changes in the prototyping processes used by this company. The 'static' assembly method allows these anticipated changes to be made over the coming years. Office space is set on these decks and is situated in opposing corners of the building to provide the spaces required for prototyping. Office and design studio spaces are fitted in around prototyping requirements. These office-based and studio-based volumes are made from prefabricated modules which arrive ready-fitted; their size corresponds to that of a transport container. Modules join together to form a flexible arrangement of office and studio space, as well as providing administration space for the prototyping parts of the building. The structural strategy for Option 1 is a single grid of columns which are 6.0 x 9.0 metres. Structural stability is provided through the structural concrete cores which are arranged as enclosures to some of the spaces. The cores are arranged off the service spine in the centre, providing additional areas of structural stability along the length of the building at its centre.

Option 2 has a similar arrangement of a grid of columns, with column supports set on both sides of the corridors, as per Option 1. Cores are set alongside both sides of the central spines, allowing for flexibility in the anticipated near-future use of space; based around the central concept of the two spines. Option 3 uses the perimeter of the building to provide structural stability by creating vertical planes of stiff-

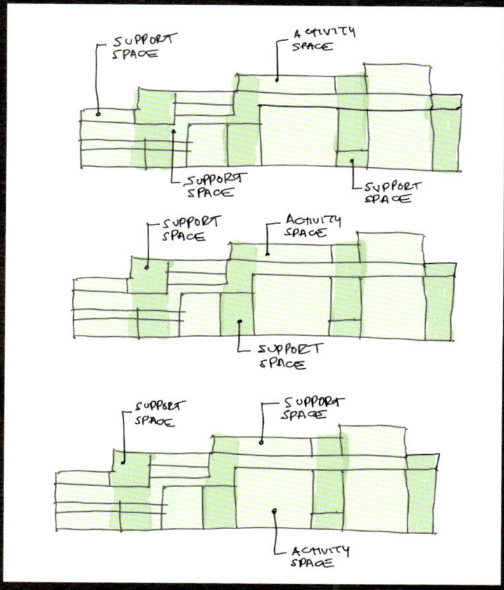

Structural strategy: Zones

Option 1 is a single grid of columns which are 6.0 x 9.0 metres.

Option 2 has column supports set on both sides of the corridors.

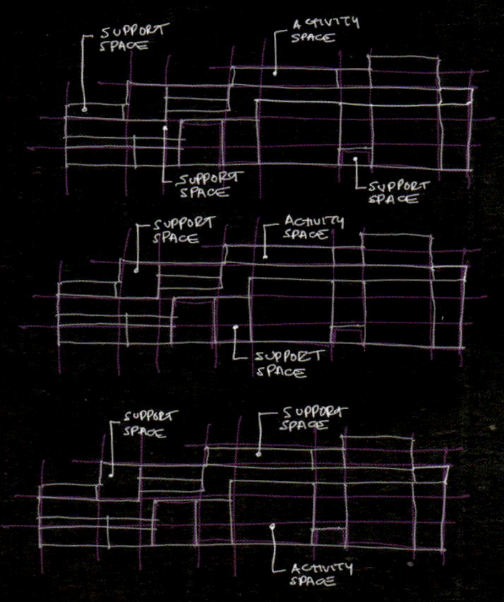

Structural strategy: Framing

Option 1 has cores arranged off the service spine in the centre.

Option 2 has cores set alongside both sides of the central spines.

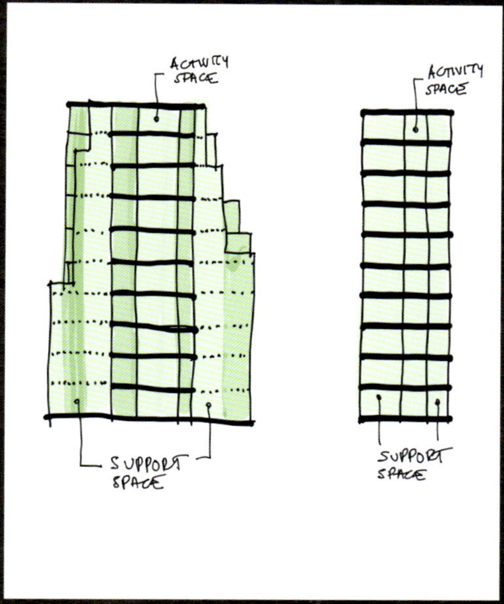

Structural strategy: Zones Option 3 creates vertical planes of stiffness around the indents.

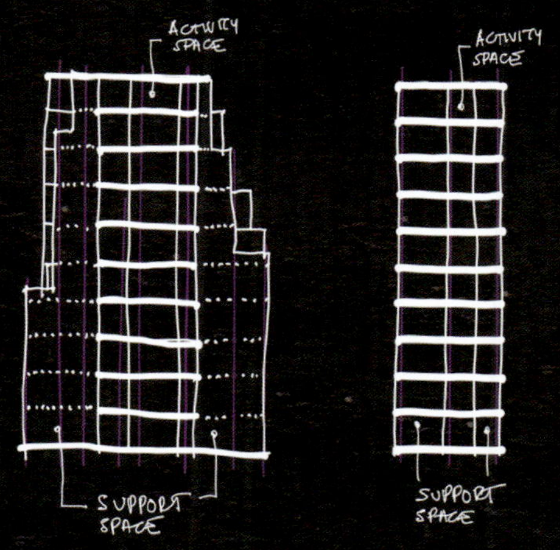

Structural strategy: Zones Option 3 has indents which are not fully glazed but have structural walls on three sides: concrete and steel.

ness around the indents, such that the indents are not fully glazed but have structural walls on three sides and beams set on the fourth side. Additional stability is provided by spaces for offices and manufacturing which have concrete shear walls to enclose them. In this option, cores provide an additional method for stability for the building through its full height.

There are two parts to the structure of this building; the central area of cores, columns and floor slabs set within the building and the outer envelope, which is a mixture of columns and structural walls.

The external envelope of glazed panels and opaque panels results in the transition from a concrete structure of walls and columns to a steel-based frame structure that supports windows. This condition exists only at the perimeter of the building and provides a more economical solution and a faster construction time than would be achieved if a single structural material were used; the steel columns would be erected more quickly than the equivalent concrete columns. Steel columns and beams are used in areas where openings in floor plates are required, such that the overall construction of each floor plate can be completed in a time to suit both openings and floor plates. This approach balances construction activities on a given floor plate in order to achieve the shortest possible construction time.

This approach is also used in internal cores, which are optimised for an economy of material as well as for the time required for their construction, which is compatible with that of the adjacent floor plates. The outer panels of the external walls are not loadbearing; they support their own weight but are fixed back to the structural walls and columns behind. The cladding panels span from floor to floor in height, which contributes to an economy of design.

At an early stage of the structural design, a roof was proposed which spanned the full width of the one side of the building, from the central spine to the external walls. The relatively low height of the space below did not add functionality to the space; instead, columns were introduced to support the roof at its mid-span, with no loss of functionality or elegance.

1A. Isometric view of environmental volumes

1B. Isometric view of environmental volumes

2A. Isometric view of environmental volumes

2B. Isometric view of environmental volume

Environmental strategy

All equipment is distributed across the roof in order to provide centralised access to all parts of the services installation, rather than distribute equipment on floors within the building. The roof level has chillers and supporting machinery as well as telecommunications equipment. The roof level equipment is accessed from the narrow ends of the building. Services are distributed down the building within blocks of ducts rather than within service cores. This approach distributes services near to where they are needed on each of the long floor plates, which avoids long runs of air-handling ducts within internal spaces. Since open, clear areas of floor plate were not required, there was no advantage in grouping support spaces of stairs, washrooms and services as volumes which are distinct from activity spaces for offices and for the prototyping spaces.

The environmental zones for Options 1,2 and 3 correspond to those options generated for spatial arrangement; the environmental zones for prototyping and offices have different servicing requirements for fresh air and mechanical servicing. Option 1 provides three environmental zones; the zones for prototyping with larger assembly areas being grouped together as a single zone. The second zone is for office space, which is clustered around the indents in the building; services are provided from the two shorter ends of the building as well as from the indents. Option 1 groups the services in order to focus on the efficient use of service space by providing the shortest distance for air distribution from the point of supply. This approach allows for further subdivision to form smaller environmental zones in which different levels of thermal insulation within external walls, as well as different requirements for air changes, can be accommodated.

Option 2 uses a similar principle, but sets the circulation of air down through the cores arranged alongside the two spines, bringing the advantage of providing services locally to smaller spaces. This approach also allows the spaces for prototyping and offices to be grouped, for example, where there are higher numbers of air changes required in prototyping spaces, and where lower thermal mass is required in office

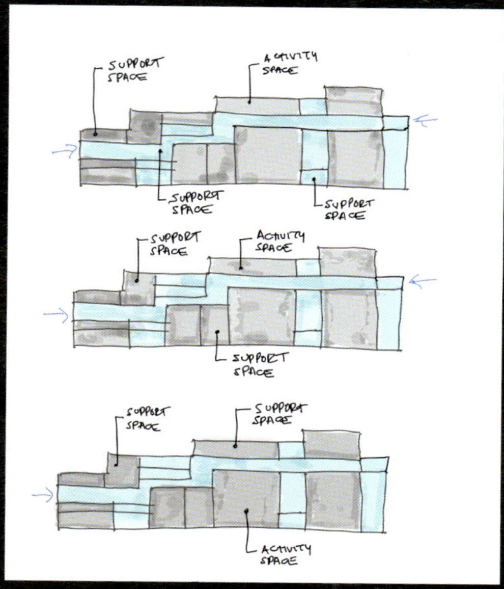

Daylight strategy for direct sunlight and cross ventilation.

Option 1 admits daylight and air within the corridors, which have a higher level of daylight and air than the spaces it serves.

Option 2 admits daylight and air across the width of the buildings, in circulation and meeting spaces.

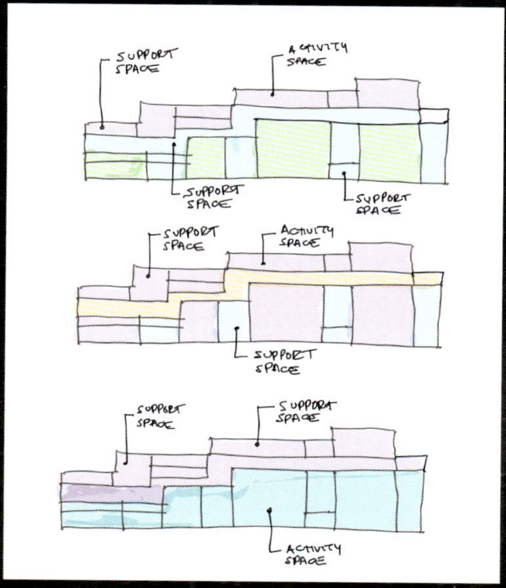

Environmental zones

Option 1 groups the services in order to focus on the efficient use of service space by providing the shortest distance for air distribution from the point of supply.

Option 2 sets the circulation of air down through the cores arranged alongside the two spines.

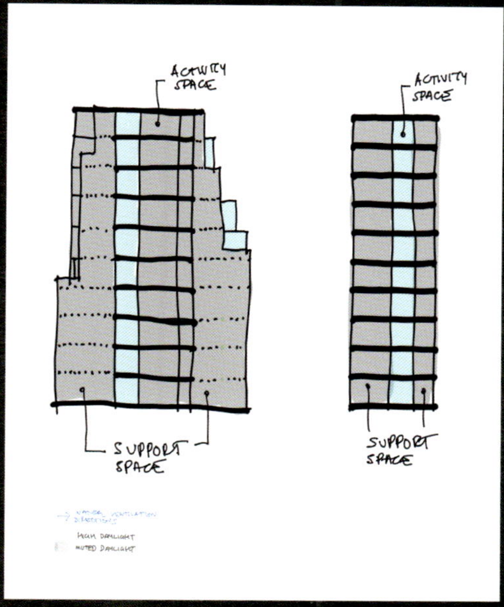

Daylight strategy for direct sunlight and cross ventilation.

Option 3 admits daylight and air in localised areas adjacent to the facades.

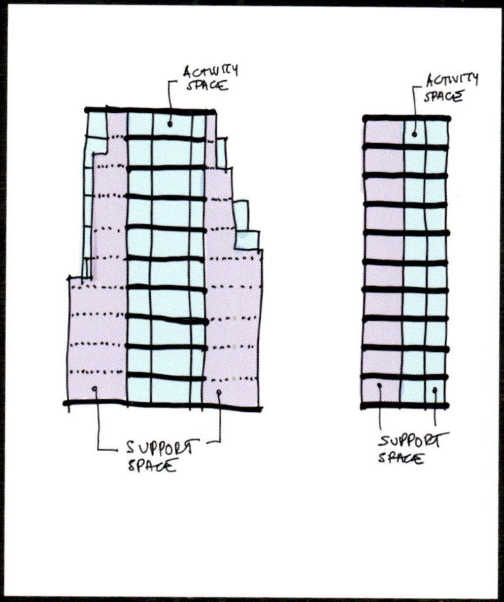

Environmental zones

Option 3 distributes services vertically down the centre of the spine.

spaces. Prototyping spaces are more open to the outside air during the working day; the offices have more air recirculated.

In Option 3, services are distributed vertically down the centre of the spine, with one of the spaces being used for services equipment which is specific to each floor. Chillers and air-handling units are set out at roof level in a way that supplies a vertical distribution of ducted air down through cores which are set on both sides of the central circulation spine. Each core services a zone within a given floor level, such that each zone has services distributed locally to its 'hub' of support space. Each service hub is independent of its neighbouring cores and vertical service runs. As a result, all environmental zones can be operated together as a complete system; almost as a separate building within a building.

Option 3 was the preferred solution to environmental zoning, which led to separating the distribution of ducted services supplied to internal areas from those which serve the perimeter of the building. There is a marked difference in this building between the environmental requirements for circulation areas and those for interstitial spaces between groups of office and prototyping spaces. Circulation areas have limited user occupation; interstitial spaces are in constant use for design meetings, and so require as much fresh air as the offices. Edges of the building adjacent to external walls are generally used for office space even within prototyping areas. Those spaces adjacent to external walls are glazed, and those for the internal prototyping spaces have opaque walls. The prototyping spaces are large volumes with small groups of people occupying them at any given time. Environmental controls are provided close to where people are working, resulting in less tempered space needed at higher level, where machinery is installed and where access is required for maintenance. As a result, each of the prototyping spaces has two modes: the first mode for full occupancy of people within the space, and the second as a standby mode where machinery is being operated and where trial runs for fabrication procedures are in progress. Office spaces do not require a two-mode operation; those spaces are tempered accordingly.

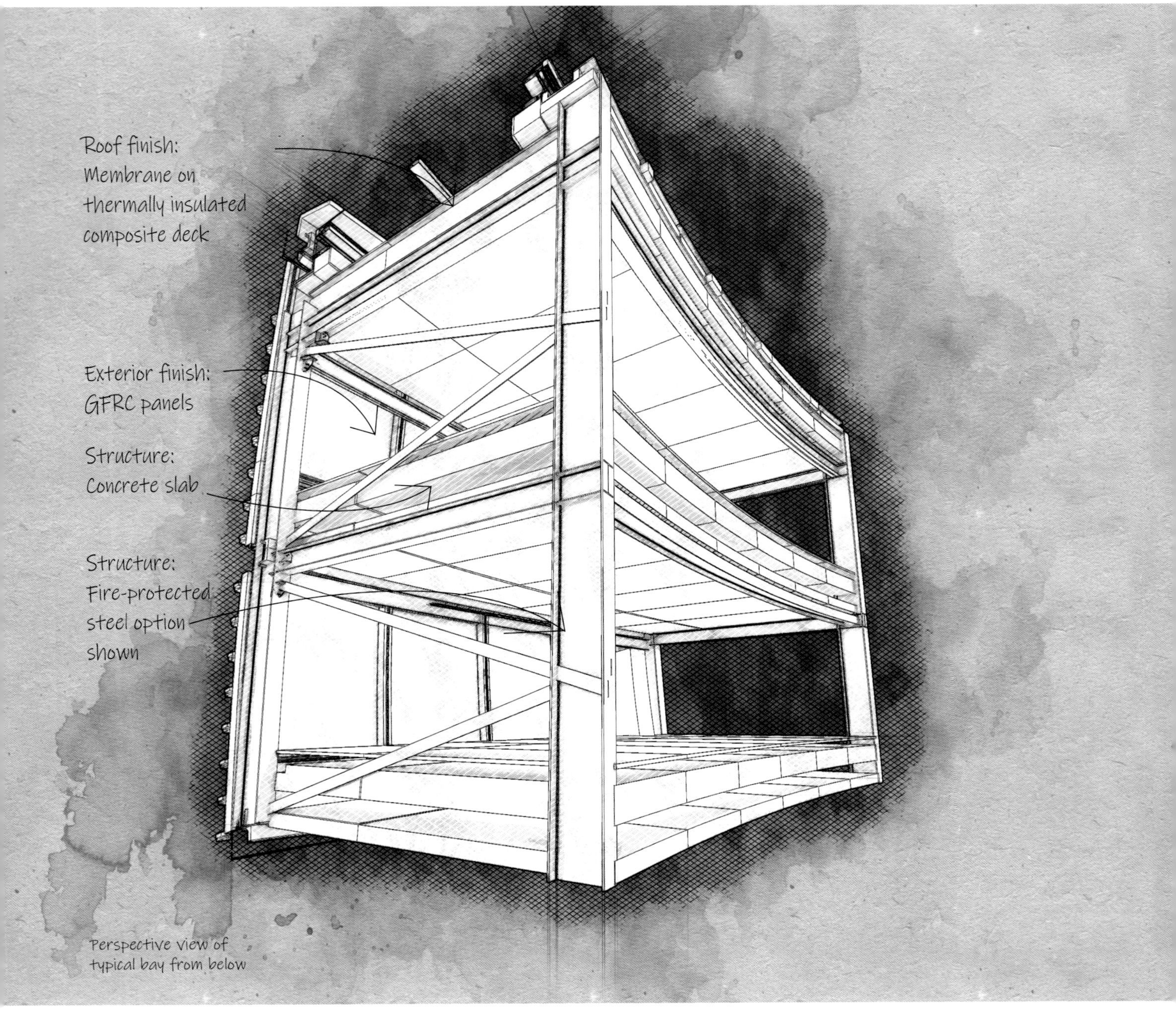

Typical bay adjacent to facade

The design approach for this building brings together structural design and spatial design by using structural components and assemblies as part of the visual design. Most structures are required to be wrapped externally with thermal insulation and additional cladding panels; internal structures are enclosed with interior finishes as their appearance is often incompatible with the need to enclose interior spaces.

Permanent formwork was used as the method of enclosing structural concrete walls with formwork that is made from GFRC (glass-fibre reinforced concrete) that is itself a finish to the outside. This method avoids the need to create additional decorative cladding by setting GFRC formwork on the outer side and an interior standard formwork on the inside face of the wall. When set around the arrangement of steel reinforcing bars, the concrete is cast into the void created by the formwork; that formwork stays in place when the concrete has cured. This method allows external structural walls to have a finish for weather protection which is integral to the process of casting the concrete instead of adding another layer of cladding. The facades aim to strike a balance between views from outside to inside the building in order to make visually transparent the activities within the building, with the need for controlled daylight within different sets of spaces. The control of daylight to glazed walls is provided by metal-framed panels which have the mechanical ventilation and electrical equipment integrated into a 1.0-metre-deep double-facade zone; the double-skin facades are set on both sides of the building in order to accommodate equipment and access to that equipment. As a result of the use of a double-skin facade, the opaque walls are suited for access to specific parts of machinery in the facade zone. Other areas of opaque facade are clad in GFRC panels, translucent metal panels and areas of full-height glazing.

Parts of the facade which accommodate services equipment and air supply ducts have cladding panels which are metal faced both on the outside and on the inside, and thermally insulated to provide the 1.0-metre-deep zone for the accommodation of equipment as well as for stairs. Lifts/elevators are also located in these perimeter areas.

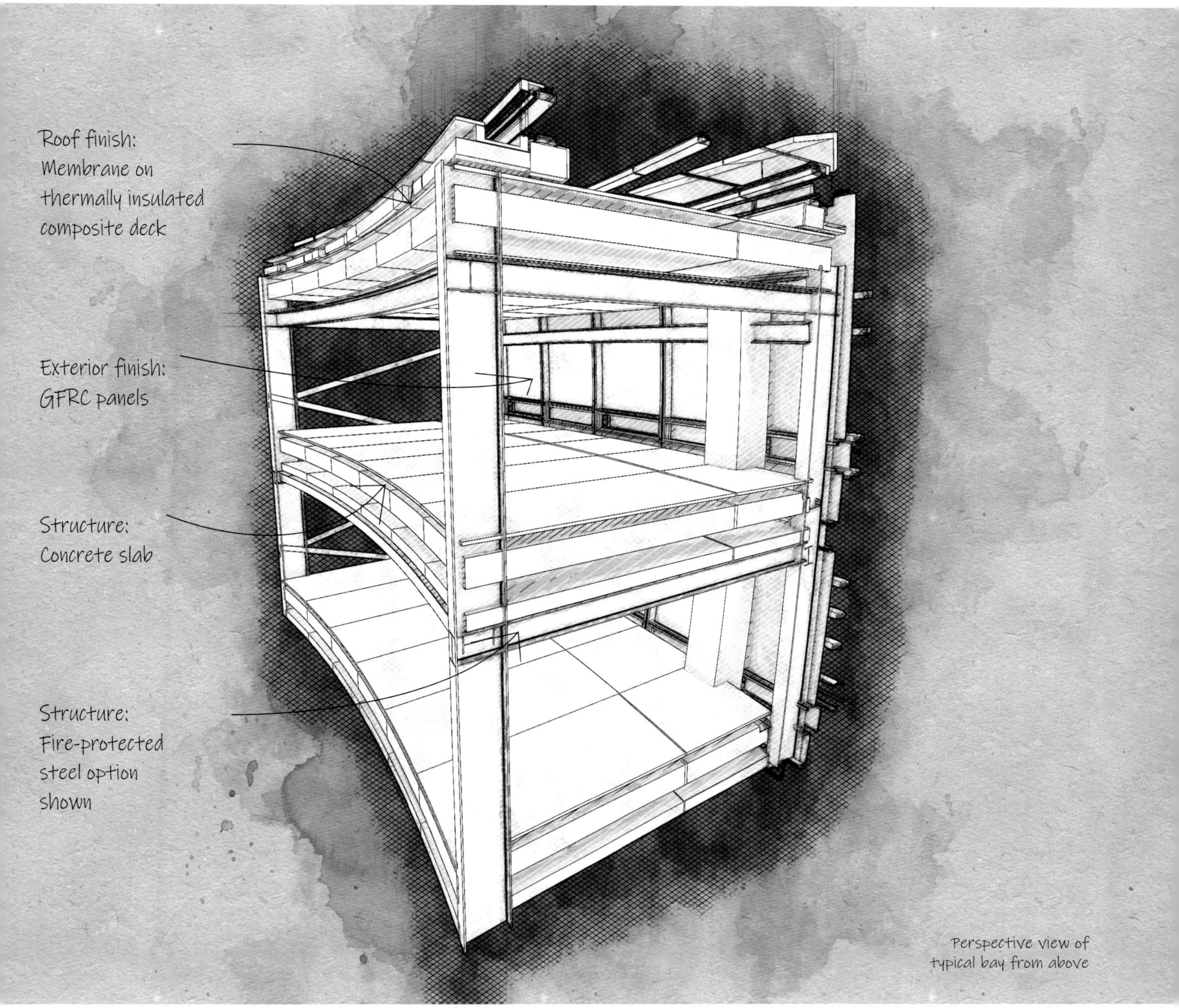

Perspective view of typical bay from above

The glazed walls comprise site-assembled curtain walling, which is inclined from the vertical plane in some areas to provide clear views out and avoid reflections within the space. The glazing uses a patch-plate glazing system in order to allow the glazing to be demounted and replaced easily with the lowest number of fixings required. Panels on the narrow end elevations are straightforward to remove in order for them to be replaced with glass or to be able to be swapped between glass and metal to suit functional requirements. Currently, the opaque walls enclose stairs; they are demountable to meet the expectation of near-future manufacturing requirements. This building is very much a 'kit of parts' where individual elements are disassembled and rearranged as facades and as part-structures while the building remains in use; similar in concept to an oil rig production platform.

The external envelope follows a strategy of providing glazed openings deep into the facades; these openings are created as floor-to-ceiling glazing with bands of opaque panels around the edges at every floor level. This approach avoids the need for a full curtain wall that would hang from each floor level and would be set in front of the floor slabs, which provides no specific advantage in this case.

The opaque panels which cover large parts of the building are formed with opacified glass, which allows for limited light transmission through the glass while still maintaining the level of privacy required in some office spaces, together with the control of solar gain that would otherwise be experienced through the glass.

This approach avoids the need for internal or external blinds with associated maintenance and provides an even level of daylight across a large floor space. The facade also holds some of the air supply ducts, which form the primary circulation of air around the building through the two side facades, rather than through the cores in the centre of the building. This approach facilitates an easier distribution of air within each floor, minimising depth required at ceiling level; benefiting from the need for opaque facades through parts of the facade. The glazed areas do not use curtain wall but instead use metal window frames to suit their complexity of form.

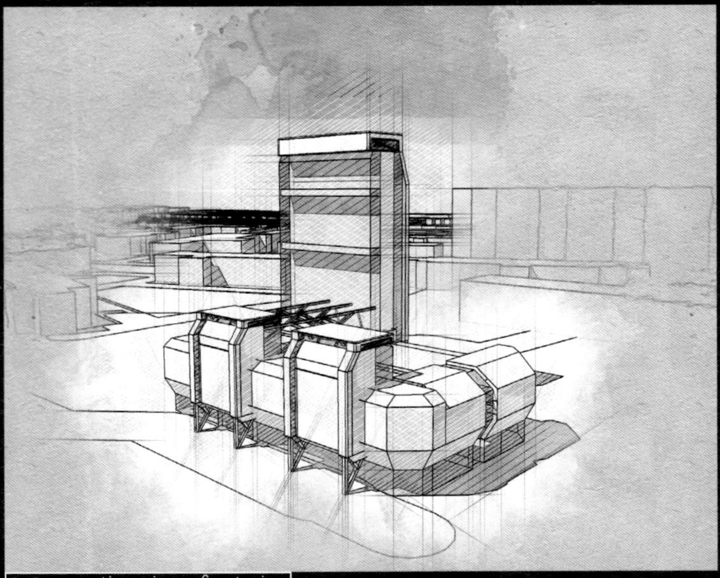

1. Perspective view of exterior

2. Perspective view of exterior

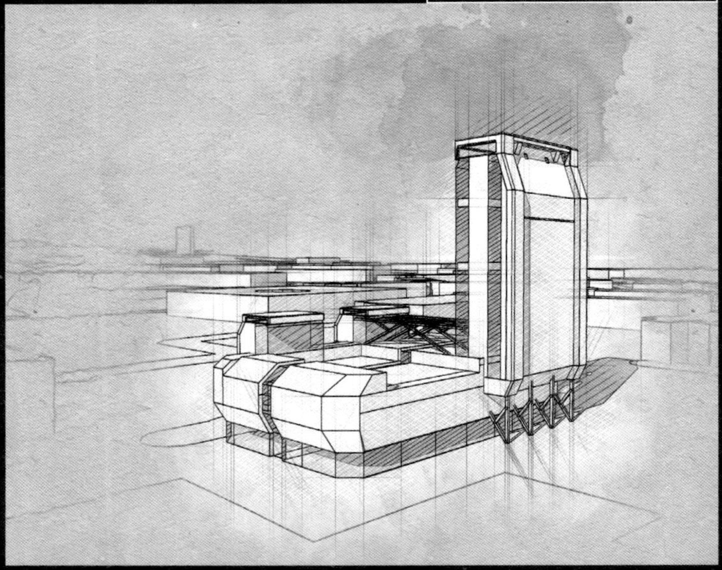

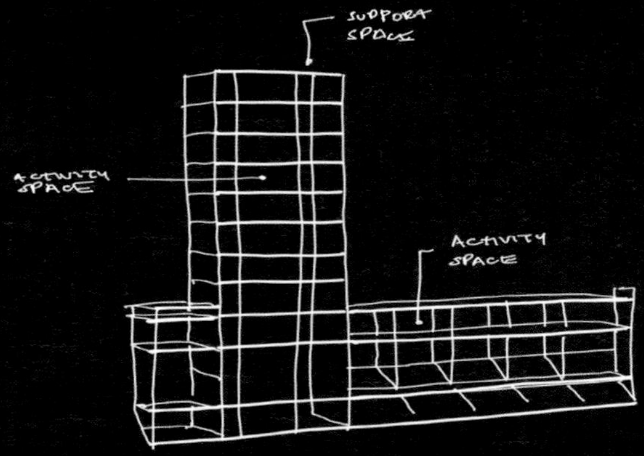

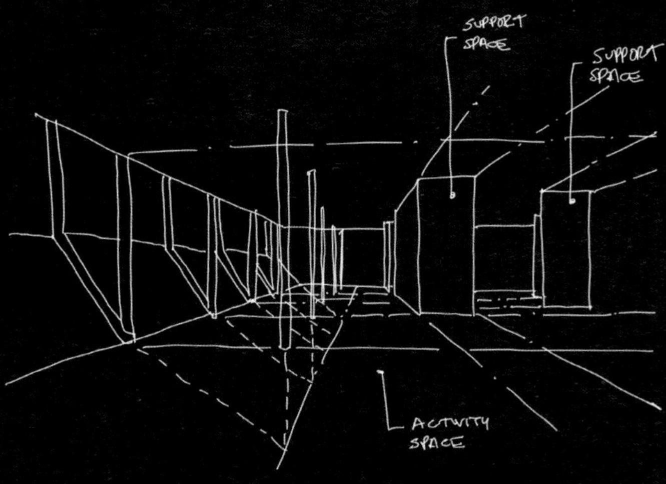

Optimisation for fabrication

In this project, a key aim of the structural strategy is to remove the need for secondary supporting structures for roof and facades. This goal is achieved through the coordination of the design of the primary structure with that of the envelope in order to ensure they can avoid duplication of components; for example, avoiding a doubling-up of facade framing to a column and beam, immediately adjacent to one another, that forms part of the primary structure. This approach enables the architectural intent to be realised for complex geometry facades as well as providing greater structural and material efficiency. The associated reduction in material allows the time and cost of manufacture to be reduced, particularly when the remaining system components are designed to be straightforward for manufacture. Panelisation methods allow the structure supporting the form of the external envelope to be efficiently created from an arrangement of members, where the geometry and size of the members are optimised to ensure their manufacture and installation is feasible as well as economic. During this process factors of material weight, fabrication and installation are evaluated in each part of the overall arrangement of members; the most efficient arrangement being based on these factors is then adopted. Commonly the solution will involve a combination of overall dimensions and a mix of flat and curved members; the number of very large or doubly-curved members is usually minimised as these are more difficult to manufacture and install. Reducing the weight of the overall structural arrangement is achieved as a result of refining component sizes using parametric design, an iterative process which is also used to achieve the reduction of secondary structure. Together these contribute to a decrease in the self-weight of the system, such that the dead load of the main structure, facade structure and roof structure is reduced. This provides the potential then for reducing the weight of the main structure when compared with standalone primary, facade and roof structures, resulting in a cumulative reduction in the material weight required for the project, associated assembly costs and environmental impact through fabrication and transport. The system design is opti-

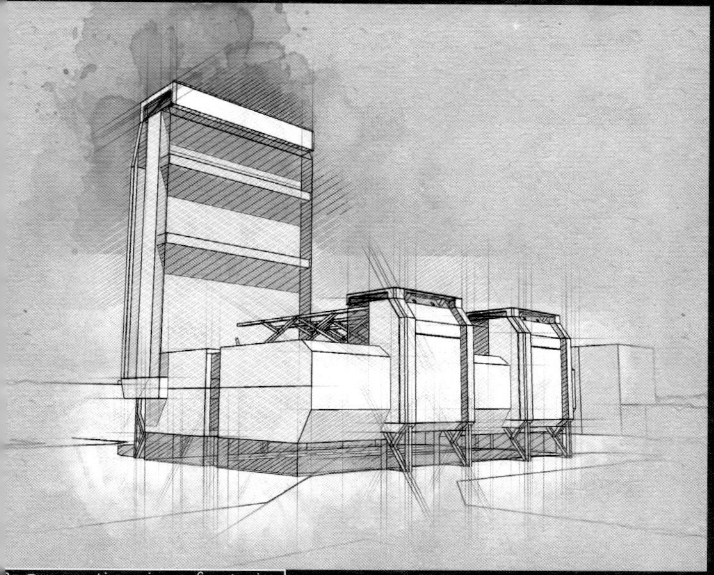

3. Perspective view of exterior

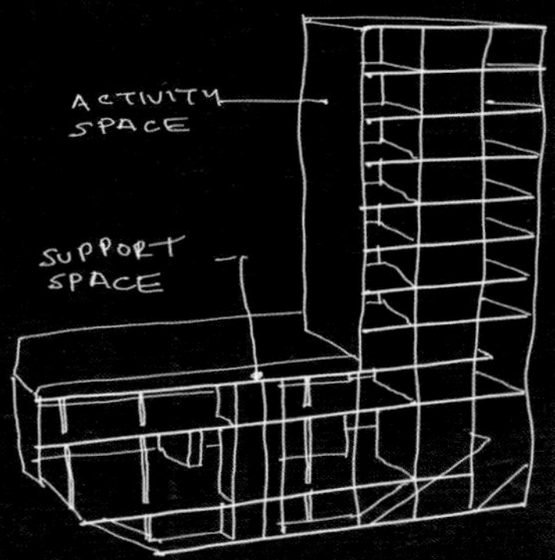

4. Perspective view of exterior

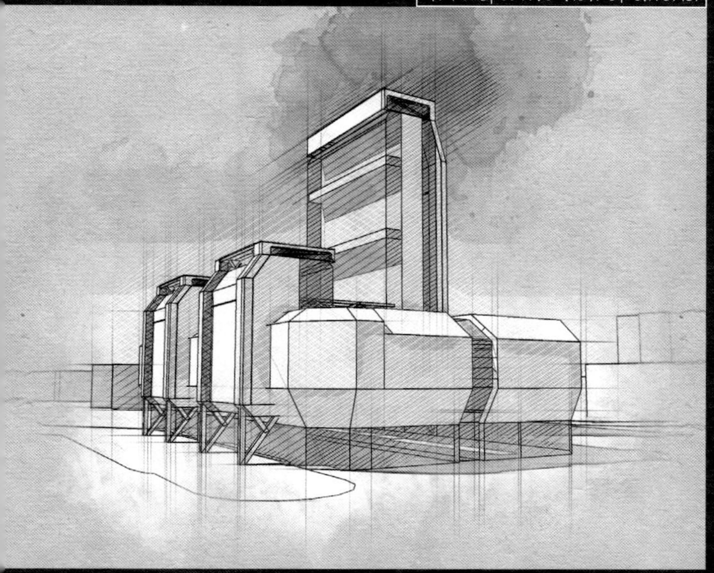

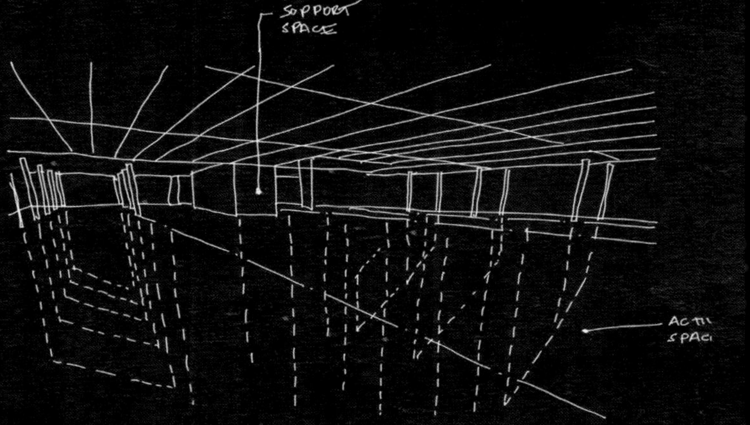

mised based on the material weight; achieved through the implementation of parametric design to explore possible component dimensions and assemblies. The parametric design approach gives the ability to define' fixed' and 'variable' metrics based on the project constraints; iterations of the design are then produced based on altering the variable metrics. Evaluation of the system variations is based on material weight and structural performance; the most efficient solution can then be selected. The method results in a reduction in construction material which inherently reduces the fabrication cost. The system's weight is reduced through the elimination of the secondary structure; this is achieved by analysis of the global and local structural response in parallel, allowing the full behaviour of the structure and envelope, as well as their interaction, to be evaluated. By evaluating the structural response of the system on a global and local scale, using the output of each to inform the other, the behaviour of complex geometry systems is more accurately modelled and evaluated. This allows the design to be refined to a high level while retaining confidence in its structural integrity. Optimising the system for weight can be expanded by considering the associated cost; this consideration of parameters in parallel creates greater structural and cost efficiency of the design. The cost metric is useful when comparing different materials for a system technology, as each material has a different strength-to-weight ratio, such that the lightest material solution may not coincide with the lowest cost option. A balance is to be found between weight reduction, which may have assembly and environmental benefits, and cost reduction, which is likely to make the project more feasible. The consideration of both weight and cost allows new materials to be considered and evaluated for system technologies to determine the most appropriate solution for the system, optimised for both the sustainability and cost of the component materials used. Emerging materials such as composites can provide particularly high strength-to-weight ratios alongside increased durability properties, but these materials are commonly more expensive. There is a trade-off to be made between the system's weight reduction and the project cost.

1. Project volume in site context

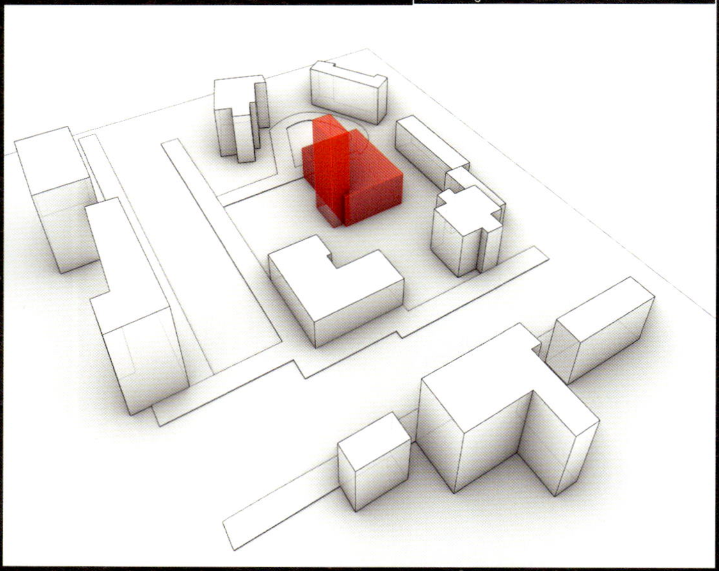
2. Project volume in site context

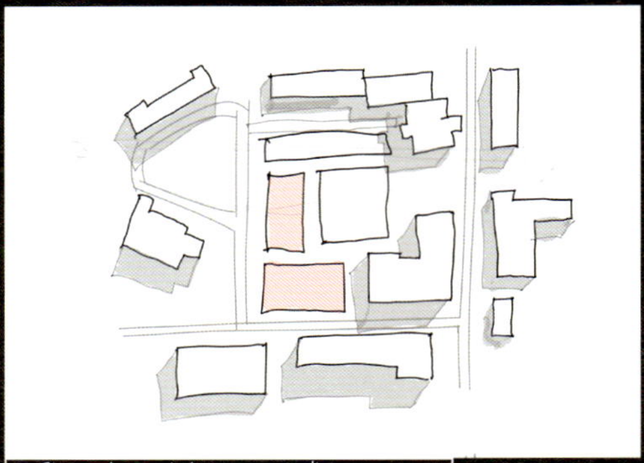
1. Site plan with surrounding context

Option 1 comprises two blocks arranged in a largely orthogonal arrangement that follows the orientation of buildings which surround the site.

2. Site plan with surrounding context

Option 2 rotates those buildings such that three small external spaces are created, all of which are rectangular in shape and are framed by the new buildings.

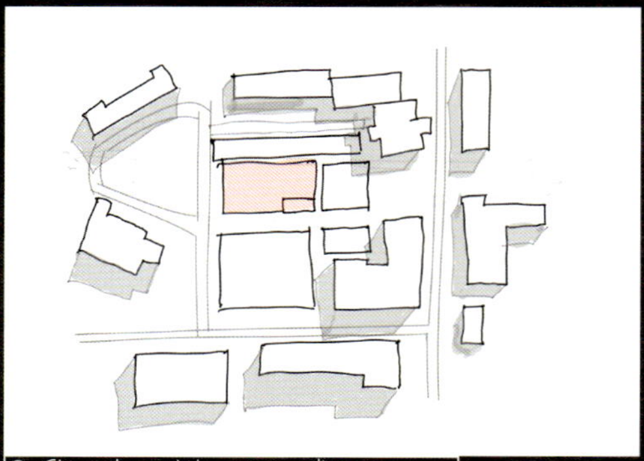
3. Site plan with surrounding context

Option 3 combines these volumes as a taller building where the office space is stacked as a mini-tower.

Site context

This building is a space vehicle refurbishment and exhibition centre that inhabits part of an industrial area undergoing renewal. Pieces of used space vehicles are available for public view. The building comprises two large-scale volumes which are juxtaposed. An enclosed tower building is used for the exhibition of components, together with associated visitor facilities and administration offices. The refurbishment of space vehicle assemblies takes place in one half of the ground floor podium; the other half of that volume accommodates large-scale refurbished exhibits. The tower is linked to the podium-level exhibition space. The podium-level exhibition has a changing set of assemblies on display that are moved periodically to a large-scale visitor attraction, where used components are joined together to make large-scale assemblies which will be visually recognisable to the public.

The main part of the assembly hall is located next to a regional rail line that provides easy access to the building. Existing rail tracks are set along both of the long sides of the building, allowing incoming components to be offloaded on one side of the podium; the other side of the building is used for the loading of refurbished items which have been exhibited onto rail wagons for despatch to their final destination at a major visitor attraction. Assemblies move on rollers from the refurbishment hall to the exhibition hall, where they await onward shipment by rail.

The assembly hall has two mini-towers that rise above the roof level of the podium, providing both access and vantage points for items which are exhibited permanently on the roof of the podium. This arrangement of tower and split-podium, together with the mini-towers attached to the podium, are at a scale sympathetic to the volumes around it. The design creates a sense of what may be possible in the near future with space travel, showing the older technologies from which more ambitious spacecraft assemblies have been developed over the last 50 years. The building is located in an area which has been deindustrialised and which is undergoing urban renewal. The site context for Option 1 comprises two blocks arranged in a largely orthogonal arrangement

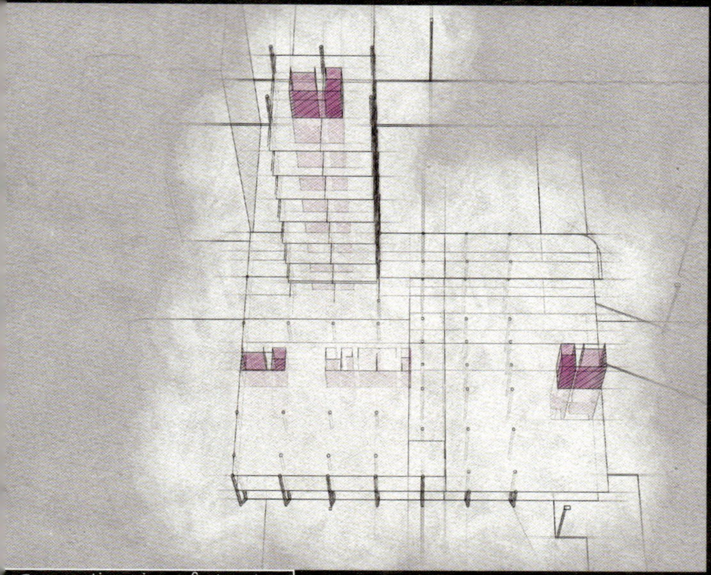

. Perspective view of structure

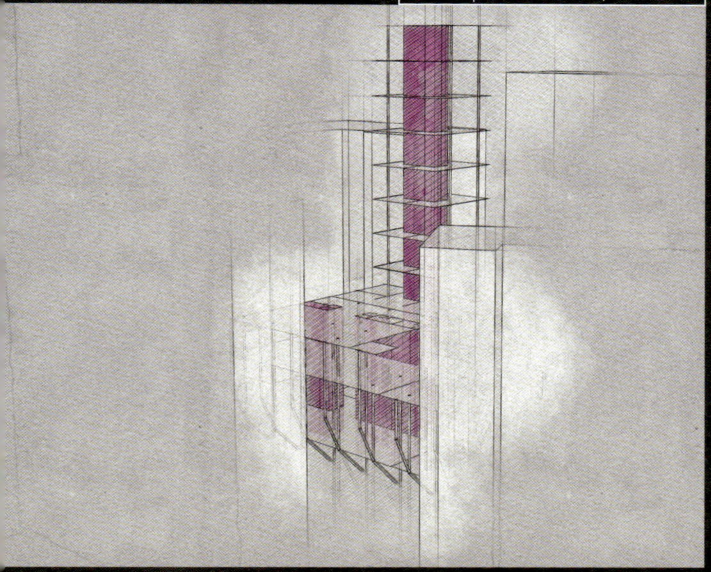

2. Perspective view of structure

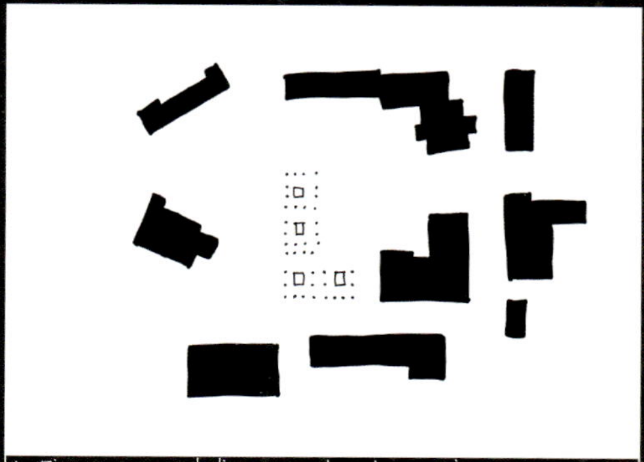

1. Figure-ground diagram showing project structure

Option 1
The structure is set out on a grid which follows that of some of the surrounding buildings.

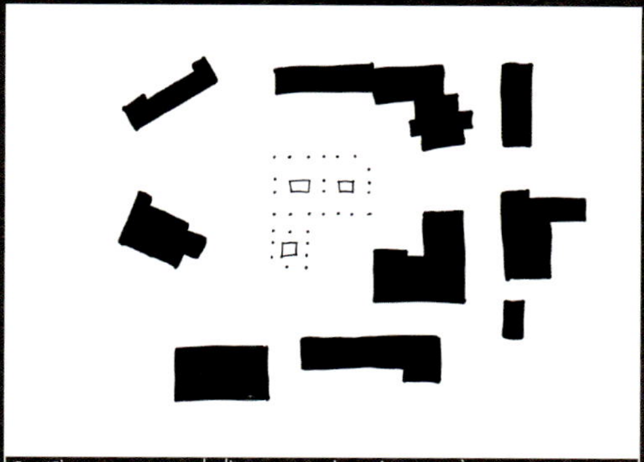

2. Figure-ground diagram showing project structure

Option 2
The structural grid integrates cores within the geometry.

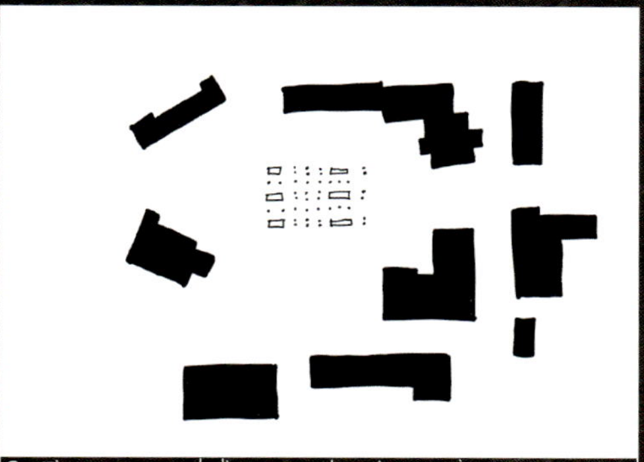

3. Figure-ground diagram showing project structure

Option 3
The structural grid is reduced in span in order to accommodate a more complex arrangment of spaces.

that follows the orientation of buildings which surround the site. Some of the newly revitalised buildings set adjacent to this project are beginning to be set out on a different grid; this project aligns itself geometrically with its older neighbours in order to strengthen the existing massing of built forms. The two parts of the building comprising refurbishment and administration result in the characteristic form of this light industrial building. As a result, the two blocks are arranged side by side. This arrangement leaves space for car parking to one side of the building, and frames an external space which can either be landscaped or can provide a site for a further building that will complete the arrangement of buildings on the site. Either an external landscape space can be created or a building can be inserted; either solution will provide an organised arrangement of internal and external spaces.

Option 2 rotates those buildings such that three small external spaces are created, all of which are rectangular in shape and are framed by the new buildings. Once again, these three sites can be developed with buildings or can be left as external spaces where both solutions are consistent with the surrounding massing of built forms. Car parking is located in the same position adjacent to the road as in Option 1. The building is again divided into the two parts of refurbishment and office-based spaces.

Option 3 combines these volumes as a taller building where the office space is stacked as a mini-tower, inserted into the exhibition space in order to integrate it more closely. The arrangement creates three larger external spaces; all three are rectangular. The podium brings smaller-scale subdivisions of the volume adjacent to the facades which are used as visitor entrance spaces as well as defining routes through the building both horizontally and vertically. This approach allows the scale of the building to be further reduced without creating arbitrary subdivisions between volumes. The subdivisions were created in order to bring the scale of the visitor experience to that of seeing a series of building forms which are of a similar scale to the adjacent buildings, allowing the functioning of this building to be more clearly visually articulated through its constituent forms.

1. Sectional perspective view

2. Sectional perspective view

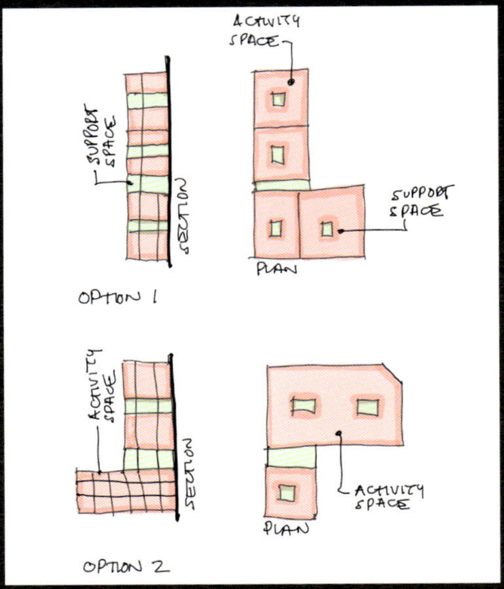

Spatial arrangement
Option 1 provides access through the link block.
Option 2 uses the same link concept, but the office space and the vehicle refurbishment space are completely separated.

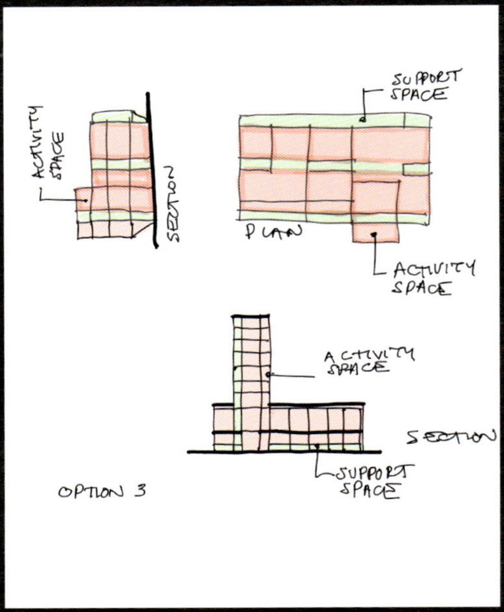

Option 3 creates a series of more differentiated spaces.

Spatial arrangement

The purpose of the building is to refurbish parts of space vehicles for permanent exhibition. Space vehicle parts are lifted from rail wagons adjacent to the rail line and are lifted into place next to the building, where they are moved into the building on rollers.

In Option 1, access is provided through the link block; circulation proceeding in a rectilinear way from that arrangement with one large single hall for vehicle refurbishment and two floors of office adjacent to it. Another part of the vehicle refurbishment hall can be converted into office space with the addition of a new floor or mezzanine floor.

Option 2 uses the same link concept, but the office space and the vehicle refurbishment space are completely separated, with that space having a single floor in a single volume arrangement; the office being arranged over four floors. Circulation is by way of the link between the two buildings, with a movement along the edges of the building and routes going into the single refurbishment space. Option 3 creates a series of more differentiated spaces as a way of reducing the envelope of the building and bringing spatial components together into a denser form, where the office space is set into the volume of the refurbishment hall. The workshop building is divided into three smaller spaces with a roof-top space for communal canteen facilities. Spaces within the podium are arranged around the structural grid onto which the towers are offset on the side of the podium housing the refurbishment hall. Five spaces are created of approximately equal size, which comprise two main spaces where work is carried out. These main workspaces are set immediately behind the mini-towers that project out from the building. Volumes on either side of that main space house support spaces for tools and equipment. The refurbished items are then rolled through to the exhibition space such that there are up to two exhibition pieces being shown at one time. These are divided by circulation spaces and information areas together with platforms from which visitors can look down onto exhibited assemblies as well as viewing them from ground level. The items exhibited are then rolled out in the same direction; moved out of the building where they are loaded onto rail trucks and

. Perspective view of interior

2. Perspective view of interior

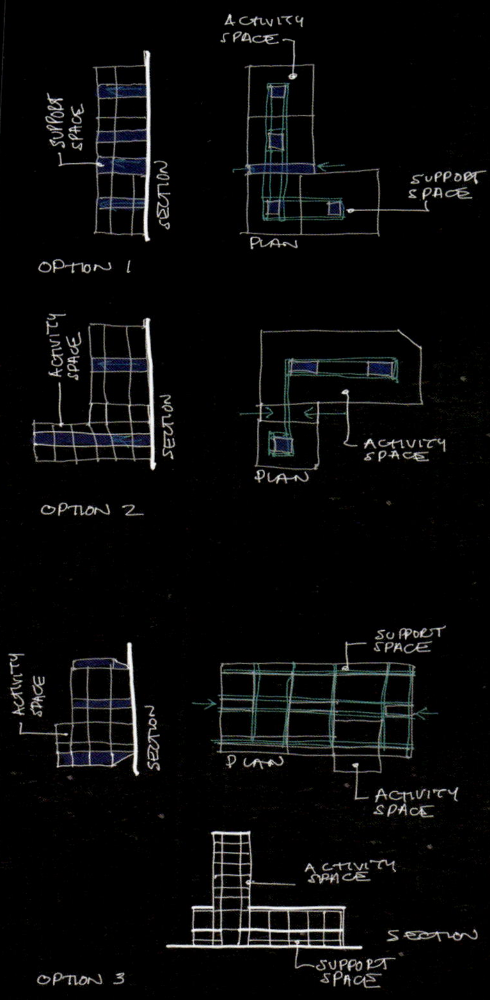

Internal circulation

Option 1 comprises routes within the building, creating a circuit around the service cores and the internal workshop spaces and meeting spaces that link those two volumes.

Option 2 links the service cores in their internal spaces.

Option 3 traces circulation at roof level by rooflights which provide continuous strips of daylight down the edges of the building.

sent onwards. The rails for the track are recessed into the ground surface to allow for vehicle access and ease of movement around the building. The main tower comprises four separate areas divided into two exhibition galleries on each of three floors above the podium level. The upper floors provide views out onto the roof where permanent exhibits are in place, which can also be viewed from an enclosed area at the opposite end of the podium, at roof level from the main tower. This projecting volume at roof level provides access to the permanent exhibits and to a light and sound show which is projected from audio-visual equipment attached to the facades of the tower.

Internal circulation

The internal circulation for Option 1 comprises routes within the building, creating a circuit around the service cores and the internal workshop spaces and meeting spaces that link those two volumes. This approach provides consistency with both the structural layout and the spatial arrangement of Option 1. Option 2 has a similar concept of corridors linking the service cores in their internal spaces, but the links between those spaces create more internal meeting space than is currently required.

The goal of having a balance of internally closed-off space for workshops, meeting rooms, cores, refurbishment space and offices is achieved in Option 3. Circulation within the building for Option 3 is traced at roof level by rooflights which provide continuous strips of daylight down the edges of the building and down the centre of the podium.

The complete building is accessed from both ends of the podium, on the exhibition side and on the corresponding edge of the refurbishment hall. The podium halls have circulation routes that extend the length of the building, which is accessed in the same way from the short ends. The tower is then accessed from the central circulation space that extends the full length of the building as well as having a separate access from the short sides of the podium.

This project creates zones for circulation by identifying them as continuos glazed slots at roof level and on facades, where daylight is needed.

1A. Isometric view of structure

2A. Isometric view of structure

1B. Isometric view of structure

2B. Isometric view of structur

Structural strategy

Structurally, the design comprises two sets of assemblies. The first set comprises a primary structure of steel sections used to create the edges of the tower, the platforms inside the tower and the projecting volumes set into the roof of the podium. These assemblies extend through the full height of the building. The second set of assemblies consists of secondary structures of smaller-sized frames that surround the primary structure but are structurally independent of the primary structure.

The structural concept is based on the use of prefabricated structures which arrive, both by rail and by road, at the project site where they are assembled and installed. The structural strategy is to accommodate a set of interlocking or adjacent forms which are highly geometric and complex in shape. The frames for each part of the building are made as individual lattice frames, to which facade cladding panels are applied on site. The need for a close fit between the parts of the building led to the solution of making the cladding from steel framing to match that of the primary steel frame supporting the elements behind. There is a requirement for large areas of opaque outer facade panels, together with areas of glazed panels and louvre panels. A raised platform for the delivery of certain items for the refurbishment process is provided on an upper deck. As a result, the design is made as a series of modular components; a structural 'kit of parts', where each part is structurally self supporting, and the elements are bolted together to create the overall forms. The complexity of the forms is a response to that requirement. The larger structures make use of portal frames, which are bolted together to form a complete podium form. These are clad in a variety of panels of large size; structurally continuous with the framing below. The cladding is continuous across the complete external envelope, supported by continuous framing in order to reduce the amount of secondary framing required, and to maximise the free space inside the internal spaces. This is very much a series of 'shrink-wrapped' building spaces created by components which are then bolted together. The building form is 'shrink-wrapped' into a set of linked forms that wrap

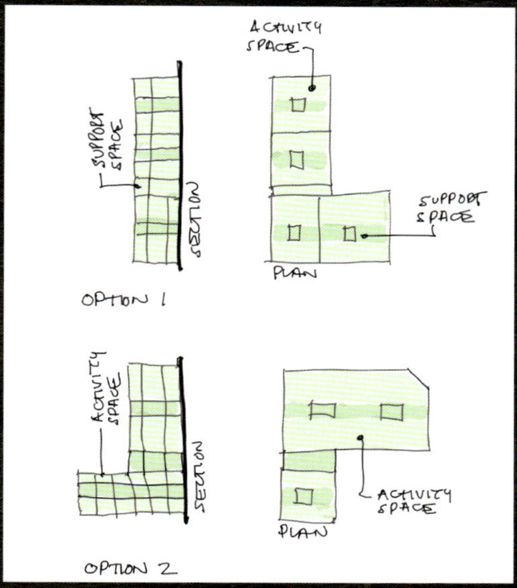

Structural strategy: Zones
Option 1 has service cores and meeting spaces arranged in the middle of each of the rectangular forms.
Option 2 provides more floors in the office tower than that of the previous option.

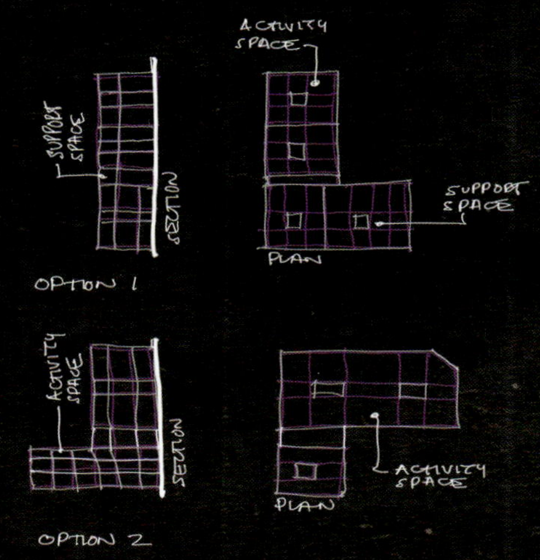

Structural strategy: Framing
Option 1 has service cores arranged at each end of the building.
Option 2 has a similar structural grid arrangement to Option 1.

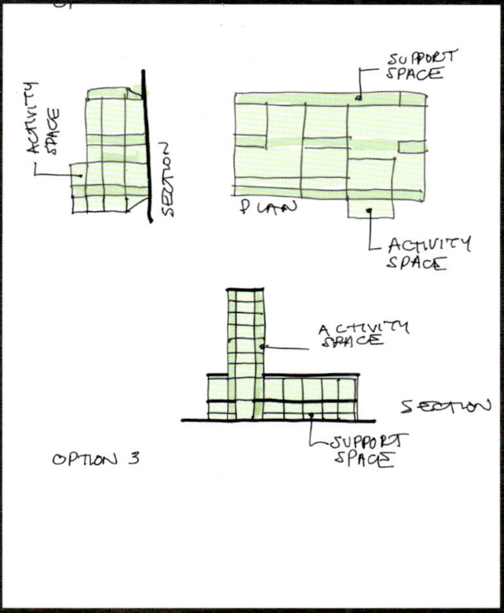

Structural strategy: Zones
Option 3 combines the individual forms into a single building.

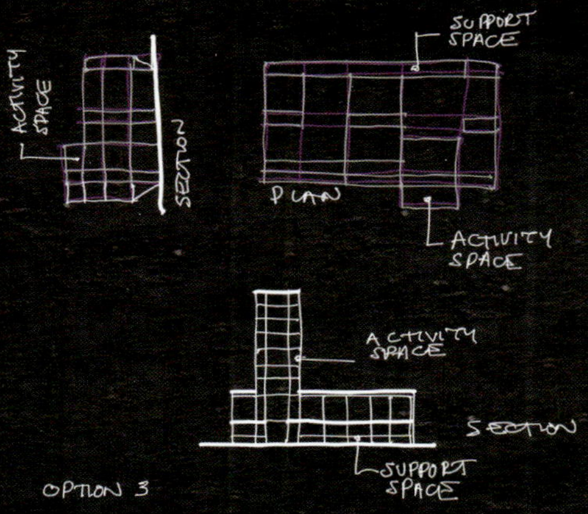

Structural strategy: Zones
Option 3 uses a single structural grid.

over a series of different spaces. In this design, each element is shrink-wrapped; the different structural elements are then bolted together, or separated to suit the functionality of the building. This approach of prefabrication with shrink-wrapped elements allows for a high degree of flexibility during the design development process and allows different parts of the building to be extended outwards without the need for dismantling any of the facades.

The structural strategy for Option 1 comprises a grid of columns and beams which is based around the need for having service cores and meeting spaces arranged in the middle of each of the rectangular forms. Service cores are arranged at each end of the building with spaces for internal meeting rooms set between the two cores. In the second block the office space makes use of a similar concept, which allows for the refurbishment block by having the internal zones between the cores removed such that there are two lines of refurbishment space. The office space has a similar arrangement in that the internal spaces can be arranged for meeting rooms. The intention of the cores in the refurbishment space is to create some closed-off workshops in the centre of the building. The structural grid is arranged around that concept to create clear span spaces from the core to the external walls, and from between the cores themselves.

The structural strategy for Option 2 uses a similar concept to Option 1, but the office tower has more floors than that of the previous option. The form of the office block becomes a tower, with a similar structural grid arrangement in order to create clear-span spaces within the volume of the tower. The refurbishment space has a similar arrangement to Option 1, but is rotated through 90 degrees. Both Option 1 and Option 2 allow for a link between those two buildings such that the internal circulation is in harmony with the structural grid.

The structural strategy for preferred Option 3 is one of combining the individual forms into a single building with a projecting tower and using a single grid; a structural grid of beams and columns arranged such that service spaces form horizontal strips across the refurbishment hall and up through the height of the office tower.

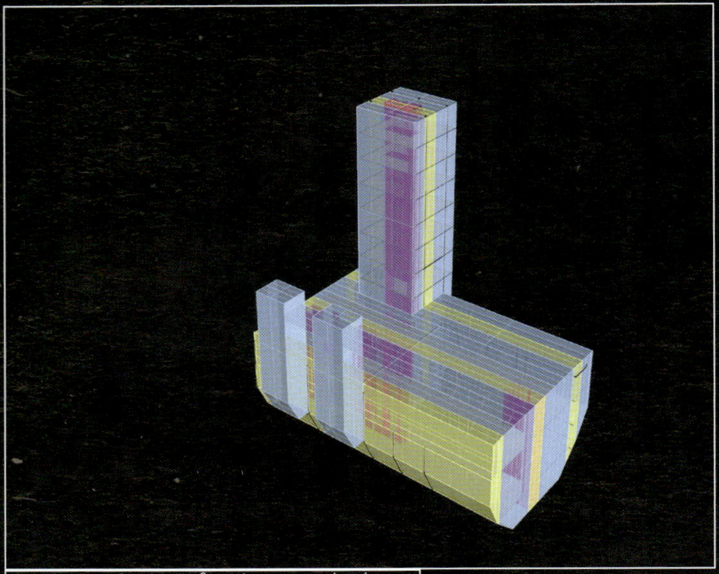

1A. Isometric view of environmental volumes

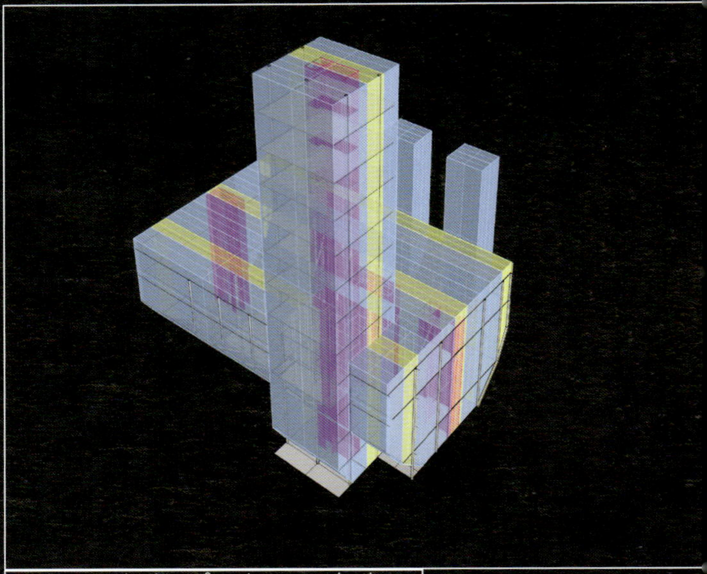

2A. Isometric view of environmental volumes

1B. Isometric view of environmental volumes

2B. Isometric view of environmental volume

Environmental strategy

A priority of the environmental strategy is that the building's owner-occupiers will carry out maintenance themselves with technologies that are available to the owner. The structures vary in scale from the single tower element to the medium-sized bolt-on opaque volumes at the side, down to small projecting volumes which accommodate stairs and entrance spaces. In addition, glazed units, together with an integration of the services components and chillers on the roof, are designed to be part of the visual expression of the building, rather than to be concealed behind panels. Where possible, the design takes the opportunity to show off the structure and structural components, which vary from medium scale to large scale. A design priority is to give as much flexibility to the owner as possible while showing that the building is continuously a 'work in progress'; one which is effectively a machine for making and doing tasks related to the refurbishment of part assemblies from space vehicles. The same system is used for the medium-sized structures and for additional elements, like structural supports for the services elements. Support platforms for additional external communications equipment, as well as external stairs, are all designed using a common system of pressed steel members which are bolted together to form lattice frames, then built up into larger platforms which are then either enclosed or exposed within the building. Only the tower has the primary structural steelwork associated with the traditional or the standard way of constructing using steel I-sections.

Both the exhibition and visitor tower, and the podium halls are naturally ventilated for three months of the year and are mechanically ventilated for the remaining nine months, when temperature control is required.

Option 1 of the environmental strategy takes the need for having a core space in the middle and an internal workshop space, and creates that as a single environmental zone surrounded by the red zone, as shown in the illustration. This single space has different ventilation and servicing requirements. The office part of the project has similar arrangements, but this block can be divided in two; each becoming refurbishment space instead of office space, and vice versa.

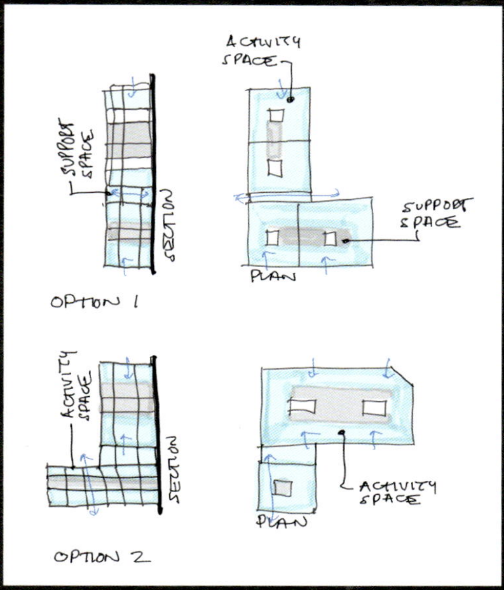

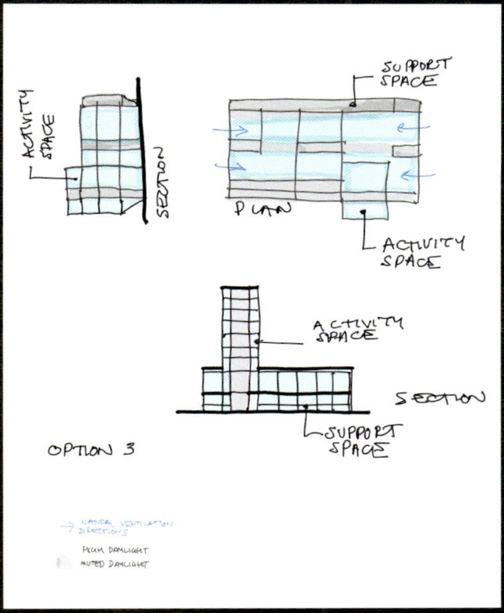

Daylight strategy for direct sunlight and cross ventilation.

Option 1 admits daylight and air from the perimeter of the building, with subdued daylight between cores.

Option 2 admits daylight and air from the perimeter of the building, with subdued daylight around cores.

Daylight strategy for direct sunlight and cross ventilation.

Option 3 admits daylight and air in bands which traverse the building, with subdued daylight between cores.

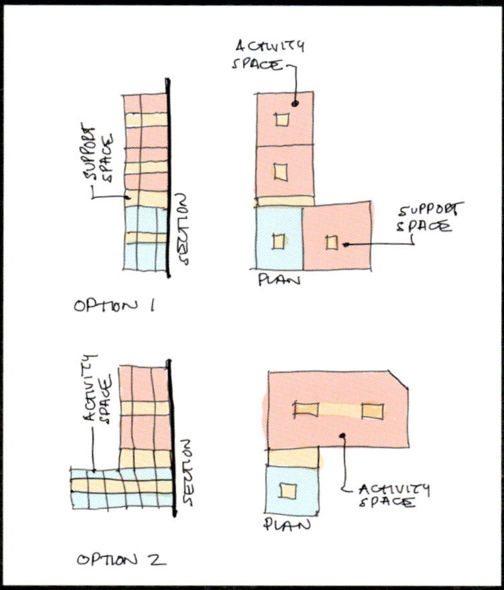

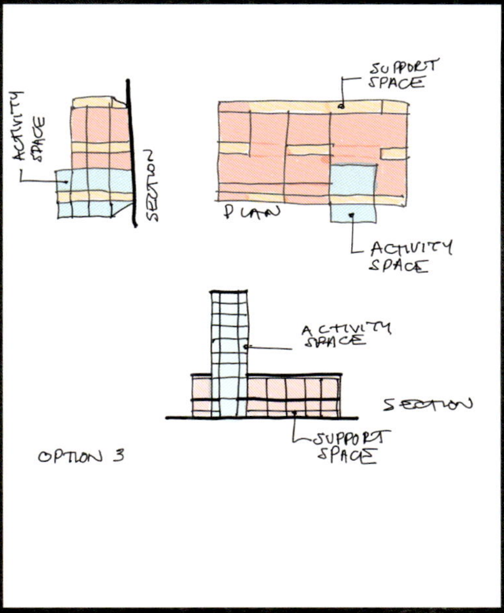

Environmental zones

Option 1 responds to the need for having a core space in the middle and an internal workshop space.

Option 2 takes the environmental zones of the Option 1 refurbishment building and rotates it through 90 degrees.

Environmental zones

Option 3 has three primary environmental zones comprising the refurbishment hall, the exhibition hall, and the mixed-use space in the office tower.

Option 2 takes the refurbishment building of Option 1 and rotates it through 90 degrees, retaining the structural grid; the overall form of the building is adjusted such that it lines through and is consistent with the structural grid of the office tower. The tower has a single environmental zone like Option 1. The link space between the two buildings is treated as a core space effectively, and its treatment as an environmental zone is like that of the cores and internal meeting spaces.

Option 3 has three primary environmental zones comprising the refurbishment hall, the exhibition hall, and the mixed-use space in the office tower. These zones are further subdivided, following the structural grid lines, into five environmental zones in the refurbishment hall and five in the exhibition hall. The zones are created by thermally insulated screens that are deployed from roof level in the manner of overhead lifting doors. This approach allows for the spaces to be subdivided in response to the changing use of those spaces. An equivalent single large space would be uneconomic to mechanically ventlate if only a small part of that space were used at any given time. The subdivision of the space in the two halls is enabled with lifting doors to create smaller spaces to ensure that the energy used in those spaces is used efficiently. As a result there are five environmental zones in the refurbishment hall and five in the exhibition hall. The tower has environmental controls provided on each floor and these are such that the tower is divided into three upper floors, each with its own environmental zone to respond to the very different needs of the types of exhibition that are envisaged. The environmental design also responds to the use of the top floor as a largely administrative open office space. The three circulation zones, which are continuous through the height of the tower, provide a thermal buffer between the exhibition halls, as well as from inside the building to outside. This zoning allows two layers of screens to extend from inside the building to outside so that lobbies are created.

A lobby space is created when the outer opening doors are lifted to allow spacecraft components to arrive or be taken out of the building. Lobby spaces have less thermal control and provision than the main spaces.

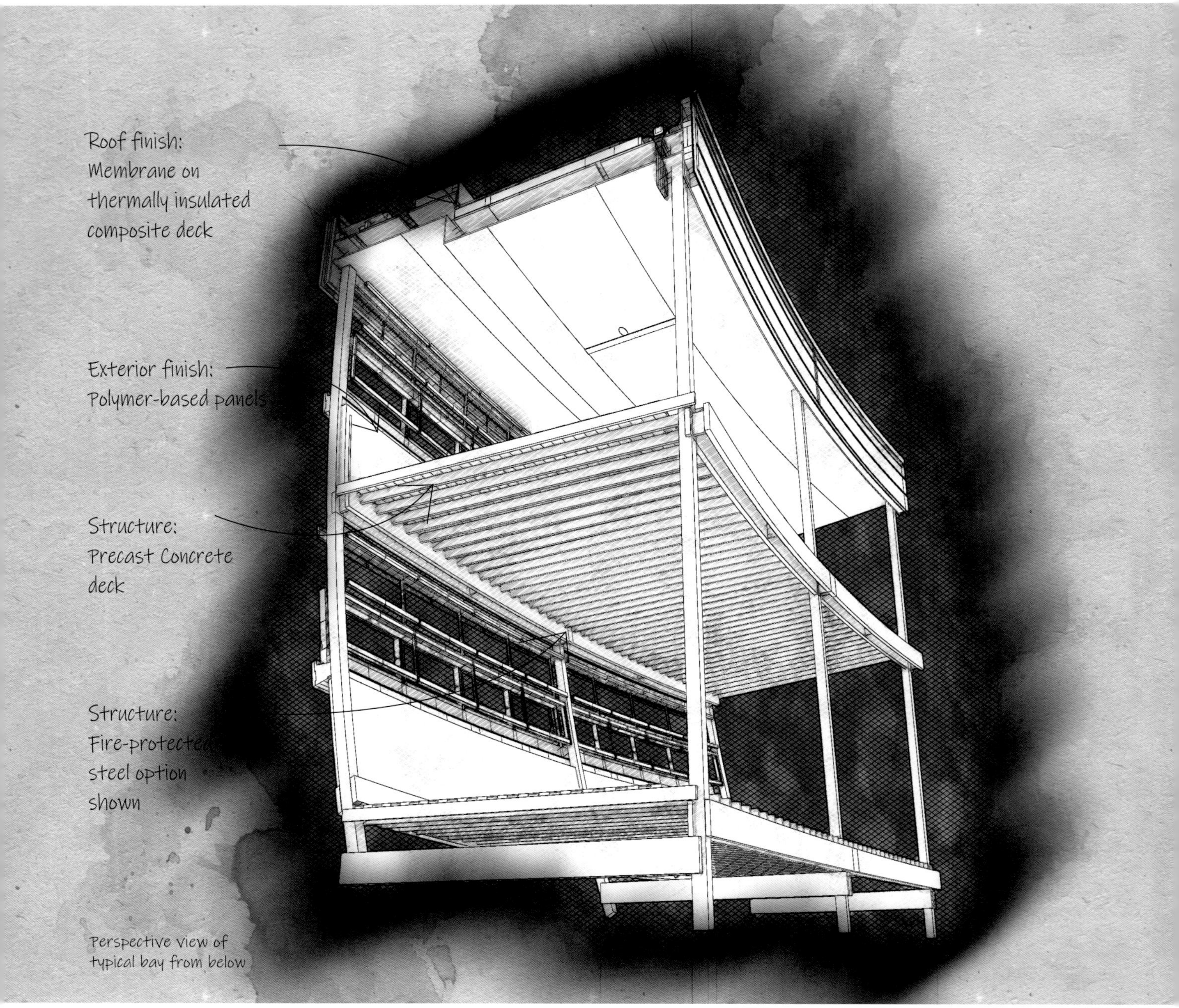

Perspective view of typical bay from below

Typical bay adjacent to facade

The modular nature of the roof and facades ensures that a minimum number of components of different types are required to be manufactured, and many of the panels used in the tower are also used in the podium. A feature of the tower is the exoskeleton of the tower, which is thermally insulated and fire protected on the outside with cladding panels fixed between them. This exoskeleton facade approach provides clear space within the building which is free of structure that would otherwise intrude into that space; internal environments are required to be kept clean and dust free. This approach gave the opportunity to put a separate roof on top of the building with a secondary waterproofing layer underneath so that the top layer is a completely opaque roof with provision for solar protection. The roof beneath is in rainscreen configuration. As a result of this structural facade concept, the building can 'morph' from having externally fixed primary structural members which are thermally insulated, to reverting to the semi monocoque style used on the podium and on the top of the tower building. The structure can switch from being outside and thermally insulated to being inside; thermally insulated on its outer face. This approach is based on the need for the buildings to be modular, and to be able to have parts which can be changed quickly and economically on the project site so that when the building is completed, elements that are required to be extended or changed can do so either in-situ, using relatively straightforward construction techniques, or can be assembled next to the building and then mounted into place.

The external envelope of the building is a series of prefabricated facade panels. Options 1 and 2 make use of polymer-based panels that provide a high level of thermal insulation. The comparative lightness in weight of facade panels allows them to be moved and rearranged as the building evolves. Option 2 has a similar arrangement of external wall panels. Option 3 takes the cores and turns them into external wall components, such that they are 4.0-metre-deep facade elements which are prefabricated and assembled on the building, so that the building becomes a set of prefabricated container-like forms stacked together

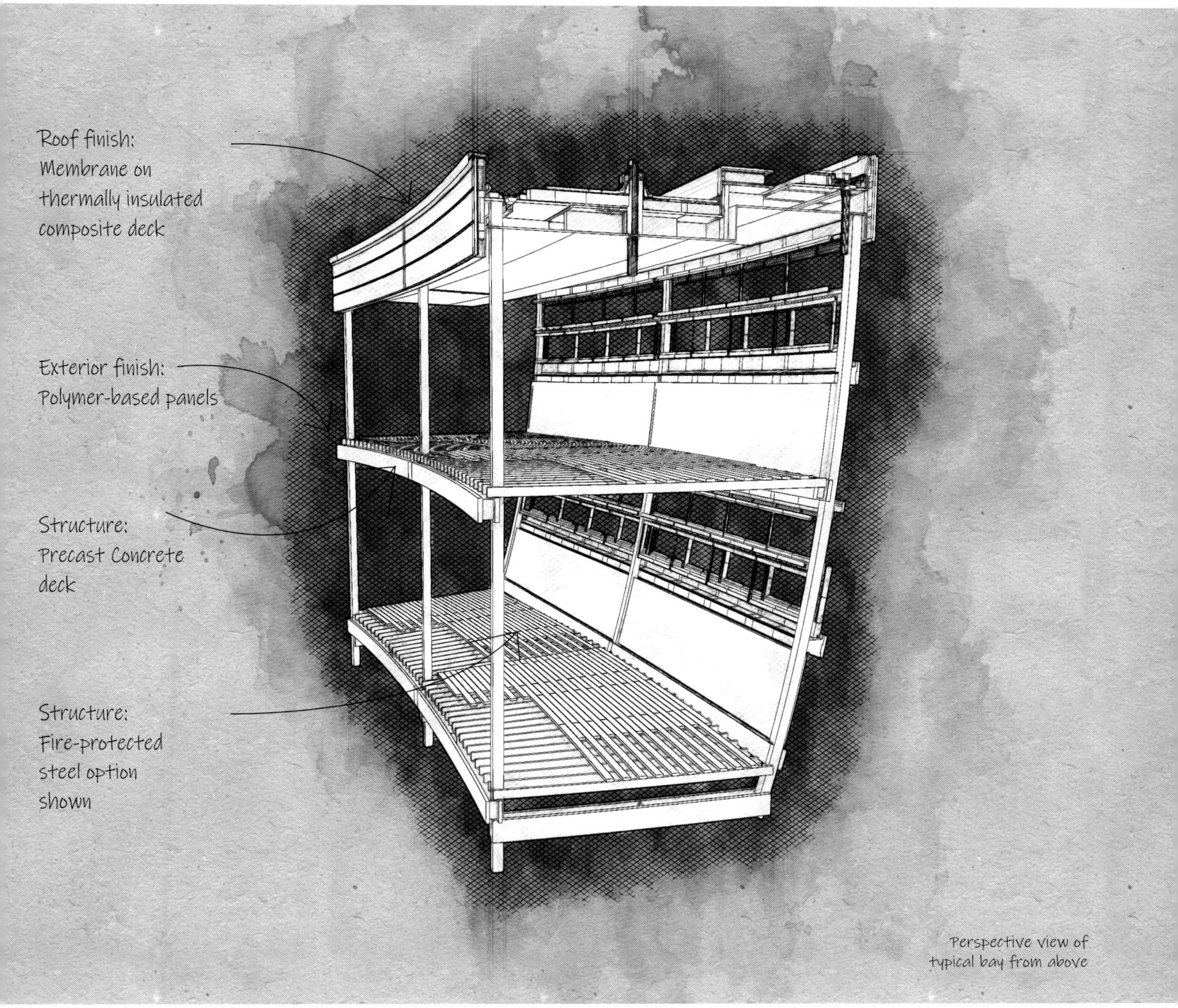

Perspective view of typical bay from above

with floor slabs that have a structural frame and floor slabs spanning between them; the external walls become prefabricated volumes. As a result, metal rainscreen panels and profiled metal sheets have been selected and applied to all three options.

The podium-level halls are divided by a glazed rooflight that extends the full length of the building. Glazing is also introduced at ground level at the base of the podium as a mixture of opaque and transparent glazing used to set the visual tone for the exhibition spaces. Glazing is introduced over the full height of the refurbishment building in order to provide filtered daylight for tasks that require a high level of natural light. Similar bands of glazing are introduced in vertical bands up the full height of the tower.

The centre of the podium volume has an opaque area of adjacent facade which encloses a daylight-controlled area for exhibitions. The two corresponding end elevations of the building have spaces for components and assemblies that are naturally lit. Most of the space adjacent to the external wall of the tower accommodates exhibitions which do not require daylight; office space and parts of visitor attraction that require daylight are arranged on the two shorter sides of the tower that look out across the adjacent parts of the site.

The glazed rooflights appear to 'fold' vertically from the podium roof to become glazed walls; at an early stage of the project, the walls and rooflights were aligned with each other to create continuous lines of circulation both horizontally and vertically.

In Option 3 these zones are moved apart such that the glazed walls are discontinuous with the rooflights to provide a very different space and a very different visitor experience from the enclosed spaces. Rooflights have been positioned to create views through to the exhibition space from roof level, as well as from the mini-towers. Glazed areas of the mini-towers provide viewing space to look out from and to provide visitor access at roof level; viewing the exhibits as well as having the ability to walk across the full extent of the areas of glazed rooflight. The facade systems used in the tower are the same as those used in the podium and two mini-towers in order to provide a visual continuity.

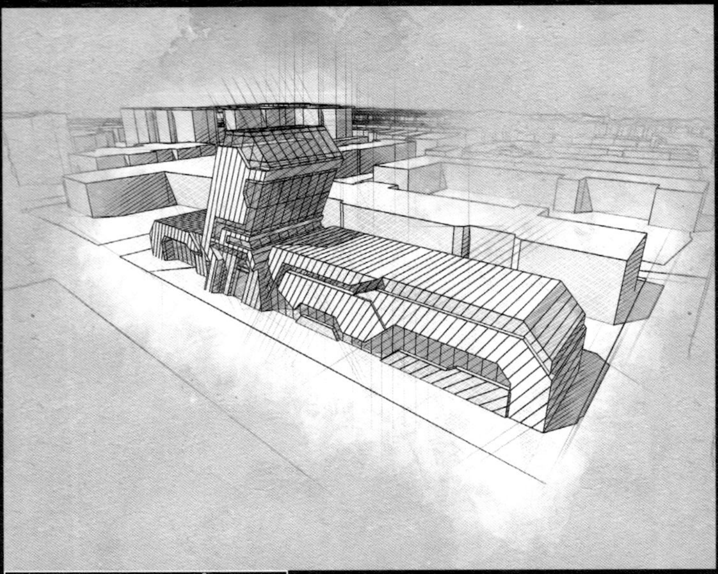

1. Perspective view of exterior

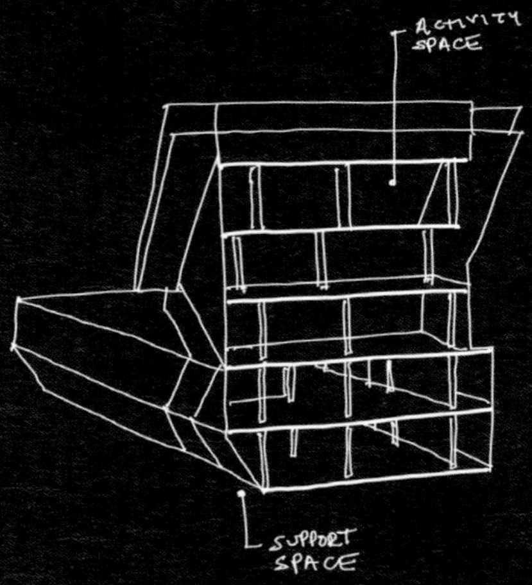

2. Perspective view of exterior

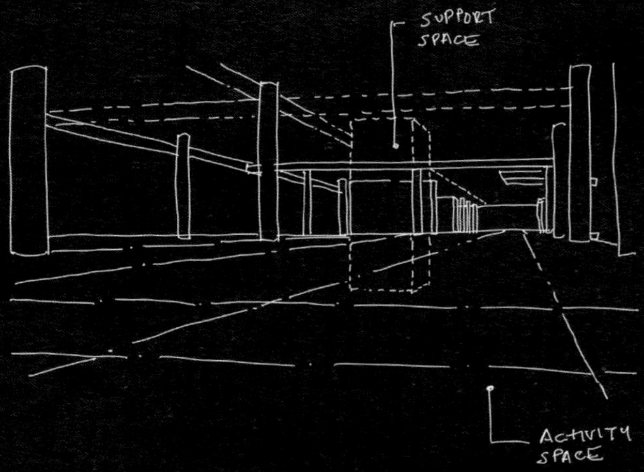

Development of an integrated structural system for envelopes

Interfaces between building assemblies are often slow to fabricate and install, and often have a performance which is lower than that of the adjacent systems for which the interface is created. In this project, interfaces between systems for structure, facade and roof are avoided by adopting a single integrated system across the project.

The design of the project suggested that five facade systems and two roof systems would be needed. Given the high cost of juxtaposing these different systems, both in the project management and in the on-site installation of interfaces it was decided to integrate all facade systems into a single type that would be able to respond to changes in facade requirements from glazed areas to opaque areas. The design led to two facade systems: the first being a framed arrangement supporting the glazing; the second system is a point-fixed panel system for the support of opaque panels. The point-fixed panel system was also applied across the podium roof; set above a thermally insulated, waterproofed deck. This approach allowed two envelope systems to be used for the complete project; the installation could be undertaken by only two companies working alongside one another. The advantage of reducing the number of systems became primarily one of achieving watertightness of the building at an earlier stage of construction; the backing walls for both opaque panels and glazed panels could be installed quickly without the need for interfaces. This method would allow the installation of interior assemblies within the building to begin when watertightness of the facades and roof had been achieved.

The integration of facade and structure as loadbearing facades allowed greater freedom in the design of the external envelope to suit the requirements of spatial arrangement of the volumes immediately behind. This was achieved by designing systems for facade and structure in parallel, using a digital model that integrated the structures of all three systems; enabling the interaction between the different elements to be accurately simulated and evaluated. The approach permitted the evaluation of key design aspects in parallel; a network of analysis to optimise performance of structural, thermal, moisture and

. Perspective view of exterior

4. Perspective view of exterior

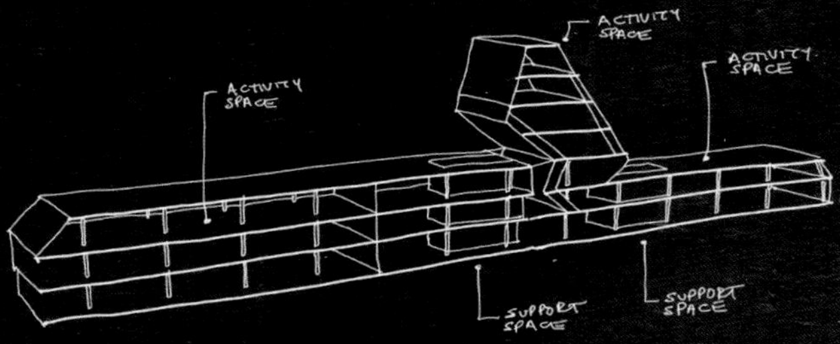

environmental considerations simultaneously. Established systems for the fabrication of structures and facades are made efficient as a result of the design of components and assemblies which are based on standard guidance and previous application. This approach would not have been sufficiently evolved for non-rectilinear geometry design proposals, as the behaviour of a structural system or a facade system of such an arrangement changes with its specific orientation; established systems are commonly designed to be used either vertically in walls or horizontally in roofs.

For this project, the system evolved a combination of structural and external envelope systems which are required to be used for the complete building: these systems were designed, at the early stages of the project, primarily through geometric modelling of structural forms and environmental volumes as well as their interfaces to ensure a close integration with supporting structure. The research of systems for structure and external envelope undertaken for this project provided a basis for the building's structural geometry. Interdependent governing aspects were considered in parallel to achieve an overall optimised design. The environmental strategy informed the evaluation of the performance of the proposed system; the internal environment and the sustainability of the operating building were essential to establish the benefits of the combined structure and facade systems, and how they could be optimised for performance and economy. The early stage integration of the supporting structures for floor plates, facades, support spaces and environmental design installations were essential to the search for economy and speed of construction. Geometric models were used for structure, facades and environmental control equipment in order to identify benefits for their integration. The overall reduction in the number of structure-based components was key to enhancing performance and optimising construction costs.

A single integrated system of structure, facade and roof was developed for this project in order to reduce interfaces between these three different building systems; creating a continuous high-performance structure and envelope that is used across the building.

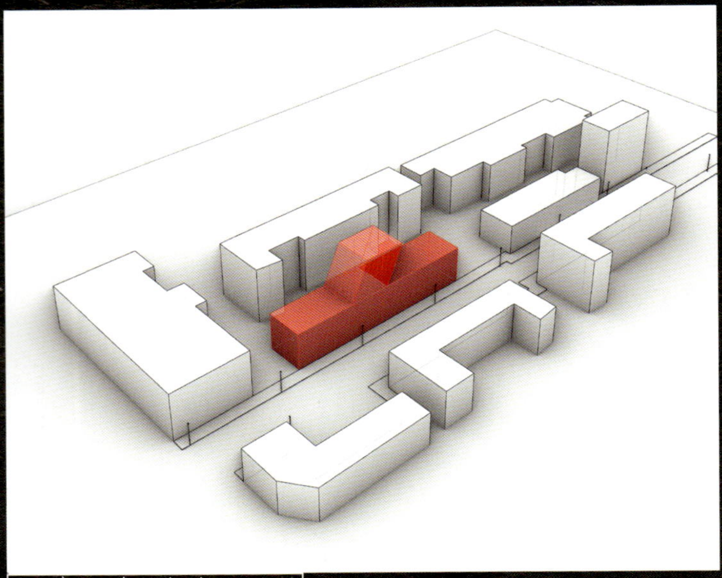

1. Project volume in site context

2. Project volume in site context

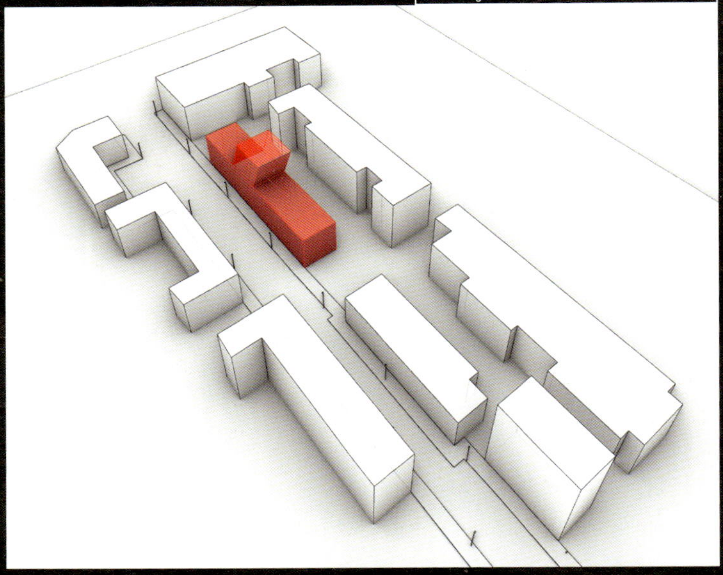

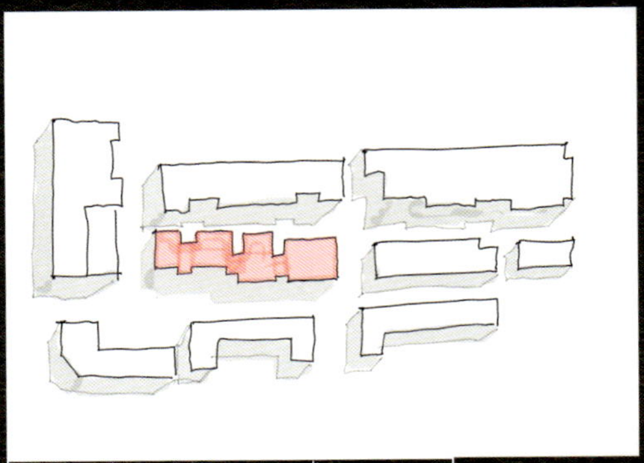

1. Site plan with surrounding context

Option 1 is based on mapping the functions set by the brief/program and arranges them to allow continuous access from a central corridor.

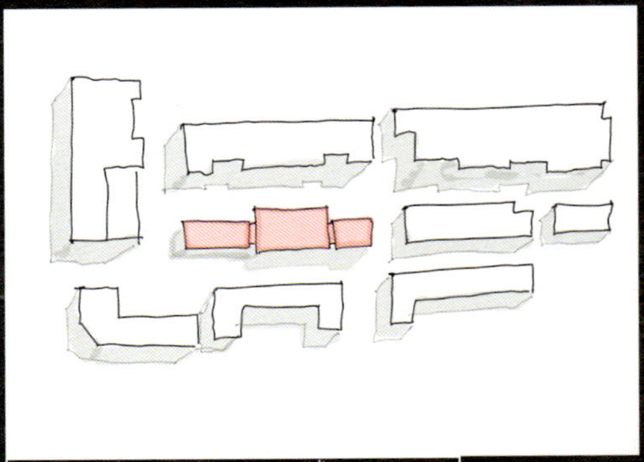

2. Site plan with surrounding context

Option 2 simplifies this concept by grouping the spaces into similar-sized volumes with compatible functions between spaces.

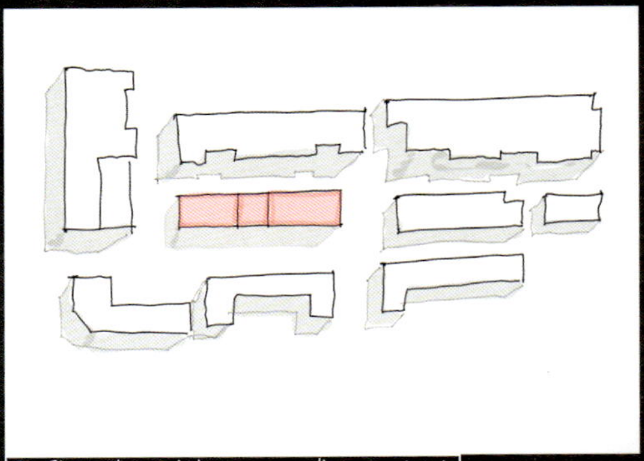

3. Site plan with surrounding context

Option 3 combines the buildings into a single volume; minimising the building envelope.

Site context

This university building is set within a college campus constructed in the 1970s. Surrounding buildings are of rectilinear design, with largely reinforced concrete wall panels, long strips of horizontally oriented windows and flat roofs. The project design followed the massing of those existing buildings, in terms of the general size of the existing forms, but adds an additional vertical form. The overall height of the building is similar to those of the existing blocks. This addition follows the general typology of two other tower buildings that adds an additional six storeys above the podium level. In this project, the use of reinforced concrete panels was applied in a way that aims to show the development of that material in the years since the campus was built. The design did not make a different use of materials from those already used on the campus; a decision was made not to move towards a design that was completely unlike any of the buildings that were already in place. A key driver in site context was to validate the design of the forms that already constitute the core of the campus, as the current buildings are a successful arrangement of teaching blocks and student accommodation. The geometry of the building was driven by environmental considerations of orientation in relation to the passage of the sun in a way that provides passive shading; making use of angled vertical geometries in order to provide oblique views up and down the height of the building. This approach was taken without introducing tall, wide spaces that are associated with the linking of spaces in other neighbouring buildings. The inclined angle of the internal linking spaces provides a point of departure for creating room shapes for lecture theatres as required by the brief/program. These spaces are expressed on the outside of the building so that they can be seen within the site context as part of an overall composition of massing. The overall context of surrounding buildings has been integrated by following the approximate floor-to-floor heights of the existing building, as revealed on the outside of this building. The horizontal lines of floor slabs in glazed areas and opaque panels are made visible from outside the building; the scale of internal spaces, as well as the external volumes of the building, are

Perspective view of structure

2. Perspective view of structure

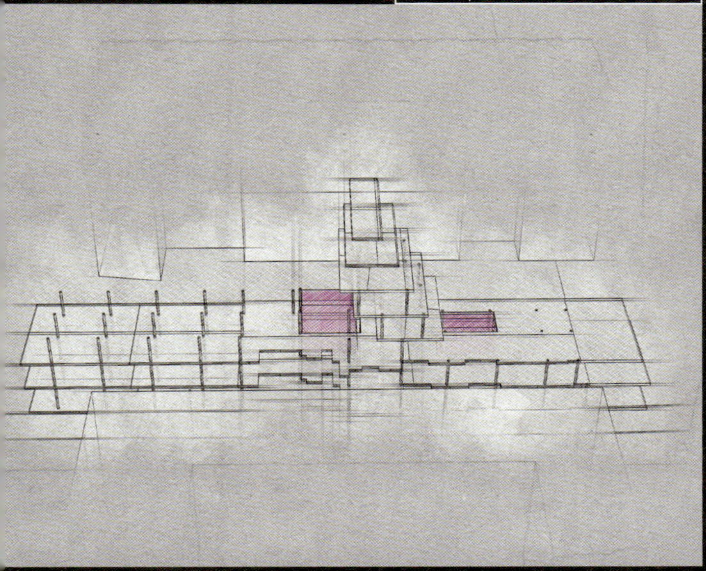

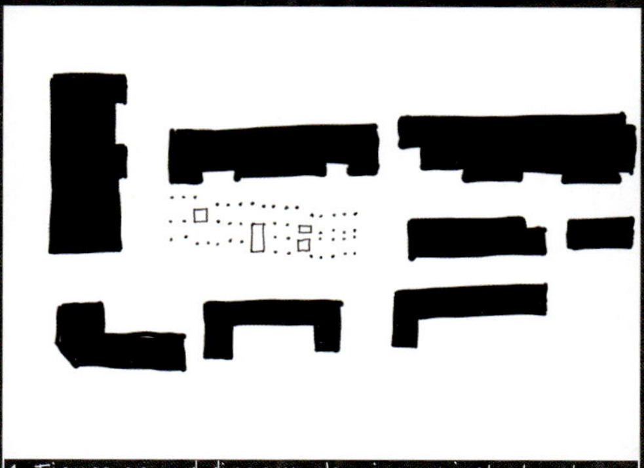

1. Figure-ground diagram showing project structure

Option 1
The structural grid follows the footprint of functions but does not respond to the geometry of surrounding buildings.

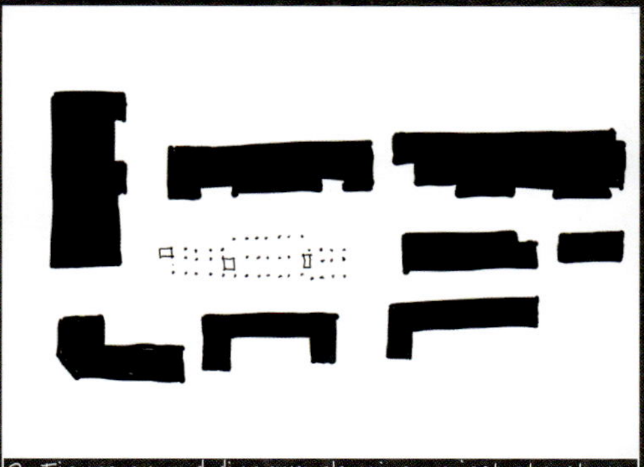

2. Figure-ground diagram showing project structure

Option 2
The structural grid follows the footprint of functions but responds partly to the geometry of surrounding buildings.

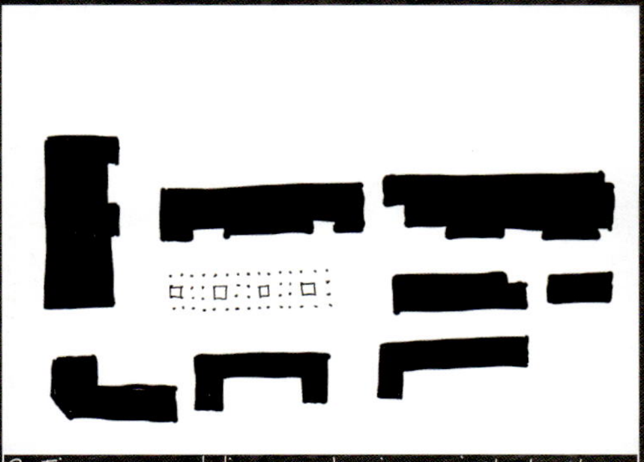

3. Figure-ground diagram showing project structure

Option 3
The structural grid follows the footprint of functions and responds to the geometry of surrounding buildings.

made similar to those of adjacent buildings. A difference internally is that the construction makes use of thinner components than those in the existing buildings. In the surrounding forms, a visual drama is created with the use of large-scale structural members to create a language of exposed structure. This building structure is not exposed either externally or internally; the external envelope is expressed as a map of internal functions, comprising three interlinked teaching blocks. These volumes include a mixture of lecture theatres, teaching rooms and communal activities including a bookstore and a food outlet. The three blocks accommodate different sets of space: a tall block in the centre of the building, a smaller block extending over three floors, and a long, low block set over four floors. The forms of all three blocks are set together as a juxtaposition of functions that are united visually and organisationally into a single language, using an integrated system for structure, facade and roof. All spaces are arranged along a central thin, canyon-like space which extends down the length of the building. Bridges set across the voids link the educational spaces with the social spaces. Option 1 for the site context is based on mapping the functions set by the brief/program and arranges them to allow continuous access from a central corridor. Both teaching space and circulation space are formed as indented links between the blocks that comprise the building. Spaces are arranged over three floors with the sports hall split between two floors, resulting in a large building footprint. Option 2 simplifies this concept by grouping the spaces into similar-sized volumes with compatible functions between spaces to create three linked blocks, where access is provided between blocks with two links.

Option 3 uses a concept developed from Option 2 of combining the buildings into a single volume; minimising the building envelope and creating a form that is compatible with the built environment around the existing building. The provision of the required level of complexity of spatial arrangement is achieved vertically, as seen in section; the building rises vertically by creating the required spaces through different heights, creating an interlocking relationship in section, but providing a clear, straightforward arrangement of spaces in plan.

1. Sectional perspective view

2. Sectional perspective view

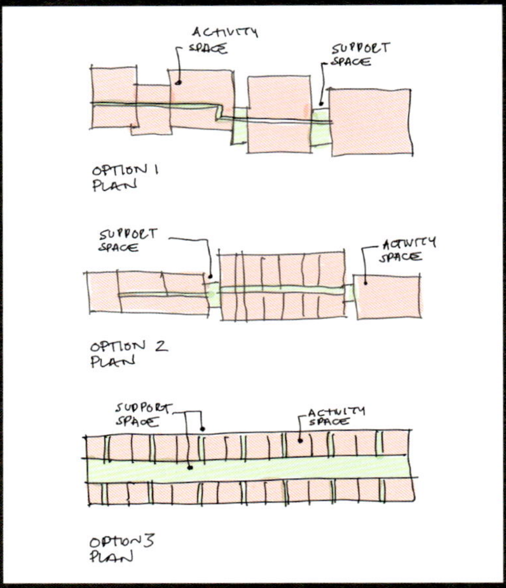

Spatial arrangement

Option 1 comprises a central corridor volume with spaces set on both sides.

Option 2 joins these elements together into a simplified form of three blocks, with two links joining the three blocks.

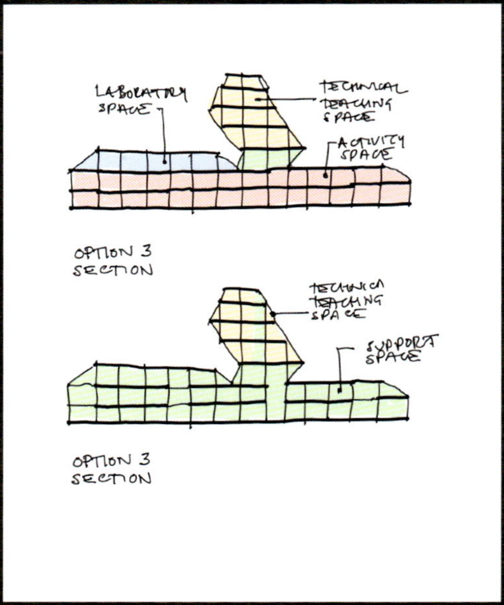

Option 3 arranges all spaces within a rectilinear box-like form in plan.

Spatial arrangement

The building comprises three blocks: A 'wedge' block which is taller than the others; a smaller 'arched' block that is immediately adjacent, and a 'cranked' block which is long and thin. The cranked block is linked to the other two blocks which are set on each side. Staff and students cross the building and pass down the canyon, which links the three blocks. Functions are stacked in layers, with the lecture halls at higher level. Service spaces are set out on the roof; they are set on both sides of the rooflights, passing around the edges of the major spaces beneath, which they service. All service spaces are accommodated in the top floor of each block. A building maintenance unit tracks around all three blocks on a continuous rail; a single cleaning unit is used to clean all three blocks. It was found to be more economic to use a single track across all three blocks than to install separate cleaning devices for each block. The roof accommodates the rails of cleaning devices; tracking up and down the inclined walls of the wedge as well as the stepped roofs; lower areas of the building are maintained from street level.

The spatial organisation for the three options is based on the creation of oblique views up and down the building on two sides of the raised portion of the building; not as a mini-tower, but as a way of 'leaning' in and out of the spaces below. As a result, no large central space is needed as the aim of the spatial arrangement is to orient building users via a visual reading of the overall forms within the building, and of views from space to space; a single unifying space is not needed.

The spatial arrangement of Option 1 comprises a central corridor volume with spaces set on both sides. The corridor is not placed in the centre, but is positioned in order to allow these spaces on either side to achieve the floor area required; the spaces are juxtaposed in order to fit into different disciplines and areas of study within the school. The two link blocks, shown in green, provide the main vertical circulation through the school. Spaces are arranged in order to maximise the amount of daylight available in each classroom which is lit naturally from one side. Option 2 joins these elements together into a simplified form of three blocks, with two links joining the three blocks. The circulation passes

. Perspective view of interior

2. Perspective view of interior

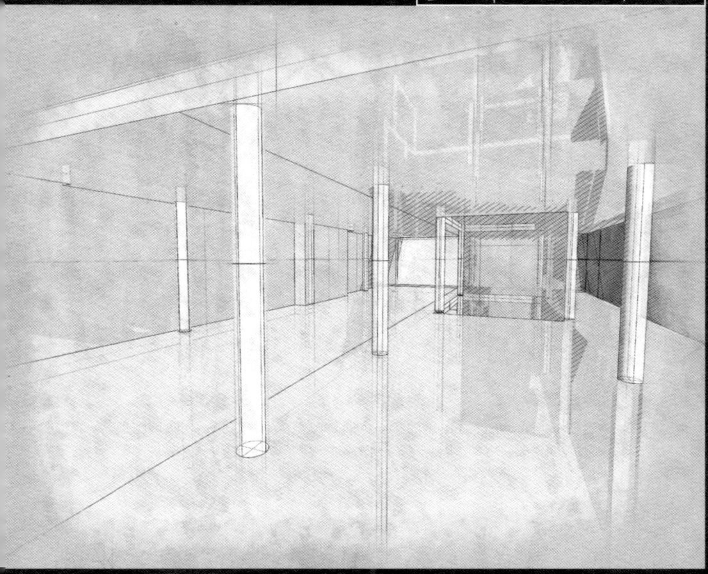

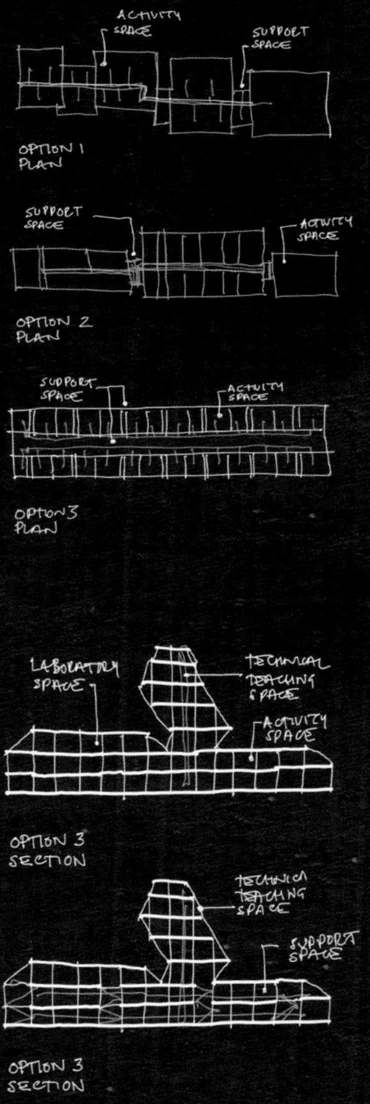

Internal circulation

Option 1 is based on a central corridor that changes direction in response to the spatial arrangement.

Option 2 creates a more evenly distrib-uted circulation by using the two link blocks.

Option 3 has most of the move-ment of people through the building occuring adjacent to the facades.

down the length of the building in steps up, creating gentle folds in the circulation flow within the link volumes. Where vertical circulation also occurs, the simplifying of the blocks creates a slightly more complex section than the simple extruded form of Option 1.

Option 3 arranges all spaces within a rectilinear box-like form in plan, where the complexity required to form the space is achieved when viewed in section, such that the main circulation routes follow the edges of the building, shown in green in the diagram, with routes that cross the top corridors, passing from side to side.

Internal circulation

The internal circulation for Option 1 is based on a central corridor that changes direction in response to the arrangement of teaching spaces and administration spaces set on either side of the corridor. A disadvantage of this circulation route is that it provides the only means of traversing the length of the building. Option 2 creates a more evenly distributed circulation by using the two link blocks as a way of providing access directly to larger spaces set on either side. This approach also creates two corridor routes which cross the width of the building in two locations. However, this option still has a central corridor in most of the building. The limitations of circulation provided by a single corridor are addressed in Option 3, where most of the movement of people through the building occurs adjacent to the facades. Teaching spaces are set to one side of this perimeter circulation route, such that the rooms have borrowed light from the glazed outer walls intc the classrooms. This arrangement allows the corridor to serve as an acoustic buffer between the teaching rooms and the outside. This arrangement has natural light at two levels that suit teaching rooms, but also has a mix of naturally lit spaces, enclosed spaces and top-lit spaces. Circulation between parts of the building is introduced by corridors that cross the width of the building in order to allow access to spaces in the centre of the building, such that access can be gained to classrooms and teaching spaces from the long corridors along the facades, or through the corridors that cross the width of the building.

1A. Isometric view of structure

2A. Isometric view of structure

1B. Isometric view of structure

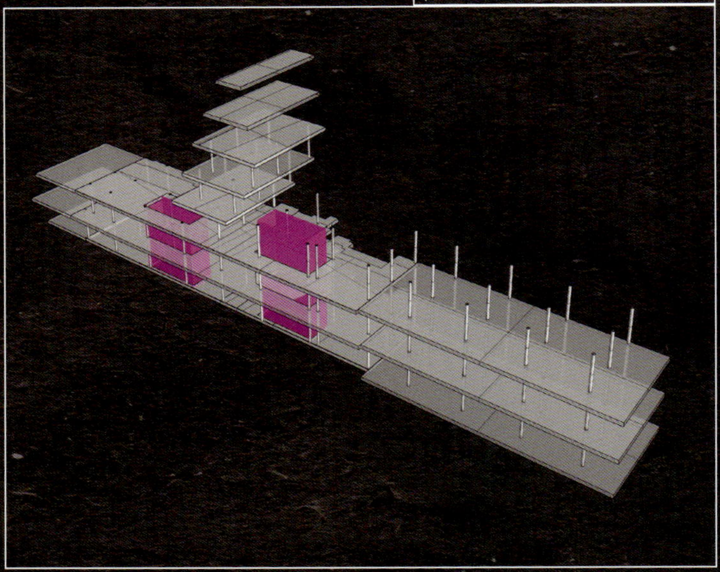

2B. Isometric view of structure

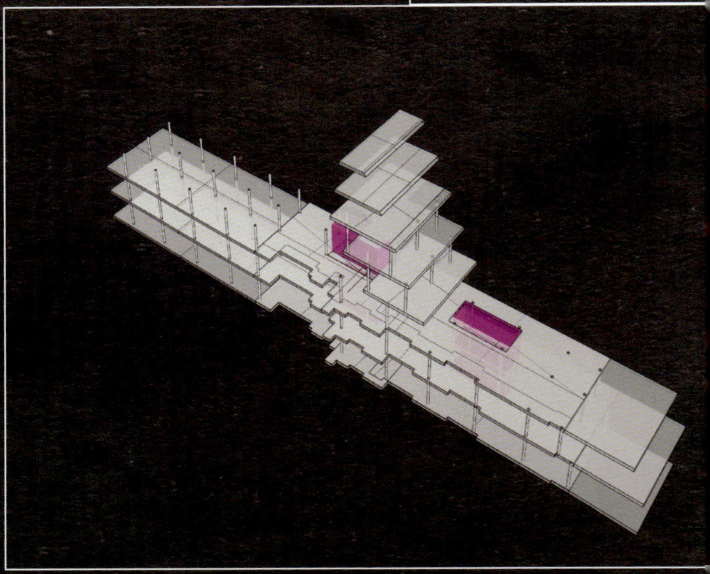

Structural strategy

The structural strategy is based on a set of reinforced concrete frames which support GFRC (glass-fibre reinforced concrete) facade panels without the need for secondary steelwork. Other external walls are built in twin-wall configuration as diaphragm walls, with thermal installation set in the void between the two layers of concrete. The facade panels follow a similar geometry to the primary structure behind in order to eliminate secondary framing internally. Rooflights are heavily shaded to provide protection from the effects of the sun.

Floor decks in the reinforced frame use flat slab construction. The frame is braced with cores which are set between the arrangement of internal spaces; a row of service cores creates a stair and food outlet, together with storage spaces. Stairs follow a linear route down from floor to floor; oriented in a transverse direction. The building is divided up into zones for fire safety, with the long 'street' divided into three separate zones. Roof-level equipment is positioned continuously along the length of the building; divided into fire-resisting compartments. Much of the daylight in the internal spaces is admitted through the central street, which is glazed, and where the heat is controlled by the solar control glass; daylight is admitted here in order to provide borrowed light which is directed to internal spaces. There are crevice-style cutouts, seen in vertical section, to ensure that daylight penetrates to areas where it is needed, as is the case with the cranked slots that are formed both in the facade and in the roof. Light is brought to where it is needed, reducing the dependence on electrical lighting as much as possible during daylight hours.

The structural strategy for Option 1 divides the building into three separate structures. Option 2 reduces this arrangement to three blocks, and makes use of a structural frame with columns set out at closer dimensional centres. Option 3 has a single large-scale frame, which creates the different forms that step in and out.

The basis of the structural design for Option 1 was to consider the different component parts of the building as separate structures. Movements expected in the different geometries were accommodated by

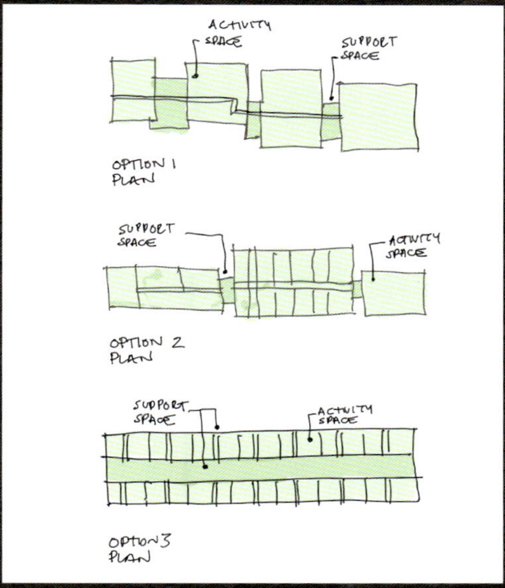

Structural strategy: Zones

Option 1 divides the building into three separate structures.

Option 2 reduces this arrangement to three blocks.

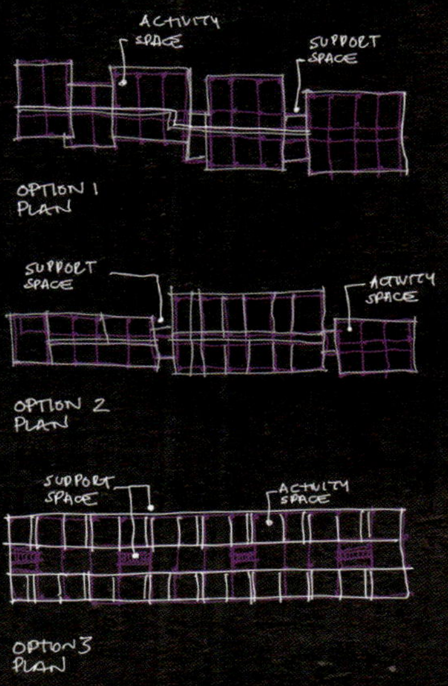

Structural strategy: Framing

Option 1 rearranges different component parts of the building as separate structures.

Option 2 has a structural frame with columns set out at closer dimensional centres.

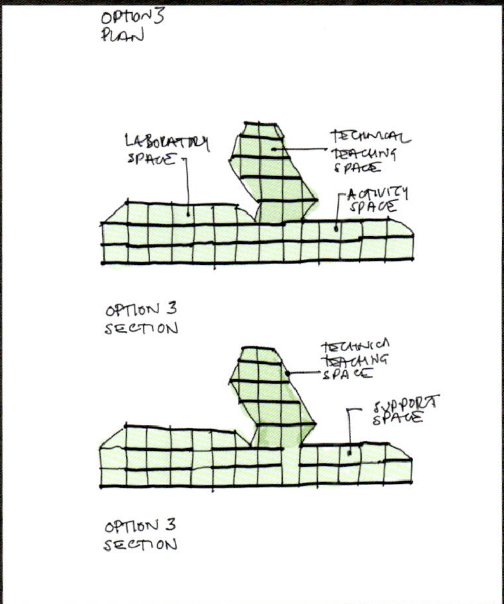

Structural strategy: Zones

Option 3 has three constituent elements which are unified as a single structure.

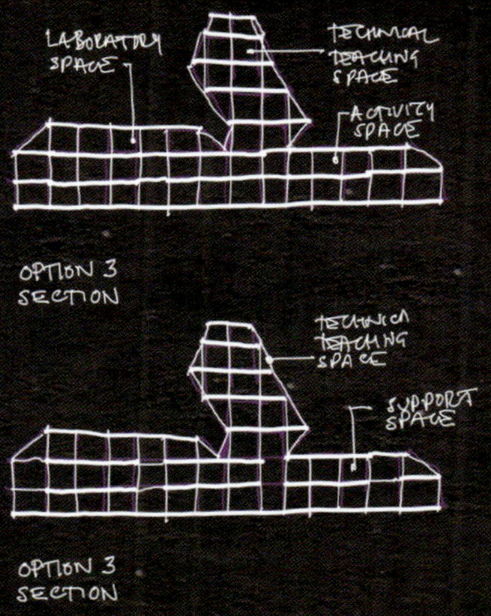

Structural strategy: Zones

Option 3 has a structure which reduces the quantities of materials required when compared with the other two options.

separating the forms as a tower shape in the middle, with two long blocks arranged on both sides of the tower. This tall central part of the building has large triangulated sections on the end elevations that take up the form seen in cross section. An arrangement of columns is repeated down the length of the tower in plan. Floor slabs are set between the structural sets of inclined columns. The triangulation provides the structural frame that supports the floor slabs. The form of the two long blocks on both sides of the tower were designed as reinforced concrete frames with vertical columns and structural cores in the centre of the building, creating long thin zones for staircases and washrooms. Circulation within the building was primarily between these three building forms such that they could be seen as being visually separate but linked by glazed facades and a glass roof, creating a glass slot between the tower and the two side blocks.

Option 2 explores the idea of making the structure with all-vertical columns, keeping the floor slabs in the same position but creating slots in the floor slabs in order to admit daylight and to provide views up and down the height of the building. This arrangement creates a sense of the building as a set of juxtaposed volumes when viewed from points along the internal circulation routes. The community within the building can be seen at a glance while moving around the structural forms of the three component parts. Inclined columns were required at the interface between the tower and the two side blocks.

In Option 3 the three constituent elements of the building are unified as a single structure. This structure reduced the quantities of materials required by reducing the number of columns and beams and allowing slots to be created; not by creating cutouts in the floor slabs but by omitting areas of floor slab within structural bays. This solution still works with the arrangement in plan of the two sets of structural cores which extend down the centre of the building; cores which are common to all three options. Option 3 provides an economy and simplicity internally that allows the facade elements, where not in twin-wall configuration, to be supported directly to the primary structure while remaining economical and allowing the building to be constructed quickly.

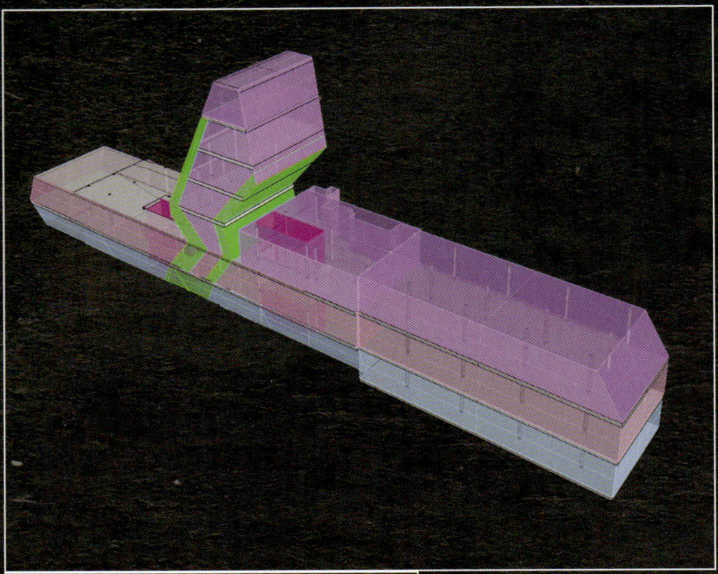

1A. Isometric view of environmental volumes

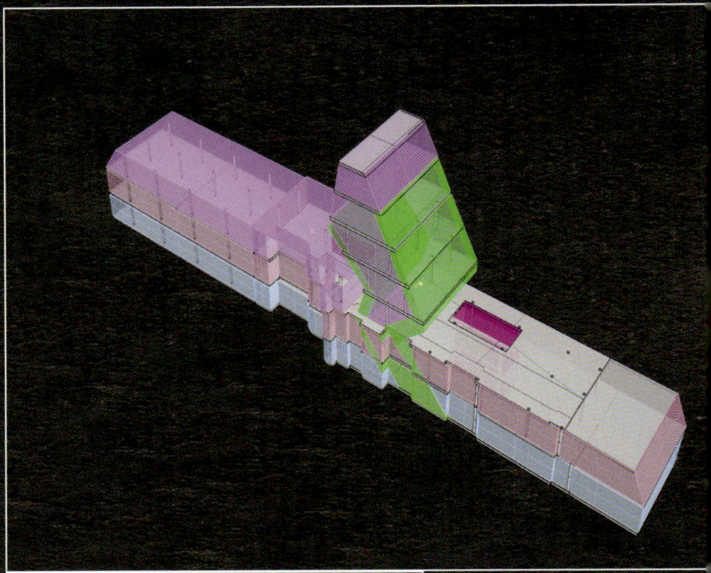

2A. Isometric view of environmental volumes

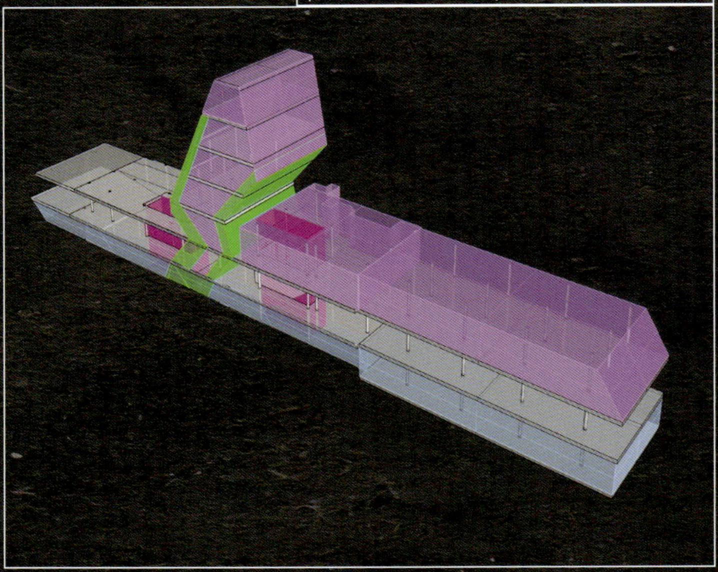

1B. Isometric view of environmental volumes

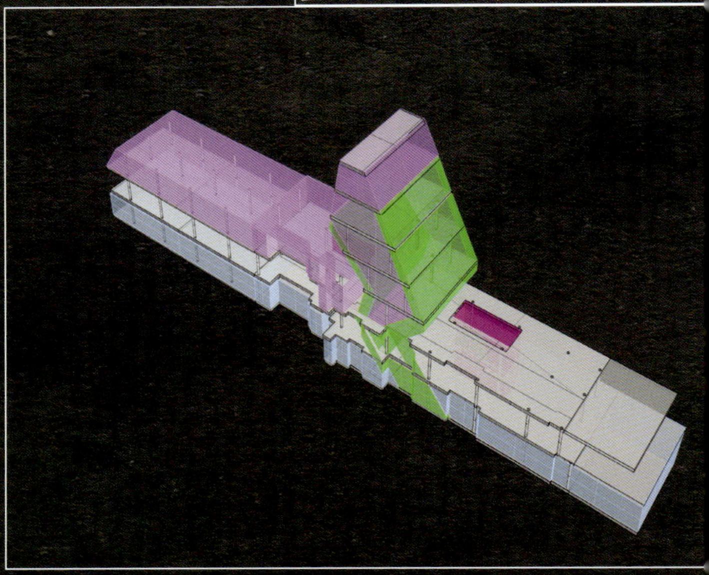

2B. Isometric view of environmental volume

Environmental strategy

All three blocks are glazed through their full height, but have stairs and services equipment in an interstitial zone between two glass walls; providing views through but using the escape stairs as solar shading devices. The environmental strategy is one of the building essentially being supplied with fresh air from roof level. The building is protected from the effects of external air temperatures as a result of the GFRC panels on the facades. Night-time cooling at the mid-season and in winter is used to cool the building fabric at night in order to reduce the need for air conditioning during parts of the day. The aim for this university building's level of energy consumption is to be similar to that of a continental European building which requires heating in winter; this building requires cooling in summer. The three blocks each constitute an environmental zone; each is split into three zones, divided between the different functions. The six environmental zones are each protected from the effects of the sun; the environmental zones have equipment to provide air at different temperatures and different speeds. Some environmental zones are under frequent use; others are used infrequently. The services equipment is provided accordingly; different equipment for each zone.

The environmental strategy for Option 1 is based on having three blocks; each with its own environmental zone. The circulation areas in each zone are formed as secondary environmental volumes; classrooms and administration spaces being part of the single red zone shown in the illustration.

In Option 2 the low level of differentiation allows the environmental zones to be consolidated into fewer volumes.

Option 3 uses a floor-by-floor layering of environmental zones rather than a vertical arrangement across several floors. These zones change from floor to floor, reflecting the different functions accommodated on each level, with a mixture of circulation spaces: administration on the ground floor, teaching spaces on the first floor and smaller teaching spaces on the second floor. In all options, the building is divided into four environmental zones: the tower, the two long blocks and the circula-

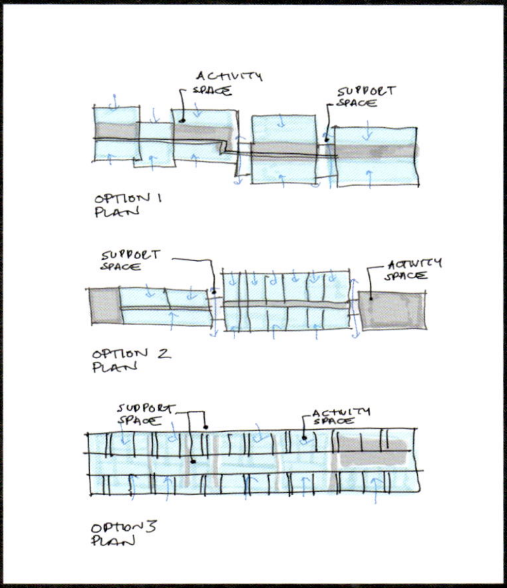

Daylight strategy for direct sunlight and cross ventilation.

Option 1 admits daylight and air from the two long facades.

Option 2 admits daylight and air in the two central blocks only.

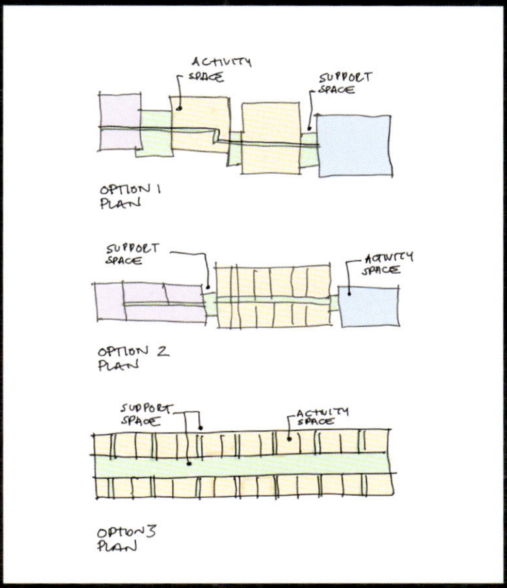

Environmental zones

Option 1 is based on having three blocks; each with its own environmental zone.

Option 2 provides a low level of differentiation to allow the environmental zones to be consolidated into fewer volumes.

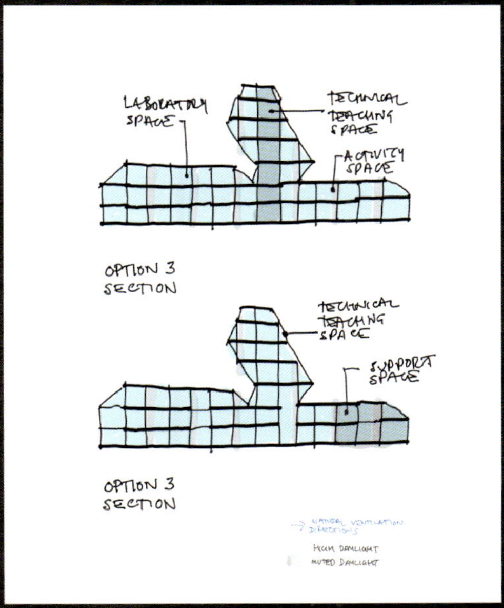

Daylight strategy for direct sunlight and cross ventilation.

Option 3 admits daylight and air from the two long facades but has subdued daylight in circulation areas.

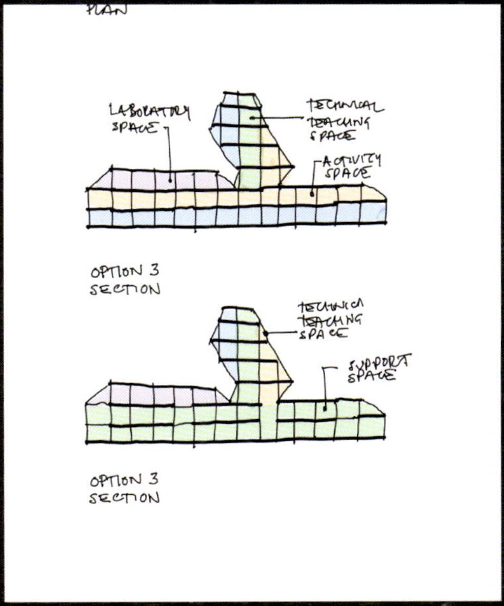

Environmental zones

Option 3 uses a floor-by-floor layering of environmental zones rather than a vertical arrangement across several floors.

tion spaces. The tower comprises mainly teaching spaces, set on one side of the tower. The environmental zone for the circulation spaces within the tower continues up to the highest floor. At the top of the building, there is a zone for building services that distributes services down through the height of the building. This service floor occupies the enclosed roof area of the tower. The internal volume within the tower adjacent to the outwardly inclined wall is a space which is tempered like that of the circulation spaces that it adjoins.

The environmental zone of circulation extends vertically and horizontally through the building. The second and third environmental zones occupy each of the blocks on either side of the tower. These volumes have ventilation which is both natural and mechanical; provided by a ducted air-handling system.

Option 2 divides each of the blocks into a separate environmental zone which constitutes a variation to Option 1, where the orientation of the two blocks on both sides of the tower were considered to have different environmental requirements and were serviced separately. The environmental design of the tower extends down to the ground floor in Option 2. The area is considered to have teaching spaces and support spaces which have a common pattern of use, reconfigured as two groups. The circulation spaces have the same arrangement as an environmental zone, as per Option 1.

In Option 3, each floor of the lower building is considered to be a separate environmental zone, allowing each to respond more closely to both patterns of use and the numbers of people using spaces – which have different functions and are of different sizes, and which vary from floor to floor. The tower has three environmental zones: The first zone is at lower ground level, where circulation space and the reception area is considered to have the same environmental requirements as the circulation spaces around it, continuing through the upper floors.

Option 2 is the preferred option for this building in its efficient combination of spaces with the least amount of ducting. In addition, functions within each of the blocks adjacent to the tower are similar at each floor level, making the differentiation of zones in Option 3 unnecessary.

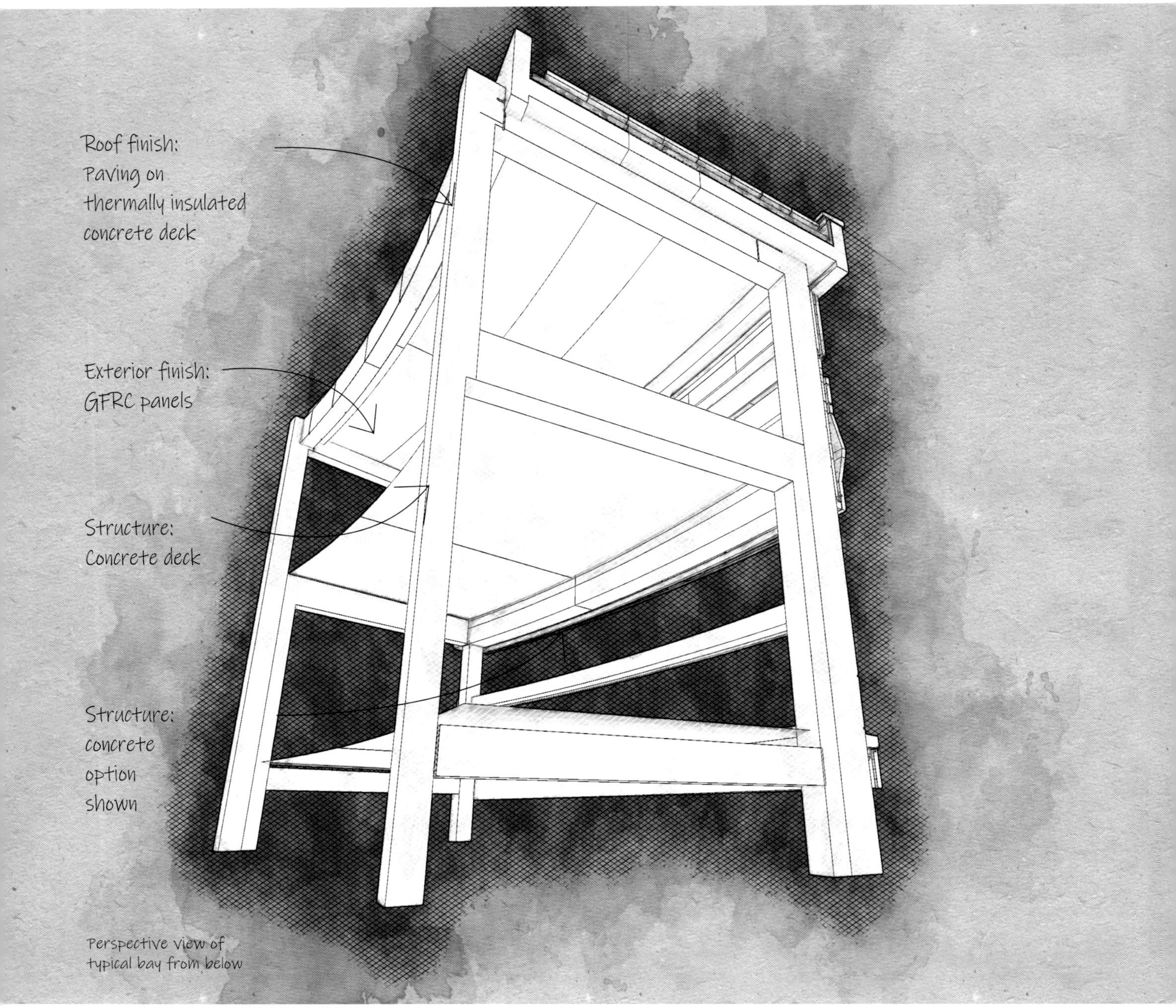

Perspective view of typical bay from below

Typical bay adjacent to facade

The building is clad in GFRC panels which are fixed onto a reinforced concrete frame with a twin-wall configuration in reinforced concrete. Gaps between cladding panels have open joints associated with large-size GFRC panels; most of the facades are opaque with slot glazing allowing lighting from outside. There is a high degree of visual sculpting of the external forms of the building in order to respond to the external envelope being formed as a 'shrink-wrap' of the functions within the building, as expressed on the outside of the building. The external forms are generated directly from the spatial relationships set out in the brief/program, which reflect the internal spaces set out on both sides of the circulation spine. The GFRC panels and areas of glazing form an external envelope with a weatherproofed backing wall immediately behind it that creates the characteristic forms of the design. The external envelope also has areas of glazed facade; translucent glass and perforated metal set adjacent to the areas of GFRC panels. The contrast between the facades of the three external forms reflects the nature of the functions that are accommodated behind the facade, allowing the building to be expressed in both elevation and section. The roof has a large area of opaque panels that partly cover the rooflights; positioned to provide top light to both laboratories and sports spaces. The external envelope, as part of a typical bay, is able to combine large-scale opaque elements with glazing in a way that maintains the general scale of the surrounding buildings from which it takes its cue in terms of the visual aspects of the materials used externally. Opaque panels are formed in reinforced concrete; cementitious materials are given the same finish and colour as adjacent buildings in order to create a visual harmony across the building and be in keeping with the general appearance of the materials used on surrounding buildings in the campus. The glazing is fixed to structural columns and beams where the columns are arranged in order to meet performance requirements of the facades, which in turn are arranged in order to frame and enclose the spaces behind. Glazing bars align with partitions; structure is aligned with the GFRC panels, which project out from the building.

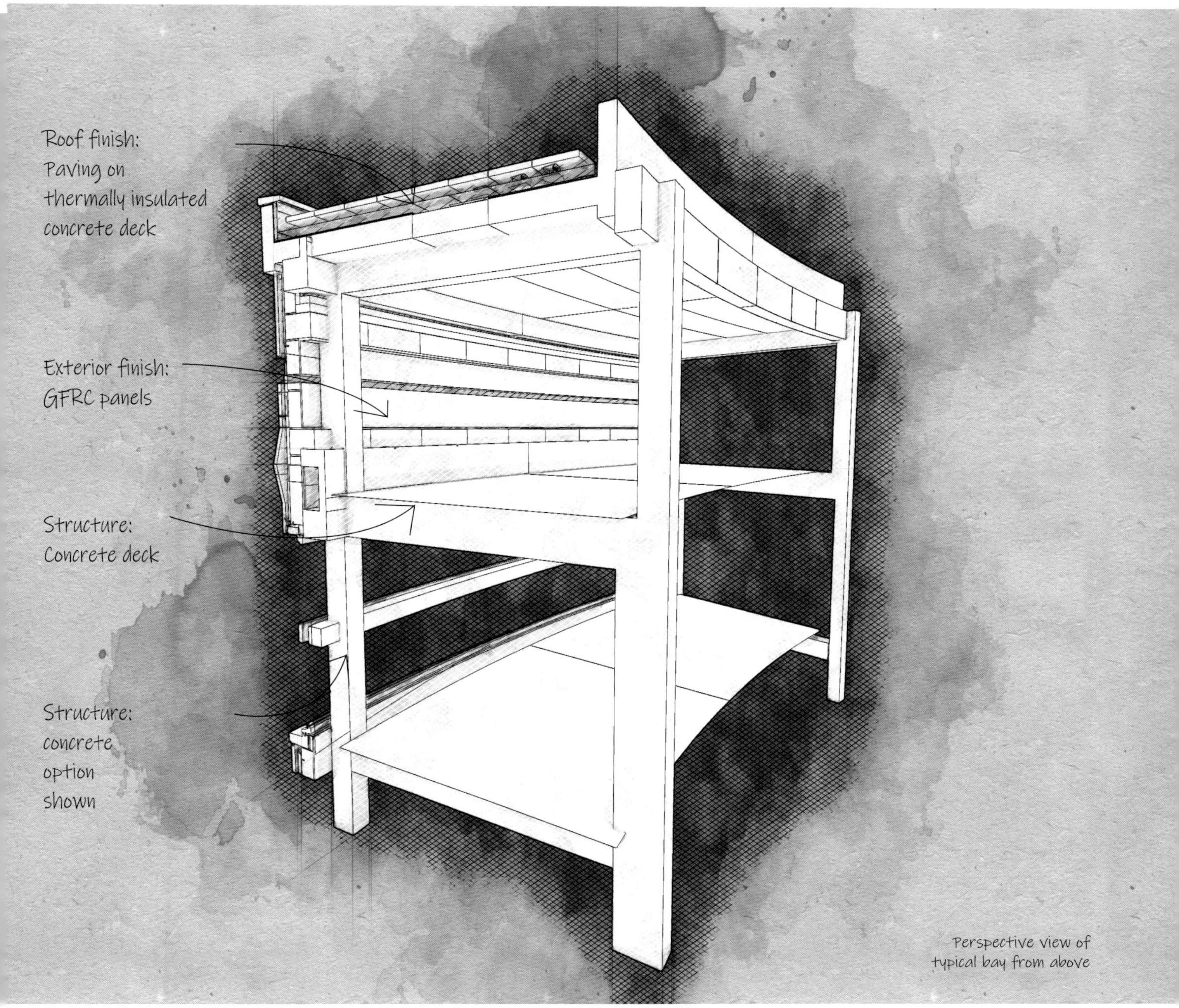

Perspective view of typical bay from above

These projecting bays have a void behind in which air-handling ducts are routed and from which fresh air is drawn through a series of filters and fans into the internal space. Some of the cladding panels that are used as covers for ducts have full structural walls behind them. Other panels are not loadbearing and are designed to serve as covers for fresh air ducts only. The structural forms of the tower respond to the needs of circulating air, combined with a structural design that is both efficient and responds to the need to support air ducts on the outside of the building. The system across the roof is one of using the GFRC panels in a rain screen configuration such that joints between the panels are open, allowing water to drain beneath while providing a durable surface to the roof of the building. Ducts and electrical supply routes are set within the roof void; picked out visually as a network of panels across the roof. Panels can be lifted for maintenance and replacement of service items in the void between the outer GFRC panels and the membrane roof below. Areas of facade have steelwork which projects from the floor slabs, and these support the external GFRC panels whilst creating a void for air-handling ducts. The introduction of fresh air through the facades into a set of ducts takes air up to where it is required in the building. This void is accessed from inside the building through openings in the drywall system, accessed on a floor-by-floor basis.

The system of projecting facade elements is used in order to allow air to be captured from below without it becoming a highly visible part of the facades and without it being integrated into a curtain wall system. Glazed facades in the tower integrate light shelves into horizontal joints. The light shelves are also used for solar shading, but reflect daylight on their upper surface, through the glass and onto the ceilings of teaching spaces. The use of light shelves is made possible by supporting them on the framing of the glazing, which would not have been economical if curtain walling had been used as the glazing system. The framing supporting the glazing is made from steel box sections with support integrated into their design. The solar shading, used on the south facade, is oriented to shade the sun at a higher angle than that experienced on the east- and west-facing facades.

1. Perspective view of exterior

2. Perspective view of exterior

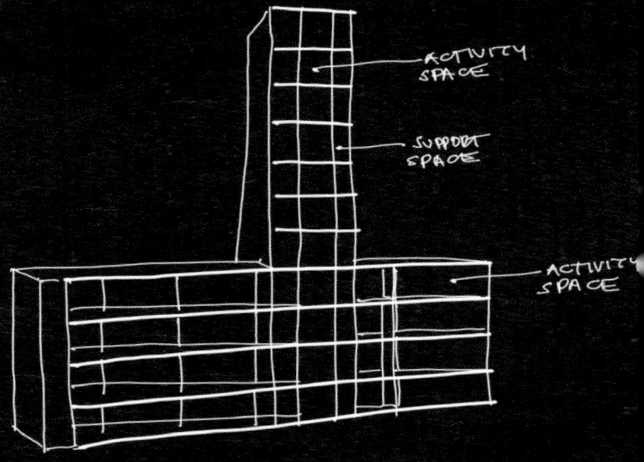

The influence of facade panelisation in structural design

Panelised systems for facades are designed such that structural movements and load paths are known and predictable, allowing the behaviour of the facade system to be understood; allowing the structure to be optimised accordingly. Each panel is designed with the same number of supports per panel; commonly four. These connections are accounted for, structurally, through the definition of a static scheme that incorporates both one-dimensional and two-dimensional rollers and hinges. Each panel connection is defined to allow for thermal expansion and predictable movement of the panel elements. For building forms where the facades deviate in angle from the vertical, this task of definition is achieved using computer-generated scripts to define and orient the different connections of the panel. This information is then input and coordinated with the structural model for global finite element analysis. Panel fixing systems are designed to provide a simplified static scheme, ensuring that load paths and structural movements are predictable.

This approach allows the primary structural system of a building to be refined while retaining its integrity. Structural connections are designed on the basis of sets of support reactions obtained from the global analysis of the external envelope system, taking into account both static and fatigue loading. This increased knowledge of the loading to be transferred by the connections enables the connections to be optimised for higher accuracy, producing increased structural efficiency in both the primary structure and the external envelope.

External envelope systems are developed from a design concept through to a physical prototype using a combination of current technologies, assisted by computer-based 3-D modelling and structural analysis at a global and local scale. The two scales of analysis are linked to enable the data from one to inform the other. This method avoids the need to consider a high number of structural inputs which are specific to shape and load. It also avoids the introduction of approximations for the geometry of the building, which would likely incorrectly represent the global structural response outputs. In contrast, this method used for the structural concept design uses structural primitives that effec-

. Perspective view of exterior

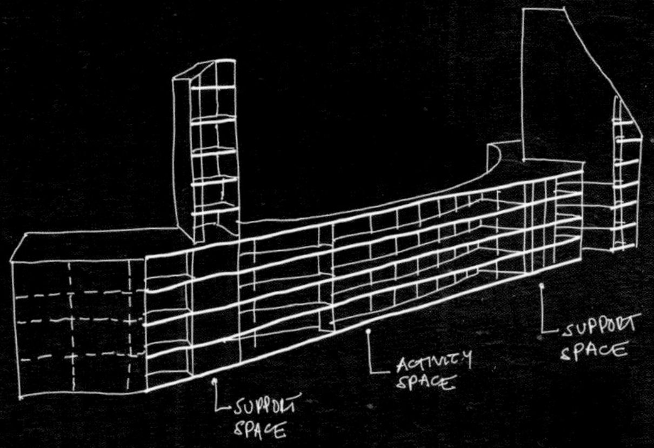

4. Perspective view of exterior

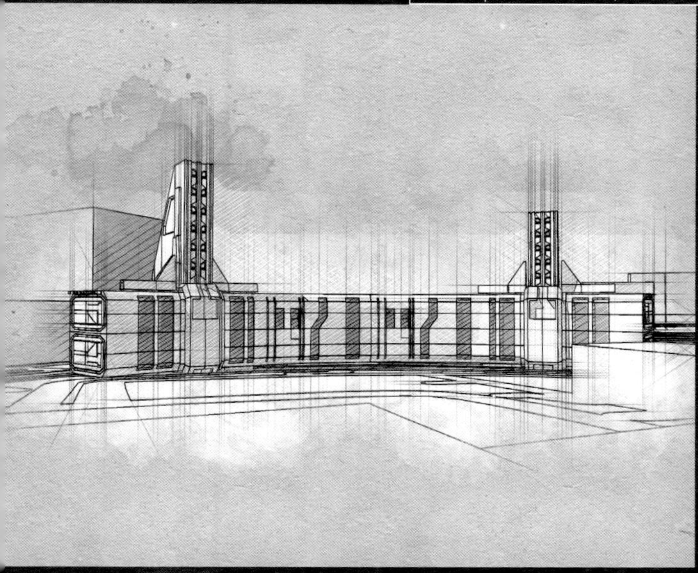

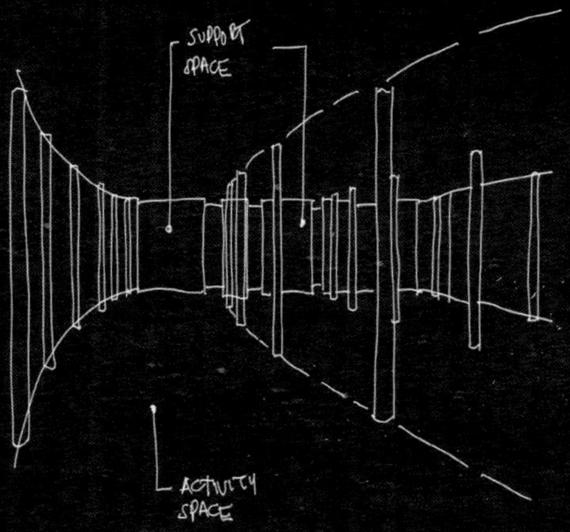

tively capture the shape and global behaviour of the structure that is studied. This method is applied to the structural design of this project. This approach is applied to the system components in parallel, taking into consideration their interaction through connections and load transfer; iterations are used to refine the design of each component. As a result, design decisions can be made quickly based on the result of successive design developments, bringing the benefit of increasing the efficiency of both the design process and the resulting facade system.

This parametric approach is part of an innovative design process supported by physical testing; parametric design allows many different component configurations to be tested with different material thicknesses and geometries.

This method allows materially efficient solutions to be found; reducing the cost and environmental impact of the system. The proposed configuration is then validated through further computed-based analysis and physical testing of the components to ensure the determined performance is achieved in practice.

The resulting panelised system is highly adjustable to allow a single system to be used across the external envelope, connecting to both a concrete and steel supporting structure, reducing the complexity and construction time of the assembly. For this project, a single bracket type is designed to be used across the whole system of connections between facade and primary structure. The multiple panel directions present in the facade system are required to connect to different supporting structures; an issue resolved through the development of an adaptable bracket.

Although many of the panels have different shapes, panels are designed to have identical corner geometry and their connection system is the same. This allows the supporting brackets to be connected to the panels at the same position with a single fixing method.

This approach to facade panels and their connection to primary structure allows the floor slab edges to follow the straight lines of beams rather than a curved edge of slab. This outcome is achieved with point fixings that can take up the curve of the facade.

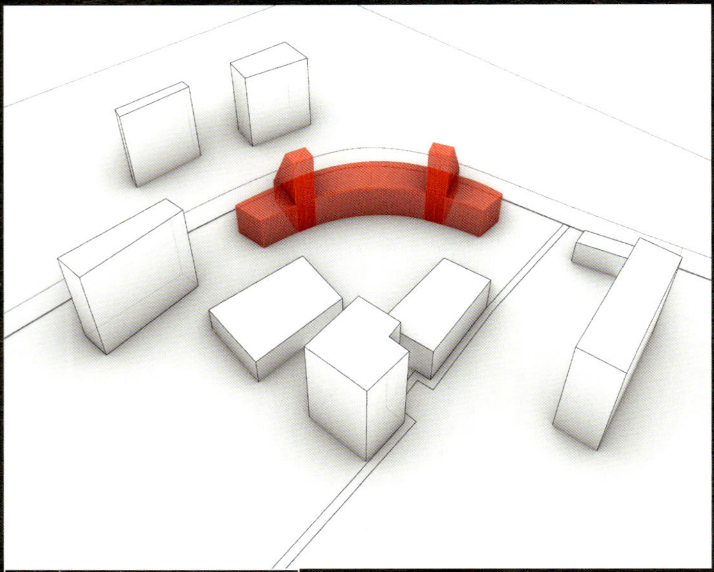

1. Project volume in site context

2. Project volume in site context

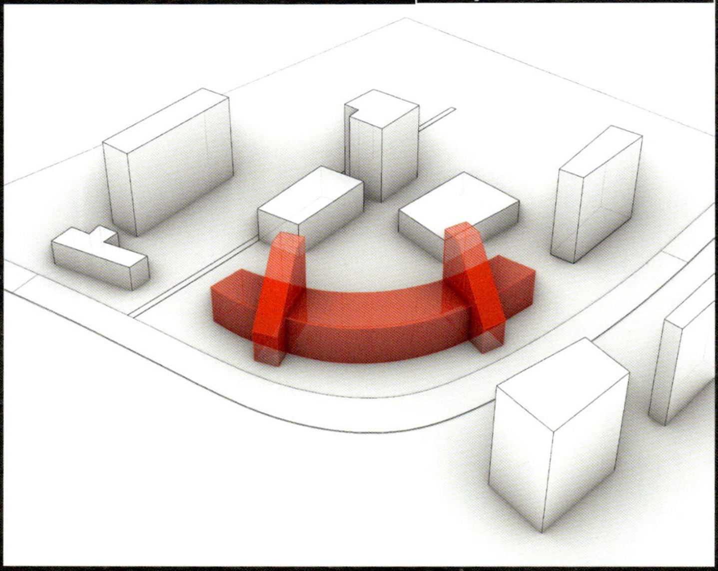

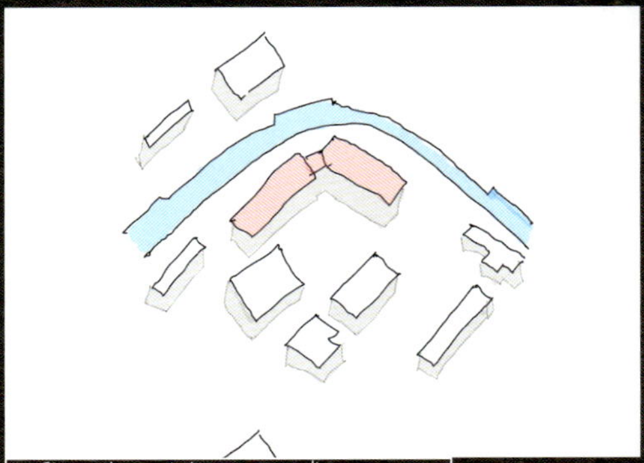

1. Site plan with surrounding context

Option 1 comprises two blocks with a link building between them.

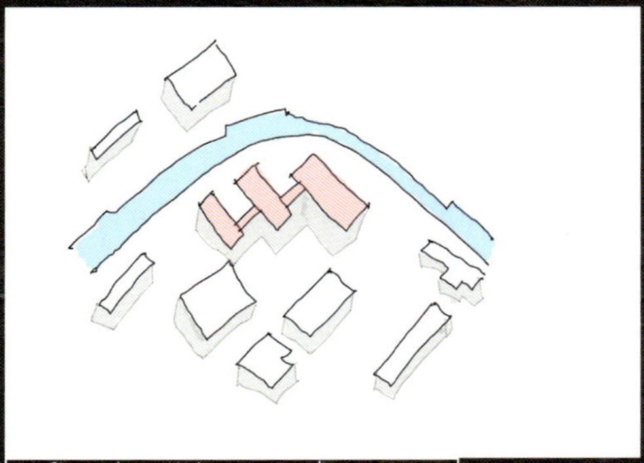

2. Site plan with surrounding context

Option 2 allows the buildings to respond to the curve of the canal by stepping the buildings.

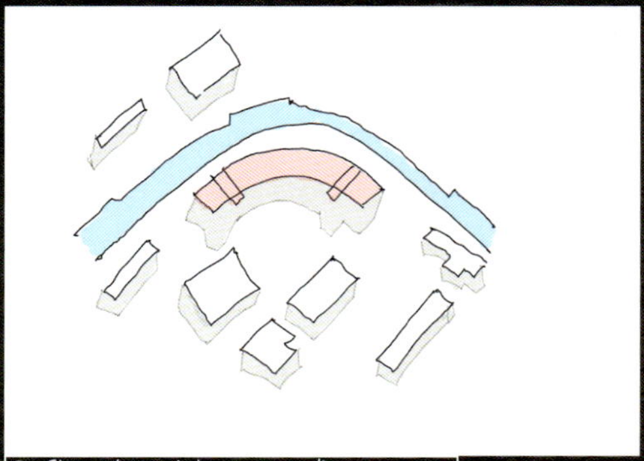

3. Site plan with surrounding context

Option 3 follows the curve of the canal.

Site context

This manufacturing building is for a technology company that makes electric transportation devices and drones. The building is, specifically, a large-scale research and development facility for tasks of design, research, prototyping and a 'mini production' assembly hall for testing production workflows for each product. The assembly hall comprises a long, curved volume in the form of a tube-like structure formed as a bridge which spans between a vertically projecting design block and an engineering prototyping block of similar vertical form. The bridge is curved in the form of a quarter-circle and is set at ground level. The design aims to provide as much flexibility as possible in how the main spaces are serviced; from above and from the side walls on several levels.

The site context for Option 1 is of two blocks with a link building between them. The building accommodates manufacturing and office space in each of these blocks. The L-shaped form is made in response to the geometry of the surrounding buildings as well as responding to the adjacent form of the canal which passes around the building, following a curved form.

In Option 2 the buildings respond to the curve of the canal by stepping the buildings, in plan, in order to create views around those blocks that form a visually staggered profile; the layout of the three blocks also responds to the orientation of adjacent buildings. Two link blocks are created between the three main blocks in order to provide a continuity of functionality within the facility. Option 3 follows the form of the canal, allowing the building to be unified into a single form where functions are more easily distributed. The building form does not follow that of the adjacent buildings but forms a 'pivot', in plan for a new set of adjacent buildings that can choose to respond to the form of the canal, which is currently not the case.

Option 3 is the preferred choice of spatial arrangement for the building and for its relationship with the site context. An advantage of Option 3 is in its response to the site context by following the curve of the canal which would have the potentially difficult issues of every space in the

Perspective view of structure

2. Perspective view of structure

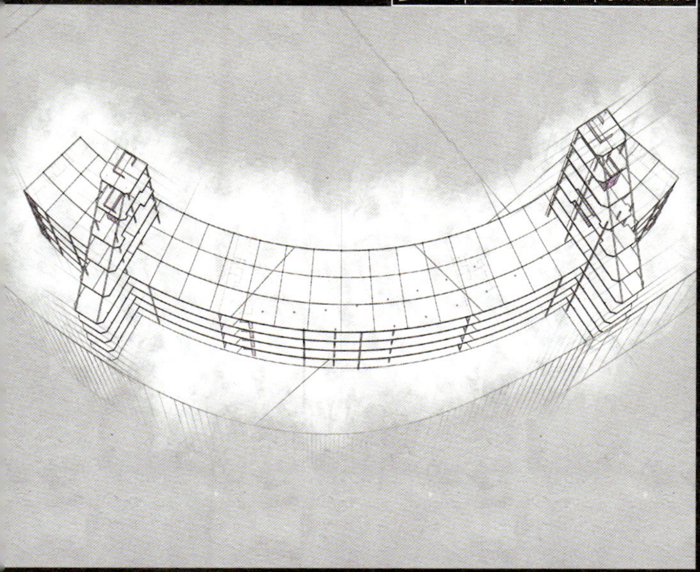

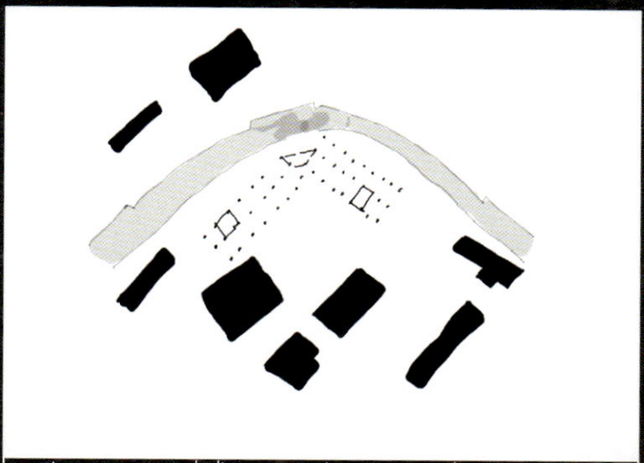

1. Figure-ground diagram showing project structure

Option 1
The structural grid, seen as columns on the outside of the building, does not respond visually to the form of the canal.

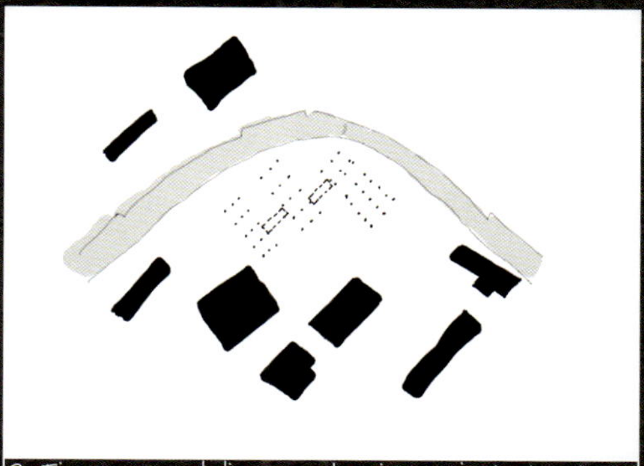

2. Figure-ground diagram showing project structure

Option 2
The structural grid responds partially to the form of the canal with a stepped arrangement of columns.

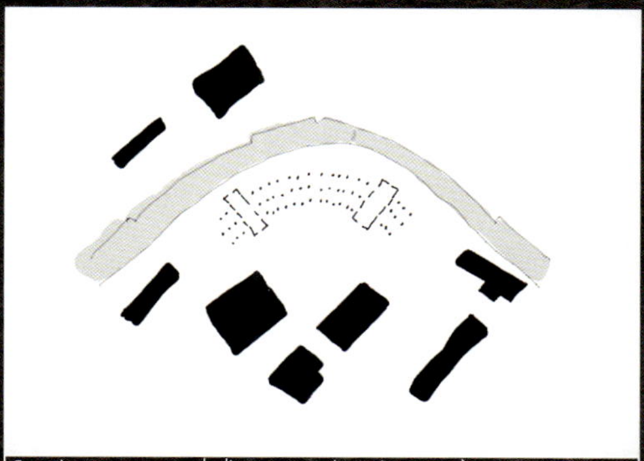

3. Figure-ground diagram showing project structure

Option 3
The structural grid recognises and responds to the form of the canal.

building being curved and consequently unable to accommodate regular arrangements of furniture and equipment. Instead, Option 3 is developed as a series of rectilinear modules which are linked together in a way that provides a continuous floor plan arrangement, with a single identity for the complete project. Option 1 is essentially two buildings with a link block entrance between them. This distributes interior space and circulation from that central entrance space.

Option 2 creates two entrance spaces, but which are still links between three blocks, each with its own identity and its own spatial organisation. The advantage with Option 3 is that everything is linked into a continuous form; a single straight rectangular block could not be accommodated on the site. Options 1 and 2 were generated for that reason. The use of the towers as roof mounted elements allow for parts of the space to accommodate equipment which is roof-mounted and is distributed horizontally at roof level, continuing vertically into the building. Both towers constitute equipment-based enclosures, arranged over three floors. The service installations are accessed from all levels of the adjacent blocks, as ease of access for maintenance of machinery used in the main spaces is a high priority of functionality. This approach brings together the need for an arrangement of the building which uses the opportunities of the site, blended with an organisation of space that benefits from that curved form.

A straight rectilinear building would not fit on the site, and did not link elements together in such a way that building users can appreciate the extent of the building either from outside or from looking outward from the windows at the front of the building; they can see the sweep of the building curve around the entrance areas.

This approach also allows the distribution of space to be sympathetic to a linked organisation both inside and outside the building, where pedestrian routes form an extension of those inside the building. Pavilions are set in the space in front of the building for use by staff. External spaces form a continuation of internal spaces, marked by pavilion buildings and staff amenities that visually anchor the building forms within the site context.

1. Sectional perspective view

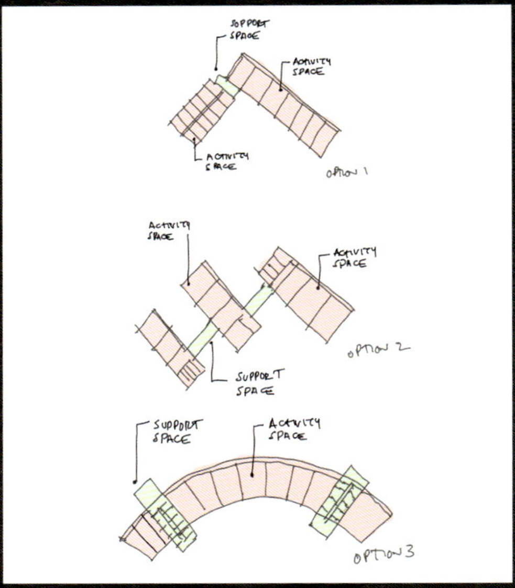

Spatial arrangement
Option 1 has an entrance in the link building between the two wings.
Option 2 has three blocks linked by two smaller hubs that provide two entrances to the building.

2. Sectional perspective view

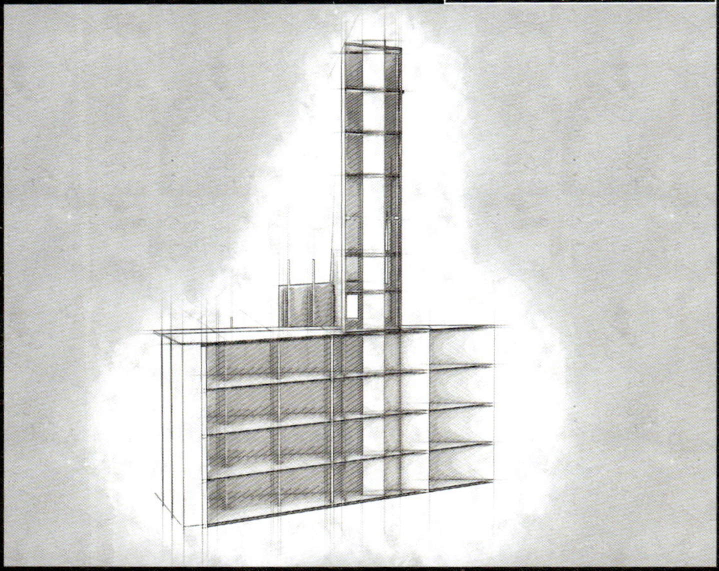

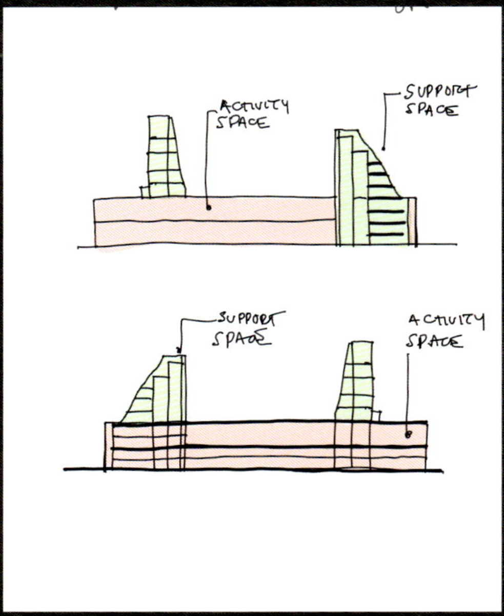

Option 3 links all volumes in a continuous curved arrangement, with office spaces set above them.

Spatial arrangement

The spatial arrangement is made in response to the way the company functions on a day-to-day basis; a close interaction between designing and fabricating products and their associated manufacturing setup. The external space around the building is not an open green space, or a planted space, but instead makes use of hard landscaping which forms part of the building fabric. An externally set testing area is formed as an extension of the building rather than as a utilitarian 'yard' space. This approach to external space allows the building to provide additional enclosures to be used for testing, storage of parts and service equipment.

These close links between 'inside' and 'outside' spaces result in the characteristic cut-out forms of the raised podium. The research and development spaces are large self-contained volumes that can be reorganised to create internal space, which is reallocated with ease. The wedge-shaped engineering block has floors arranged in vertically grouped sets of three floors. These groups of full floor plates and intermediary mezzanine floors provide accommodation for pre-design teams and design development spaces within a single engineering team studio. The spatial arrangement of Option 1 is one of an entrance in the link building between the two wings. This approach allows access from both sides of the link building; to the manufacturing spaces and to the offices. The solution reduces links between each block as a result of the need to pass through the link area, which would extend in height above the entrance level, in order for it to meet the expectations of an entrance that provides a welcoming space for visitors.

The entrance space is shared in Option 2: The three blocks linked by two smaller hubs provide two entrances to the building, one of which is the primary visitor entrance and the other for staff entry. The two entrances allow each block to be organised independently of any other, which is not the case in Option 1.

Option 3 unifies the office space and manufacturing space by linking all volumes in a continuous curved arrangement, with office spaces set above them.

Perspective view of interior

2. Perspective view of interior

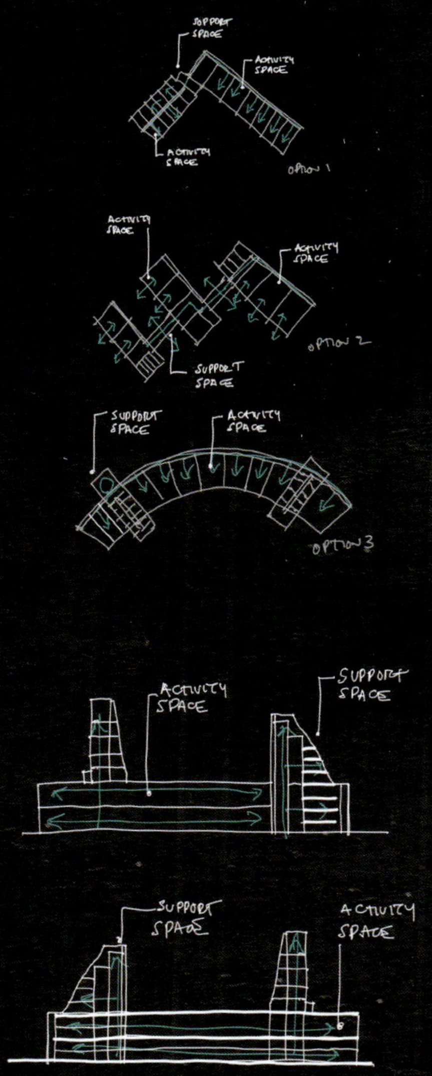

Internal circulation

Option 1 comprises a corridor set along one edge of the manufacturing spaces on the canal side.

Option 2 has three blocks which are relatively independent and have a common circulation route through the building.

Option 3 uses a circulation route along the canal side of the building.

Internal circulation

The internal circulation of Option 1 comprises a corridor set along one edge of the manufacturing spaces on the canal side; the corridor continues down the middle of the second block of office space, which is set on both sides of a central corridor. The link blocks between buildings create the entrance and exit, providing hubs for movement around the building. In Option 2 the three blocks are relatively independent and have a common circulation route through the block down one long side of the building, providing access into adjacent clear space. Circulation is shared between three buildings, with access spaces which also provide ancillary support space for washrooms, meeting rooms, canteens and reception.

In Option 2, these two link buildings remain as separate buildings which provide independent circulation routes that fit into space that serves the three blocks on each side of the link blocks.

Option 3 uses a circulation route along the canal side of the building, extending in the form of a ribbon around one side of the perimeter of the building. This route links to vertical circulation spaces that serve the offices above. Movement of goods in and out of the building is by way of the side of the building that is away from the canal, allowing circulation within the building to be independent of the circulation of vehicles and goods. This pattern is repeated in Options 1 and 2 where pedestrian access, and the movement of people around the building, is independent of goods and the movement of vehicles.

The internal circulation of Option 3 aims to be matched by an external circulation space which follows the path of the canal immediately adjacent to the internal circulation, providing a parallel route to traverse the building.

A second external route has pavilions set against the canal with seating areas and an outside cafe for staff. The inner curve, which faces the front of the building, has a different-shaped external route which forms an elliptical shape to follow the form of the building; creating an external walkway for staff. This space has pavilions with seating and space for exercise as well as outdoor meetings.

1A. Isometric view of structure

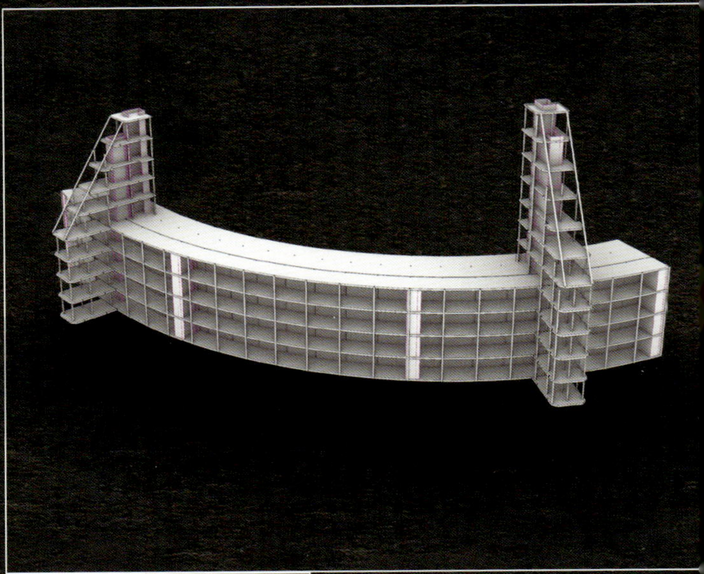

2A. Isometric view of structure

1B. Isometric view of structure

2B. Isometric view of structur

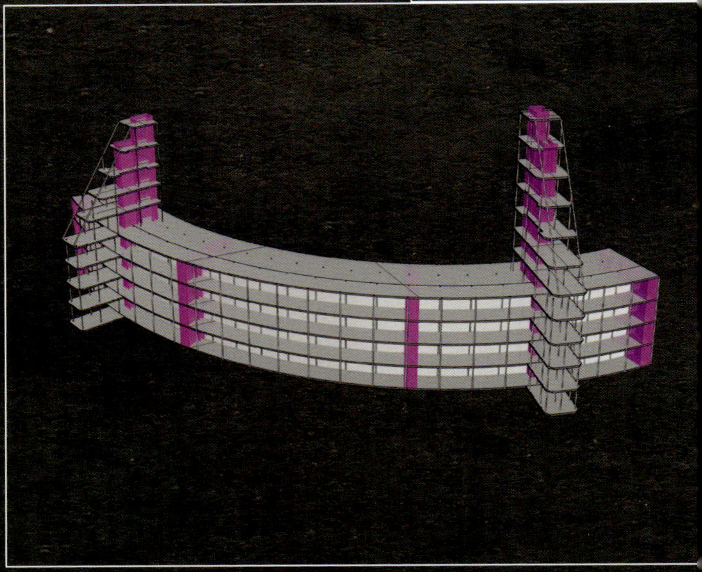

Structural strategy

The structural strategy is to have the podium block constructed as a steel-framed, tube-like structure, or a 'trussed hull'. This approach to the podium block allows openings of different sizes to be formed within the 'tube', rather than limiting opening sizes by a structural concept of regular reinforced concrete floor slabs supported by a frame of columns and beams. This approach allows structure, facade and roof to be 'wrapped' in a continuous external envelope; an envelope which forms an integral part of the primary structure.

The bridge structures are supported on steel-framed consoles. This concept ensures that all floor plates are completely clear of columns. All core spaces, escape stairs and washroom blocks are accommodated in a series of 'zigzag' forms within the structure of the tube. In some cases, stair cores with washrooms and storage spaces pass through the framed tube-like structure. Access is maintained along the circulation route down the centre of the building. Rooflights are set at the top level of the tube, providing daylight that is controlled externally with mechanical baffles; directing daylight to the floor levels arranged within the tube. The tube is separated into zones in order to create a set of clear, open spaces. The wedge-shaped tower is designed as a steel frame with metal deck composite floors. The perimeter of the structure is formed by lattice frames and geodesic frames. The vertically set service cores form a 'spine' that is located at the centre of the tower, extending through the height of the building, dividing the wedge shape into two halves of space on each floor.

Structural stability is provided by diagonal bracing in the steel frame that forms the structure of the cores. Slot windows extend up the height along the long sides of the wedge block such that the diagonal cross-bracing can be accommodated within opaque walls set on both sides of these slot windows. Individual platforms are set on the facade which provide an access deck for goods being lifted into storage and to the roof-mounted locations of service equipment. Air ducts pass up the vertical slots on one side of the windows. The shorter end elevations accommodate different-sized slots for the controlled admission of

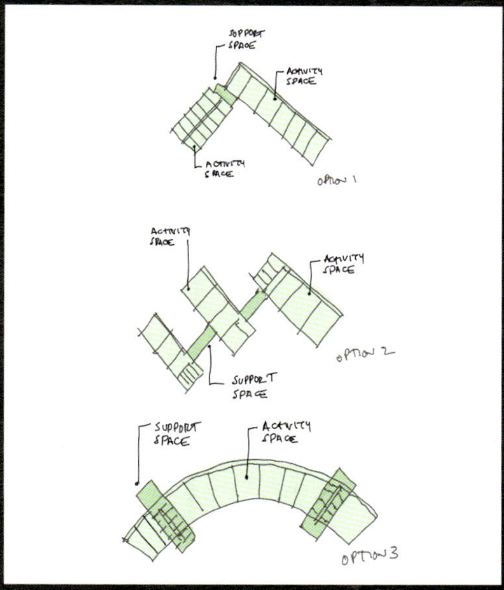

Structural strategy: Zones
Option 1 is of a rectilinear arrangement of steel portal frames that support both industrial unit and office space.
Option 2 provides three independent structures.

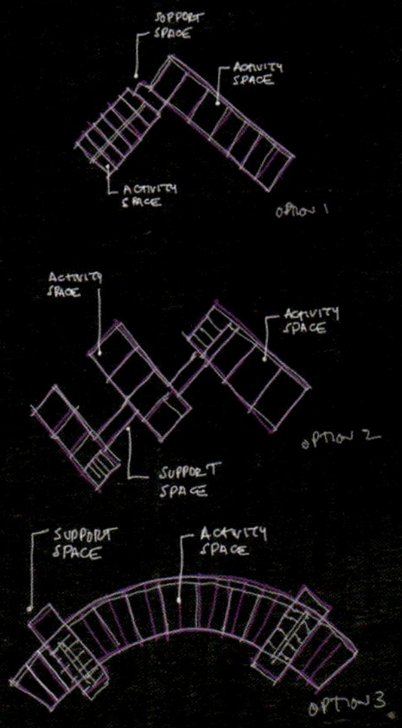

Structural strategy: Framing
Option 1 uses a common structural system throughout the building.
Option 2 has two link buildings joining them to create a total of five structures.

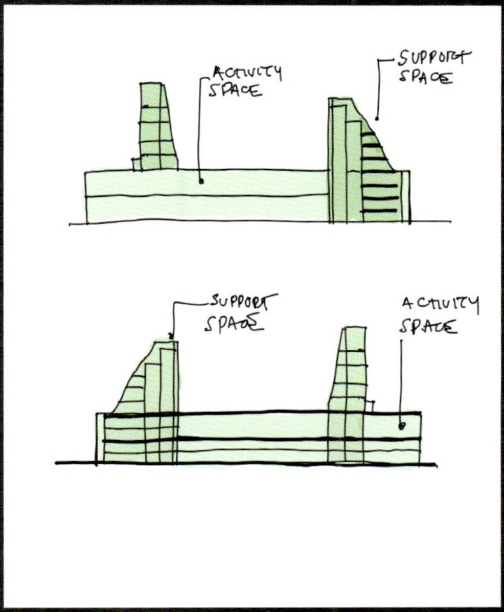

Structural strategy: Zones
Option 3 allows for the expansion of the industrial spaces by expanding out from both sides of the building.

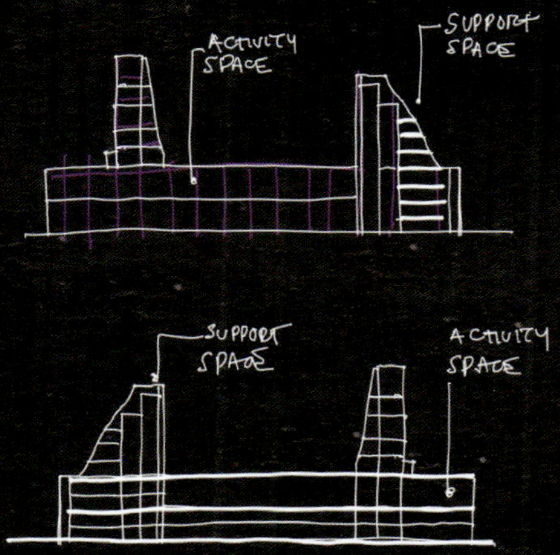

Structural strategy: Zones
Option 3 has flexibility of structure to allow the building to grow and change in response to evolving spatial requirements.

daylight. The engineering building comprises two large modules, set on top of the bridge, using the same construction of steel frames formed at their perimeter by large-scale lattice frames and geodesic frames, which create the enclosures. Floor decks are formed as composite decks as per the 'wedge' tower. This building has its service cores located adjacent to the outside curve of the podium block, on the opposite side from the projecting enclosure that cantilevers out at the front of the building.

The cantilevered elements of the building direct light and shade to provide solar shading to the exposed glazed end of the building. The two modules themselves are supported on steel-framed consoles that provide the support for both modules. The structural strategy for Option 1 is of a rectilinear arrangement of steel portal frames that serve both industrial unit and office space. These have a common structural system; the complete building allows for those functions to change in the near future, allowing office space to become industrial space and industrial space to expand further into the building. Option 2 provides three independent structures, with two link buildings joining them to create a total of five structures. In each building the structural form is the same. Between one of the buildings the form is rotated through 180 degrees in relation to its neighbour; the two link buildings use the same structural arrangement.

All five structures are independent of one another, allowing link buildings to be dismantled quickly and for the manufacturing buildings to be enlarged. This layout provides considerable flexibility for an increase in size of the units as well as an increase in their length; but is more difficult to apply to Option 1.

Option 3 allows for the expansion of the industrial spaces by expanding out from both sides of the long facades of the podium in order to increase their size; the loop can be increased in length. The office spaces are set above the industrial spaces, allowing them to extend either upwards or across the roof, or to be increased in size. The flexibility of the structure allows the building to grow and to change in response to evolving spatial requirements.

1A. Isometric view of environmental volumes

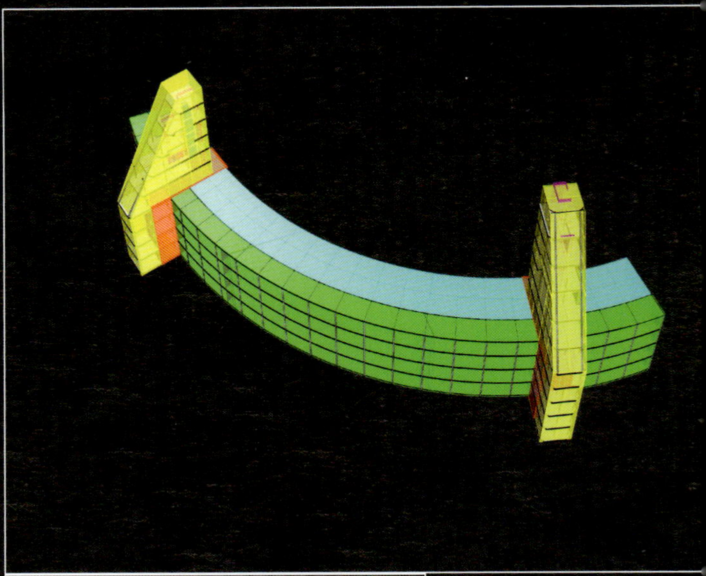

2A. Isometric view of environmental volumes

1B. Isometric view of environmental volumes

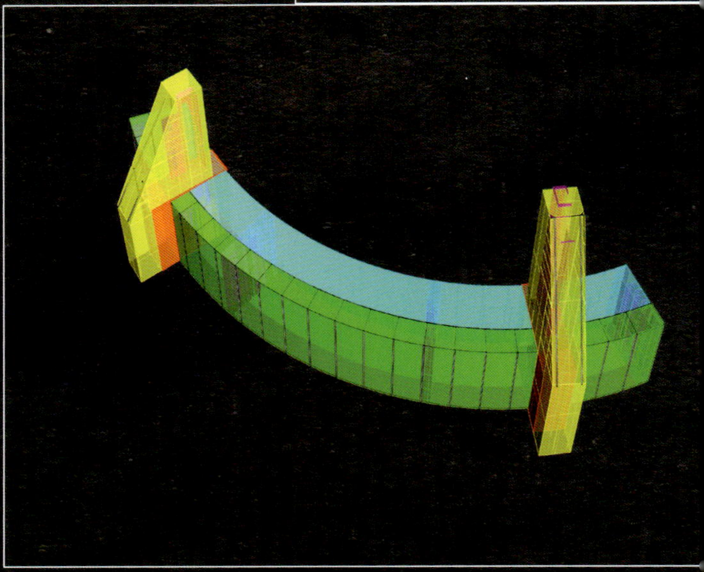

2B. Isometric view of environmental volume

Environmental strategy

The environmental strategy is based on treating the podium as a set of environmental zones formed from each of the 'bridge' structures which are set above ground level. The wedge-shaped tower, positioned at one end of the bridge structures, comprises three separate environmental zones; a podium level zone, a mid-tower zone and a zone set within the wedge form at the top. Each of the research and design development spaces in the 'bridge' zones has specific environmental performance requirements set by the brief/program; spaces are semi-external, used for the raising and lowering of equipment to and from the podium block.

Other spaces have prototypes lowered to the outside from parts of floors which can be lowered to ground level on winches – sometimes fully exposed to external conditions – and are closed at other times. As a result, each part of the building is an environmentally independent zone. The same principle applies in the wedge-shaped tower, where each of the three-story blocks that form the tower are self-contained units; formed as separate environmental zones. The two modules in the research and development and engineering blocks above the bridge are formed as independent environmental zones, but share the service provision of mechanical and electrical equipment modules located at roof level.

A similar principle applies in the second tower, where ducted air is directed and redirected around the different environmental zones. Only the bridge has separate zones where their respective service provisions do not interact with one another.

One key environmental design criterion was a strategy for the admission of daylight, limiting use of electrical lighting inside as much of the building as possible during daylight hours. Most spaces in the podium bridge are toplit; daylight in these spaces is directed by baffles around the building. The environmental strategy for Option 1 is of having separate environmental zones for manufacturing and for offices. The third environmental zone is created by the entrance space and the associated circulation spaces of corridors. Services are distributed from 'envi-

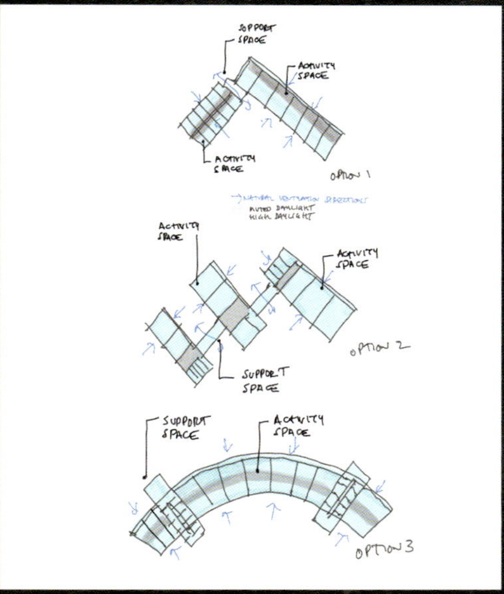

Daylight strategy for direct sunlight and cross ventilation.

Option 1 admits daylight and air from one side of each block. Circulation areas have subdued daylighting.

Option 2 admits daylight and air from both sides of each block. Circulation areas have subdued daylighting.

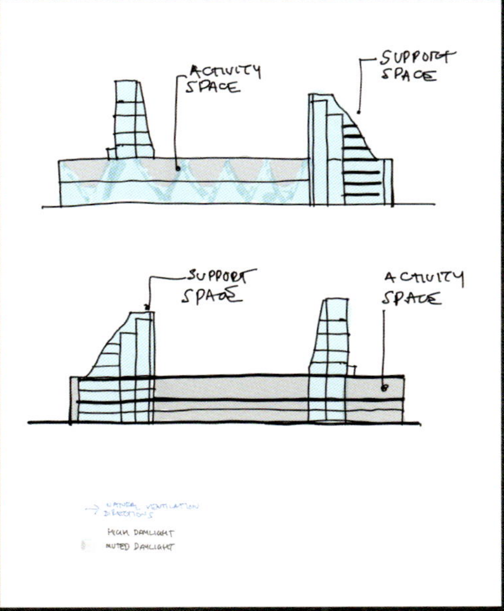

Daylight strategy for direct sunlight and cross ventilation.

Option 3 admits daylight and air from both sides of each block. Central areas within activity spaces have subdued daylighting.

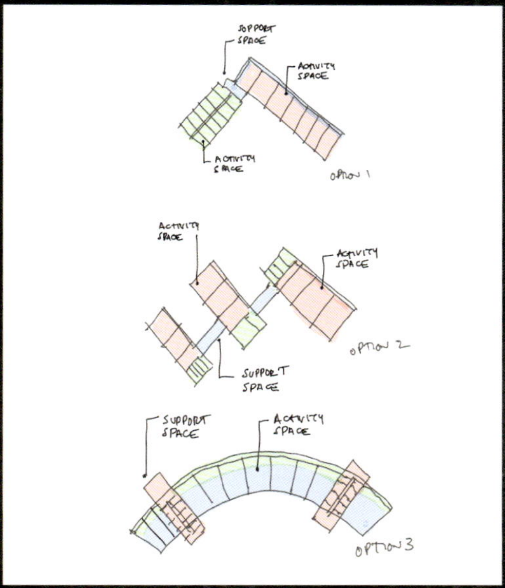

Environmental zones

Option 1 has separate environmental zones for manufacturing and for offices.

Option 2 provides three separate buildings with separate air-handling units and services equipment.

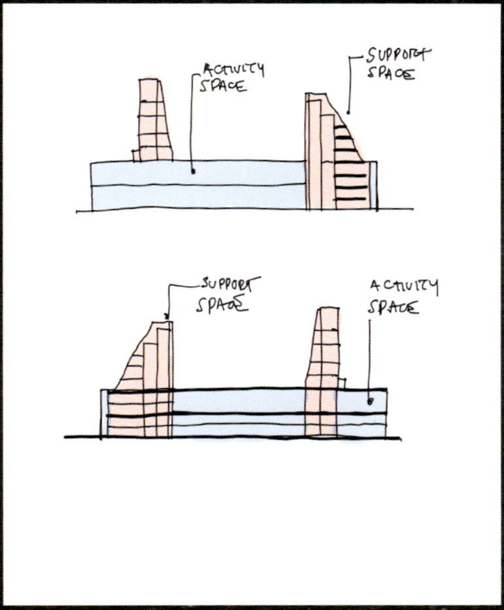

Environmental zones

Option 3 has environmental zones which are created within each unit within each zone.

ronmental units', which include air handling units, which are set on top of the link building; services are distributed at roof level down the length of both blocks, with air and electrical supplies being drawn down into the spaces beneath. Option 2 provides three separate buildings with separate air handling units and services equipment. The air handling units are set on top of the office spaces, with air distributed across the roof and down into the spaces below. All three blocks use a similar arrangement. The link buildings use ventilation taken from each of the blocks adjacent to them. Option 3 has environmental zones which are created within each unit at the back of each zone, facing the canal; each unit is a tightly controlled environmental unit where office space and ancillary functions can be added.

Large spaces within the bridge, with lifting doors that open frequently during the day, have a less closely controlled environment; highly thermally insulated but where heating or cooling energy can be provided rapidly. Office spaces set above the curved manufacturing spaces comprise a third environmental zone which provides air for associated circulation spaces. The environmental strategy for this building is focused on how environmental supply services move around the building, then creating separate zones with different performance criteria and different occupations at varying times. The strategy is based on the need for fresh air that is filtered through equipment at roof level and therefore the most efficient distribution of air and electrical services is the primary consideration.

As a result, the rooftop is used as the means of distribution of services, and this is achieved using the the least amount of duct runs when balanced against the need for additional sets of equipment to supply that air. The balance between those requirements was found by having two areas for services equipment at roof level which distribute first horizontally, then downwards into the building. All spaces have approximately the same requirements as environmental zones, since similar tasks are being carried out in each space with a similar number of people. For this reason, the same level of thermal insulation is provided in the external envelope.

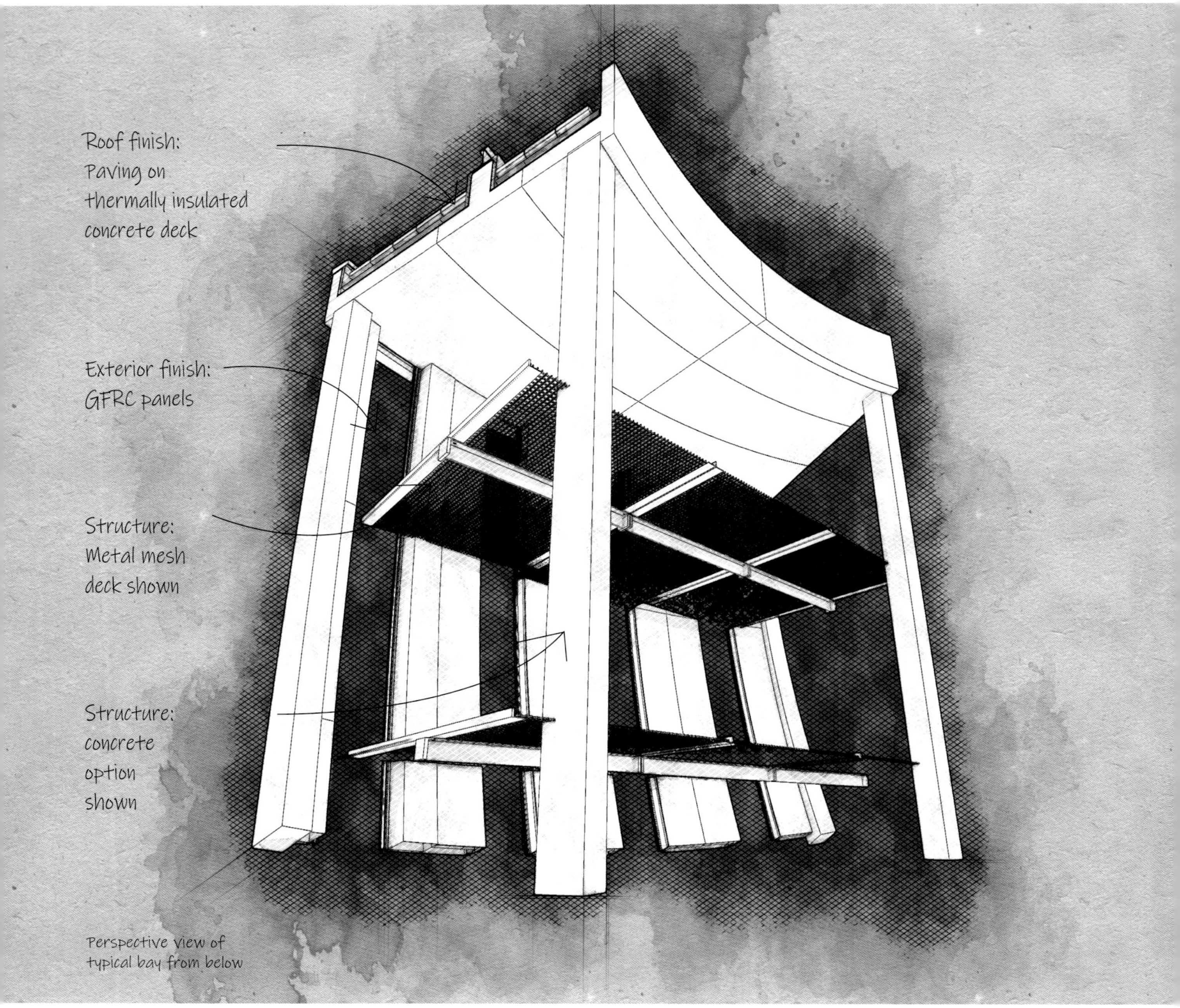

Typical bay adjacent to facade

The building is constructed in modules which are partially assembled on site and partially off site. Components for the external envelope arrive as a kit of parts which are part-assembled locally; the remaining components are assembled on site. The project is structure led with cladding panels used to enclose the primary structure which is external in some locations. This externally mounted structure is thermally insulated and waterproofed.

To ensure its durability and weather-tightness, the exoskeleton bridge is formed as a single structure which is supported on framed consoles that are clad in thermally insulated metal panels. The consoles form part of the structure of the two towers mounted at each end of the building. Prefabrication methods are used for the steel window frames and for the mixture of GFRP-faced (glass fibre reinforced polyester) panels and metal-faced panels which enclose the opaque areas of wall. The bridge is formed with large-scale openings for large-scale doors and glazed panels along its length; the supporting structure is framed as a mixture of geodesic frames and lattice frames. The wedge tower has more opaque facades than the podium; these are constructed from curtain wall. The facade panels have air-handling ducts and electrical supplies fixed to the inside face of the panels. The internal face of the panels, within internal spaces, has fixed furniture items of shelving and drop-down desks. The facade is very much a combined envelope and service space, accommodating equipment that runs from the building to be fixed to the walls. Part of that shelving is integrated as part of the framing of the external wall and serves as open shelving and cable trays. The system integrates cupboards, as the spaces are required to be clear of service equipment as all electrical supplies are floor-mounted, set within the depth of the floor. Electrical supplies for portable equipment such as laptop computers and office-based support devices are set within the facade zone. The edge of the space is part of the circulation zone down the length of the building, and it is here where the equipment is accessed. Facade panels are thermally insulated and weatherproofed externally. These tall storey-high panels

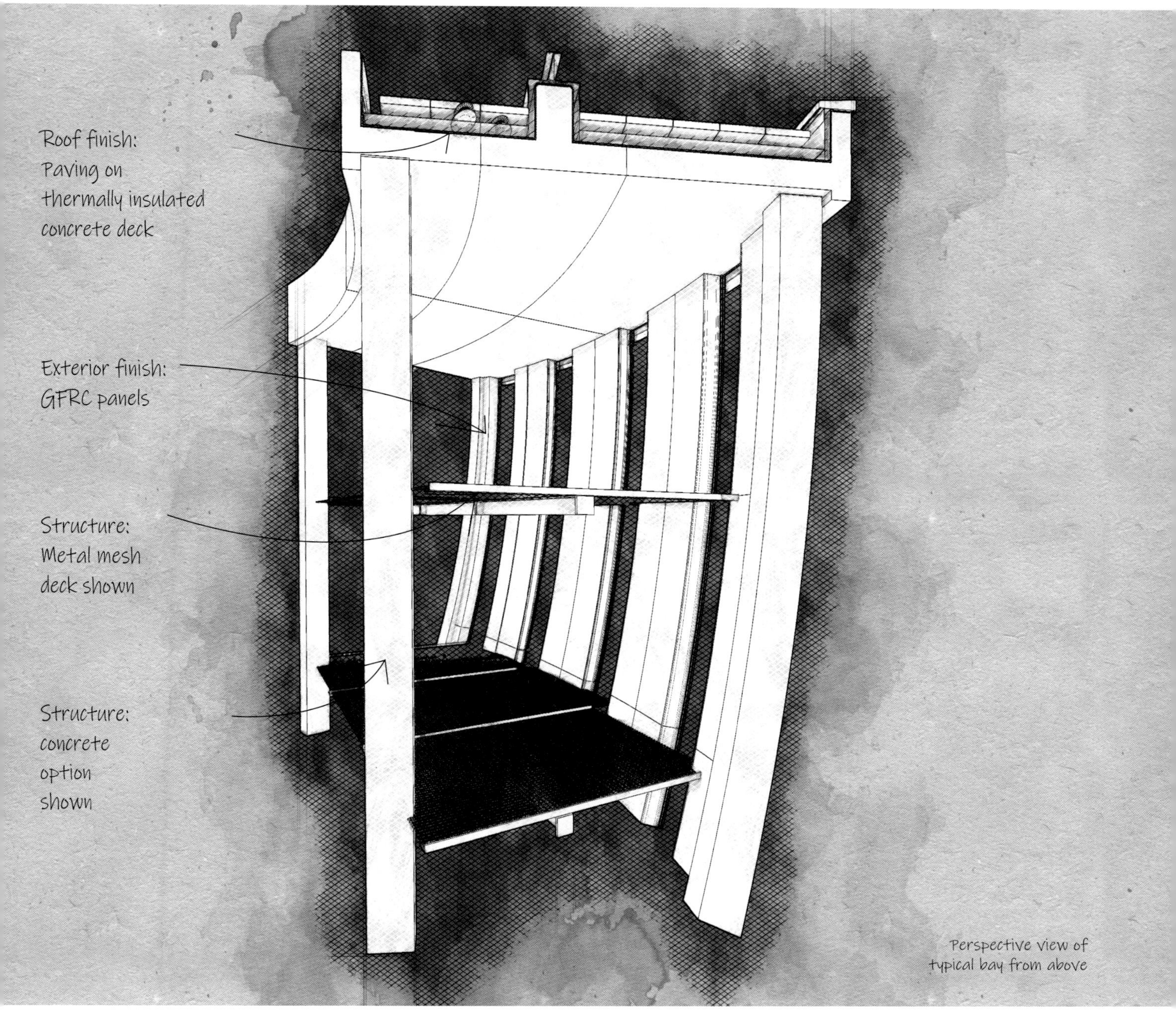

Perspective view of typical bay from above

use the framing to support electrical services and air-handling ducts in some areas. The facade zone is 1.5 metres deep in order to accommodate a 0.5-metre-deep facade panel in which the shelving and drop-down desk space are located.

The 0.5-metre-deep facade panels are required to be fixed back to the primary structure as storey-high units, in some cases extending the full height of the bridge extending over two floors. Panels are required to be fixed back at the top and bottom of the unit. At the top they form a wide parapet zone as they are connected back to the roof immediately behind. At the base, panels are connected back to the lowest level of floor slab set above ground floor level. This results in the characteristic chamfer at the top and bottom edges of the panels in response to the need to connect back to the primary structure. For this reason, a lightweight frame is used within the panels; the frames are used as a support for the GFRP outer layer.

An advantage of this approach is that the glazing can follow corresponding lines of ductwork rather than needing to be set as continuously vertically framed elements. A benefit of the large size of these panels is that this approach is also used where panels are cladding structure which projects from the building as a result of the need for the structure to support other assemblies in the building. This language of the typical bay behind is based on the need for close integration between structure, services and daylighting in the building.

The typical bay in this building can be regarded as extending across the full width of the building, to include both facades and the full extent of the supporting structure between.

The servicing of the building has been designed on this basis, with the possibility of full cross ventilation from one side of the building to the other, and the possibility of staircases and washrooms being integrated within each typical bay. This approach would extend to the roof, where a single mechanical ventilation installation would service the extent of a typical bay and its associated stairs, washrooms and technical equipment.

1. Perspective view of exterior

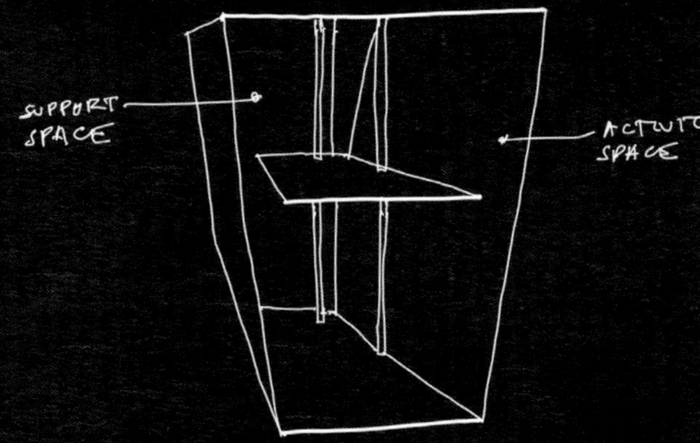

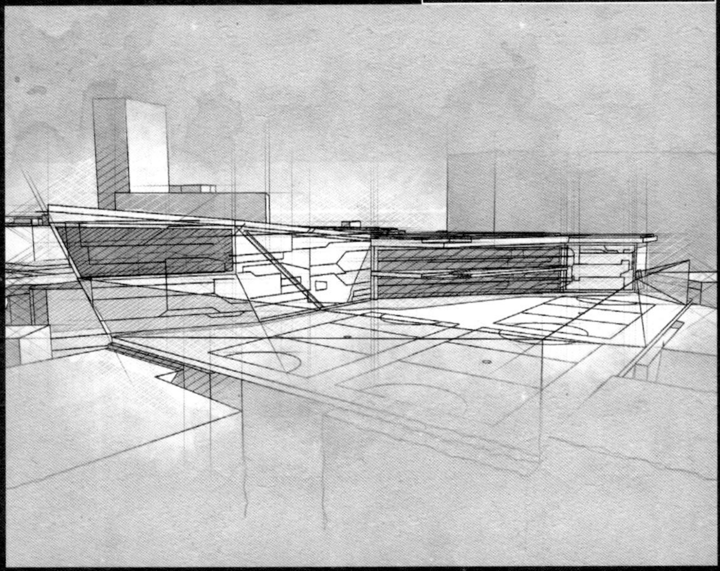

2. Perspective view of exterior

Visual considerations of structure

A key consideration of this project is to provide visual integration of structure – expressed both internally and externally – with the spaces that the structure encloses, while minimising the energy required to fabricate, transport and install the materials in the building. This consideration sits alongside the importance of ease of construction, the quantity of materials used in the construction of the structure, and associated issues in the fabrication of components.

This aim is achieved by permitting a high level of differentiation of the assemblies that form the spaces both visually and technically, with articulated (jointed) forms that express how the structure is working both as support and as enclosure. This approach may take the form of expressing structure that projects out from the building, beyond the facades. Alternatively, structure may be expressed in forms that are visible within spaces, such as that of the three blocks set adjacent to the rooftop sports courts. As a result, a single steel or reinforced concrete frame would not satisfy the requirements of a separation between all these forms; each element is designed to be constructible in a phased sequence as partly prefabricated sets of facade assemblies. Facades that are prefabricated as large-scale assemblies can be transported by road as pieces of preassembled building, complete with a partial supporting structure. A set of small buildings set side by side results in internal space-specific design considerations becoming more important than the potentially more economical solution of a rectilinear braced frame. In this project, some structural and volumetric elements are projecting, with columns which are made visible in areas of the facade where they do not suit the visual requirement of a set of interlocking separate building forms. As a result, the technology used is one that provides structural bays of different height, width and depth; some structural bays appear to be stacked one on top of another, and appear to be separate from one other. This approach allows phased construction to take place on site and avoids the spatial limitations of a rectilinear frame coupled with a single facade system, which suggests that only a single volume exists in the building behind, rather than the

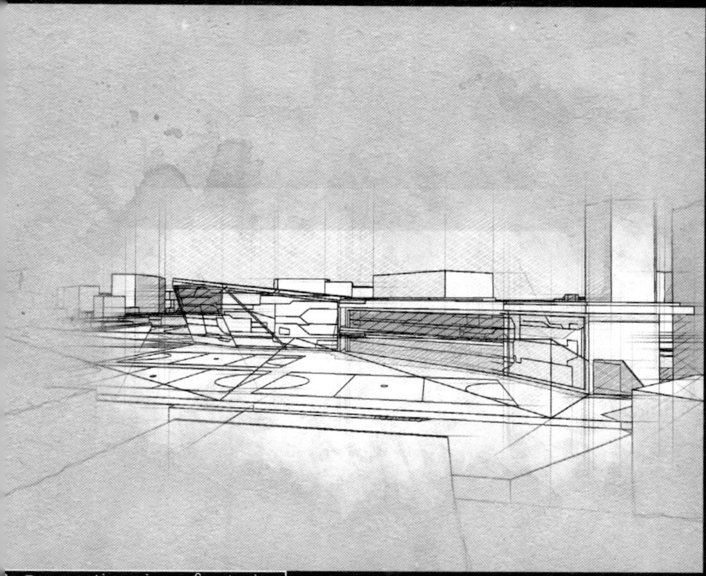

3. Perspective view of exterior

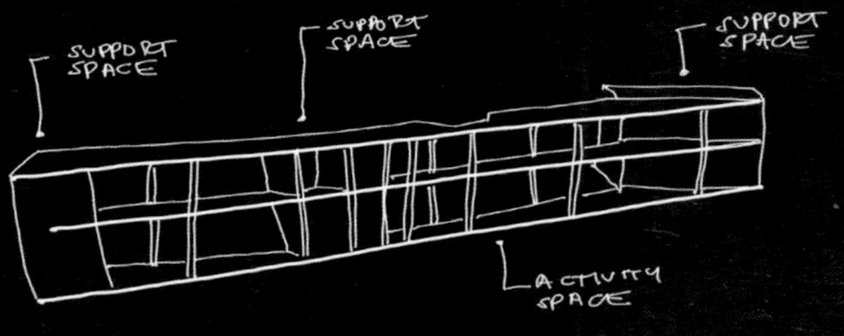

4. Perspective view of exterior

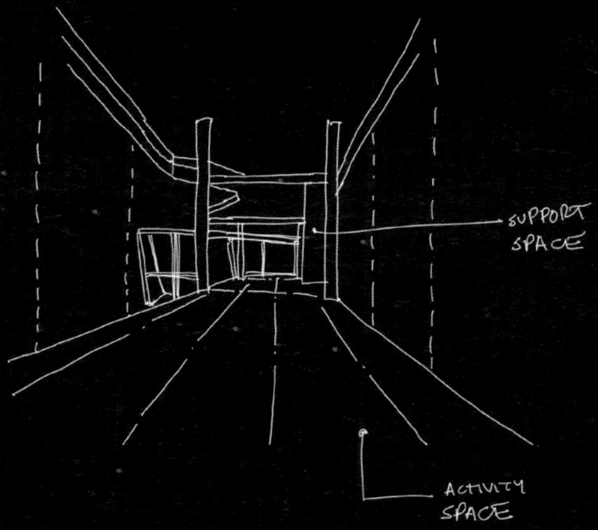

sequence of spaces that is a key part of the design approach. With the solution of structural bays of different size and arrangement, the facade form appears to be projected, or extruded, through into the inside volumes.

The design is also based on a regular structural system used for braced frames, but the setting out of the structural members, offset and cranked away from each other, together with varying thicknesses of walls internally, provides the required expression of differentiation of facade and associated internal spaces behind the facade; the facade layout is driven by the spaces required to be created, not by the structural grid. This approach results in structural elements, set next to each other, which are independent of one another but have structural continuity within the building. Internal spaces are made visible externally, just as they are appreciated and viewed when standing within those spaces.

Lines of structural frame are picked out as panel joints in the external envelope. Panel joints are created along the lines of structure that support those panels This approach allows the joint lines to be authentic, as the points of support are immediately behind and follow the lines of structure. The large facade panels do not exceed a size and weight that would require their subdivision. Roof panels also follow the lines of framed structure beneath; panels project upwards in different areas to allow for drainage to follow the falls of the membrane layer beneath. Facades and roofs are able to accommodate glazed panels which are of a size that can be accommodated as triangular-shaped units, based on the module size that is used on the roof, and which can be manufactured as double glazed units.

The triangular panels are part of a folded structure that provides strength to the facades, avoiding the need for a regular vertical division of glazed panels that is more associated with a commercial environment than with a sports facility. Edges of the roof extend outward as framing to enclose the sports courts area in order to create not just a framing to support the perimeter fence, but also to support the pavilions used for the storage of equipment.

1. Project volume in site context

2. Project volume in site context

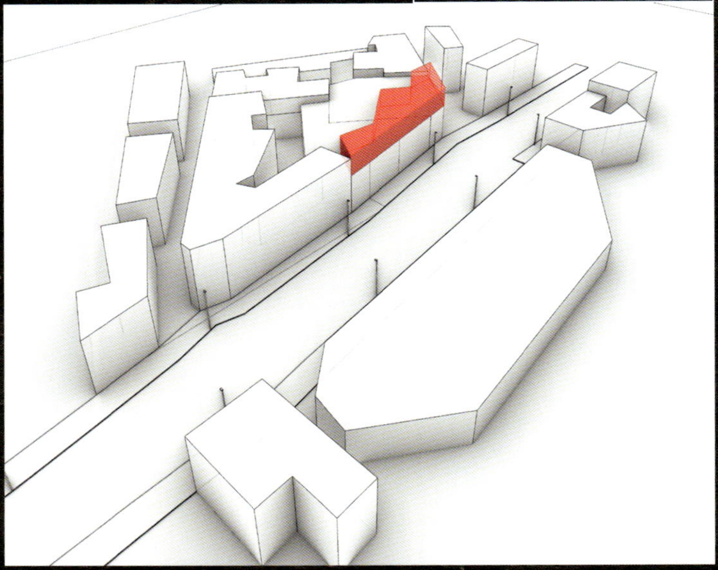

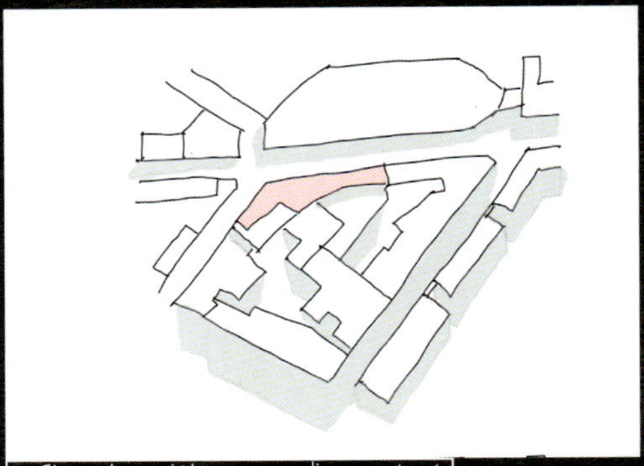

1. Site plan with surrounding context

Option 1 comprises a single building form with movement joints between three component parts of the new building.

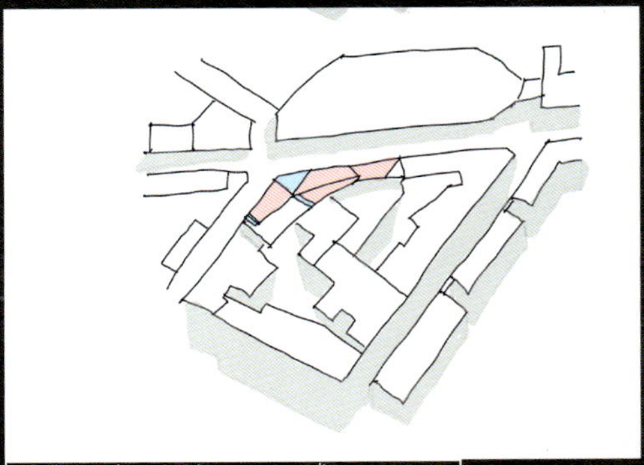

2. Site plan with surrounding context

Option 2 is separated as three buildings set side by side.

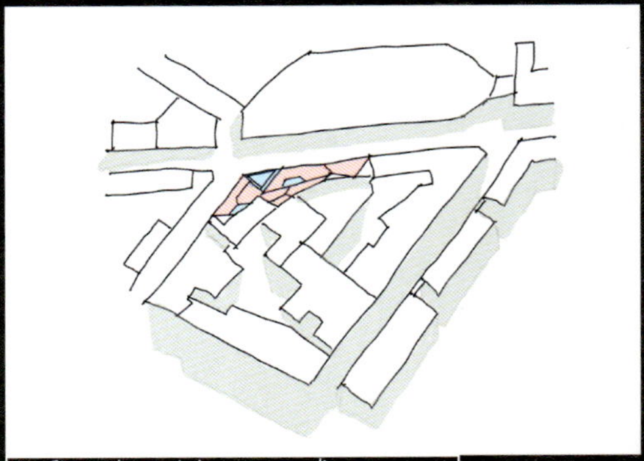

3. Site plan with surrounding context

Option 3 provides a greater differentiation between the spaces of the three blocks.

Site Context

The site context is an urban location where a single-storey addition is being inserted onto the top of an existing building; the current building form is three storeys high. The existing building is divided into three separate blocks which are connected with movement joints. The new addition is an extension to that building which is set above the existing flat roof.

This project is for a sports centre operated by a university, accommodating sports courts on a site that had originally been conceived for indoor sports only. The centre has multipurpose courts for basketball, netball and tennis. The design evolved from an indoor centre into creating external sports courts surrounded by a safety fence that allows more courts to be provided, allowing the indoor spaces to be used for more activities than sports only. The courts were moved outside; indoor spaces were repurposed as changing rooms and shower facilities, rooms for meditative pursuits and fitness classes. The approach taken by the owner is one of being able to reach out to the community to provide a sports facility that can be shared with local residents, as well as being used by university students and staff. As a result, the boldness of the roof deck which stretches over several existing roofs of university buildings formed a key part of the design intent, creating the context for interaction with the immediate built environment. The roof deck has canopies which extend out from the building both from the edge perimeter fence and from the building. These spaces are covered in order to avoid an all-netting protection that would have been required as an alternative to the canopies. It was suggested that netting be introduced to canopies at street level between university buildings; this was considered to be too much of a built intervention outside the scope of the project. Enclosure to the courts has been created visually by making the canopies cover the sports courts; becoming a key architectural element in the design as a result of them covering an area greater than the building itself.

Canopies are designed as an extension of the building. In form, they do not adopt the curved surfaces associated with tent-like structures, but

. Perspective view of structure

2. Perspective view of structure

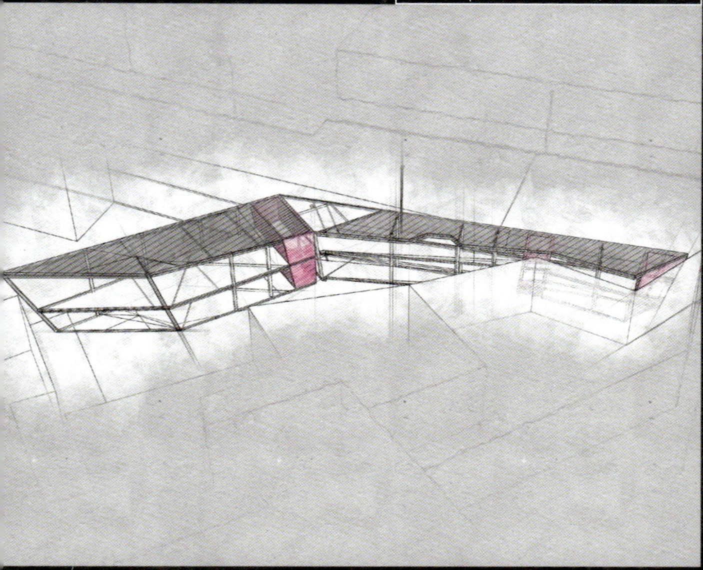

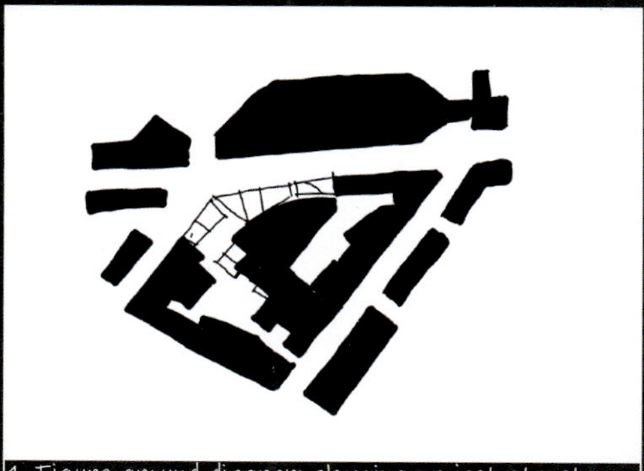

1. Figure-ground diagram showing project structure

Option 1
The structure is supported at points that can provide support from the building below.

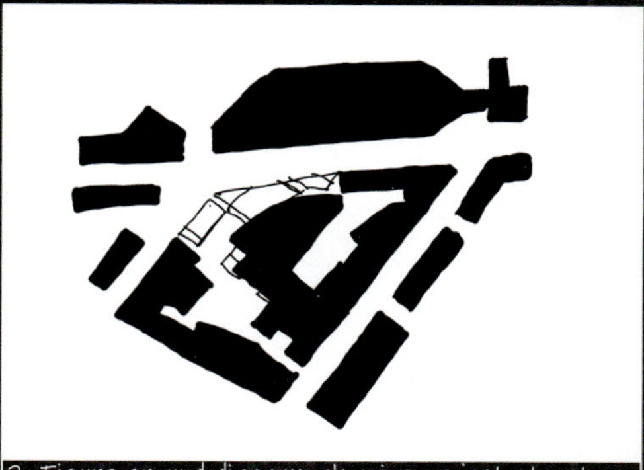

2. Figure-ground diagram showing project structure

Option 2
The structure is supported at points that can provide support from existing load-bearing walls.

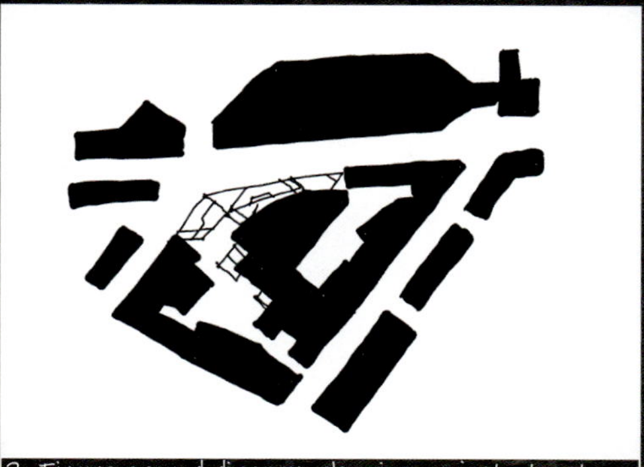

3. Figure-ground diagram showing project structure

Option 3
The structure is supported both at points that can provide support from the building below and from existing load-bearing walls.

instead are of triangular form. The reason for this is to allow a multi-functionality of the canopies in that they can accommodate secondary elements such as smoke extractors adjacent to their edges as well as supporting floodlighting for sports activities on the courts during hours of darkness. The night-time lighting is hung beneath but is high enough in the canopy that the fabric serves as a light shade, ensuring that light pollution does not extend to neighbouring buildings but instead provides a gentle level of illumination through the canopy material.

This approach avoids the effect of glare created by harsh floodlights, which were considered to be the most contentious part of this design proposal, both for the university and for neighbouring businesses and residents. The sides of the external areas are translucent and are not made from the regular chain link fence; translucent sheet materials are used that allow light to pass through but which avoid creating an effect of glare that would have been experienced in surrounding buildings. This effect of shading from lighting is achieved without a loss of performance or efficiency of the lighting installation.

Option 1 for site context is to have a single building form with movement joints between three component parts of the new building which correspond to the movement joints in the supporting building below. These joints are not immediately visible within the new spaces being created. Option 2 works with the fact that the building is in three parts – which is expressed in the form of the building, as well as in the internal circulation routes – in order to draw people into the building through the middle block, towards each of the two end blocks. The corner part of the building is identified as a separate volume; rooflights are introduced in order to provide daylight adjacent to areas where there is an opaque wall. Option 3 takes Option 2 and develops it further in order to provide a greater differentiation between the spaces of the three blocks in the building and create rooflights which emphasise the route along which users are walking in the circulation spaces beneath. The roof deck is adjacent to the building, which is flat and has additional structures added to partially cover two courtyards in order to provide a surface for a set of small sports courts for basketball, netball and tennis.

1. Sectional perspective view

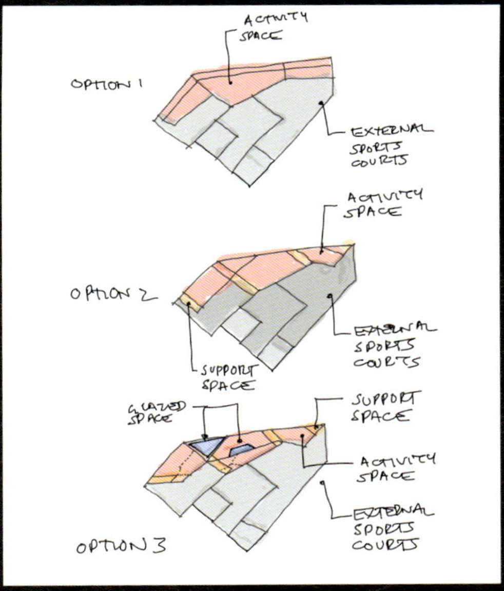

Spatial arrangement

Option 1 positions the service spaces along the elevation facing the street.

Option 2 positions the service spaces to create transition zones between volumes.

2. Sectional perspective view

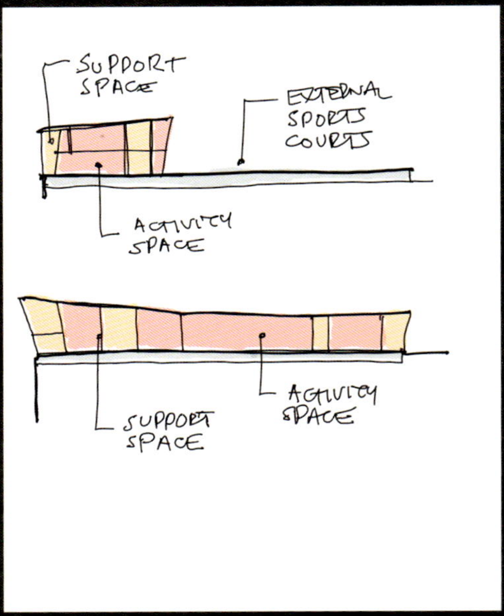

Option 3 allows the support spaces to become external forms and internal spaces in their own right.

Spatial arrangement

The spatial arrangement of Option 1 is to position the service spaces along the elevation facing the street, forming a 'deep wall' between the street and the new building. Internal spaces form a single volume that faces the rooftop rather than the street below; the inclined roof planes of the rooftop extension direct the focus of the space towards the new roof deck. Views out to neighbouring buildings, and from neighbouring buildings to the new rooftop extension, are framed in order that the new building contributes to the surrounding urban forms and external spaces. The single volume creates a set of sub-spaces through changes in the direction of the wall planes and changes in height of the roof planes. Option 2 takes the idea of the support spaces as transition spaces rather than as a separate zone for support activities; a step beyond the arrangement of Option 1. Option 2 positions the service spaces to create transition zones between volumes by taking the three spaces and creating a unique spatial experience within each space, which has its own form and its own framing of views to the outside roof deck and the urban environment beyond. The creation of three identifiable volumes allows different activities to take place simultaneously, providing more flexibility in functionality within the building than is provided by Option 1. This approach allows the spaces to work a little harder for their occupants. Support spaces form a set of intermediary volumes between activity spaces rather than forming a continuous deep outer wall as set out in Option 1, forming an acoustic barrier between each of the three activity spaces. Option 3 allows the support spaces to become external forms and internal spaces in their own right, rather than being interstitial elements between activity spaces. This option differentiates further between both the activity spaces and the support spaces by introducing a more complex building form into the site; subdividing the roof into smaller zones in order to create a strong spatial differentiation between each of the spaces which are now linked dynamically. The outside sports courts can now form a continuity with the internal spaces rather than being adjacent to them, which is the case with Option 2. The support spaces are adjusted in their position accordingly.

. Perspective view of interior

2. Perspective view of interior

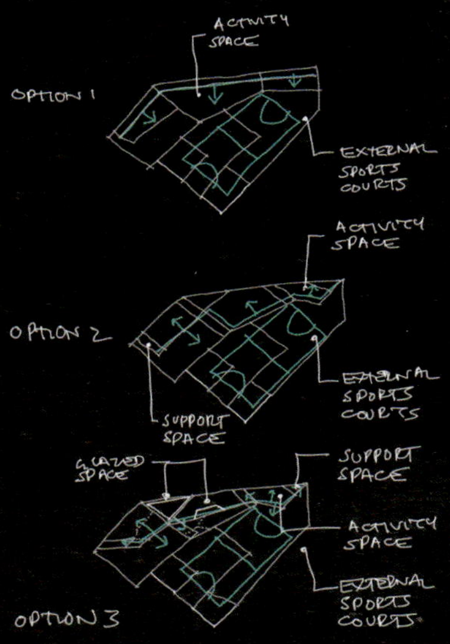

Internal circulation

Option 1 is based on a corridor space along the 'deep wall' of the service space wall on the street side.

Option 2 moves the circulation space route to the sports-court side of the building.

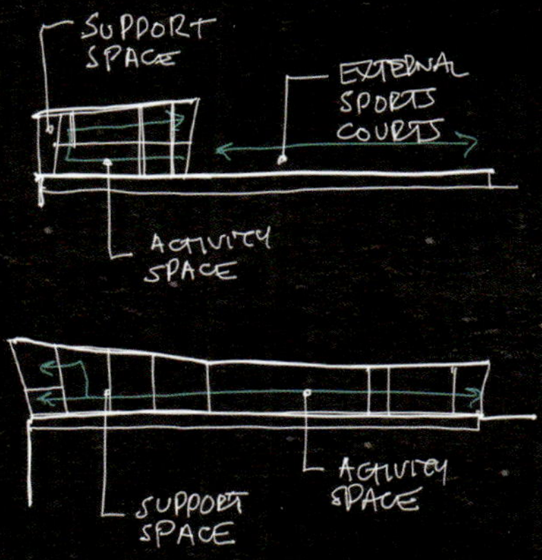

Option 3 moves the circulation to the centre of the building, with the two end spaces retaining the circulation route from Option 2.

Internal circulation

An essential part of the internal circulation in this project is that of the direct links between spaces in the building and the adjacent sports court, together with a need to be able to have spectators standing around the edge as well as seated on two rows of benches on the side of the court opposite the enclosures. This circulation is based on groups of people moving from space to space, and from the interior space to one of the sports courts. This project aims to reduce the distraction to players caused by movement around the courts of those not taking part. The internal circulation for Option 1 is based on a corridor space along the 'deep wall' of the service space wall on the street side, allowing the main space to be accessed by way of a lobby which forms an integral part of the service space. The single spatial volume can be subdivided with sliding or folding partitions, or may retain its open plan arrangement such that the three juxtaposed spaces are defined only by their form and their different ceiling heights, which can be accessed from different points along the deep wall corridor. Each space then has a link through to the service space behind which changing rooms, vertical circulation and storage are housed. Option 2 moves the circulation space route to the sports-court side of the building in response to a dispersed arrangement of service areas; now differentiated as spatial forms in their own right. The service spaces each accommodate a different function: changing rooms, vertical circulation and sports equipment. This circulation route along the roof deck side of the building provides more engagement with the sports facility outside at roof level than is the case with Option 1, allowing spectators to see the sports court activities as well as provide space for supporters during a match. Option 3 moves the circulation to the centre of the building, with the two end spaces retaining the circulation route from Option 2, along the wall facing the courts, such that there is a larger zone for spectators to gather in front of the main sports court while circulation takes place behind the viewers, making easier use of the internal spaces. The centre of the large space has roof glazing introduced to mark the presence of the circulation route.

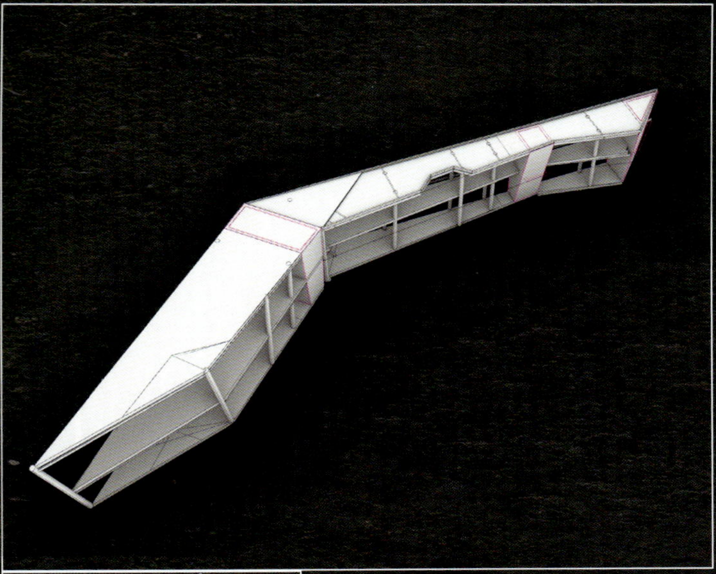

1A. Isometric view of structure

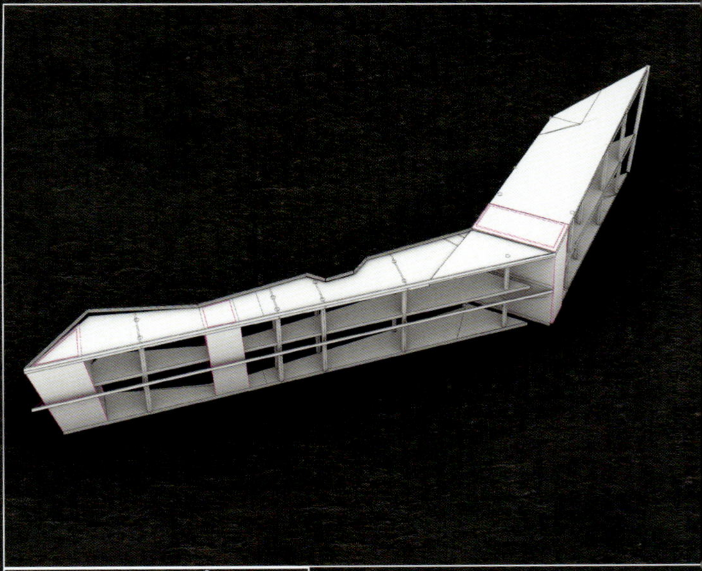

2A. Isometric view of structure

1B. Isometric view of structure

2B. Isometric view of structur

Structural strategy

In response to the building being supported on different existing buildings beneath, the structure of this rooftop sports centre is divided into three component pieces, with a movement joint between each part. This approach allows the structure of the new building to be compatible with movements of the existing structures beneath. The structure is a steel frame, with lightweight cladding panels forming the external envelope, making the building as light in weight as possible.

The structural strategy for Option 1 is to create a perimeter steel frame whose column positions coincide with points of support in the structure of the buildings below. These points of support are load-bearing walls and columns that can carry the additional loads imposed by the superimposed new structure. Column positions in the new structure are irregular but provide an opportunity for an irregular geometry of structural frame that responds to the geometry of the exterior forms and interior spaces of this rooftop extension. The structural frame of the extension includes both the enclosed part of the building and the roof deck; both take points of support from the buildings below, interacting with these buildings only at points. Consequently, the steel frame follows the perimeter of load-bearing walls and columns beneath. The distances between structural supports are kept as similar as possible in order to distribute loads as evenly as possible from the rooftop extension to the buildings below. The structural steel frame comprises primary and secondary members that form walls as vertical trusses and roof decks formed as composite metal floor plates supported on a steel frame.

Option 2 responds to the changes of spatial arrangement from Option 1, with corresponding changes in the orientation of some of the structural members in order to respond to the different arrangement of internal walls which form part of the load-bearing structure. Option 2 has structural members with spans that reflect the load-bearing capacity of the points of support beneath, while ensuring that both the enclosure of the rooftop extension and the roof deck are as economical as possible. Both Options 1 and 2 require a steel deck set externally that

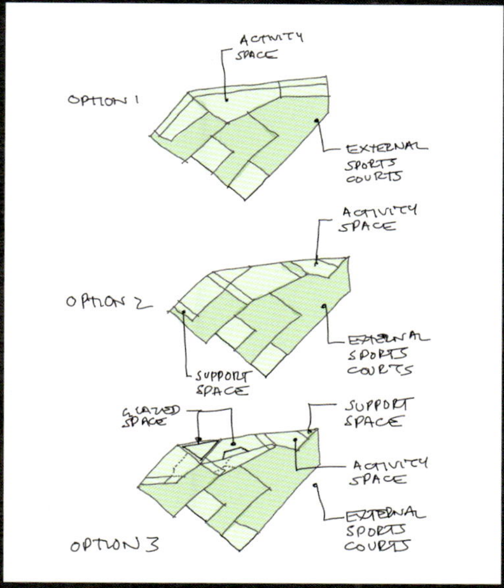

Structural strategy: Zones

Option 1 responds to the different disposition of internal spaces and external volumes.

Option 2 has a frame that provides structural stability to the concrete cores.

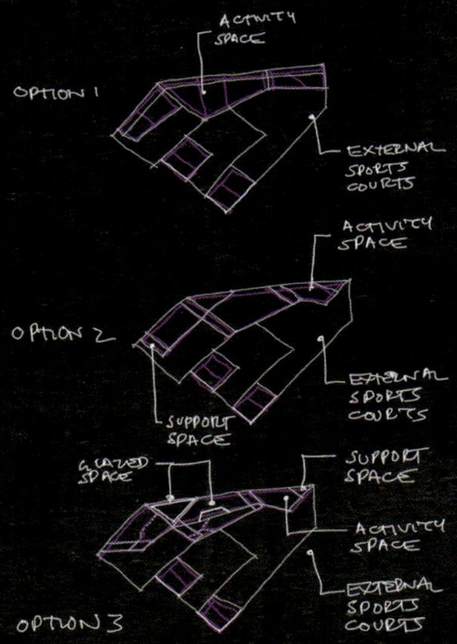

Structural strategy: Framing

Option 1 has points of support from existing load-bearing walls and columns.

Option 2 has structural members with spans that reflect the load-bearing capacity of the points of support beneath.

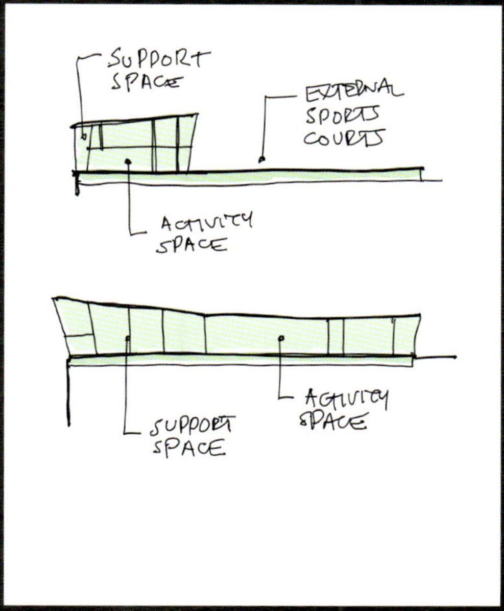

Structural strategy: Zones

Option 3 has framing members increased in frequency in order to avoid the need for secondary steelwork.

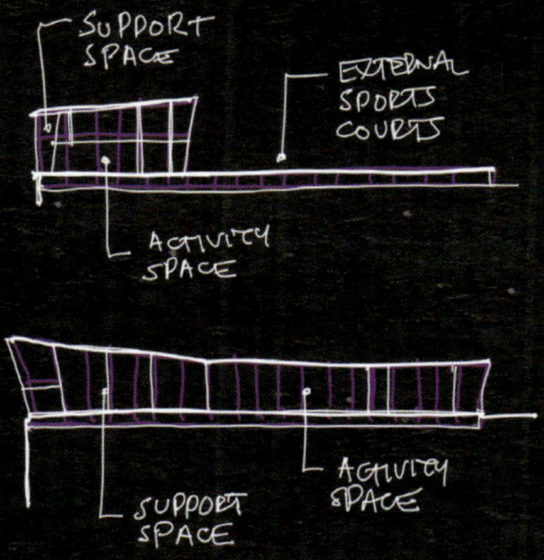

Structural strategy: Zones

Option 3 allows the spans to be optimised for prefabricated floor decks and prefabricated roof deck elements.

spans across parts of the courtyards in order to provide a playing surface for the sports courts. Option 3 takes Option 2 one step further, where framing members are increased in frequency in order to avoid the need for secondary steelwork, allowing the spans to be optimised for prefabricated floor decks and prefabricated roof deck elements to be used.

The advantage of the prefabricated approach, in addition to providing external roof decks, is speed of construction in an environment surrounded by buildings inhabited both as workplaces and as residential accommodation. As a result of the need to prefabricate as much of the building as possible, elements are lifted into place in similar-sized assemblies, for bolting together in place. These assemblies are delivered to site pre-painted and, where possible, pre-finished in order to reduce the local environmental effects of painting and coating. Bolting rather than welding is also preferred, for reasons of safety for those constructing the building as well as for neighbours in adjacent buildings. This approach of prefabrication and pre-assembly ensures that construction activities are focussed on fixing items together rather than the rooftop environment being used partly as a site-based workshop. A key aspect of the structural design is to create a series of assemblies which can be lifted into place with a minimum of disruption to the functionality of surrounding buildings owned by the same client, with a view to minimising construction activities in the construction of this building.

The most significant part of the structural design is the introduction of a rooftop deck which spans two existing rooftops, with a deck extending across a void between two buildings.

The roof deck is formed as a series of beams set above the roof level; the courts are supported on space frames beneath in order to provide a continuously rigid surface for playing. The court deck comprises a build-up of a rigid deck surface which is fixed onto secondary steel members beneath. The complete construction is weather protected and the decking is a fire-resistant set of interlocking panels with a sports court finish. An acoustic layer ensures that noise generated by the deck is absorbed in its construction.

1A. Isometric view of environmental volumes

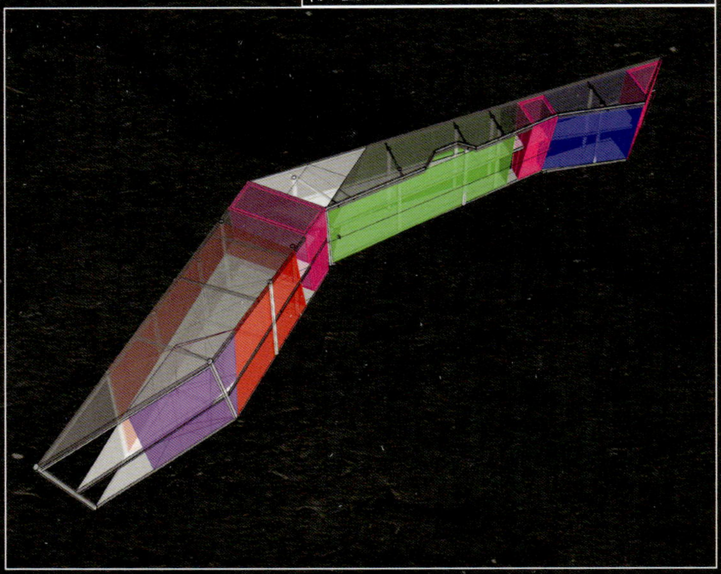

1B. Isometric view of environmental volumes

2A. Isometric view of environmental volumes

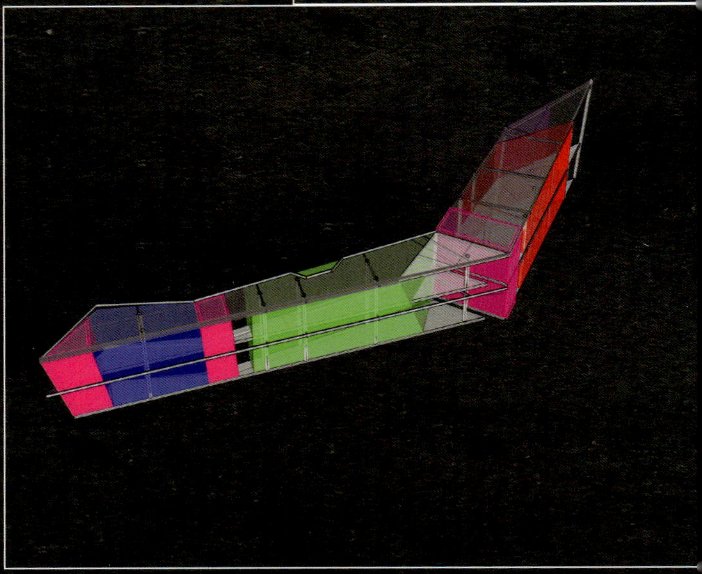

2B. Isometric view of environmental volum

Environmental strategy

The environmental approach to this project is of considering the enclosed part of the building that is positioned on the edge of the roof and the outdoor courts as part of a single building; a mixture of indoor rooms and the courts considered as being 'outdoor rooms'. During mid-season and summer months, the doors to internal spaces can be opened and wall panels can be lifted along joints formed as folds in the triangular-shaped panels in order to allow internal and external spaces to interact with one another. It is anticipated that small tournaments for netball, basketball, tennis, badminton and table tennis will take place. There is an associated need for organised events to be conducted with a certain ceremony; matches can take place in parallel, allowing tournaments to be held both inside and outside the building. The spaces also provide spaces for drawing people together for presentations and refreshments. As a result, the environmental zones are really in two parts: the interior parts of the building are closed in winter and are heated during those months; the mid-season and summer configuration has the outdoor areas made continuous with the indoor spaces. The complete extent of the sports centre becomes naturally ventilated at mid-season, with canopied space over courts being the benefit of the rooftop location. Spaces can be open to each other without the need for additional enclosures to separate the functions from adjacent street activity. The external canopies provide translucent light from the side screens as well as framing views across the university in other parts of the external screens; these views are set from vantage points within the enclosed part of the building.

The approach of providing geometric folded canopies allows them to be part of the dynamic urban setting; creating neither a visually 'heavy' appearance nor a visually light tent-like appearance. The folded planes of the canopies are similar to those of the new building and form a continuation of that language of construction, despite the canopy membranes being light in weight and not thermally insulated. The canopies are a covering against rain and ensure that balls stay within the confines of the courts. The introduction of a fixed roof deck for sports that

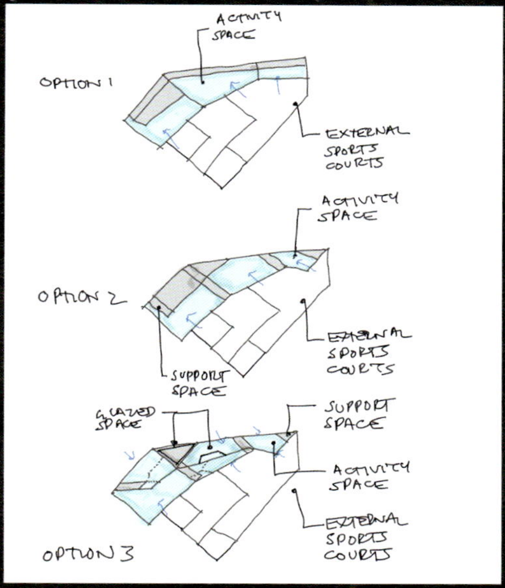

Daylight strategy for direct sunlight and cross ventilation.

Option 1 admits daylight and air from one side of the building.

Option 2 admits daylight and air from one side of the building; some spaces have subdued daylight.

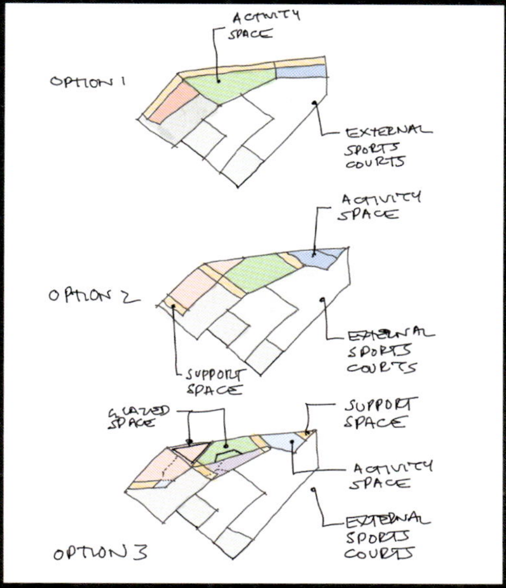

Environmental zones

Option 1 is based on considering each of the three spaces to be linked as a single volume.

Option 2 works with the modified spatial arrangement and provides similar environmental zones to Option 1.

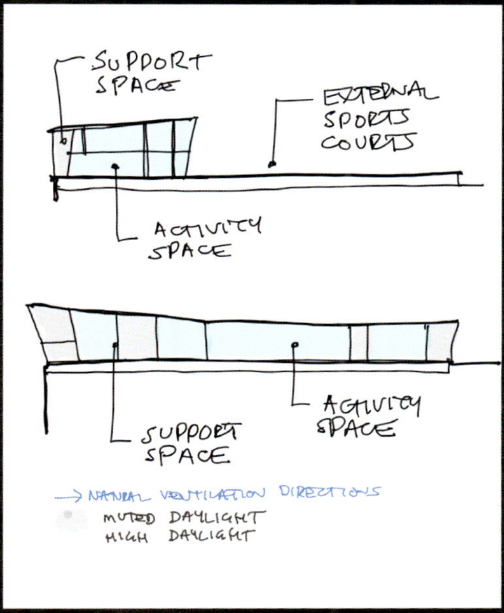

Daylight strategy for direct sunlight and cross ventilation.

Option 3 admits daylight and air from both sides of the building; circulation spaces have subdued daylight.

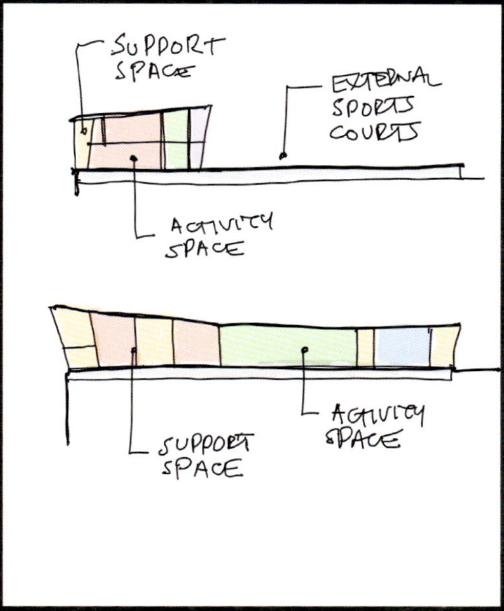

Environmental zones

Option 3 creates four main environmental zones, each of which has different requirements for daylight and natural ventilation.

spans across two existing roofs resulted in a loss of daylight to part of the space beneath, which serves as a ventilation area between buildings only. This residual space has no openings or windows that open onto it. The roof covering creates a canopy for this previously unused space which now forms an entrance space for one of the adjacent buildings within the university.

The three options look at how the enclosed part of the building can be arranged; the playing areas have a fixed arrangement of courts which can be marked out differently for different sports activities. It is envisaged that these courts will be reassigned on a regular basis in order to accommodate different activities. The outdoor space is suitable for group exercise activities including yoga and Pilates.

The environmental strategy for Option 1 is based on considering each of the three spaces to be linked as a single volume, but each space having an individual set of environmental controls; all within a single environmental zone. Each of the three spaces has a different mechanical ventilation system and different lighting arrangement to suit its expected pattern of use. All three spaces are naturally ventilated from one side, but have different levels of ventilation to suit the different uses of each space. The two smaller spaces are used as a changing room and facility for specific events; at other times as meeting rooms. These are flexible spaces; flexible in their function and in the way they are used within the context of a single function.

The second environmental zone is the service zone formed as a deep wall that forms an acoustic buffer between the main volume and the street below. These areas contain shower facilities, changing rooms and associated environmental control equipment, and a lift/elevator to take users up from the ground floor. Option 2 works with the modified spatial arrangement and provides similar environmental zones to Option 1; service areas have been moved into dividing spaces. Option 3 creates four main environmental zones, each of which has different requirements for daylight and natural ventilation. Like Option 2, the service spaces comprise separate environmental zones of washrooms, changing rooms and showers.

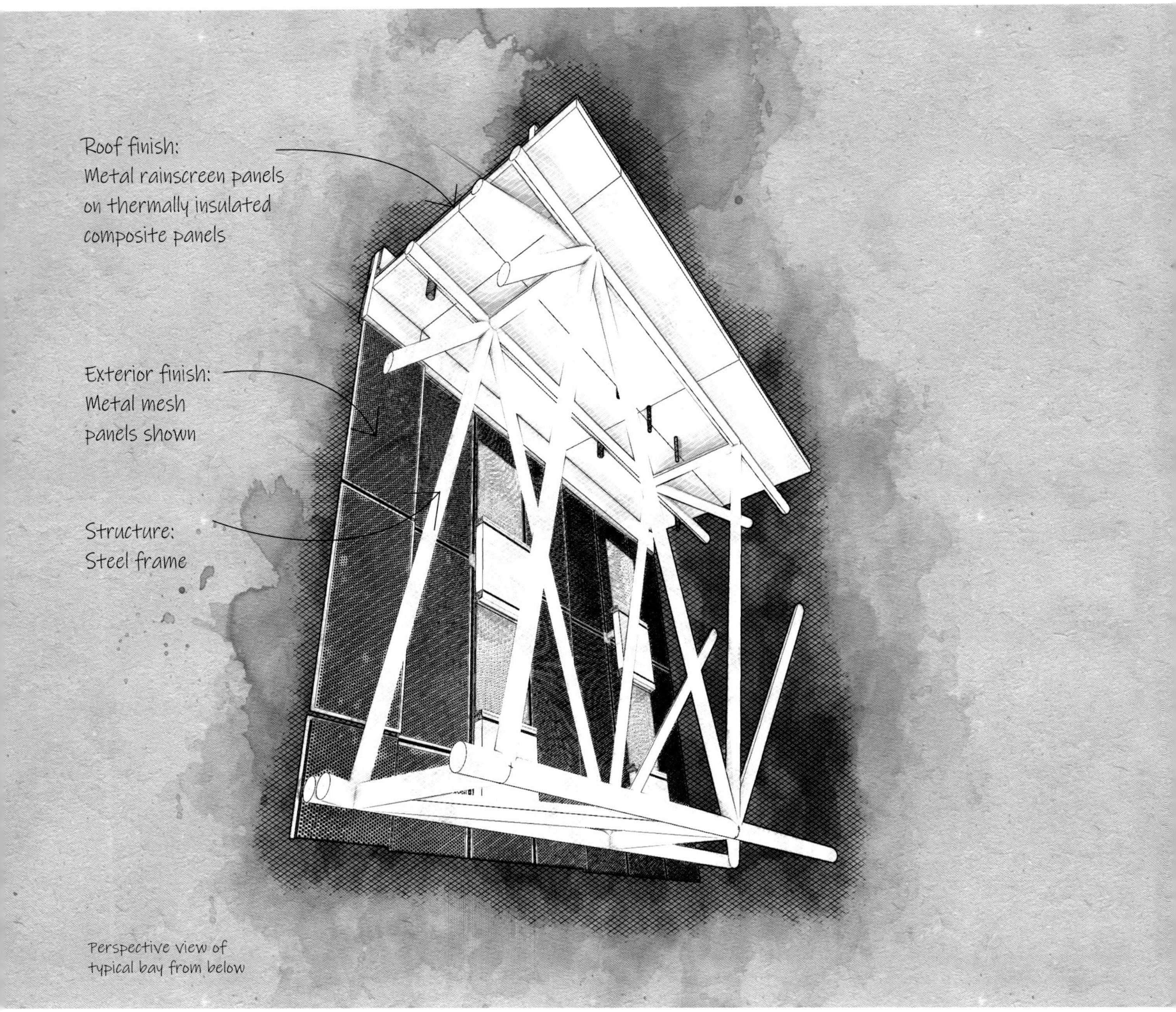

Perspective view of typical bay from below

Typical bay adjacent to facade

The enclosure is made from prefabricated panels to suit the geometry of the structure; there is a close fit between the supporting structure and the panels that are fixed to the structure. Panels are formed in both opaque metal and in solar-protective glass.

The design concept for the external envelope, which forms a key part of the typical bay, is to combine elements of structure and envelope which can be used both as enclosure in internal parts of the building and as translucent fencing in the sports court areas. An essential aim was to avoid a strong visual difference between the building on the roof and its its fence enclosing the sports courts in order to avoid it looking like a building that would be more comfortable at ground floor level. This is a rooftop building that is designed to fit comfortably both technically and visually into its environment, without appearing to be out of place or alternatively as something which would dominate the spaces that it rests upon. The envelope is designed to sit slightly above the existing buildings, while still forming a visual continuity with that building. The envelope and its typical bay do not appear to float above the existing buildings, but instead are resting on top of it, forming a continuation with the walls below. This is a building that engages with its built context, of which it is an extension. The materials used for the enclosed parts of the building, together with its fences, are of a cementitious composite material that picks up on the stone-faced elements of the building upon which it forms an extension. The envelope is articulated to follow the high level of differentiation of the forms of sports courts and the internal activity spaces and associated changing rooms and refreshment areas. Pavilions are set opposite the enclosed spaces such that they are at the scale of the smaller elements beneath and engage with the geometry of the existing building below, but with a bolder geometry of triangulation. The triangular theme is used in order to create structural rigidity within the depth of the of the enclosing walls and fences; the simple bay is actually more of a typical module within a building. Elements used for facade, roof and canopy are of a common language which can be used in different configurations. The configurations used for the roof

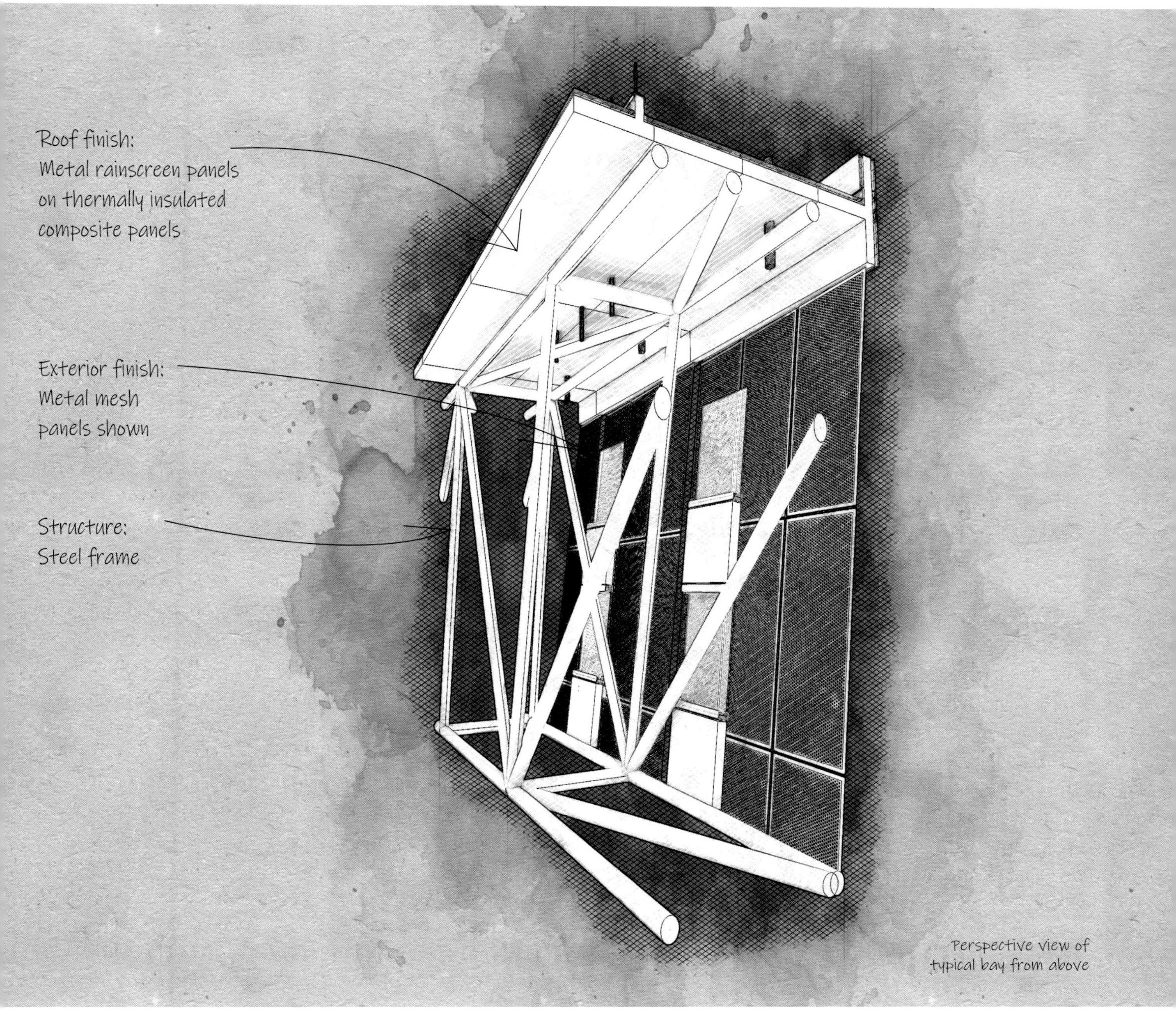

Perspective view of typical bay from above

are used on the pavilions located on the other side of the sports courts away from the building. Canopies are used in parts of the vertical enclosure, the fenced enclosure, as well as being used to enclose two of the enclosing walls. The typical bay which serves both external courts and the enclosed part of the building has electrical lighting set at ceiling level to illuminate the courts and sports halls. The same language of lighting is used internally and externally, informing the typical bay, which is that of a sports hall. The external envelope comprises cementitious composite-material panels which are of a matte finish in order not to pick up reflections during sports activities. The glazing is formed from double glazed units which are set into the spaces created by the window frames. The glazing is set independently of the steel frames, but their framing forms part of the primary structure; glazing is required to accept more movement than is the case with regular glazing frames. As a result, the glazing is set slightly forward of the steel frames to allow that movement to take place. A key feature of the external envelope is the sliding panel which can also be rotated on wheels to form a part of the arrangement of projecting screens. Rows of seating can be deployed for the convenience of spectators who are generally standing at the edges of the courts. The screens are deployed from the pavilions and from the side opposite the building.

A key aspect of the folding screens is to be able to deploy them in different configurations, to provide amenities for visitors and to create an event, or tournament, with its sense of occasion and procedures specific to each sport. This atmosphere is created by opening up the building and its adjacent pavilions, forming stalls, display boards and servery spaces; a building which can be deployed for a sense of ceremony. This approach will make the periodic replacement of panels necessary; the effect created by the folding panels will in part be achieved by using less durable assemblies than for a regular building. This mix of long-term durable building assemblies and shorter-term furniture-like assemblies is established by setting out the design life of each of the 'deployable' parts of the external envelope. This approach will bring more flexibility for the building occupier, beyond that of a single fixed envelope.

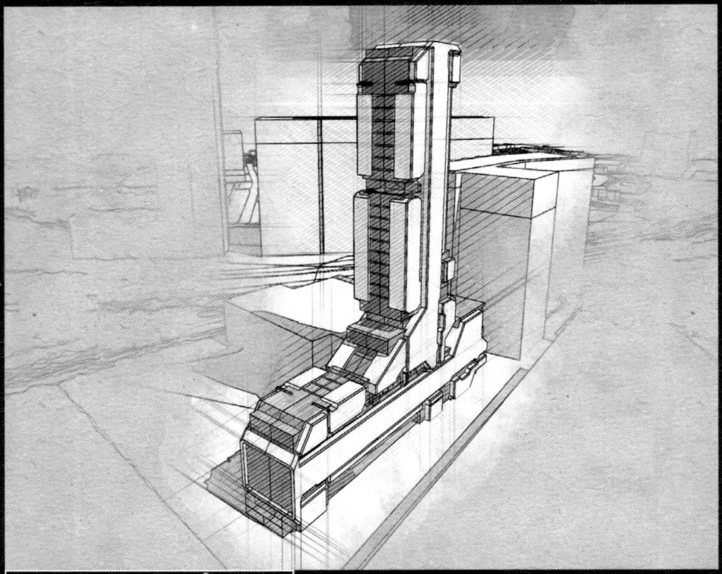

1. Perspective view of exterior

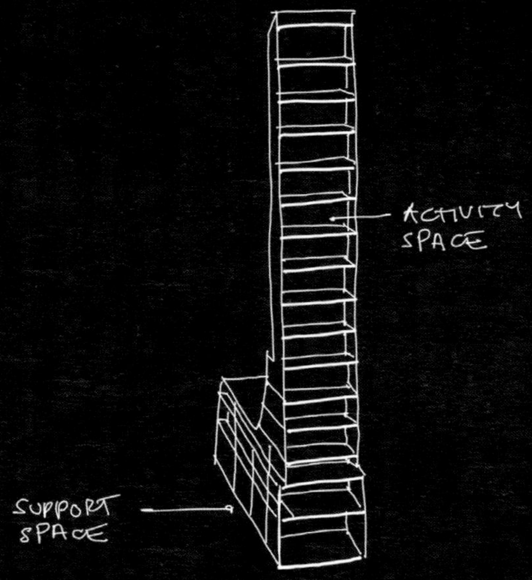

2. Perspective view of exterior

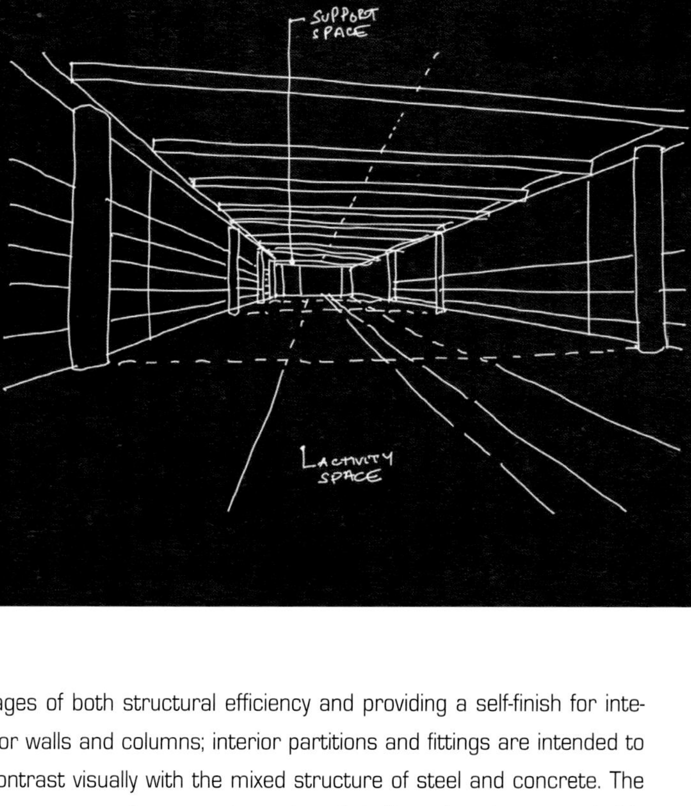

Use of single structure type across a range of scales and spans

The structural approach taken in this project was to seek to use a single structural system for the complete building; avoiding the need for additional secondary steelwork elements to support balconies and vertical runs of services that are fixed to the sides of the building. This approach is taken in order to avoid a doubling-up of structural components where a single assembly is sufficient. This approach maintains the 'loose fit' that is necessary between primary structure and service installations which allow services to be updated periodically, independently of structure, while bringing the benefits of the economy of their partial integration. This approach of a partially integrated structure allows the projecting floors at the upper levels, for example, to be integrated within the primary structure, contributing to the overall stability of the building, rather than being 'added on' to a structural design. A 'mixed' structural frame has been used; a mix of reinforced concrete and steel. Concrete is used for columns, slabs and shear walls; steel is used for service-related parts of the project. This mix has the advantages of both structural efficiency and providing a self-finish for interior walls and columns; interior partitions and fittings are intended to contrast visually with the mixed structure of steel and concrete. The arrangement of structural members is indicated on the external side walls with GFRC (glass fibre-reinforced concrete) cladding panels that allow the building to be thermally insulated and weather-protected while reflecting the exposed finishes of the reinforced concrete inner walls. This approach creates a close relationship between facade and structure while providing the required environmental performance of the external envelope. A single technology of connections between two materials of hot rolled steel and reinforced concrete was used in this project; fixing brackets are designed to be partially embedded in the concrete and partially bolted to adjacent steelwork.

This project is for a mini-tower; a typology that has appeared recently which provides the advantage of a taller building with light admitted on three sides of a modest-sized floor plate that is divided into two parts. The arrangement provides enclosure and a differentiation between

3. Perspective view of exterior

4. Perspective view of exterior

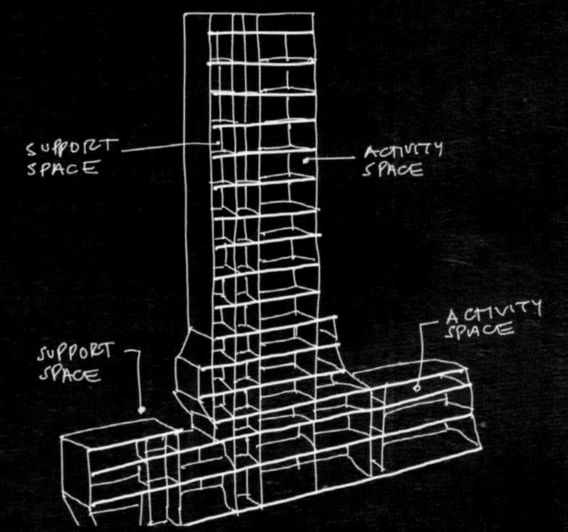

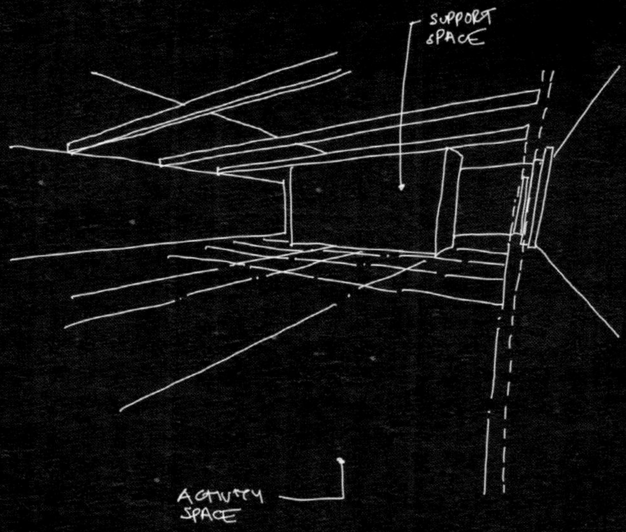

spaces at the lower level while remaining economical. As a result, the spaces which have more complex form such as workshops and assembly spaces are accommodated on a podium which is almost equal in length to the height of the mini-tower that sits above it. The tower has efficient floor plates which are relatively small and open, being divided into two equal halves. The floor plates in the tower are serviced by service modules, which are attached to the sides of the building and extend the full height of the building. This installation is fixed back to shear walls that form part of the primary structure. This approach allows the structure to be more varied than a regular braced frame in that the bracing provided by the frame is not continuous, but is part of a series of smaller-scale structural assemblies which are nested together to form the complete structure. This approach provides a high level of differentiation between spaces, answering the requirements of the brief/program with a form which is different from the more typical rectangular block-based building, while providing a spatial differentiation in the mini-tower that is generally not achievable in larger-scale towers, where most floors need to have a continuity of spatial arrangement and structure. This building as a combined mini-tower and podium building has the advantages of admitting daylight to larger numbers of spaces while occupying a small plot of land in the city. The building has a mixture of opaque facade areas which are generally service areas and glazed areas which enclose the 'activity' areas. The glazing is set deep into the external envelope profile in order to provide passive solar shading in this environment, which experiences high levels of solar gain and solar exposure for most times of the year. The structure meets service and program requirements by weaving elements of frame and bracing elements within a single form that remains economical. This structure-based solution provides the opportunity for parts of the building to be environmentally self-contained, with equipment for air-handling ducts and electrical supplies arranged in small groups throughout the height of the building. This installation is set externally, on the outside face of structural concrete walls, in order to avoid sound transmission from the equipment back into the building.

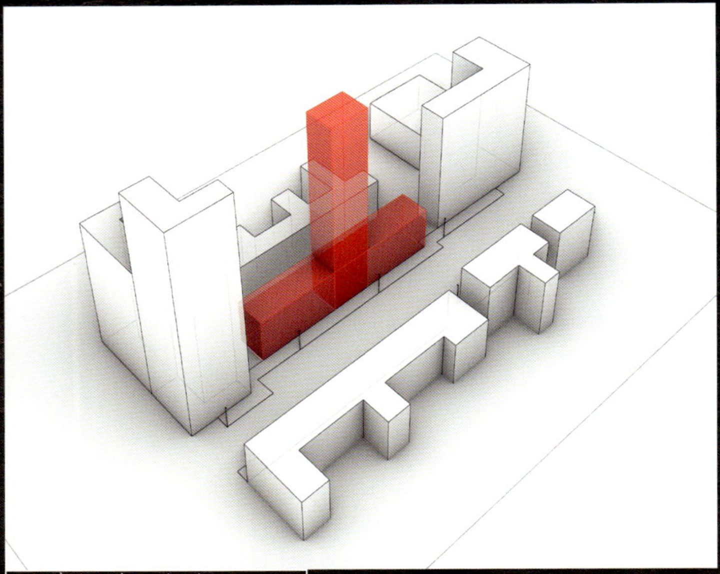

1. Project volume in site context

2. Project volume in site context

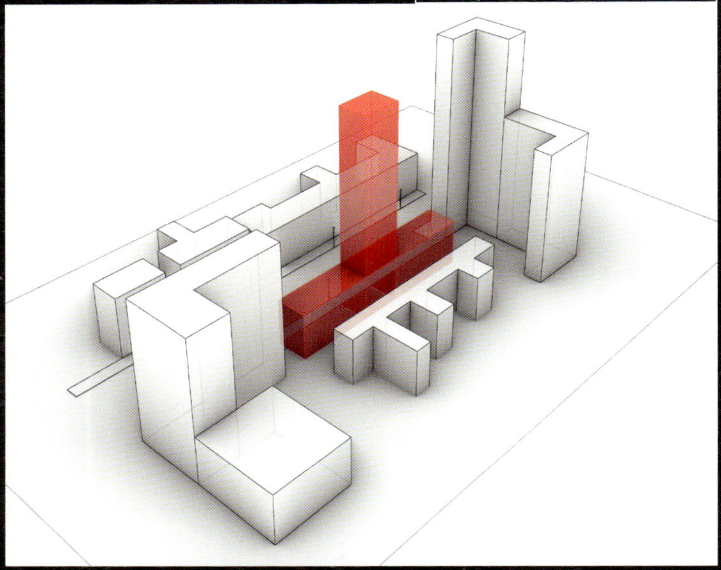

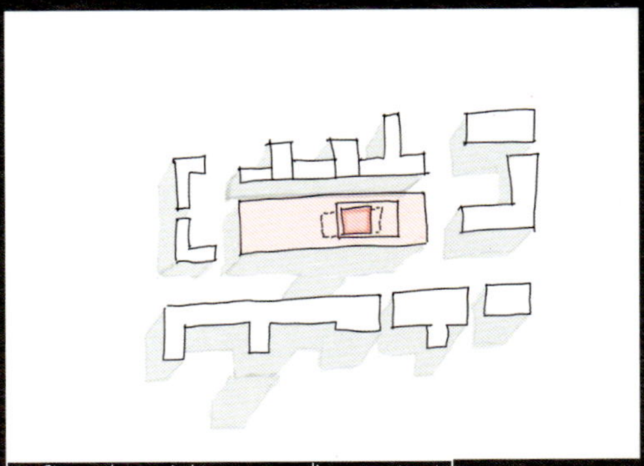

1. Site plan with surrounding context

Option 1
is for a tower which is relatively small in size, set onto a podium which is the same width as the tower.

2. Site plan with surrounding context

Option 2
has the same floor plate for the podium, but moves the position of the tower for an increased length and is shorter in height than in Option 1.

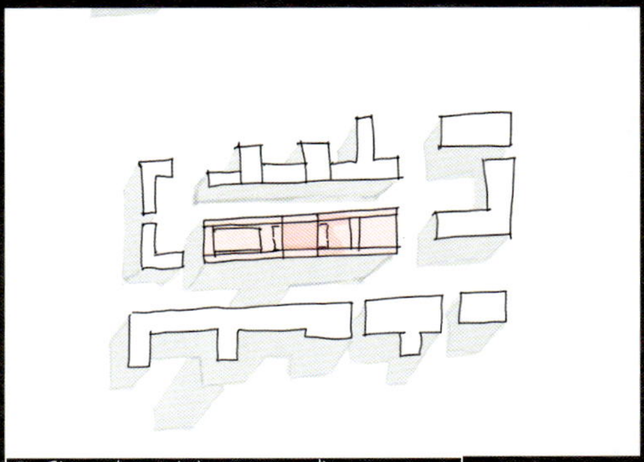

3. Site plan with surrounding context

Option 3
reduces the volume of the tower by reintroducing the outer service modules from Option 1.

Site context

This project is for a 15-storey tower set onto a four-storey ground-level podium. The tower provides a corporate headquarters for a research and development-driven (R+D) manufacturing company. The building provides space for design and R+D tasks, together with administrative office space. Office-based tasks are carried out in the tower; prototyping-based tasks of large-scale assemblies are carried out in assembly halls located within the ground-level podium. The tower comprises a set of externally fixed 'support' spaces or service support modules, including staircases, washrooms and lifts/elevators which are located adjacent to 'activity' space, or office floor plates. The support spaces are fixed to both the sides of the tower and to the side walls of the podium. The site context for Option 1 is for a tower which is relatively small in size, set onto a podium which is the same width as the tower. The overall form of the podium fills the footprint of the site, providing an urban block which is similar in size to neighbouring building forms. The office tower occupies a relatively small footprint, with its externally mounted service pods attached to it that reduce the size of the floor plate on each level to that required for activities; ancillary space is omitted as a result. Option 2 has the same floor plate for the podium, but moves the position of the tower such that it has an increased length and is shorter in height than in Option 1; incorporating a four-storey step down at the base of the tower down to the podium. Option 3 reduces the volume of the tower by reintroducing the outer service modules from Option 1; separating the fabrication facilities and service spaces into small groups of spaces, where each group has an identity that is in keeping with neighbouring blocks that surround the building.

The site context is one of an open environment in a city developed in the 1970s in this dry desert climate. The organisation of the building is driven by a need for daylight in office spaces and mainly opaque spaces for research and manufacturing. The tower is divided into two blocks, one set above the other, arranged with the aim of reducing the scale of the constituent parts of the tower in order to match the scale of the urban surroundings. This approach is strengthened by the structural walls

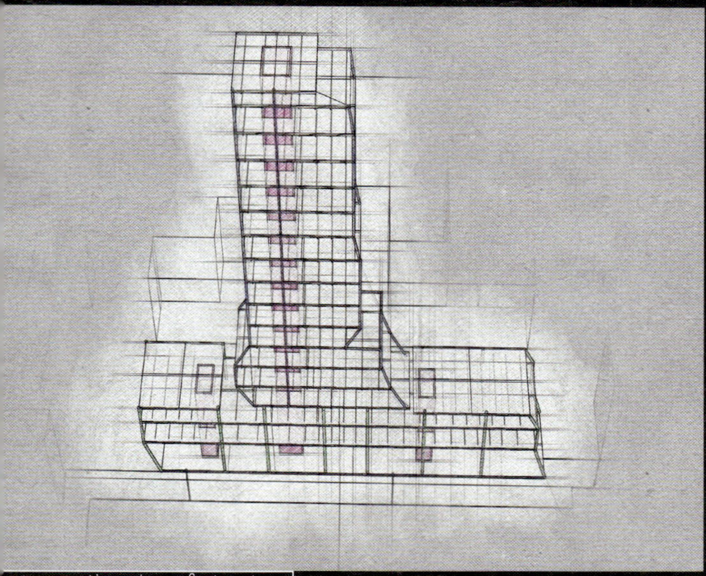

. Perspective view of structure

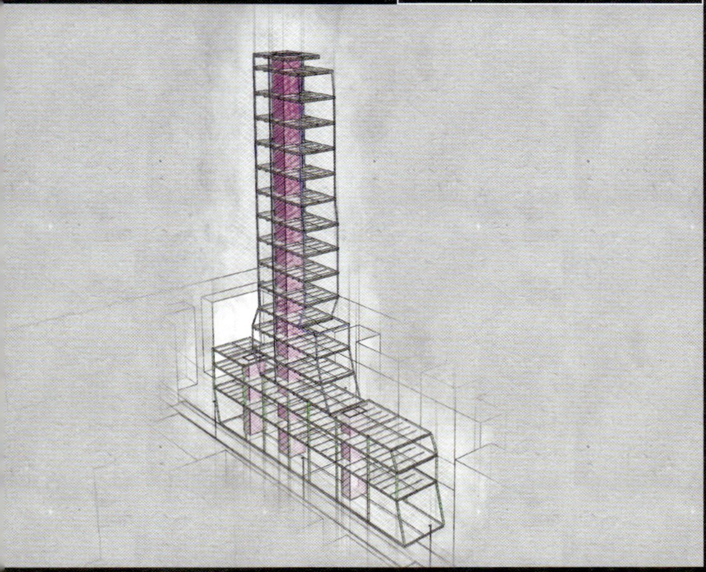

2. Perspective view of structure

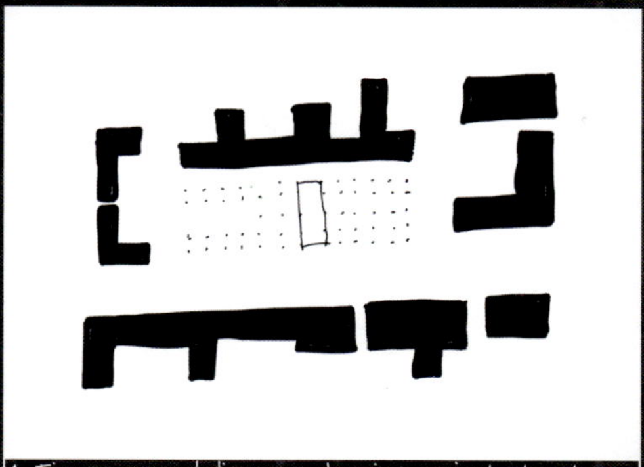

1. Figure-ground diagram showing project structure

Option 1
The column grid aligns itself with adjacent buildings. The tower is conceived as a fully load-bearing wall structure.

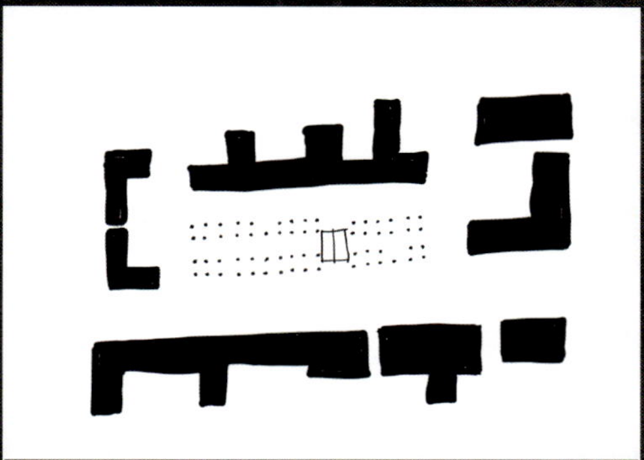

2. Figure-ground diagram showing project structure

Option 2
The column grid aligns itself with adjacent buildings. The tower is conceived as a load-bearing core structure with columns supporting floor plates.

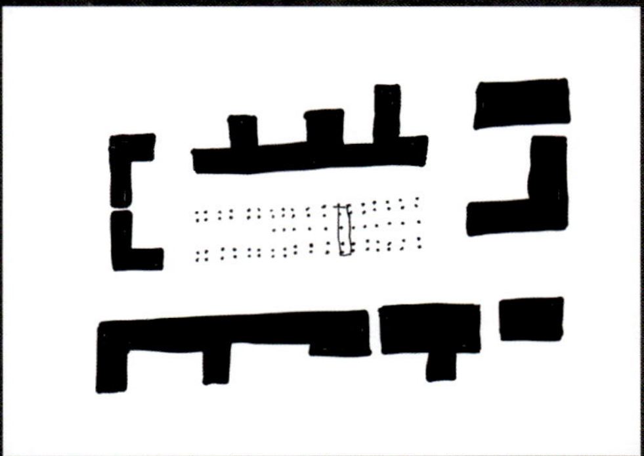

3. Figure-ground diagram showing project structure

Option 3
The tower is conceived as a load-bearing core structure, divided into superimposed structures, with columns supporting floor plates.

which support the building visually. The structural design of this building is arranged around the orientation of the constituent elements of the structure such that north-facing glazing is more generous than that of the east, west and south facades. The roofs of the podium are opaque and have glazed slots which are set beneath overhanging elements of structure and service installation, together with other protruding elements such as breakout spaces for the conference hall.

The edges of the conference hall roof provide an overhang for rooflights set one floor below.

This shading is provided passively by the forms of the buildings above; the roofs also form canopies which project out to enclose external terrace space and semi-external winter gardens. These enclosed glazed spaces set within conference spaces are oriented towards the sun of the early morning and late afternoon, and are semi-enclosed such that they trap winter sunshine to provide a buffer zone of warmth in the building, whilst in the summer they are enclosed. Winter gardens are enclosed to exclude the effects of solar gain and the high external temperatures of summer. The podium also has structure which supports air-handling ducts fixed along the length of the building and which extend across the roof of the podium, secured to the beams that span across the width of the building. Chillers and air-handling units are fixed to both the roof of the tower and to the roof of the podium in order to integrate the structure with enclosure of the equipment. The chamfered edge to the roof enclosure accommodates ducts which are set out in a circuit around the top of the tower, and which are distributed downwards in front of the continuous vertical shear walls of the tower, but behind the cladding.

This project creates a link between the form of the tower and the form of the podium; each element can be considered to be an extension of the other. The tower does not consider the podium to be a visually wide and tall volume whose role is to provide a base onto which the tower is placed. Instead, the two work together; the tower appears to have been unfolded from the podium; a part of the podium which has been disengaged and rotated upwards to create light and airy spaces.

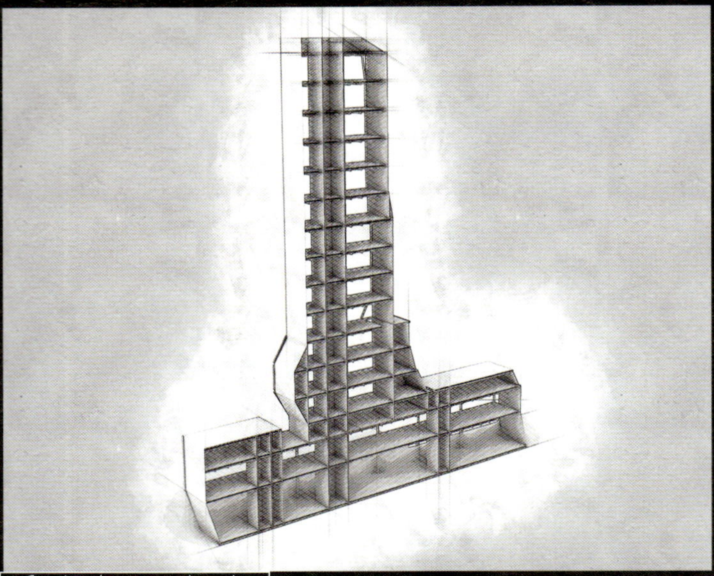

1. Sectional perspective view

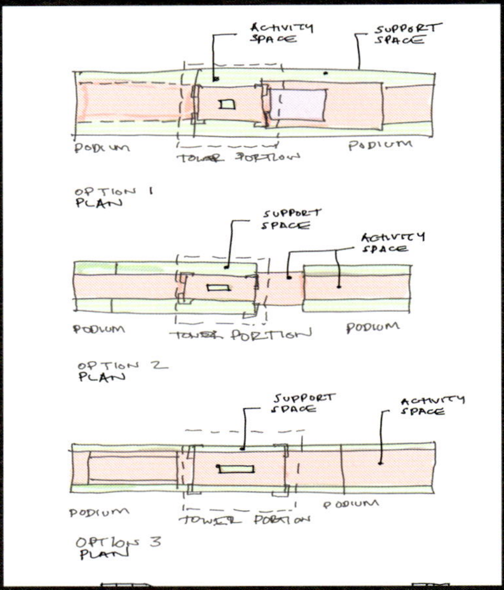

Spatial arrangement

Option 1 has a ground floor lobby that extends through the full height of the podium.

Option 2 has prototyping halls which are moved to the opposite end of the building.

2. Sectional perspective view

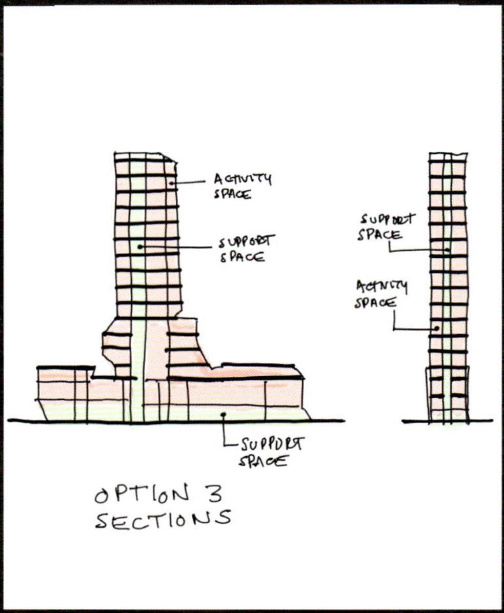

Option 3 positions the tower to be accessed externally on both sides of the building.

Spatial arrangement

The podium accommodates the company's R+D prototyping halls for the manufacture and assembly of new products. Space required for access to the tower, within the podium, comprises an entrance lobby, reception spaces and lifts/elevators. The assembly hall facilities comprise a large-scale workshop space, storage, a product presentation area and associated service space. Car parking is provided over four floors below ground; spaces are primarily for electrically powered vehicles and their recharging. This set of different functions in the podium is accommodated as a set of linked blocks which are defined by their external form and their internal space. The workshop space has a series of large rooflights that can be opened for natural ventilation, as the workshop has a secondary function as a temporary exhibition space for new products, with controlled natural light and natural ventilation provided during winter and mid-season months; roof glazing is closed off during summer months. The four-storey space at the base of the tower is the reception space for both tower and podium. A second reception space is located near the top of the tower. The spatial arrangement of Option 1 has a ground floor lobby that extends through the full height of the podium, creating a three-storey void up to roof level; the lobby is accessed from the sides of the building. The lobby and tower core are surrounded by meeting spaces, breakout spaces and associated service facilities at ground floor and first floor levels. The lobby is separated from the prototyping halls behind it; the halls are accessed at the end of the building, on its short side. The production halls have a separate entrance; there is a link to the tower, but the prototyping halls function independently of the tower. In Option 2 the prototyping halls are moved to the opposite end of the building, such that the office tower can be positioned away from manufacturing activities in the podium. The office tower is accessed from a central lobby, which is open from one side to the other, creating two blocks in the podium with the lobby between them. The overall volume of the podium is sub-divided into smaller 'linked' blocks which fit in scale with the adjacent cityscape. The tower lobbies are accessed from the long sides of

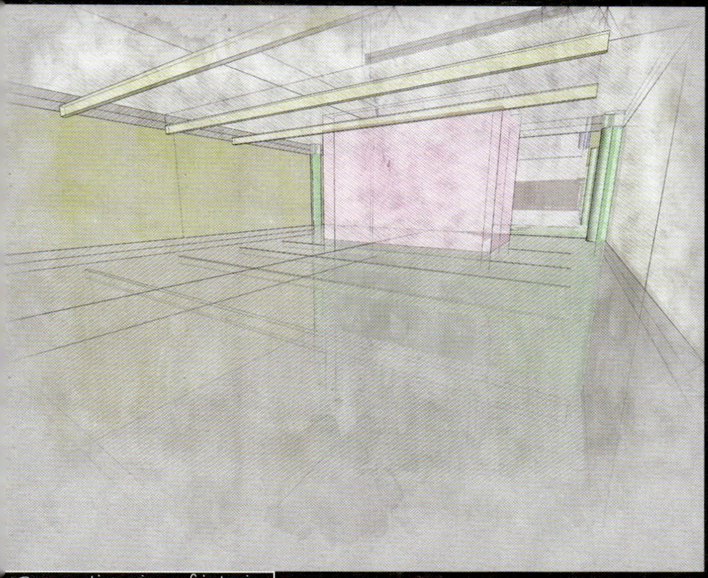

. Perspective view of interior

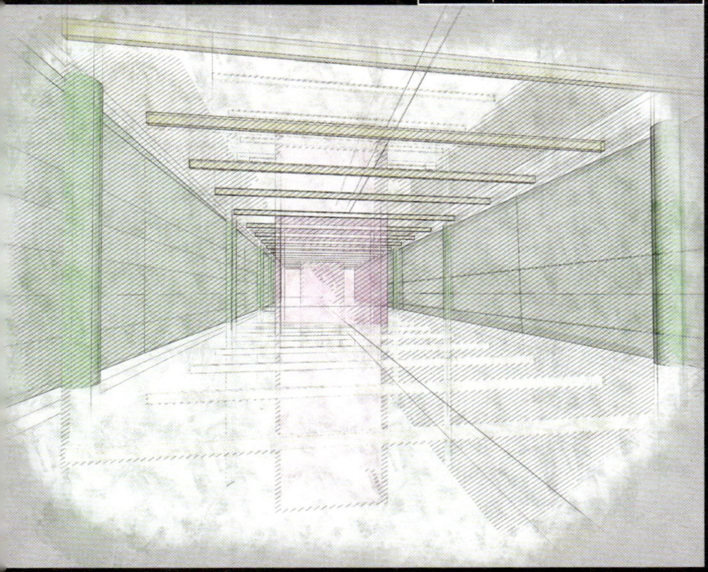

2. Perspective view of interior

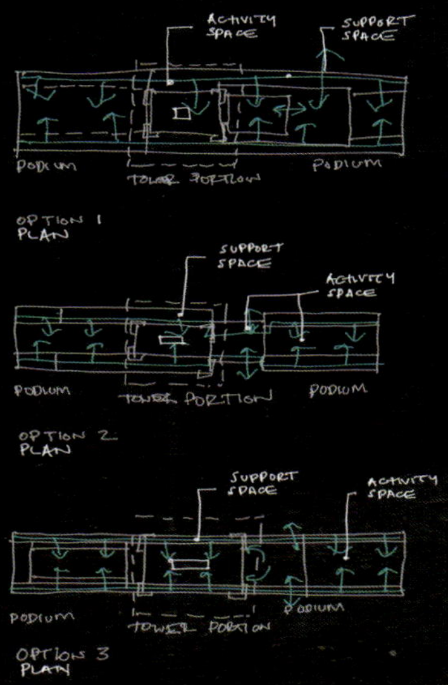

Internal circulation

Option 1 is of galleries extending the full length of the podium, providing access to the functions within.

Option 2 creates a stepped route of galleries, traversing one end of the building to the other.

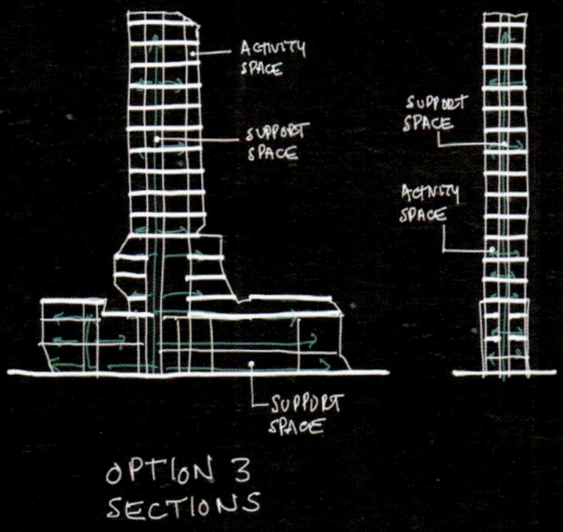

Option 3 provides a straight gallery that extends from one end of the podium to the other, from which access is made from the assembly halls to the office tower.

the podium at each end, allowing the central lobby to be used for gaining access to meeting rooms that form part of the office tower. This arrangement allows the podium to function independently of the office tower. Option 3 integrates the spatial components of Options 1 and 2 by accommodating meeting spaces and service spaces in the entrance lobby to the office tower. The tower can also be accessed externally on both sides of the building. Set adjacent to the office tower are the prototyping halls; all volumetric components of the project are linked by access corridors extending the full length of the building on both sides, providing a link directly between the tower and the prototyping halls. Option 3 integrates tower and podium while allowing for vertical stacking of spaces inside the podium, and providing a sequence of links between all spaces within the podium.

Internal circulation

The internal circulation of Option 1 is of galleries extending the full length of the podium, providing access to the functions within. The main circulation route is by way of the entrance lobby, with routes set along the edges of the building in the zone between the facade and the inner space. Option 2 has a similar concept but creates a stepped route of galleries, traversing one end of the building to the other. The central lobby is inset, creating a narrow link between the lobby and the long galleries on either side of elevations along the remainder of the podium. Option 3 has the most compact arrangement of all the circulation options; a straight gallery extends from one end of the podium to the other, from which access is made from the assembly halls to the office tower, and on to the breakout spaces. In this option the central lobby is the main point of access into the building, but additional entrances along the length of the podium can be made to allow visitors to enter the manufacturing hall display area directly, or the office tower, or the commercial spaces adjacent to the service spaces.

The prototype manufacturing spaces are arranged as a set of discrete volumes with the circulation space expressed externally as recessed spaces set between these volumes.

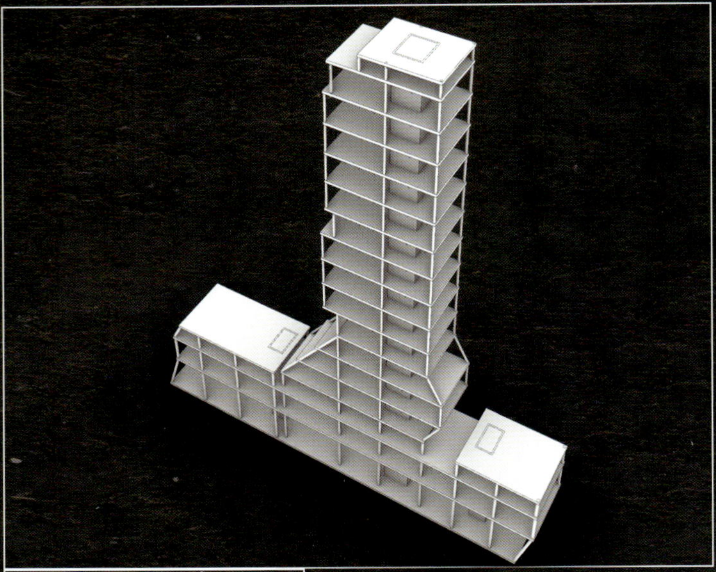

1A. Isometric view of structure

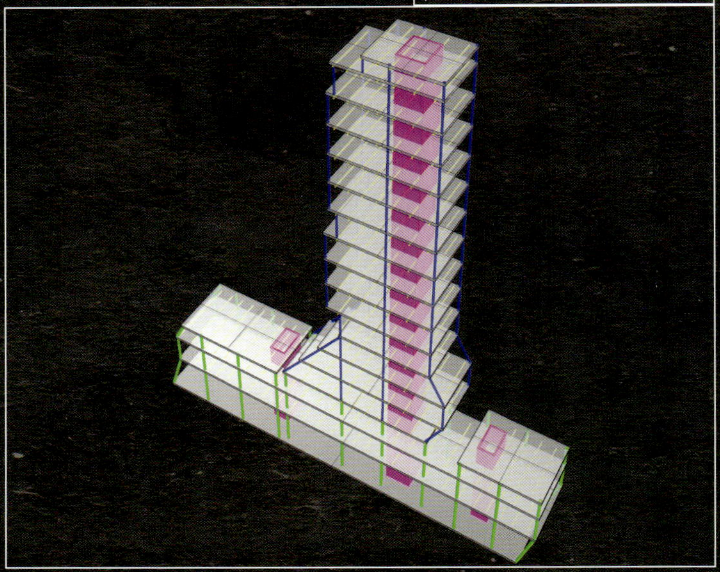

1B. Isometric view of structure

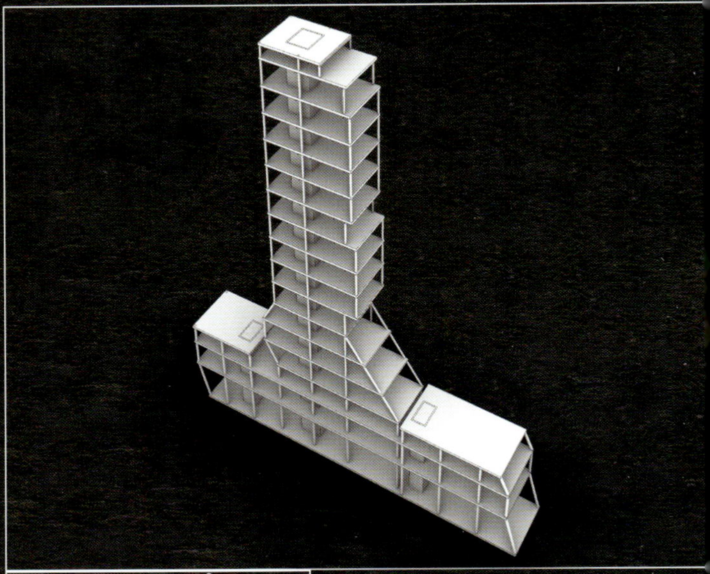

2A. Isometric view of structure

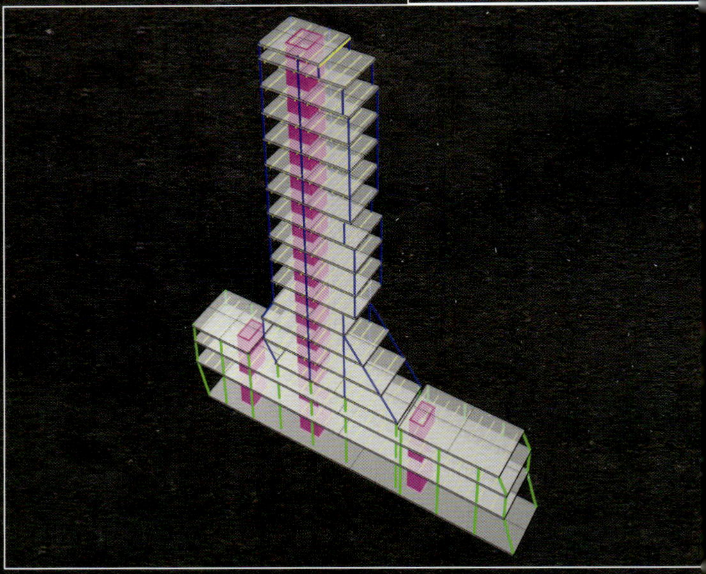

2B. Isometric view of structur

Structural strategy

The side walls of the tower are designed as shear walls; these reinforced concrete walls are clad in GFRC (glass-fibre reinforced concrete). The cores and shear walls form a wrap around the complete building with glazed slots, releasing the complete floor plate as a column-free 'activity' space. The structural strategy is one of a reinforced concrete tower with post-tensioned slabs which extend the full height of the building, spanning between the shear walls and the 'backpack' service pods set on each side of the tower. The podium is formed as a continuation of the reinforced concrete structure, with a movement joint between the perimeter of the tower and the surrounding podium; this creates the characteristic visual forms of the podium attached to the tower. The podium is also a reinforced concrete structure with post-tensioned slabs and concrete columns.

The structural strategy for Option 1 is based on the podium having a structure in reinforced concrete with the tower rising through it; a mixture of reinforced concrete and steel frames. The prototyping halls have steel trusses that span the full width of the central space of the prototyping halls; galleries set on both sides of the halls present the opportunity of a double row of columns onto which the steel trusses are supported. The column grid of the office tower extends down into the podium; it forms a separate structure to the podium, which has columns around the edge and columns in the centre. The lobby for the hotel is an open space that rises to the level of the roof glazing. The structural grid of columns and beams continues through the meeting rooms and breakout spaces to the end of the building. Option 2 considers the three volumes as three separate structures; each with its own column grid that suits the smaller spaces which are enclosed. At the other end of the building, the service spaces have a regular grid of columns; the prototyping halls have long-spanning trusses at roof level that form the wide, column-free spaces beneath. Option 3 combines Options 1 and 2 with a similar structure, but here the galleries on both long facades create a double row of columns that provide a clear route down from one end of the building to the other, and internal

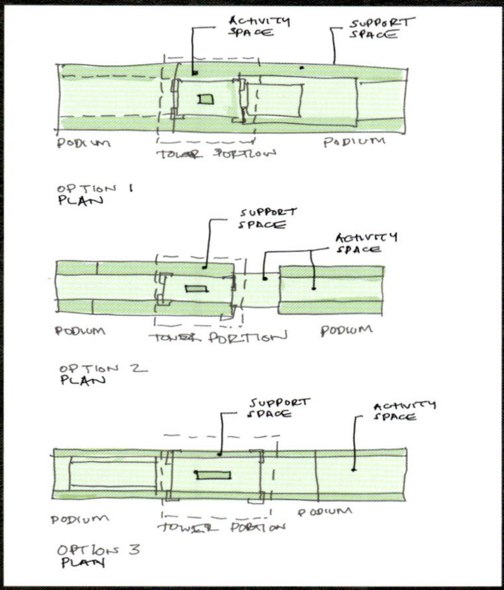

Structural strategy: Zones
Option 1 has a mixture of reinforced concrete and steel frames.
Option 2 considers the three volumes as three separate structures.

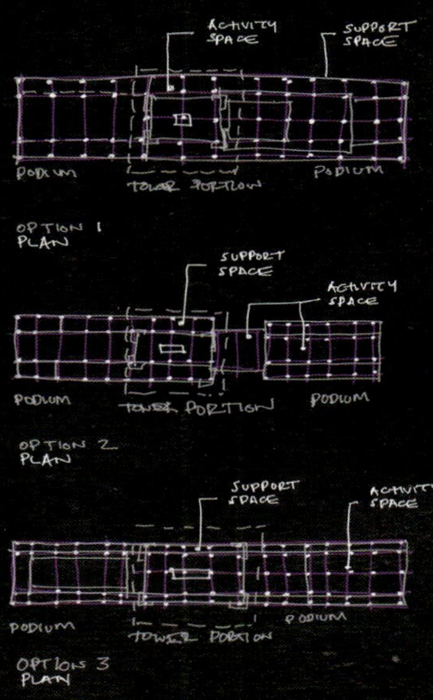

Structural strategy: Framing
Option 1 has steel trusses in the prototyping halls that span the full width of the central space.
Option 2 has a regular grid of columns in the service spaces.

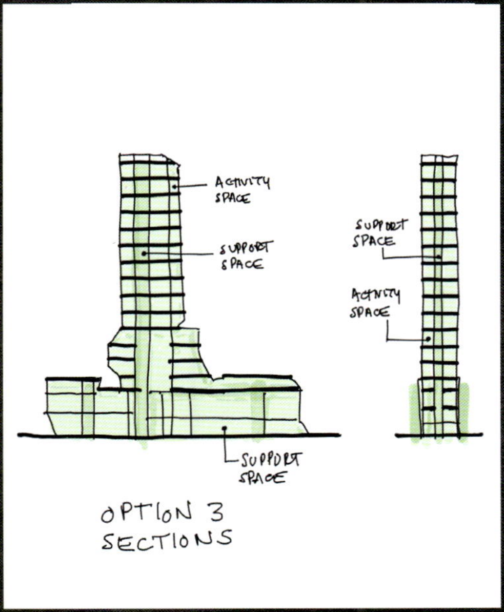

Structural strategy: Zones
Option 3 combines Options 1 and 2 with a similar structure.

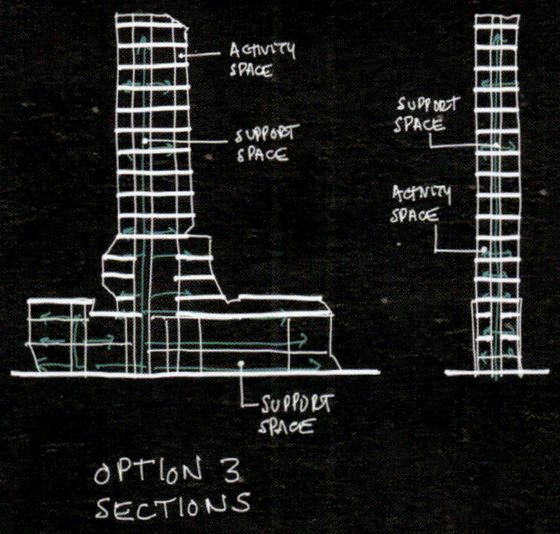

Structural strategy: Zones
Option 3 positions galleries on both long facades creating a double row of columns.

spaces have columns to suit their specific functions. Structurally, the podium does not extend beyond the sides of the tower. The tower and podium form a single continuously aligned form. As a result, the structure is aligned between tower and podium with the same structure already continuing up the tower from the podium beneath. Although the structure uses regular columns and floor slabs, the expression of the structure as seen both from outside and within the building is one of presenting the opaque elements as a visual expression of primary structure. The external structural walls in reinforced concrete provide a substrate for services that run vertically through the building. These walls extend vertically as large-scale elements, expressed as individual volumes; as two visual elements extending up the height of the tower. These forms are in addition to the three rooftop elements of the podium building beneath. The visual expression of the tower and the podium is one of showing where daylight is admitted and where it is excluded. This differentiation of form gives a sense of a visually deep structural zone that provides opportunities for passive protection from solar gain and glare resulting from direct sunlight. The building contains 'structures within structures' that are supported independently of one another. This approach avoids the need for transfer structures within a single structural form. As a result, the podium is designed as a series of nested independent structures to create the characteristic form of the building. The tower has a continuous structural frame which is consistent from floor to floor, constructed in three parts: a central structure of side-facing shear walls and two glazed volumes set on either side of this central form. Two large pod-like elements which appear to be attached to the central core are standalone elements, supported with continuous columns. The cladding creates a gap mid-way up this part of the tower to spatially differentiate the two halves of the facade on each side. The three parts of the tower constitute a set of visual 'bookends' flanking parts of the tower with a central opaque structure set between them. Although these three structures are set alongside one another, they form a single structure which is visually differentiated to visually express the organisation of space within.

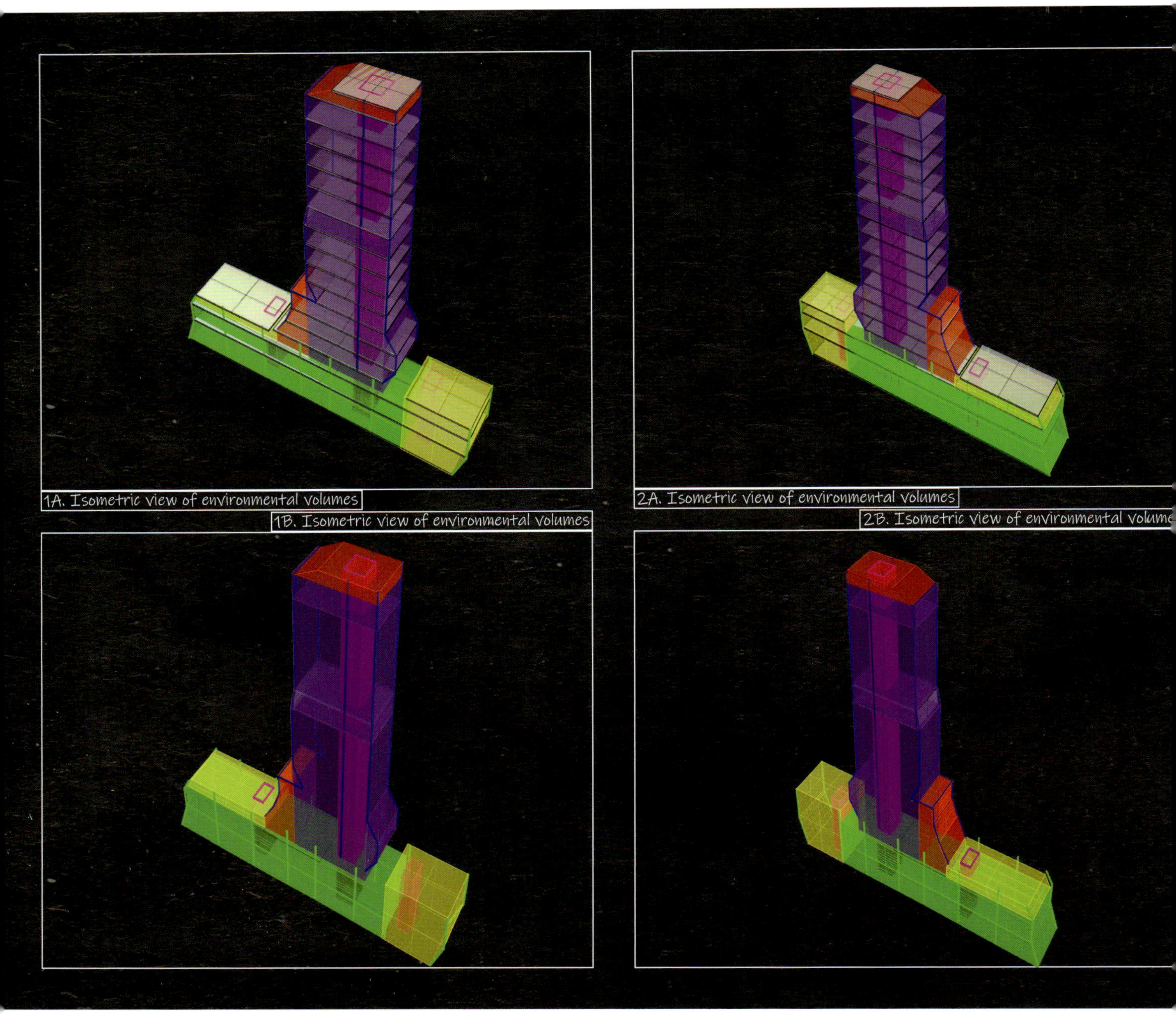

1A. Isometric view of environmental volumes

1B. Isometric view of environmental volumes

2A. Isometric view of environmental volumes

2B. Isometric view of environmental volume

Environmental strategy

The building is divided into two environmental zones: the podium and the tower. The podium has naturally ventilated, naturally lit assembly halls and product display areas with both passive and active environmental controls. The cores support mechanical and electrical services provided to the main floor plates within the tower. Externally mounted mechanical and electrical equipment 'backpack' modules of machines, ducts and pipe runs that supply all floors of the tower are fixed to the external walls of the tower. This approach allows the provision of natural daylight to both cores and the activity spaces of the main floorplates. The two reception spaces set above the podium at the base of the tower, and at the middle of tower, have a mixture of mechanical ventilation and natural ventilation through a double-skin facade, where the natural ventilation is introduced as night-time cooling during the cooler months of the year. The tower has two environmental zones, each with its own mechanical and electrical equipment, and is entirely mechanically ventilated. The lower zone and the upper zone both benefit from the slots in the facade that provide night-time cooling during the cooler times of the year, but these spaces are primarily mechanically ventilated. Chillers are mounted at the top of the building in the volume formed between the shear walls that extend through the full height of the tower. This roof also accommodates telecommunications equipment; all externally mounted, for which no environmental controls are required.

Environmental zones formed by the assembly halls share their fresh air supply; this air is moved around between the halls to suit occupancy levels. Part of the environmental strategy is to introduce limited amounts of daylight into areas where people congregate, such as entrances, receptions and the areas around the assembly halls. This reduces energy consumption as a result of a reduction in the surface area of facade and rooflight glazing, while avoiding the need for electrical lighting during hours of daylight. Daylight is admitted in the upper and lower podium, and with it, solar gain is introduced in small amounts; absorbed by the surfaces onto which it is incident, mainly the floors. Solar heat

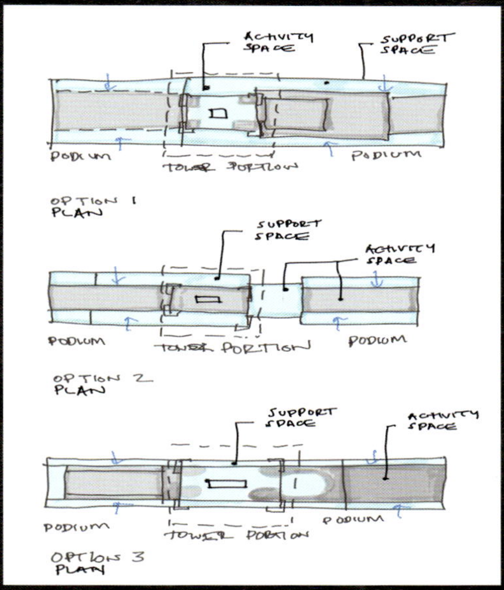

Daylight strategy for direct sunlight and cross ventilation.

Option 1 admits daylight and air into two sides of the podium and two sides of the tower.

Option 2 admits daylight and air into two sides of the podium and two sides of the tower.

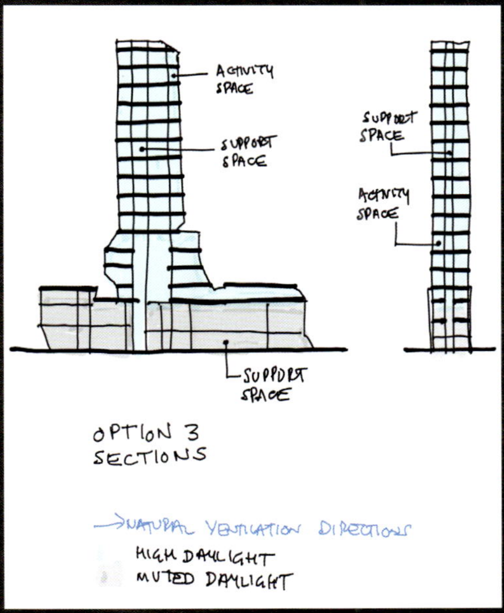

Daylight strategy for direct sunlight and cross ventilation.

Option 3 admits daylight and air into two sides of the podium and all sides of the tower.

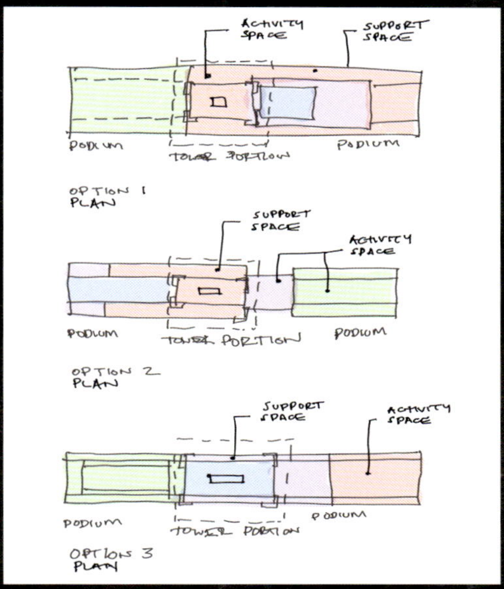

Environmental zones

Option 1 has a separate environmental system for each of the functions of assembly hall, office tower lobby and breakout spaces.

Option 2 has a more integrated set of environmental zones with assembly halls forming a single zone.

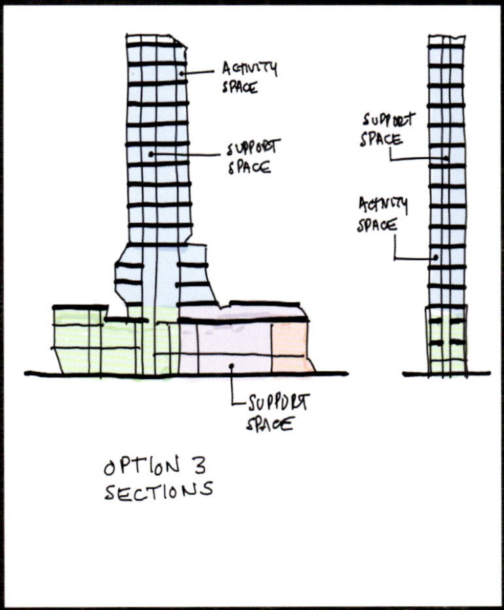

Environmental zones

Option 3 provides a separation between the environmental zones for the podium and the office tower.

gain is absorbed during the day, then removed by night-time cooling via fresh air from outside. Voids in the floors and ceilings of the podium and the office tower are used to distribute air, electrical supply and chilled water from the 'backpack' service areas. Primary electrical services are provided on the ceiling for air in ducts and electrical lighting. In some areas, electrical lighting is provided through recessed uplighters in the floor panels to provide a light which is reflected off the ceiling. The environmental strategy of Option 1 is to have a separate environmental system for each of the functions of assembly hall, office tower lobby and breakout spaces.

Option 2 has a more integrated set of environmental zones with assembly halls forming a single zone. The entrance lobby forms a second environmental zone, which wraps around the office tower core as a third zone; the office tower forms a fourth zone. The tower lobbies share the same conditions environmentally as the central lobby. Option 3 provides a separation between the environmental zones for the podium and the office tower. The main lobby, which wraps around the tower, is a third zone. A fourth environmental zone is the breakout meetings space and associated service spaces, located at the end of the building next to the entrance lobby. The galleries behind the long facades of each building are divided up into environmental zones that are linked to the performance of the space immediately adjacent. Each gallery is linked environmentally to the space it serves.

The environmental strategy is based on a mixture of semi-external spaces created around the edges of the tower and the podium, together with enclosed areas which can be naturally ventilated in winter months, but which are closed for other months of the year. The strategy for environmental control of these spaces is one of having small centralised service areas that are distributed relatively locally in the building. Services equipment areas, located on the roof of the mini-tower and the roof of the podium, distribute ducted air and electrical supplies to the interfaces between semi-external spaces and enclosed volumes. Air is distributed in air-handling ducts which are set behind opaque panels, fixed to structural concrete walls.

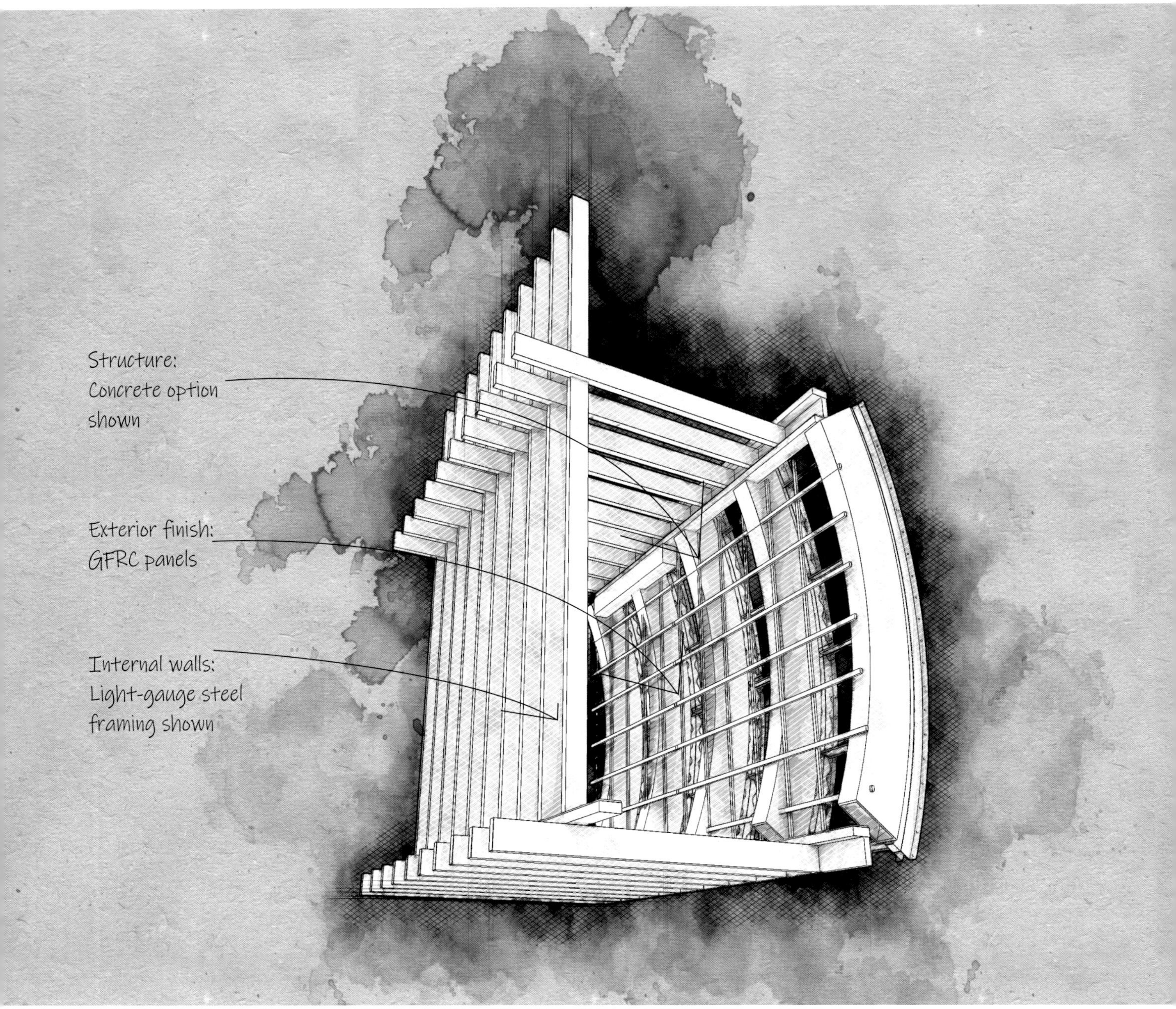

Typical bay adjacent to facade

The strategy for the external envelope is based on positioning equipment for environmental control outside 'activity', or main spaces, allowing the floor plates to be clear of ancillary spaces and equipment. This approach has led to the vertical riser ducts for mechanical ventilation being accommodated within the forms of the shear walls, which span the full height of the sides of the tower. The sides of the podium also have the services installation fixed across the full extent of the facades, leaving the roof clear of services equipment; the roof is used to admit controlled levels of daylight and natural ventilation, which is less practical to achieve through the facades. The podium is arranged such that most spaces are top lit and top ventilated; glazing and opening vents on the facades provide natural light and air to the perimeter spaces only. This approach has the effect of creating a visual 'map' of how the building is serviced, both on the facades and on the roofs. In addition to rooflights, a set of terraces are arranged on the top of the podium, with shrubs that reduce the wind speed of air currents which enter through rooflights. A second level of control of natural ventilation is provided by mechanically operated baffles fixed directly to the roof lights. The tower has opposite facades which are opaque as a result of the accommodation of services equipment and routes for environmental controls. The other two facades, set opposite one another, are largely glazed; enclosing floor plates clear of ancillary spaces. Opaque facades for both tower and podium are clad in GFRC (glass-fibre reinforced concrete) panels, which are made locally by several companies. This cementitious material is relatively light in weight, is durable and can be moulded to fit the geometry of the services installations associated with the facades and roof. GFRC cladding is used to enclose the backpack service modules, staircases, lifts/elevators, washrooms and routes for services. One side the structure is conceived as a primary structure of concrete, with a separate element attached to it which is visible on the facades and with no division between primary and secondary steelwork. The steelwork comprising the structure includes secondary steelwork that would otherwise be added at a later stage of project development.

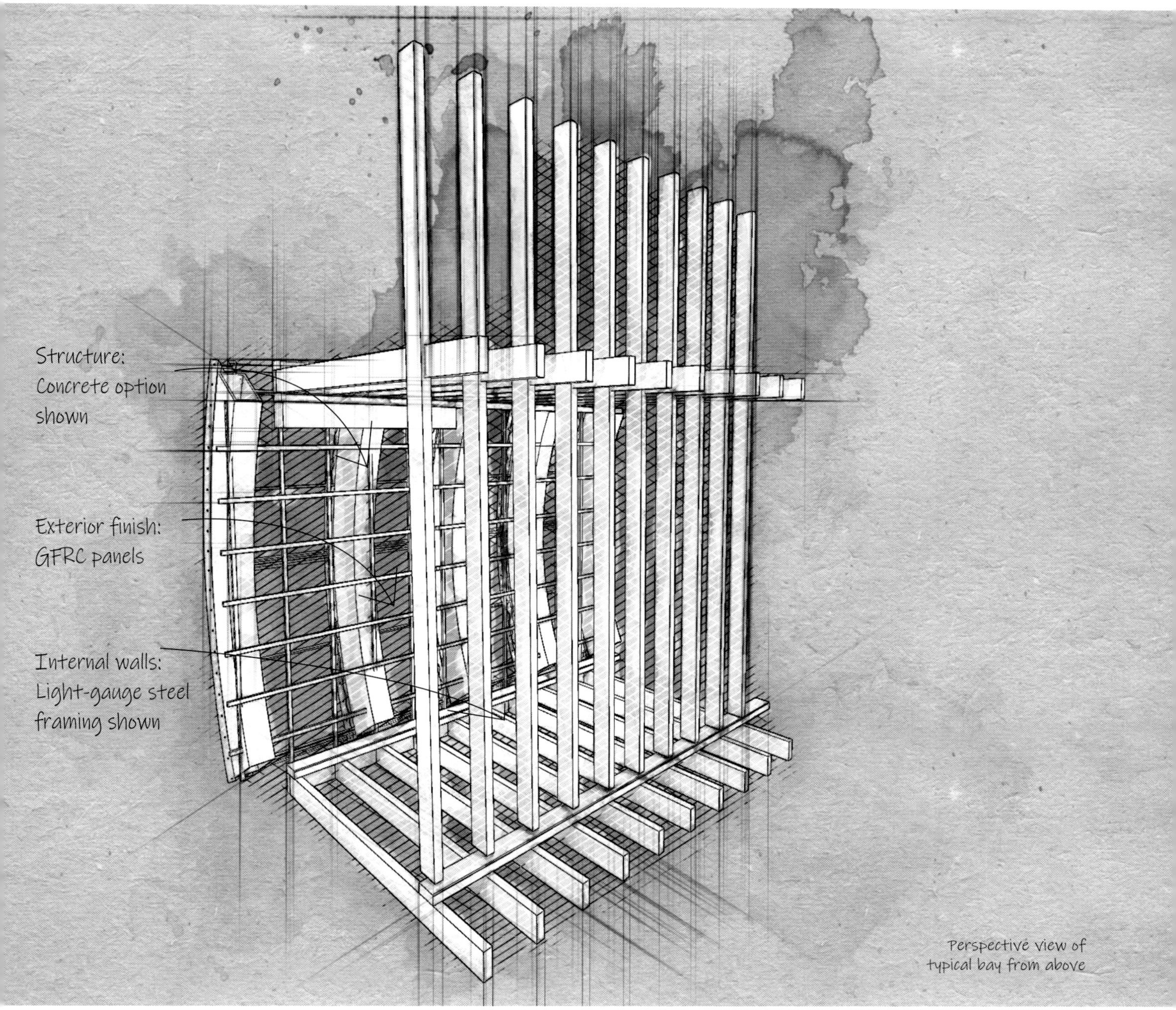

Perspective view of
typical bay from above

This approach avoids the need to make assumptions about the future needs of secondary steelwork, which may be greater than those actually required. This approach requires a close integration of the structural requirements of the detailed mechanical and electrical design as well as the interior design. In this project, primary and secondary steelwork is integrated for economy, resulting in the characteristic appearance of this building, driven by the integration of structural form.

The external envelope comprises areas of opaque cementitious panels made from a fibre composite material. The facades use solar control glass of 50% light transmission; the glass is relatively transparent when compared with other glasses used on glazed facades in the region as a result of the solar shading provided, which is largely horizontal in form to block the path of direct sunlight from entering interior spaces from mid-morning to late afternoon. This approach has the benefit of reducing or eliminating the need for electrical lighting to be used during hours of daylight. Opaque areas of facade have deep reveals, both in order to accommodate services equipment as well as to allow daylight to enter the building while protecting from the effects of solar gain. The podium has long bands of opaque cementitious panels arranged as spatial modules that are projected forward of the adjacent glazing. Additional solar protection for glazing is provided by overhanging elements. The use of horizontally oriented facade panels to provide solar shading is a key part of the facade strategy to protect glazed elements by ensuring that the shading is effective for as much of the day as possible, in response to the path of the sun at different times of day and different seasons of the year. This approach provides a visual contrast with elements that require little daylight. The elements for areas of little daylight have horizontal or vertical strips of glazing to illuminate walls internally. This approach is used both within the mini-tower and the podium. Daylight is reflected off the horizontal shading elements and into some spaces. This creates a light shelf effect in larger spaces which is however not used in smaller office spaces where it provides no specific performance benefit, but instead contributes characteristic effects of light and shade projected across areas of floor.

1. Perspective view of exterior

2. Perspective view of exterior

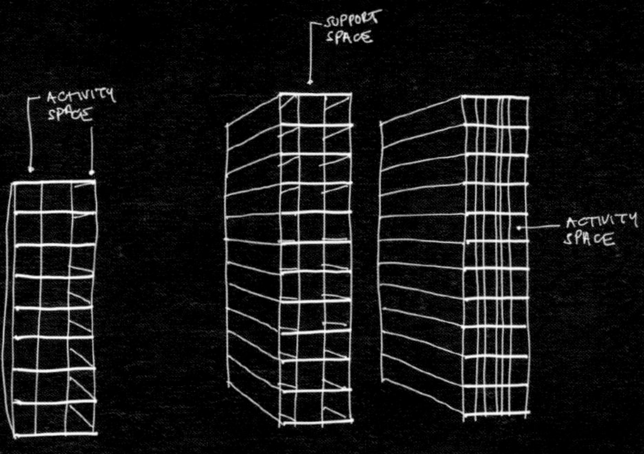

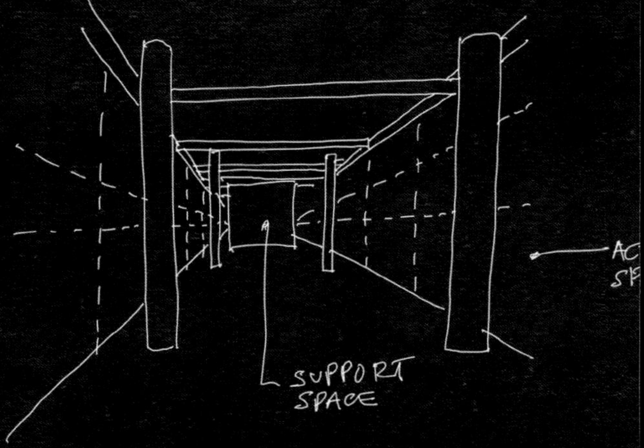

Establishing a project-specific structural design typology

In this project, the early stages of the building design process determined design strategies which were combined to establish a 'typology', or 'general type' that could be deployed over three separate buildings, and would also allow each building to be differentiated as a result of its specific functions. A typology is typically a combination of spatial arrangement, structure, environment and envelope, regardless of the function of the building. In the meaning conveyed here, a typology is a combination of those characteristics, seen across a project that requires a set of similar building forms. A typology is not established as a result of the specific functions carried out in the set of buildings, but instead seeks to find what is common to the buildings; differences in application of a typology are considered as 'add-ons' to that typology.

In this project, a typology was established in relation to specific technical performance and methods of construction, making common the technical strategies and design requirements across all three buildings, while meeting the requirements of the brief/program.

A part of the approach taken in this project was to bring together design strategies for internal space, structure, environment and facades, and link those strategies to specific construction technologies.

The structural design typology was established for this set of buildings as a result of them needing to be built as three separate blocks in order to accommodate the space required in the brief/program. A range of options was investigated based on the volumetric requirements of spatial organisation within each building. Different volumes and sizes of building 'block' were established from the preferred options. It was seen that the structural design had the greatest influence over establishing an economical building design, as elements of staircase, services and facade were largely secondary to the cost of the structure. The focus on the structural design as part of the strategy for establishing an overall set of common forms for these buildings became the key driver in the design because of the associated cost of taller buildings. The choice of three buildings, as opposed to a two- or three-storey arrangement of a single form, was the preferred option from the point

Perspective view of exterior

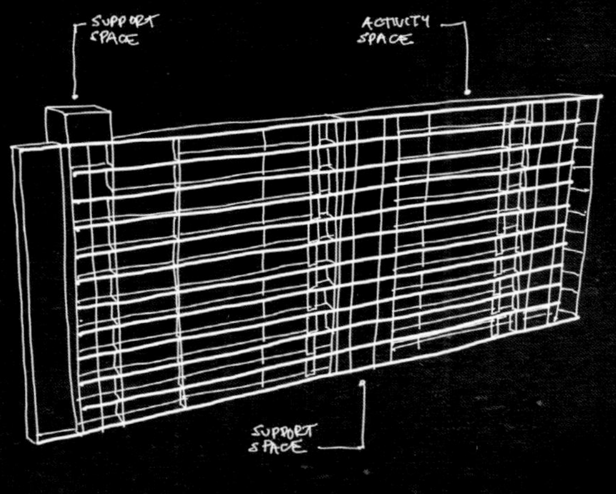

4. Perspective view of exterior

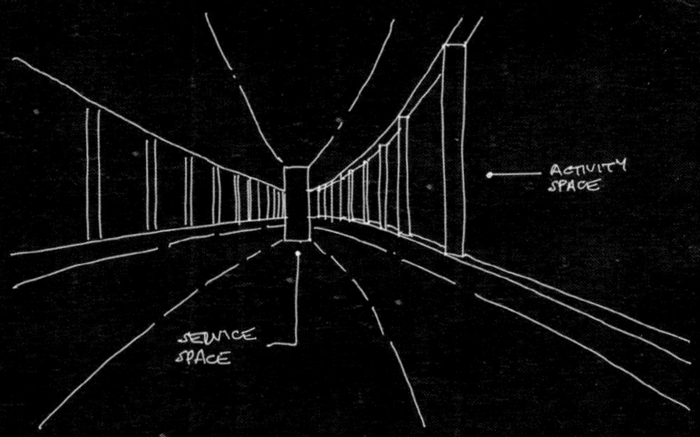

of view of functionality and circulation between groups of building users. The structural design was investigated to see how the three buildings could become a repeated form in order to give them a common identity and to simplify the design and its associated costs. As a result, key aspects of the spatial organisation were investigated to find out what was common to the three buildings in terms of floor-to-floor heights and overall dimensions of the floor slabs. In addition, the mechanical and electrical servicing of these spaces was investigated to ensure that each of the three buildings presented an economical solution as well as a highly functional solution. The design incorporates a relatively narrow width of floor plate that allows rooms on each side of a central corridor to vary in dimension without using steps in the facade. This approach resulted in the characteristic forms of the building, as seen in plan with the structural arrangement emerging as a typology. The structural arrangement considered issues of spatial arrangement, including the spacing of columns which were set on a structural grid onto which internal partitions and walls could be aligned. It emerged that the preferred location for service spaces was at the ends of each building. The service spaces comprise staircases, lifts/elevators, washrooms and associated environmental control space. It was also identified that each building would have a different arrangement of service spaces, set at the ends of each building; service spaces being fixed to the primary structure in different configurations. The typology for this building excludes these service spaces, which are added on to each building block as a set of specific spatial arrangements. Each service block is different, in response to the requirements of the brief/program for the spaces served. A set of secondary typologies were established for lifts/elevators, staircases and washrooms, each identified as a floor-to-floor-based module. This approach did not suggest that the modules were to be prefabricated; they were seen as a set of secondary typologies which ensured an economy of structure associated with an economy of external envelope that could be fixed to each of the service spaces. The resulting design is dimensionally modular, but is not conceived to be a set of repeated elements.

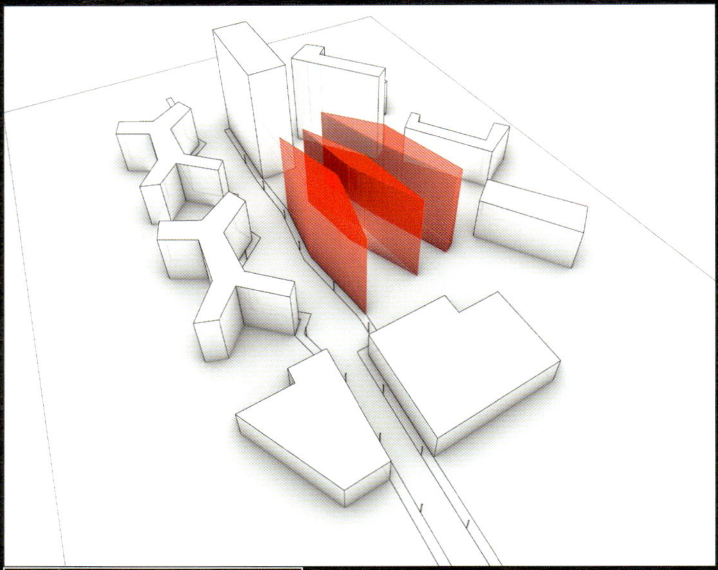

1. Project volume in site context

2. Project volume in site context

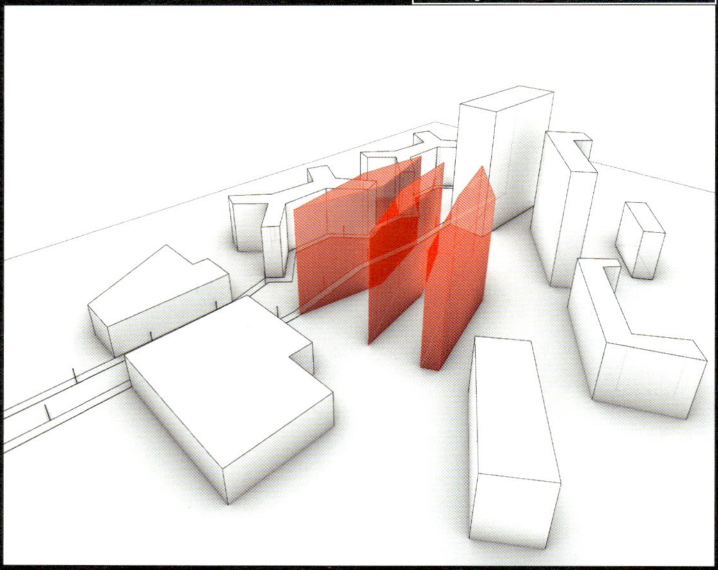

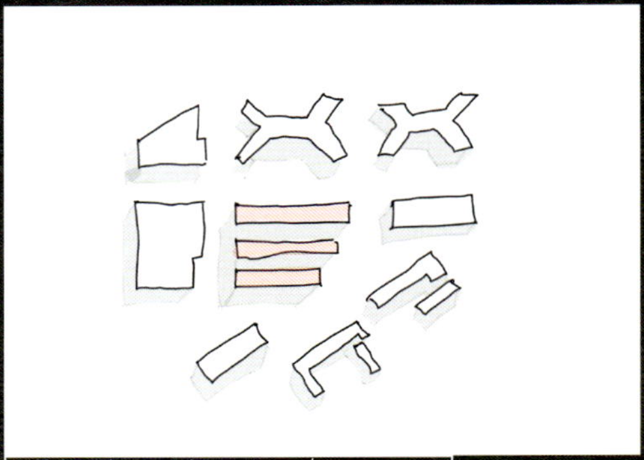

1. Site plan with surrounding context

Option 1 ties together the surrounding buildings by creating a stepped profile in plan.

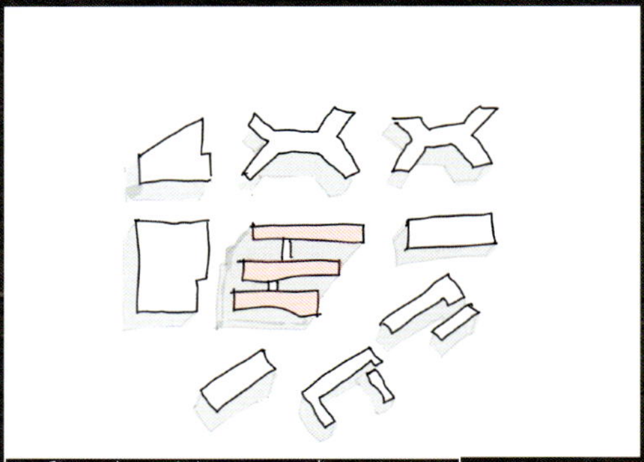

2. Site plan with surrounding context

Option 2 allows each of the three separate blocks to have a specific geometry to create a range of different-shaped spaces.

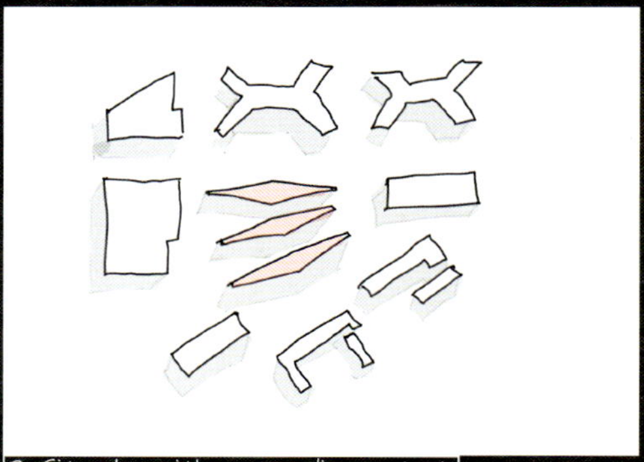

3. Site plan with surrounding context

Option 3 forms external spaces between the three blocks that serve as 'outdoor rooms'.

Site context

This set of buildings is a healthcare project set on the edge of a large university campus in a temperate climate; warm in the summer and mild in the winter. The project comprises three blocks which have mixed use of space within each building; for care, research and administration. The floor-to-floor heights are consistent from building to building, allowing blocks to be connected by bridge links and interstitial buildings within the near future. The aim here is to provide as much flexibility for repurposed use during the lifespan of the buildings, while ensuring that the reconfigured space will maintain its high levels of daylight for care, research and administration.

Once the typology had been established that was common to all three buildings, the forms could be arranged in a way that was as visually harmonious as possible with one another, within the context of the site. The height of two of the three buildings exceeds that of the neighbouring buildings. The lowest of the three buildings is approximately the same height as neighbouring buildings, but those surrounding buildings are of greater size in terms of overall massing. The three buildings developed for this project have a slenderness which allows them to sit comfortably within the site, from a visual point of view and from an access point of view.

Three options were investigated, which considered the buildings to be set either parallel to one another, or arranged in a radial format. The first two options have buildings of different length in order to create a stepped arrangement of forms that required some functions to be moved from one building to another. In Option 2, a staggered arrangement was introduced that required more link bridges to be formed between buildings. Option 3 rearranged the spaces as a single building size and typology, with different arrangements of service elements added to the ends of each building. This layout allowed the three parts of the building to function independently of one another, with access at ground floor level providing the only link between the buildings. From the point of view of site context, it was agreed that Option 3 satisfied the requirements of maintaining external space for staff and patients.

Perspective view of structure

2. Perspective view of structure

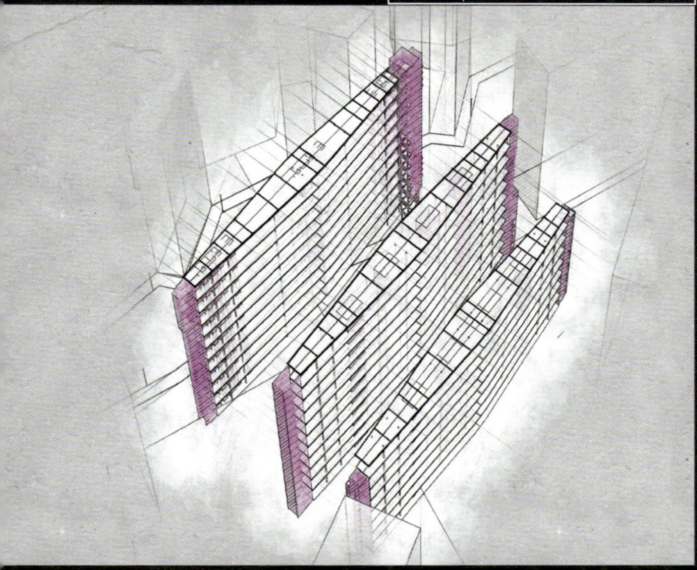

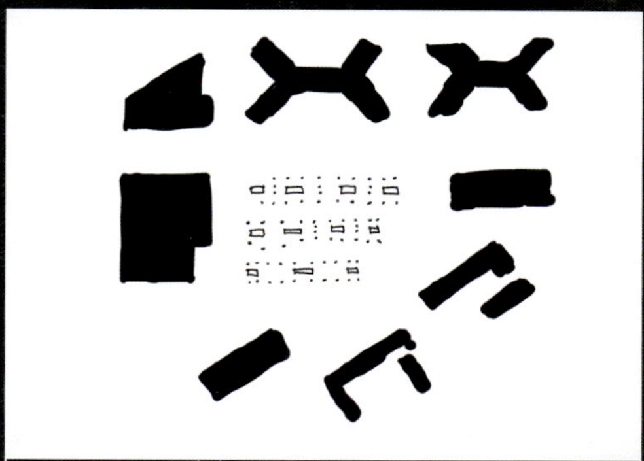

1. Figure-ground diagram showing project structure

Option 1
The structural grid is aligned with neighbouring buildings on two sides.

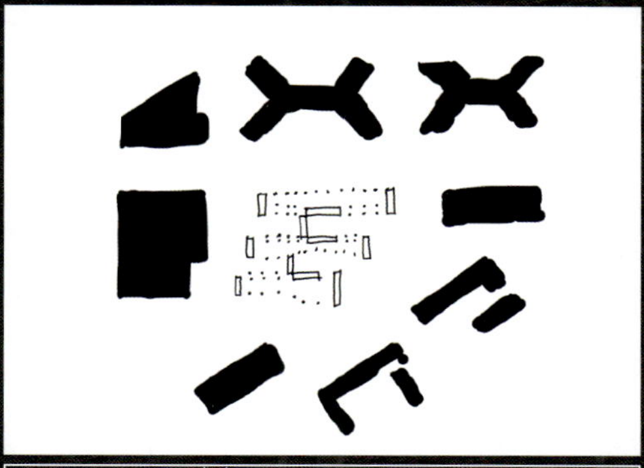

2. Figure-ground diagram showing project structure

Option 2
The structural grid is aligned with neighbouring buildings on two sides; positioning the support spaces at the ends of each building.

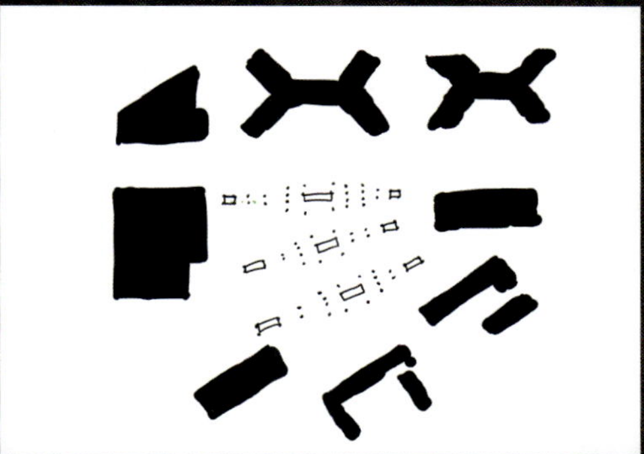

3. Figure-ground diagram showing project structure

Option 3
The structural grid is aligned with neighbouring buildings on all sides; positioning the support spaces at the ends of each building.

The site context of Option 1 is of tying together the surrounding buildings by creating a stepped profile in plan; the blocks respond to the orientation and spatial arrangement of adjacent building forms, aligning the three blocks with two neighbouring buildings. A disadvantage of this approach is the self-imposed geometric rigidity of the forms; making options for environmental strategies and spatial arrangements more difficult to implement. Option 2 allows each of the three separate blocks to have a specific geometry to create a range of different-shaped spaces; bridges are introduced to allow facilities to be shared between buildings. The shape of each building as seen in plan is different in order to respond to the needs of the brief/program as a literal interpretation of immediate user requirments, but without considering the longer-term plans for use. The three buildings are linked together at positions close to the corresponding service cores of each in Option 2 of the spatial arrangement. This approach to the site context still makes little reference to the form and orientation of surrounding buildings; this option is more of a 'machine' that fulfils its own requirements but makes it difficult for future links with neighbouring buildings to be created. Option 3 develops the arrangements of Options 1 and 2 to create a group of buildings which is more suited to the distribution of spaces in the building, and to the needs of creating usable space between the buildings. In Option 3 the three blocks create external spaces between them that serve as 'outdoor rooms' at one end of the site, while creating an entrance space into each of these 'rooms' at the other end of the site. This approach creates a radial geometry of form that is compatible with the different orientations of the surrounding buildings. The outdoor rooms are hard landscaped for use by both those occupying the building and the neighbouring university buildings of different departments and schools. Options 1 and 2 regard the spaces between the buildings as being more 'residual space', as a result of the geometry having created external spaces that do not consider the geometric proportions of the facades of adjacent blocks. At the scale of urban design, Option 3 provides a focus at each end of the building and within each of the four outside spaces created between the buildings.

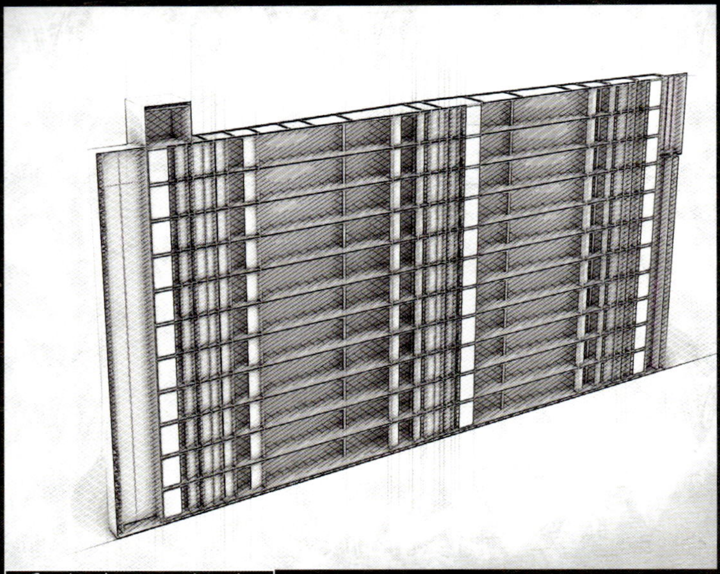

1. Sectional perspective view

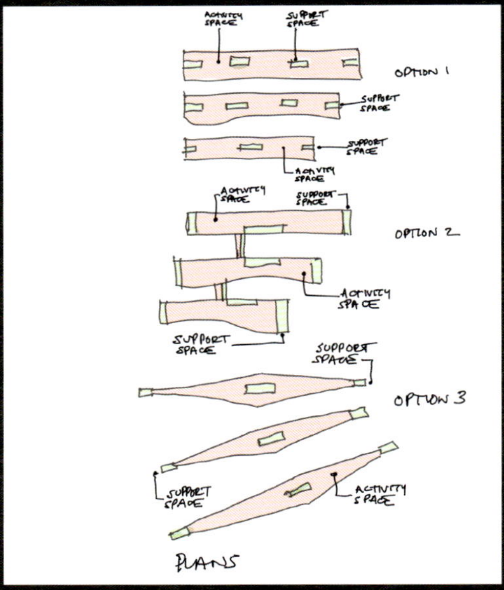

Spatial arrangement Option 1 makes use of a set of rectangular forms which results in the three blocks being arranged in a staggered arrangement. Option 2 positions more of the cores for staircases and washrooms within each building.

2. Sectional perspective view

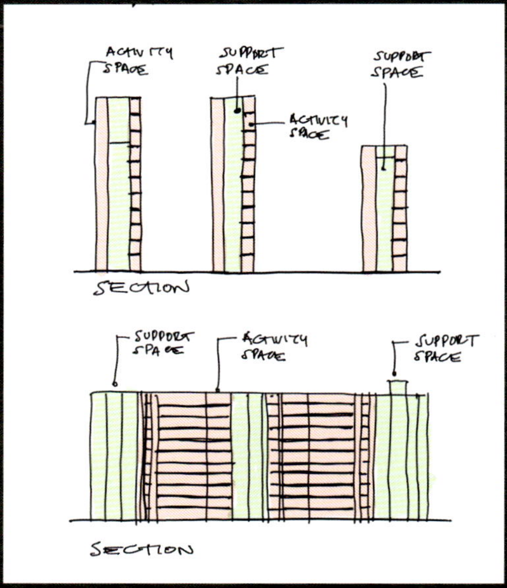

Option 3 has identical building forms that are developed to meet the requirements for different space sizes.

Spatial arrangement

The spatial arrangement of Option 1 positions the cores within the zone occupying the centre of each block such that stairs and lifts/elevators form a spine within the building, and washrooms, patient lifts/elevators and mechanical and electrical equipment are accommodated at the ends of each block. This arrangement of support spaces is able to accommodate the long, narrow floor plates of the 'activity' space. Option 2 puts all the service space into cores adjacent to bridges that link the three blocks, being linked at positions adjacent to their service spaces. The angled geometry of the facades, on each side of each block, responds to the required sizes of spaces within each building. The inflected facades have the benefit of creating external spaces and direct views of those external spaces from the bridges; buildings visually 'reach out' to one another as they are set closer together adjacent to the bridge links. Option 3 puts all the service space for vertical circulation for washrooms, and support spaces for mechanical and electrical equipment, in the end blocks which are attached to the ends of the building. Stairs and lifts/elevators are positioned in the centre of the building's 'activity' space. This arrangement provides a clear set of floor plates of linked spaces; the most flexible of the three layouts in terms of near future use. The general spatial arrangement of all three options is the same; each is established from the point of view of creating a set of repeated structures that is economical. In addition, spaces are required to meet the need to admit daylight into rooms that are to be occupied by patients and staff. Options are differentiated by taking another approach to the geometry in plan, in order to consider different-sized rooms while maintaining a single form for each building that is economical. This aim is achieved by reverting to a simple rectangular form where possible. Option 1 makes use of a set of rectangular forms which results in the three blocks being arranged in a staggered arrangement, set parallel to one another to allow the construction time to be reduced. Option 2 positions more of the cores for staircases and washrooms within each building. This results in three different building forms which do not provide an economical typology. This option sets the

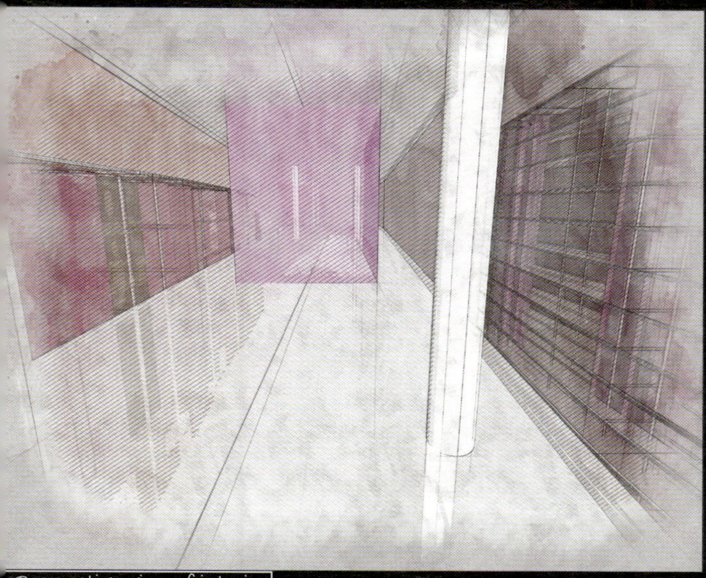

. Perspective view of interior

2. Perspective view of interior

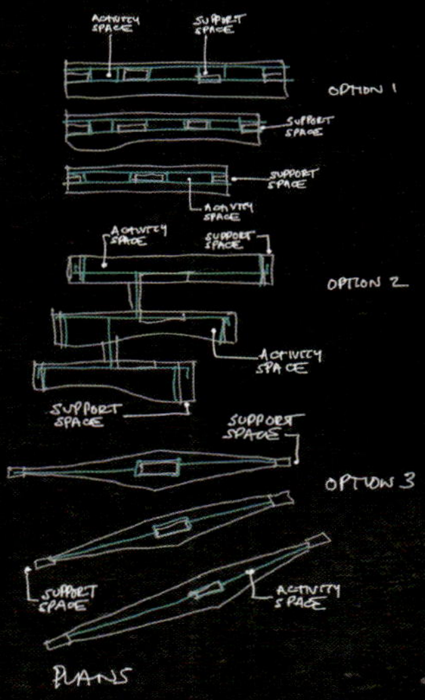

Internal circulation

Option 1 has corridor access on both sides of the line of cores.

Option 2 has a single corridor adjacent to the service core located on the sides of each block.

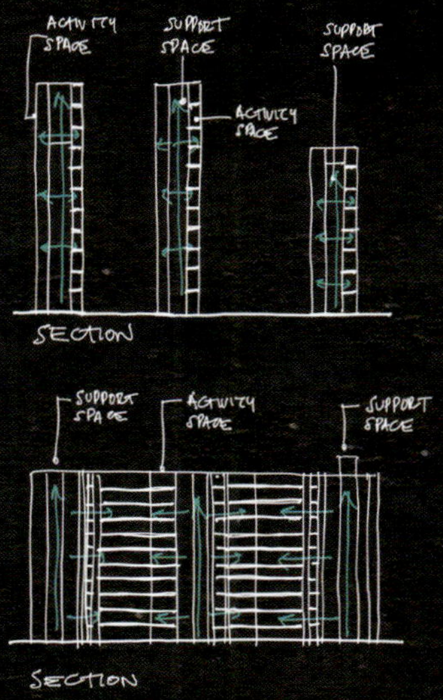

Option 3 links the two service cores which are attached to the ends of each block to the central core.

service spaces that would create floor plates which are mainly clear of support spaces at the ends of the building. As a result, each form varies in order to accommodate different room sizes; these three building forms also became more expensive than Option 1. In Option 3, identical building forms are developed that meet the requirements for different room sizes, which was achieved for support spaces at the ends of each building to suit their specific needs.

Internal circulation

Circulation through each block is provided by the core areas which are arranged differently in each of the three options. Each block is divided into two equal halves that can be accessed from the core at each end of the building. An entrance to each block is positioned at ground floor level in the core at each end; a total of six possible entrances. This arrangement provides separate access to the different constituent functions accommodated in the building with the minimum of internal circulation space. In each of the three options, this arrangement of circulation provides the shortest route between cores. Circulation between blocks is either by way of the bridges in Option 2 or by corridor located at the first basement level. Links at first basement level avoid the need to form an enclosed connection between blocks at the level of the garden spaces, which would make the 'outdoor rooms' difficult to use as an open public space.

The internal circulation of Option 1 is one of having corridor access on both sides of the line of cores that are set along the centre of each block; the corridors provide access to meeting rooms and ancillary spaces located in the space between the cores. Option 2 reduces the need for corridors by setting a single corridor adjacent to the service core located on the sides of each block, as well as those cores positioned at the ends of each block. Corridors in each block are linked by bridges in a configuration that minimises the total length of the corridor across the three blocks. Option 3 reduces the need for circulation to a single central corridor, linking the two service cores which are attached to the ends of each block to the central core.

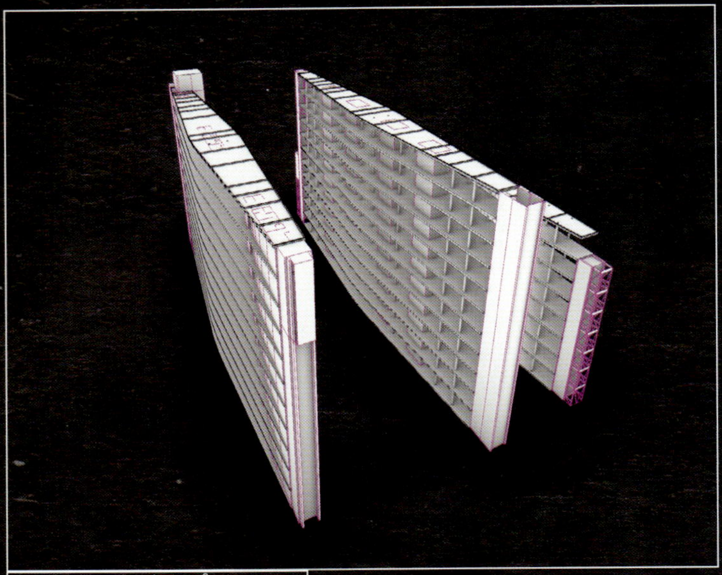

1A. Isometric view of structure

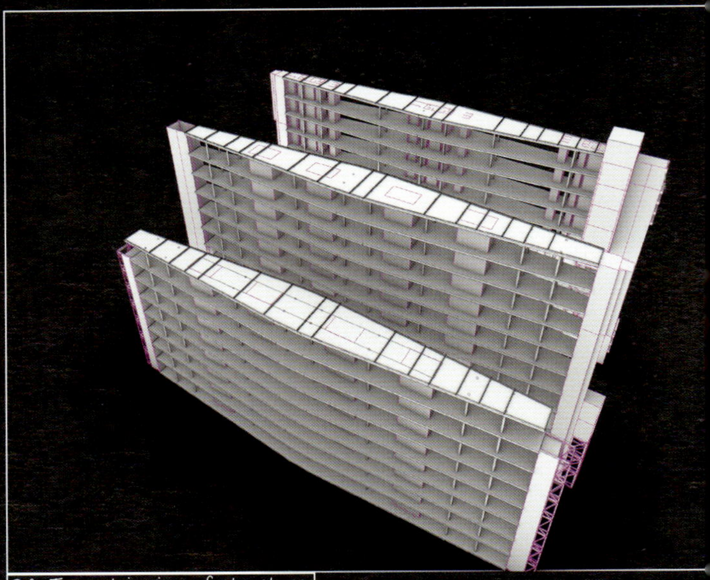

2A. Isometric view of structure

1B. Isometric view of structure

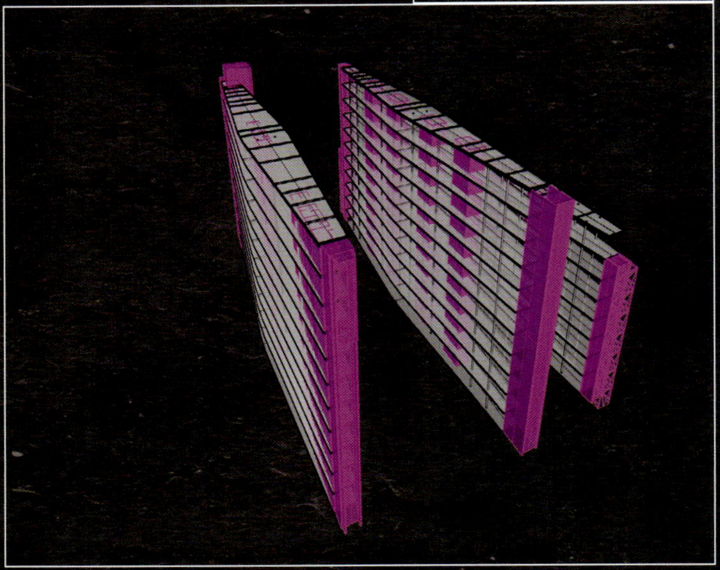

2B. Isometric view of structur

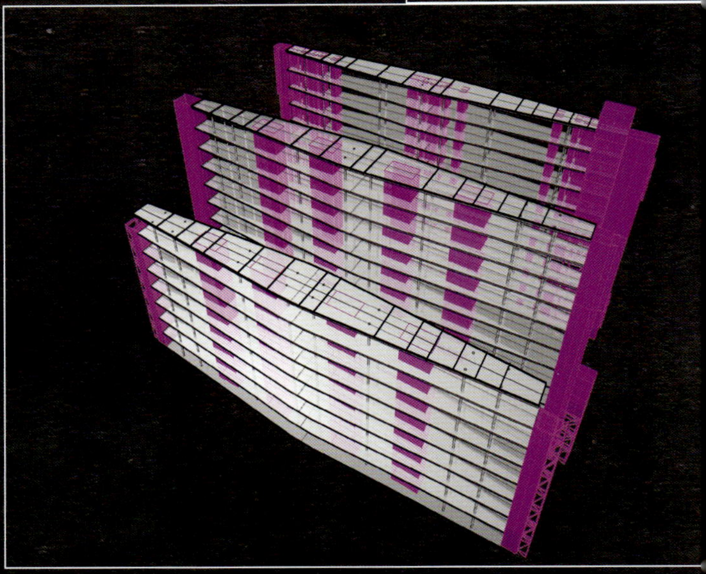

Structural strategy

The structural system for each of the three buildings is of columns set around the perimeter of floor slabs; downstand beams, spanning the width of each block, support the floor slabs in turn. A framed structure for service cores is fixed to each end of each block; there are no structural cores in the centre of the building. The internal space within the wedge-shaped 'activity' space is clear, with only a single core for vertical circulation set in the centre of each block.

The structure for each block comprises a braced frame of columns joined by beams that span the width of each floor plate. Reinforced concrete floor slabs are set between these beams; prefabricated concrete floor planks are used in the centre of the building, set between cores, in order to allow the cores for vertical circulation to change position in the near future. A key consideration is to provide clear space which is as flexible as possible; for this reason, the 'support' space modules at each end of the clear floor plates contribute to the stability of the structure of each block.

The structural strategy for all three options is based on the use of a single typology which was developed to suit spatial organisation requirements in the building, balanced with the volume occupied by each of the three blocks and its associated height. The height of each block is determined by the need to arrange them in a long narrow plan that rises through between eight and 10 storeys in height. The single structural typology that was considered required different arrangements of service space added to the ends of each building. These service spaces are secondary elements attached to the primary structure and are continuous with it. This typology needed to consider both the different heights of the building and the different arrangement of service modules attached to the ends of each block whilst using a single typology, where the only changes from building to building, in structural terms, were the sizes of columns, the depth of floor slabs and the thickness of shear walls. Otherwise, the general arrangement of each building is, from a structural point of view, very similar. Floor slabs were required to support different facade arrangements to suit the varied functional

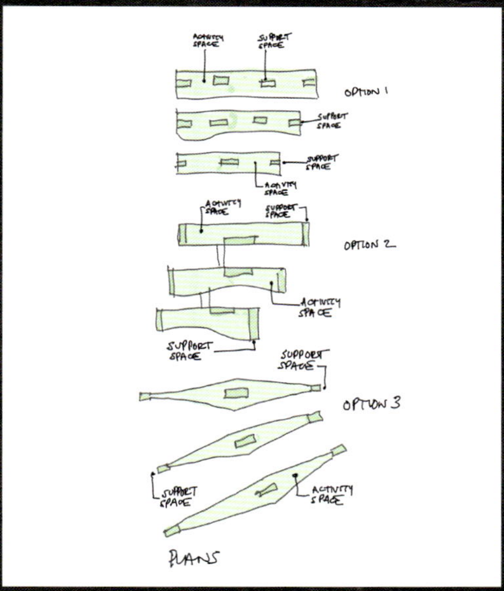

Structural strategy: Zones

Option 1
has clear floor plates where the structure spans from one side of the building to the other without intermediary columns.

Option 2
has beams used to support floor slabs that are set closer together in order to accommodate the varying shape of the floor plates.

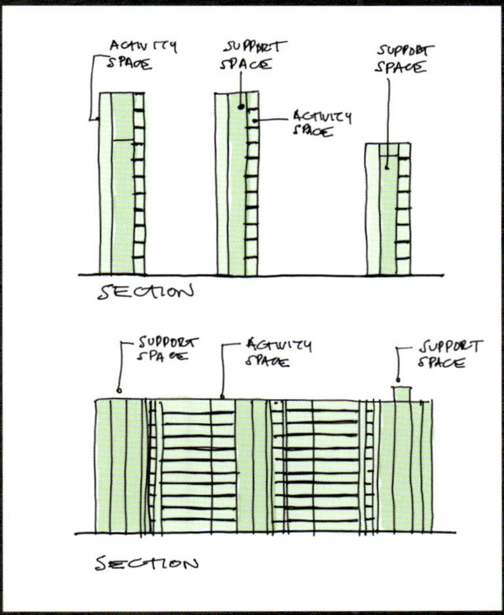

Structural strategy: Zones

Option 3
has a structure of downstand beams which span from one side of each floor slab to the other.

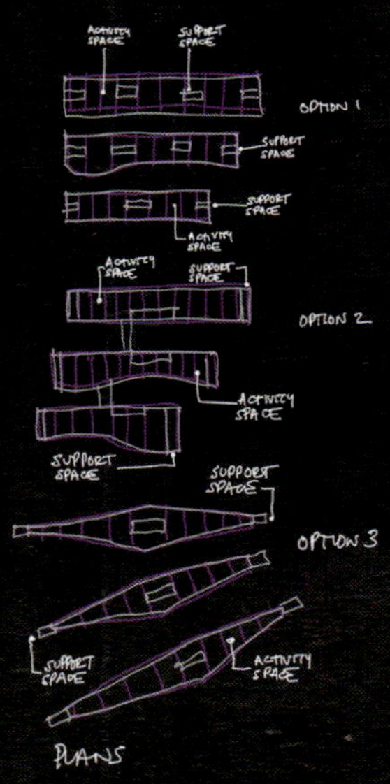

Structural strategy: Framing

Option 1
has a structural frame designed to be constructed in steel, for speed of construction on site.

Option 2
uses the cores to contribute to the stability of each block.

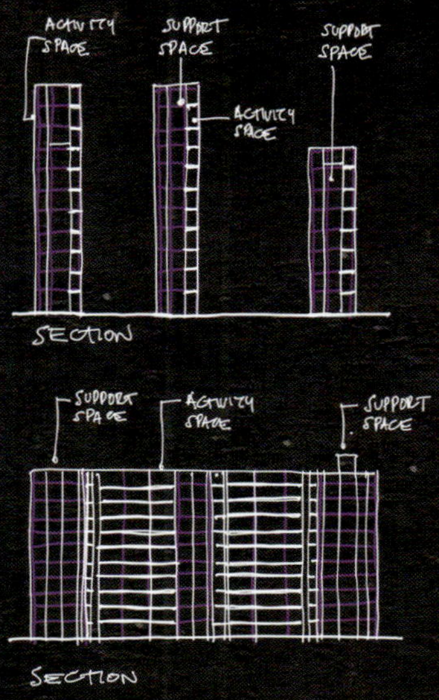

Structural strategy: Zones

Option 3
columns are positioned only at the edges of each floor slab, adjacent to the facades.

requirements of each block. Floor slabs are strengthened locally within the structural grid of each building in order to optimise the structural design. The use of a typology did not lead to a single solution of structural member sizes that suited the most onerous condition. Instead, the sizes of beams, columns and slabs were optimised to suit the specific requirements of the envelope and the service modules at the ends; elements which are not expected to require a change of design, but will be replaced in their current form during the lifespan of the building. However, the facades are all capable of of having the secondary external elements continued across the full extent of the envelope should this become a near-future requirement. Structurally, the three buildings are independent of one another. There was no requirement for spaces below ground. Options for link bridges to be inserted at different levels in different parts of the facade were all based on single-storey designs. The resulting design reduced the use of bridges by providing a connection with the ground floor level of canopy structures which allow unimpeded passage. The structural strategy for Option 1 is one of having clear floor plates where the structure spans from one side of the building to the other without intermediary columns. This option has the advantage of flexibility of the arrangement of spaces that can vary in size. The structural frame is designed to be constructed in steel, for speed of construction on site. In Option 2 a similar principle is used; beams used to support floor slabs are set closer together in order to accommodate the varying shape of the floor plates when viewed in plan. The cores contribute to the stability of each block. The bridges span from block to block without intermediary support; intermediary columns would reduce the spatial effect of the gardens when viewed from within these 'outdoor rooms'.

Option 3 has a structure of downstand beams which span from one side of each floor slab to the other, with columns positioned only at the edges of each floor slab, adjacent to the facades. All three options have columns set adjacent to the facades; none are positioned within the floor plates. In Option 3, the benefits of clear floor space are maximised by putting the service cores at the end of the building.

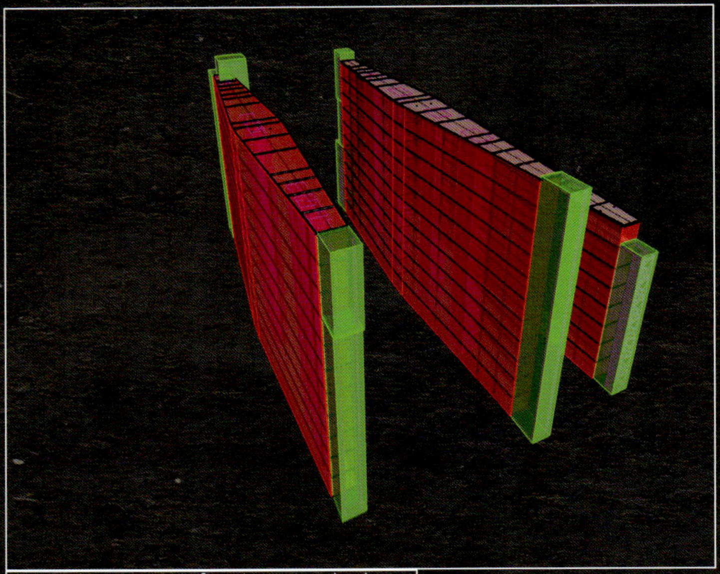

1A. Isometric view of environmental volumes

1B. Isometric view of environmental volumes

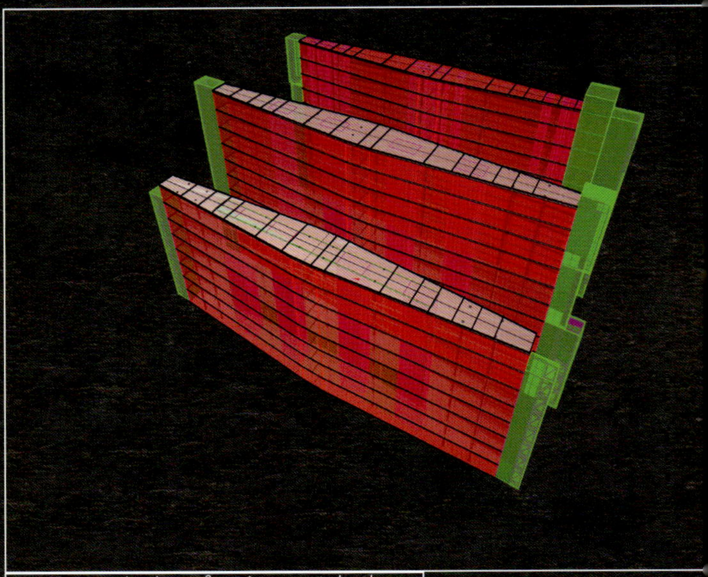

2A. Isometric view of environmental volumes

2B. Isometric view of environmental volume

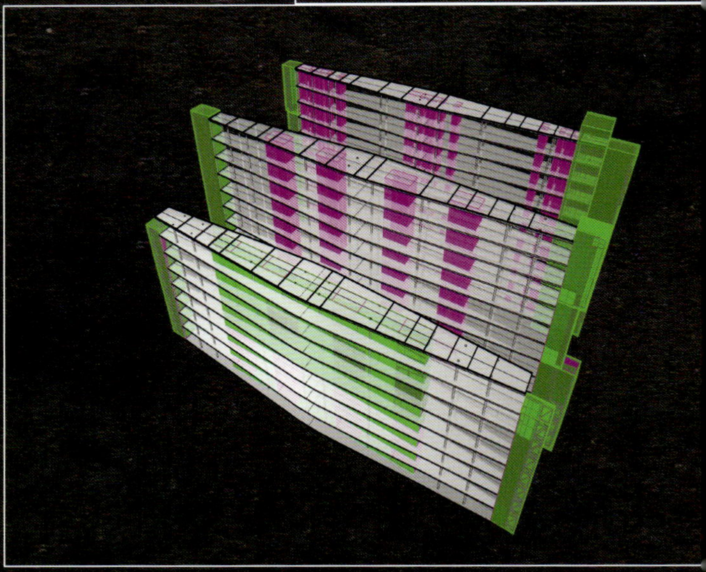

Environmental strategy

The building is serviced by air-handling ducts and electrical supplies mounted along the facades. The main mechanical equipment is housed in spaces at the ends of each building. Rooftop chillers and air-handling units are used in all three options. The environmental controls required in the building made necessary a sealed external envelope with mechanical ventilation supplied to all spaces, including communal spaces used by both patients and staff. As a result, the environmental approach is one of contrasting the external spaces with internal spaces, where internal spaces are highly controlled and regulated and external spaces are used to form links from one block to another with external canopies which are not temperature-controlled. Reception spaces and a cafe are introduced into the garden areas, naturally ventilated for most of the year, and have enclosed areas which are deployed in the colder months. This design decision avoided the need to have these facilities inside the building, which would have become part of the air conditioned and temperature-controlled spaces within the building. These single-storey reception and cafe 'pavilions' avoid the need for visitors to enter the main buildings if their visit does not specifically require it.

The environmental strategy is one of arranging separate environmental zones for each of the two end cores; both cores are mechanically ventilated during the day and are naturally ventilated at night for night-time cooling at mid-season, and moderately cool temperatures at other times of year. The central zone forming the 'activity' space in each block is the second environmental zone; mechanically ventilated with night-time ventilation but with the possibility of daytime ventilation from opening windows. The third environmental zone, formed by the roof block, is essentially a mixture of naturally ventilated spaces with supplementary mechanical ventilation and entirely mechanically ventilated spaces. This third environmental zone is isolated from the other two zones as a result of the medical research activities within. In common with other campus-based university healthcare developments, the buildings accommodate a significant research component in addition to the

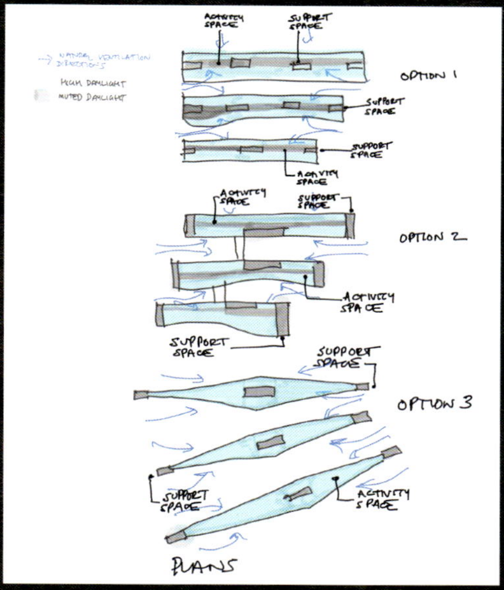

Daylight strategy for direct sunlight and cross ventilation.
Option 1 admits daylight and air from the two long sides of each block.
Option 2 admits daylight and air from the two long sides of each block; subdued light is provided in areas adjacent to facades.

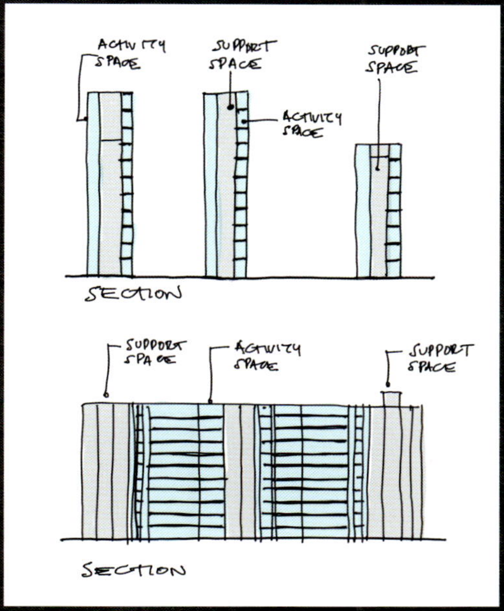

Daylight strategy for direct sunlight and cross ventilation.
Option 3 admits daylight and air from the two long sides of each block, including circulation spaces.

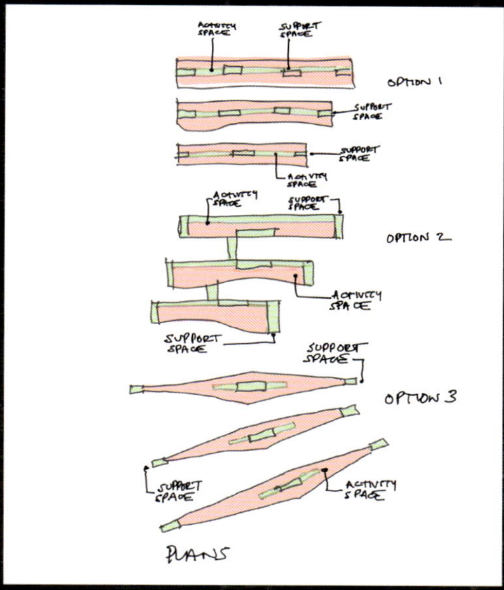

Environmental zones
Option 1 creates an environmental zone in the centre of each block which links the service cores, meeting rooms and ancillary spaces.
Option 2 sets the service zones on one side of the building in a continuous strip in the centre of each block.

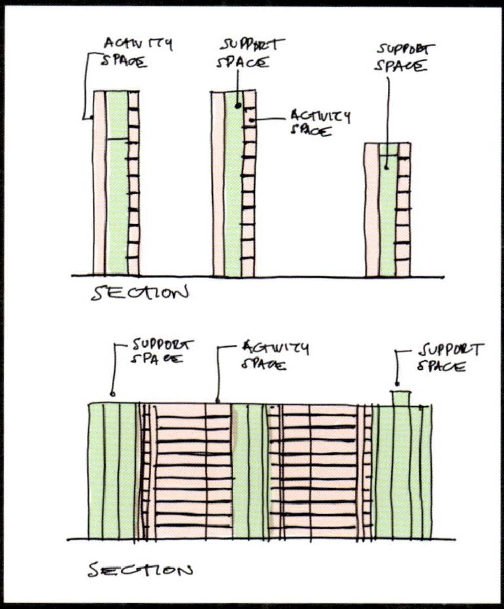

Environmental zones
Option 3 comprises a single environmental zone, in the centre of the building, for most of the floor plate.

provision of healthcare. In this project, the research spaces, occupying the top two floors of each block, require only low levels of daylight and natural ventilation. Spaces are daylit by corridors only on these levels; the corridors being set adjacent to the facades on both sides of each block. Solar shading panels are provided in front of all glazing in order to reduce the risk of glare within internal spaces. Vertical circulation is set on the ends of the building with modular stairs, lifts/elevators, washrooms, storage spaces and mechanical and electrical equipment fixed directly to the building. The modular approach is taken in order to allow the provision of air and electricity to be flexible; equipment can be moved within the structural frame in which it is positioned.

The primary benefit of arranging the building as three individual blocks rather than one single large building is to be able to provide natural daylight to as many spaces as possible; providing solar shading where required. This group of three buildings has the benefit of creating external garden spaces between the buildings for use primarily by those working in the building; each green environment being set in a semi-enclosed 'outdoor room'. The environmental strategy for Option 1 is one of creating an environmental zone in the centre of each block which links the service cores, meeting rooms and ancillary spaces. Set either side of this zone is a second and third environmental zone which face opposite directions and have different exposure to the path of the sun; solar shading is provided to the facades to suit each orientation. Each block follows the same principle of mechanical ventilation; air is distributed differently at various times of the day in response to building use and weather conditions. The environmental strategy for Option 2 sets the service zones on one side of the building in a continuous strip in the centre of each block; linked to service cores at the ends of each block. The open space forms a second environmental zone in each of the three blocks. Option 3 comprises a single environmental zone, in the centre of the building, for most of the floor plate; a second set of environmental zones occupies each of the service cores, all of which have similar environmental requirements for air and supply of electrical services, supplied both from the ends of the building and from the roof.

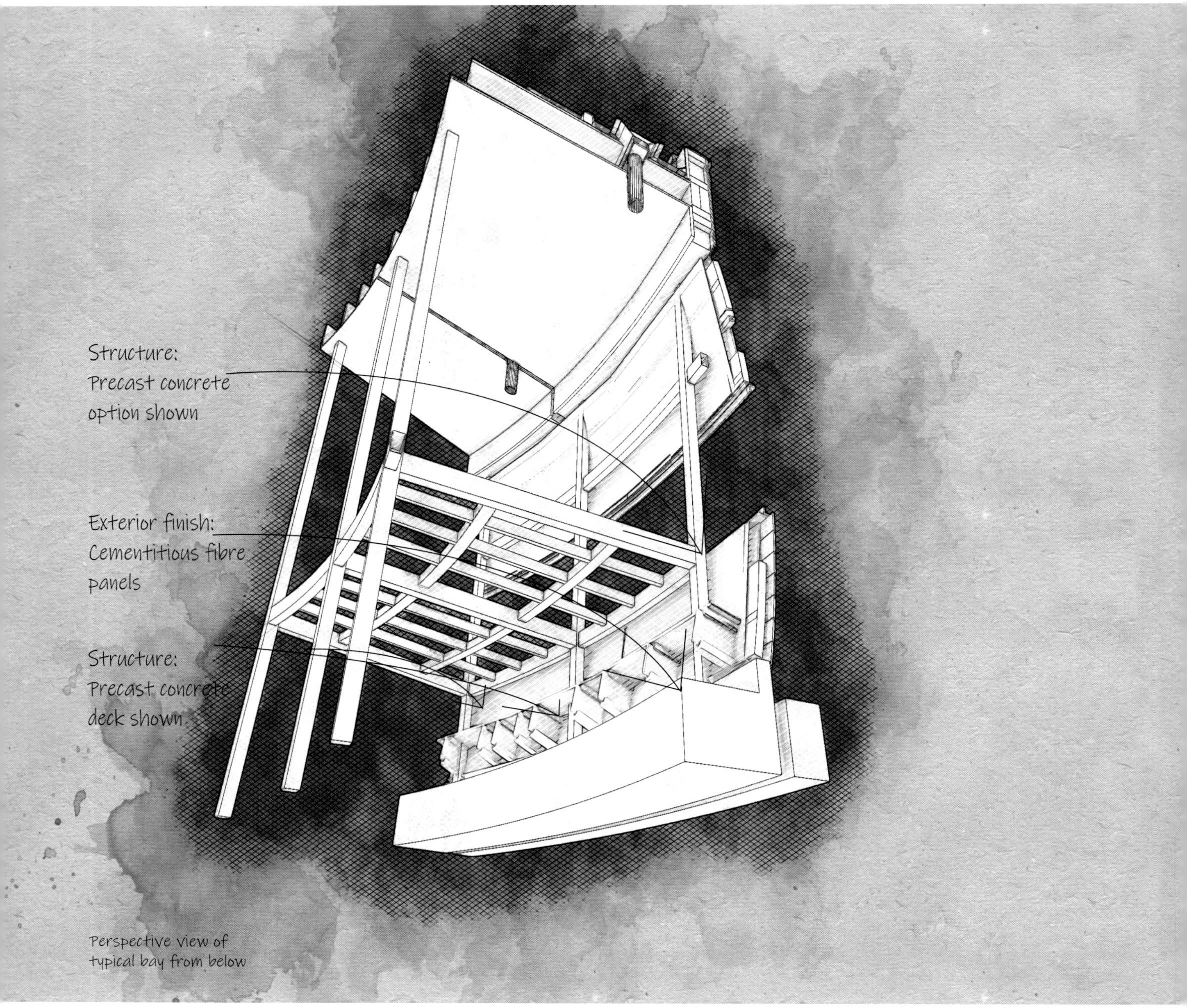

Perspective view of typical bay from below

Typical bay adjacent to facade

The three blocks have a general orientation of east to west, resulting in long south-facing facades which provide partial self-shading as a result of the juxtaposition of the buildings; each volume provides passive shading for its neighbour. However, the south-facing facades all have solar shading panels fixed externally in order to create as much solar protection as possible from early morning through to early evening. Facade panels are formed in translucent glass in order to provide views outward, below the level of shading.

This configuration works effectively in excluding the effects of the sun while allowing the glazing to be formed in clear glass, which provides uninterrupted and unfiltered views to the outside from part of the window opening in each room. Windows are all arranged in a horizontal strip configuration set at 1.1 metres above floor level and extending to just below the ceiling level. These panels are formed as continuous visual bands along the facades, including where opaque panels are used. These non-glazed panels within horizontal bands use the same material as that for the solar shading; panels are inclined slightly outwards to continue the visual language and identity of these buildings along their entire length. The arrangement of panels is repeated on the long north facades where solar shading is not required for much of the time, but shading is still introduced to provide privacy between windows that face one another, such that towards the ends of the building, views are provided of beyond the building; but the external panels provide privacy from building to building. Views are also framed to create views to the gardens below; the bottom edge of the glazing facing the external internal spaces has projecting glazed windows with a horizontal ledge providing real views directly down to the external areas. An additional glass horizontal ledge forms a triangular-shaped window.

The external envelope for the service modules added to the end of each building are formed as cementitious fibre panels, as are the opaque panels that surround the blocks of glazed windows These panels are large-scale and reduce the amount of secondary steelwork and secondary fixings that would be required to support the cladding panels.

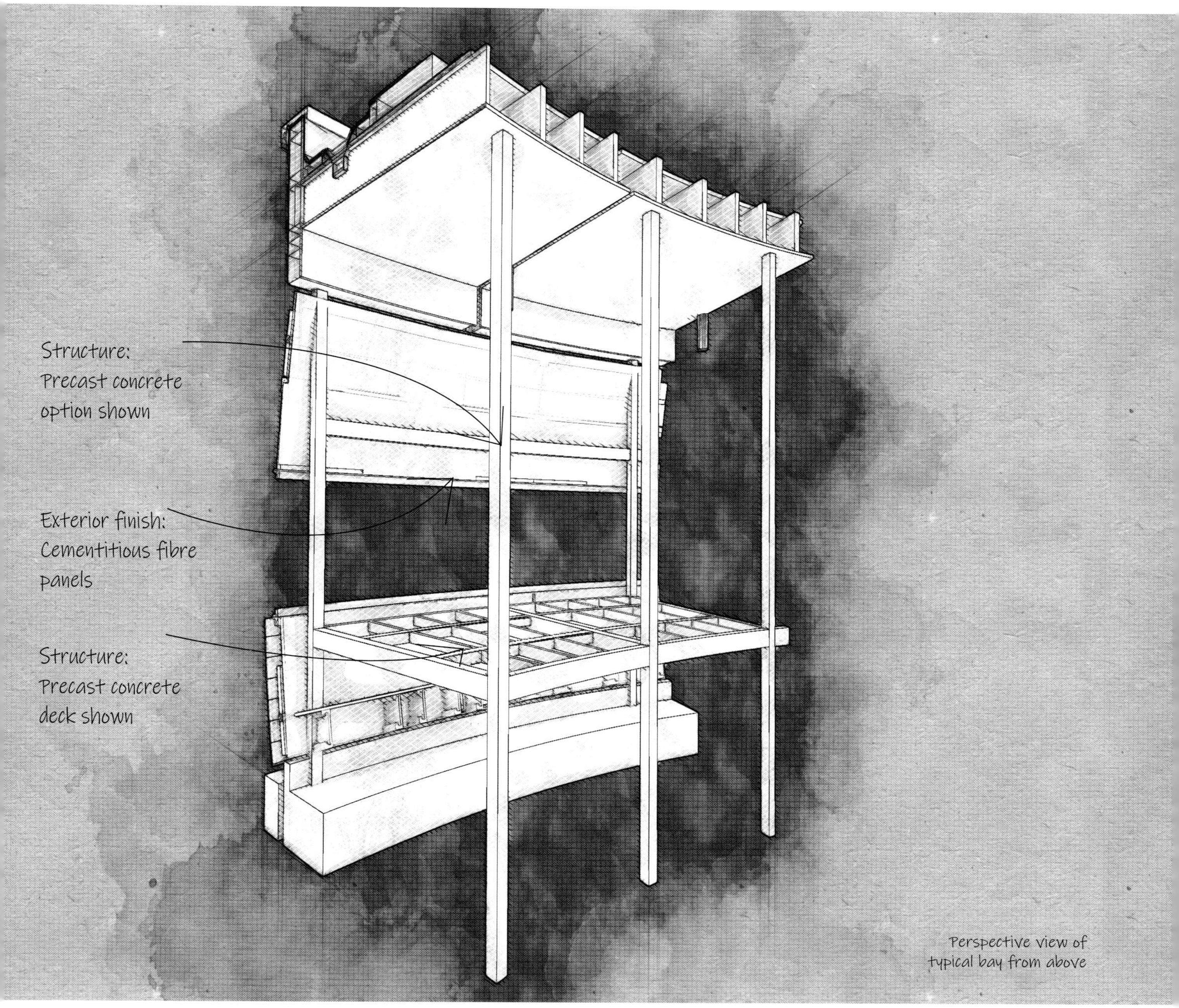

Perspective view of typical bay from above

The modular nature of the service buildings ensures that the facade elements are fully modular in size, but do not necessarily need to be prefabricated and delivered to site as finished assemblies. As a result, typical bays and the service modules are more complete volumes than the typical bay inside each of the buildings, which extend the full width of each block from facade to facade.

The general internal arrangement of spaces is repeated on all floors in each building. The long north elevation of each of the three blocks has full-height glazing inserted into the facades for the common spaces in each building and administration offices on the floor above, as used by staff and patients. This full-height glazing is not introduced onto the south facades, as the solar shading would reduce its effectiveness significantly. The use of solar control glazing would have removed the transparency that is essential in the use of glazed areas on this project. The two blocks which are shaded by the outer block in a passive manner have a mixture of opaque panels and glazed slots that provide views through, but create the privacy required in these administrative spaces as well as the ground floor spaces facing the two external spaces which are linked visually.

The facade system allows for panels to be moved and repositioned to suit the current and near-future requirements of each of the three spatial arrangements of 'support' and 'activity' spaces; support spaces for lifts/elevators are positioned at the ends of each block.

Each block has 45-degree fixed glass shading in order to provide solar control without the associated loss of daylight. In addition to the glass-based shading, some facade areas have perforated metal shading which is arranged as vertically sliding panels. The metal shading is used to shade office space. Each block has an opaque band at the top, which continues down the vertical faces of each block. This facade area accommodates meeting spaces, which benefit from the wedge-like shape of the building to form spaces suitable for giving lectures and presentations. The roofs of each block accommodate additional mechanical and electrical equipment, as well as a continuous track for a cleaning cradle set at roof level on each building.

1. Perspective view of exterior

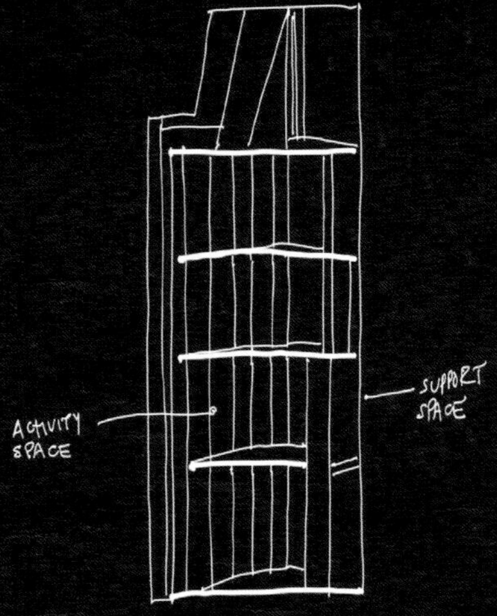

2. Perspective view of exterior

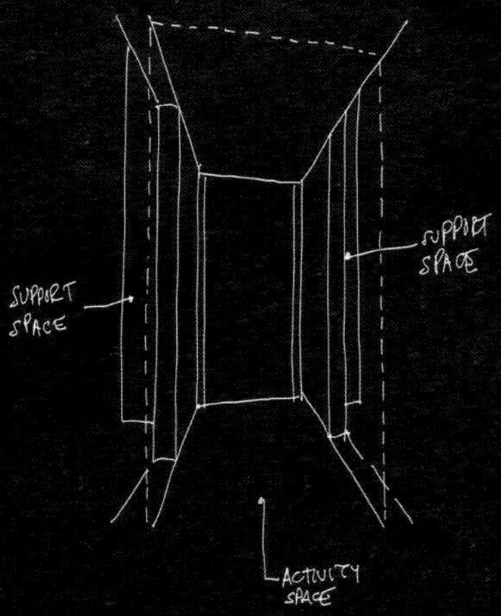

Integration of additions into existing fabric

A key theme of this project was to create a small cluster of buildings, in a hospital campus, as an alternative approach to demolishing a corner building. This medical research facility comprises four groups of activities; each group requires approximately the same floor area and each requires isolation from the other activities. The initial proposal as set out in the brief/program was to demolish the existing corner building on part of the site and replace it with a four-storey building; each group occupying a single floor, with separate administration for each group accommodated on a fifth floor. The three options developed for this based the design on retaining the existing building and adding in four smaller buildings; each building to accommodate a group, including its own administration. All options increase the requirement for external walls and structure to enclose each building when compared with replacing the existing building on its existing footprint; the extra facade and structure allows a cluster of small buildings to occupy the residual space between existing buildings. This residual space cannot be used for larger-sized buildings than those shown in the options, as access to existing buildings would be blocked and would change the character of this well-established campus setting. The environmental strategy is to ensure that a cluster of four separate buildings do not consume more energy when in use than an equivalent single building.

In terms of the built environment, this project seeks to retain and revitalise what has already been constructed; adding small additions as infill in ways that allow the site to be utilised effectively without loss of existing patterns of use for those occupying the neighbouring buildings. The key environmental aim is to add to, and modify, what is already constructed; to retain what is useful, and to create what otherwise could not be created for this site.

The project makes use of the existing infrastructure of water supply, drainage and utility services that are used by the existing building; their common ownership avoids the need for a separate system of utilities. This approach is based on working with what is already in place instead of starting anew; working with the site context of adjacent existing build-

. Perspective view of exterior

4. Perspective view of exterior

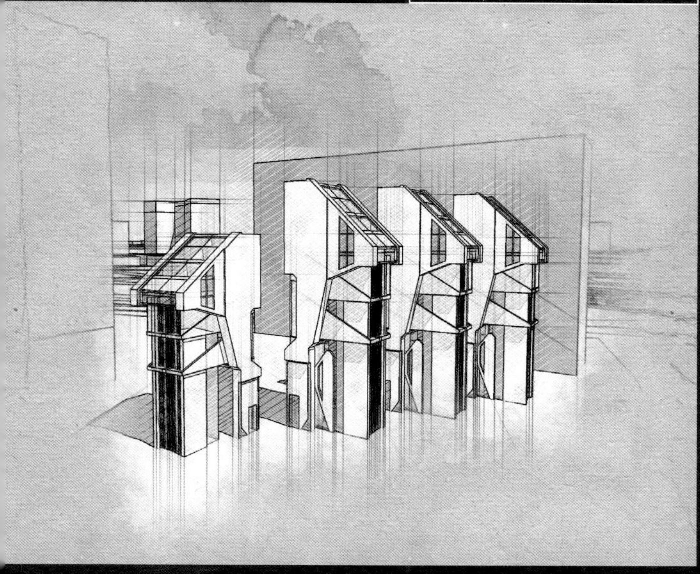

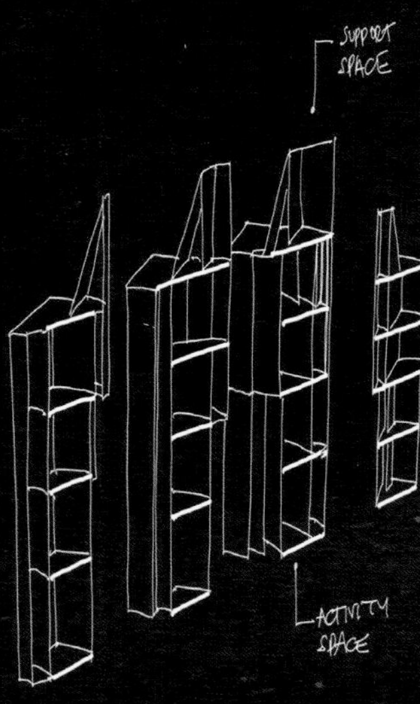

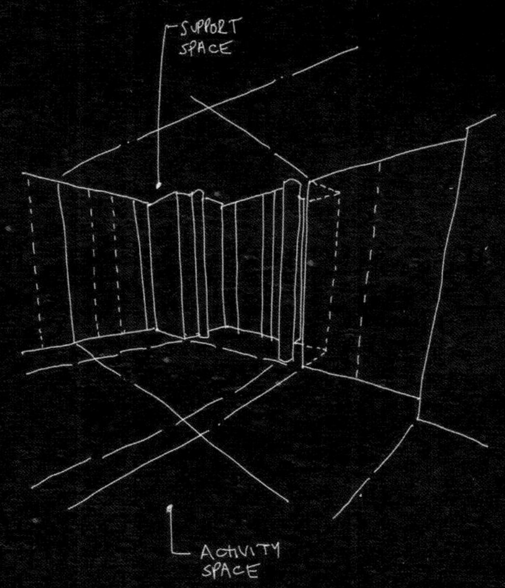

ings and landscaping. Options could have been developed for extending the existing building to be demolished, but this would have blocked access routes and would have prevented the new building from having the required identity of the four groups that are working towards a common set of goals. This approach is based in the introduction of built forms which are dependent on the context into which they are introduced. Location and site context is a key to design development; the preference here was to retain as much existing built fabric as possible rather than demolish and start again. Much is lost in demolition, both socially and environmentally: This project aims to avoid demolition where possible; showing how new buildings can be created with specific additions and changes to this urban context. To remain economical, this approach to design is enabled through reduced quantities of materials and systems for facades, structure and environment control, allowing the design options to be scalable. This advantage of scalability is made possible in this project by using an evolved 'kit of parts', from small-scale additions to the scale of a group of buildings of similar size.

The massing of all three buildings in this context ensures that the buildings appear to step down to fit into their surroundings, geometrically aligned within the streetscape. In order to fit into the current massing of neighbouring buildings, the four new buildings are aligned with buildings which are immediately adjacent. This aim is achieved by use of modular construction, where spaces between buildings can be linked; built volumes can be 'stacked' together in different configurations, providing a high degree of control over how the external forms and internal spaces are arranged while remaining economical. This prefabricated and modular system provides sufficient flexibility to provide complex arrangements of internal spaces that would be difficult to achieve with more traditional construction methods.

The existing corner building has windows and doors replaced using the visual language of the original fittings. The new additions do not seek to integrate the existing building visually within the language of the new buildings. Internal spaces are refurbished and a new ventilation system is introduced to meet contemporary requirements for building users.

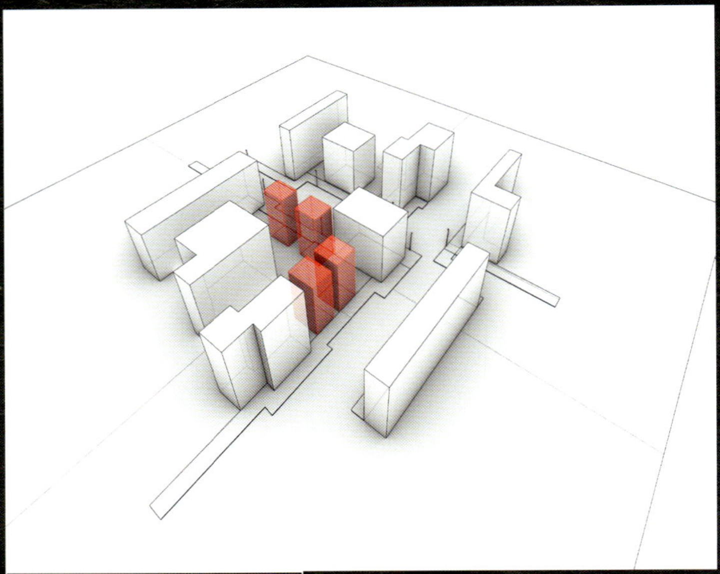

1. Project volume in site context

2. Project volume in site context

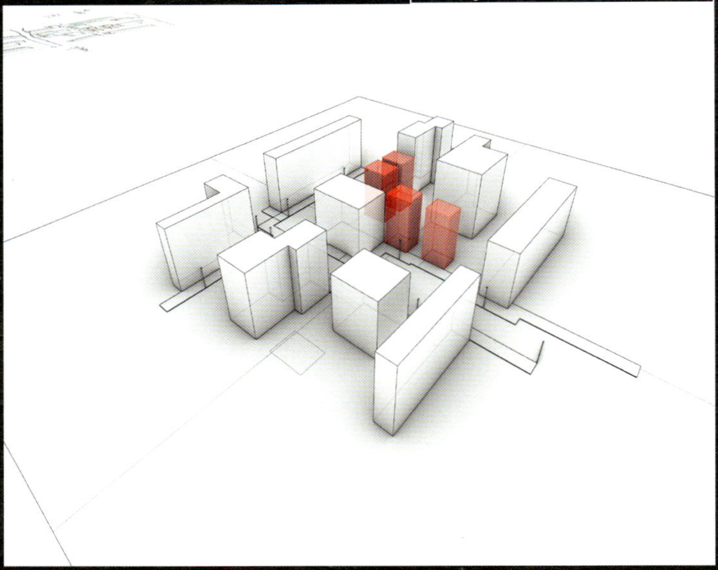

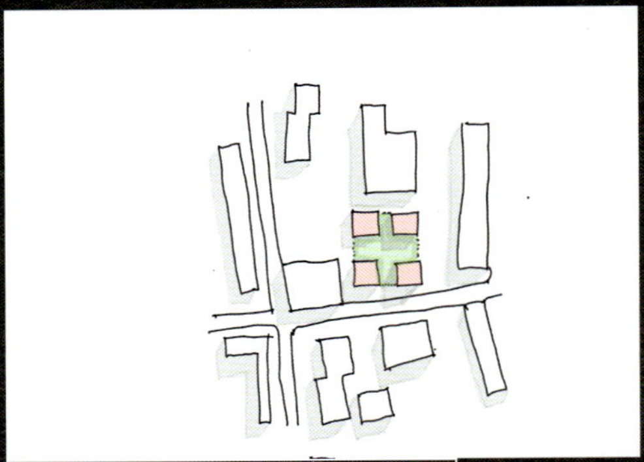

1. Site plan with surrounding context

Option 1 places all four blocks in a square in order to create a space between the buildings.

2. Site plan with surrounding context

Option 2 sets the buildings up on 'stilts' that visually separate the upper floors from the ground.

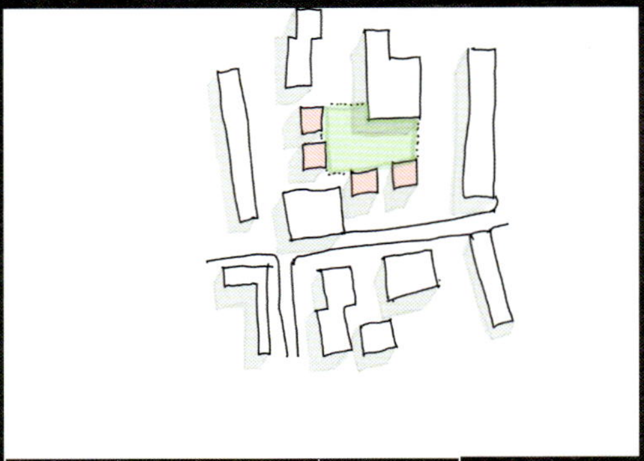

3. Site plan with surrounding context

Option 3 sets the buildings in two straight rows such that the pavilions appear to wrap around other neighbouring buildings.

Site context

This medical research cluster accommodates laboratory spaces for short research cycles, resulting in the need for small teams to occupy self-contained sets of linked spaces. Each 'pavilion' building comprises a set of mainly opaque walls with the design driven primarily by the need for mechanical ventilation and low levels of daylight in the laboratory spaces; requirements in common with similar research facilities. The building accommodates a mixture of laboratories on the upper floors, and research and development on lower floors. The functions carried out within the building can be clearly understood by a visitor to the building, as the spatial organisation is seen in the forms comprising each building.

The buildings are arranged as a set of four 'pavilions' arranged in a configuration to suit the irregular spaces available for their construction, while retaining the existing corner building; the pavilions are arranged in a rectilinear geometric arrangement in order to fit well with neighbouring existing buildings. The building at the corner of the site is retained since it continues to serve its purpose well; the environmental cost of demolishing this building and replacing it with something that would accommodate the same space was considered to be uneconomical. Within each mini-tower, the use of these spaces is slightly different from the other three pavilions but all are based on an arrangement of research laboratories, testing facilities, teamwork, whiteboard spaces and administration. Due to the sensitive nature of the work being carried out, the organisation of buildings is something of the reverse of what might be expected elsewhere; all spaces are set above ground floor level as vertically stacked external forms and internal spaces, instead of occupying a one- or two-storey building of larger footprint. The laboratories are set above the ground level access space; administration and meeting rooms are set on the first upper level, with laboratories being set above this floor level. Chillers and small air-handling units are located externally at the top of the building.

The design of each of the four buildings is driven primarily by a need for inward-looking spaces that still provide an outward expression of the

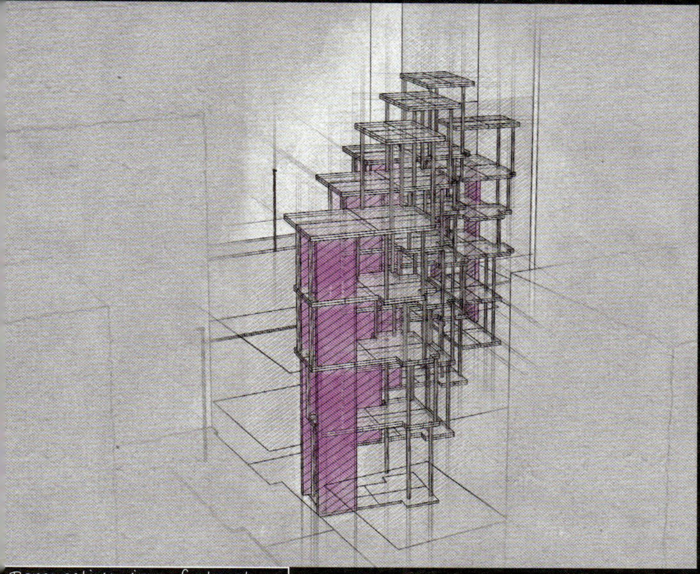

Perspective view of structure

2. Perspective view of structure

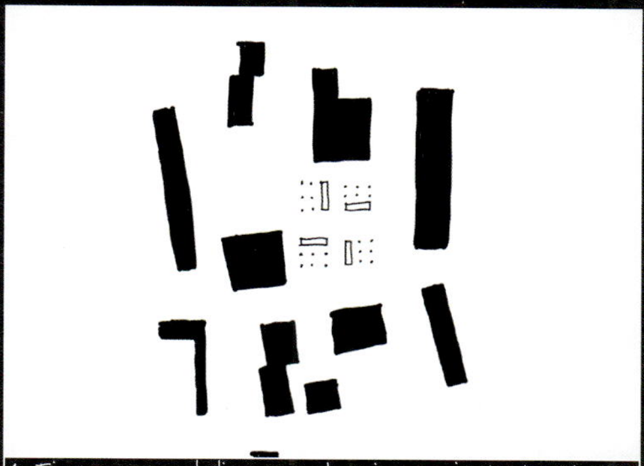

1. Figure-ground diagram showing project structure

Option 1
The structural grids of each pavilion are set at 90 degrees to one another; arranged within a square plan.

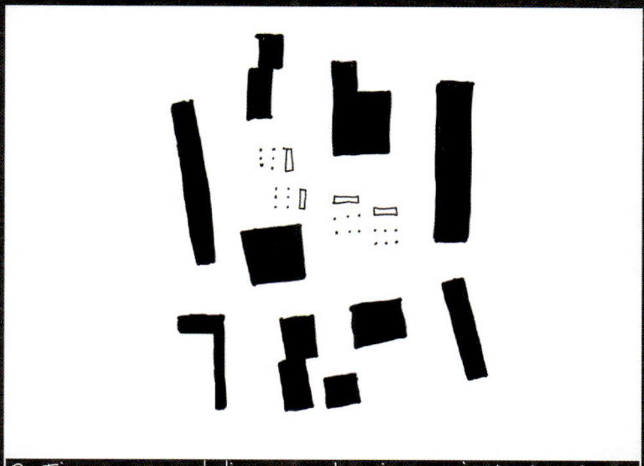

2. Figure-ground diagram showing project structure

Option 2
The structural grids of pavilions are set in two staggered rows that face the street.

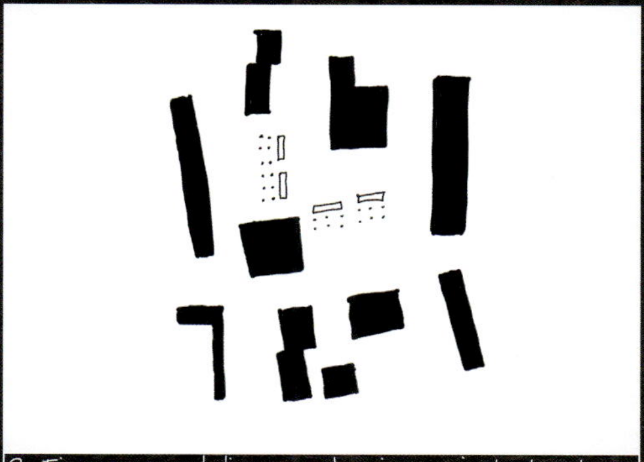

3. Figure-ground diagram showing project structure

Option 3
The structural grids of pavilions are set in two aligned rows that face the street.

technical tasks that are performed within. Functions within the building are internally focused, requiring little natural daylight. Ventilation and natural ventilation are provided to meeting rooms and administration offices; there is no specific need for natural ventilation in the laboratories. Each building is arranged over five floors; building 'high' rather than 'wide'. The form of each building is a consequence of the brief/program as interpreted in a set of pavilions; meeting the need for larger spaces to be accommodated at higher level and the smaller spaces to be within easy reach of the ground floor access facilities. The buildings are arranged very differently from a regular set of buildings; floor plates are provided at high level for reasons of privacy only.

Option 1 places all four blocks in a square in order to create a space between the buildings, both for access and external public space; they are all framed by the four buildings. The close proximity of the buildings, and the fact that the volumes are elevated, creates a space underneath the buildings which can be occupied; allowing the external space to extend under each of the four buildings. Option 2 sets the buildings up on 'stilts' that visually separate the upper floors from the ground; allowing the external space to continue under each pavilion. Pavilions are set perpendicular to one another, forming two stepped rows of two buildings, each with an external space in front; this external space extends partly under the buildings in order to provide a welcoming point of access to the main body of the spaces set above ground level. Option 3 sets the buildings in two straight rows such that the pavilions appear to wrap around other neighbouring buildings. The space in front of the pavilions is created by setting a constant distance from this neighbouring building; the space is formed by seating and hard landscaping, creating a semi-enclosed 'outdoor room' between the buildings. This external space creates a scale of space that can be inhabited through geometrically arranged seating; being formed as either a garden or as a grass lawn. With Option 3, the pavilions occupy most of the space available between the existing buildings. The arrangement of the pavilions in each option responds to that of adjacent buildings, while maintaining a distance between each pavilion.

1. Sectional perspective view

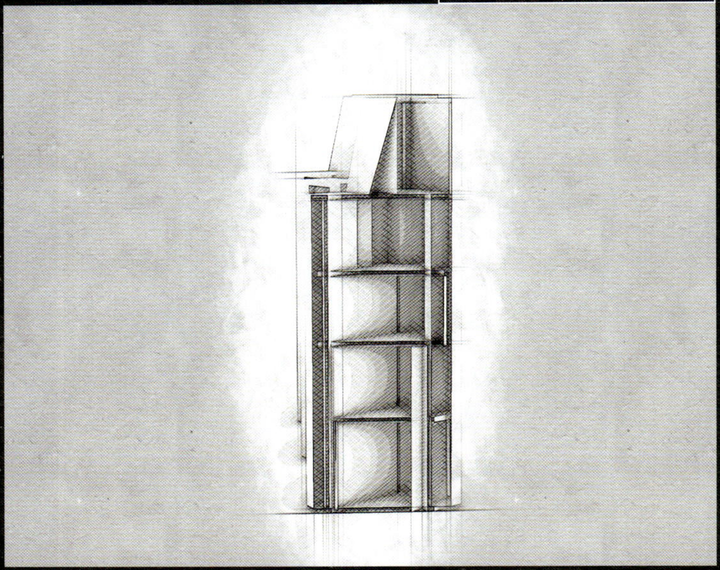

2. Sectional perspective view

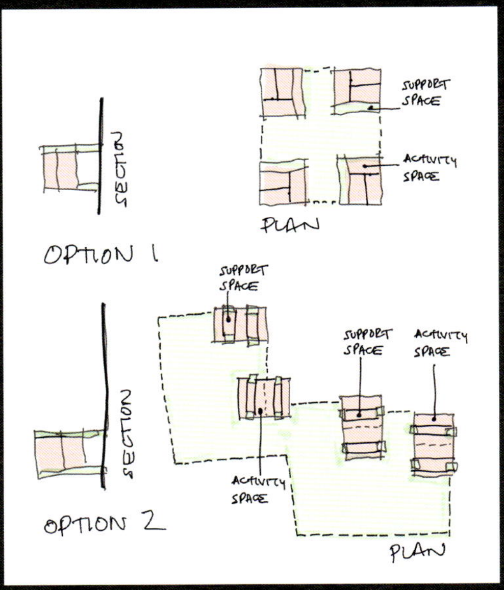

Spatial arrangement

Option 1 puts the service spaces along one side of the building, with the activity space set alongside that space.

Option 2 rotates each pavilion form through 90 degrees in relation to its adjacent pavilion.

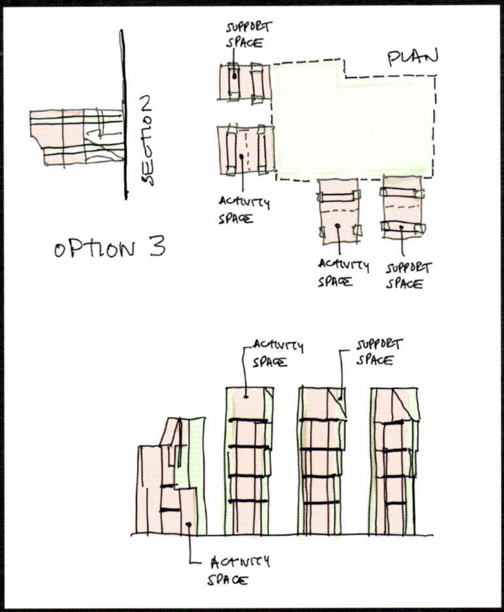

Option 3 takes the service cores and sets them partially externally.

Spatial arrangement

The spatial arrangement within each pavilion comprises two sets of space arranged parallel to one another; the 'support space' or service space supports the 'activity space' or main floor plates, which are set immediately adjacent. Given that each building is laboratory based, the relationship of support space to activity space is not as clearly defined as is the case in buildings in which less technology-driven activities are carried out. In these four pavilions, most of the space is actually service-based. Laboratories are set out as closely arranged spaces; meeting rooms are much more open and fluid in terms of their distribution, which can be changed to suit the changing needs of each team. Stairs, lifts/elevators, washrooms and vertical service runs comprise the main parts of the support space; arranged in each of the mini-towers with a slightly different arrangement to suit the laboratory equipment being accommodated.

The spatial strategy for Option 1 is to put the service spaces along one side of the building, with the activity space set alongside that space. All four pavilions are arranged in a similar manner; each pavilion is rotated through 90 degrees in relation to its immediate neighbour; placed together to form a square-shaped arrangement. The creation of four separate pavilions in place of one single building creates external spaces which are as significant in their arrangement as that of the internal spaces. This approach forms external spaces which transition from an outdoor room, to a sheltered portico for access, to a plinth for the positioning of vertical circulation core walls. These walls are always set behind the projecting structural 'stilts' that support the upper floors.

Internally, the spatial arrangement is straightforward, with access from the ground floor leading up through a stair or lift/elevator to service cores at the upper floors, with washrooms on each floor and an adjacent clear and open floor plate. Options 2 and 3 have the same spatial arrangement within each pavilion, but each pavilion form is rotated through 90 degrees in relation to its adjacent pavilion. Views between buildings are created as oblique views to the two buildings set at 90 degrees to whichever space is creating the viewpoint. Option 2

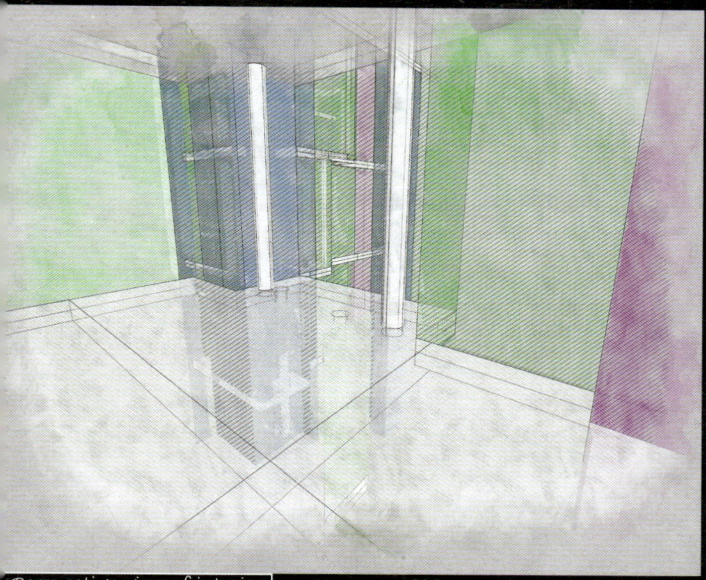

Perspective view of interior

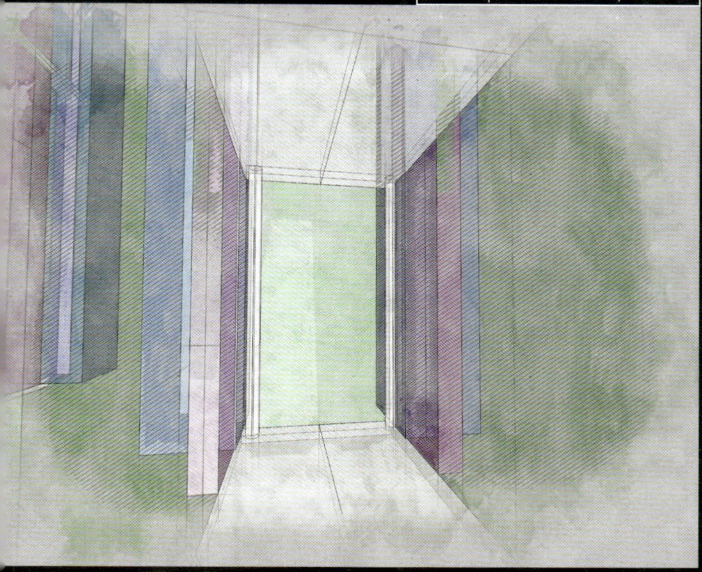

2. Perspective view of interior

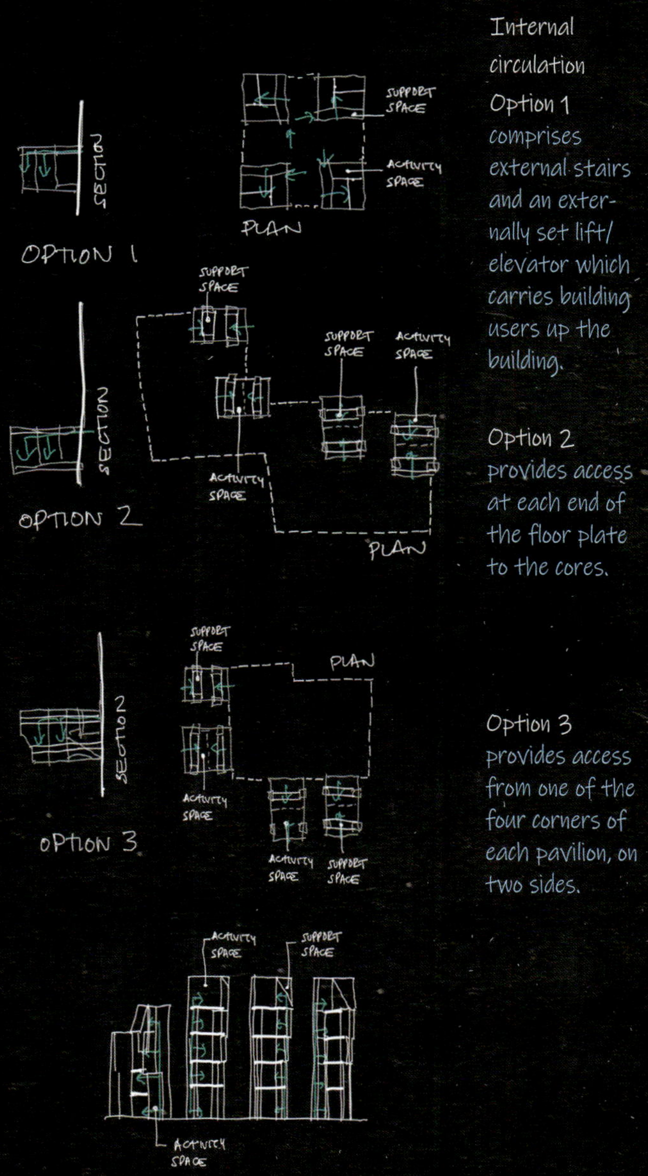

creates a stepped arrangement of volumes in order to distance each pavilion as far apart from its neighbours as possible. Option 3 takes the service cores and sets them partially externally, forming an enclosed volume at the 'back' of the pavilion, away from the facade facing the newly created external space.

Internal circulation

The internal circulation for Option 1 comprises external stairs and an externally set lift/elevator which carries building users up to the research labs and rooms on the upper floors. Research work is carried out away from the public gaze. There are no reception spaces at ground level in each building; security is carried out on a floor-by-floor basis. As a result, no reception space is needed at ground floor level in any of the pavilions. In Option 1, internal circulation is provided through the access courtyard or 'outdoor room'. The edges are the end walls of the courtyard, at ground level, rising directly to the floors above. Access is by stairs which are set at one end of the circulation core; alternative access by lift/elevator is provided at the opposite end of this core. In Option 2 a similar provision is made, with access at each end to the cores for stairs and lift/elevator respectively, which allows for a single point of access into each pavilion; a secure lobby is created at each floor level. In Option 3, access is provided from one of the four corners of each pavilion, with access provided on two sides; circulation is created around the edges of the floor plate in order to join the four service cores, which contain washrooms and storage areas, in addition to the two corner cores that provide staircase and lift/elevator access.

The arrangement of space was considered through three options; the options vary from a cluster of four buildings to two terraces, to two stepped arrangements that invite the public to use the external space. Option 2 is the preferred option as the arrangement creates an external space that is protected from the adjacent roads. The arrangement within Option 2 is of all blocks being set slightly out of alignment with the building's neighbour, creating oblique views up and down the street to reveal the pavilion forms and create a visual dynamism on the street.

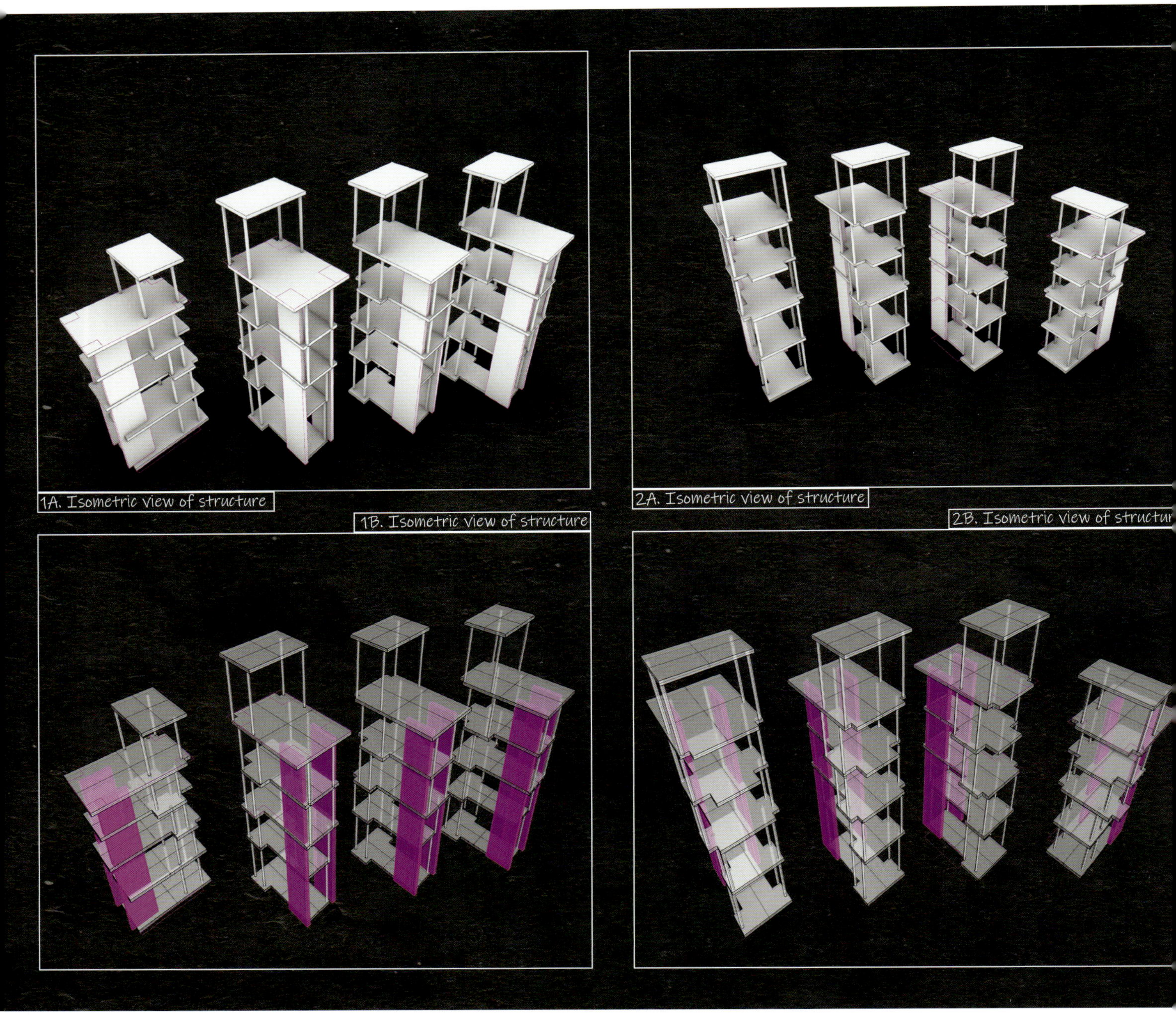

Structural strategy

The structural strategy is to use a steel frame that supports the prefabricated modules that enclose interior space. This strategy is based on a triangulated exoskeleton frame supporting prefabricated light gauge steel floor cassettes and prefabricated services assemblies, together with a facade system that encloses the complete set of assemblies. Stair enclosures form part of the set of structural components. Laboratory modules are installed as prefabricated and preassembled modules which are slotted into the building frame and left in place so they can be removed and replaced or refurbished with some ease, without disturbing other parts of the building construction; this is very much a 'plug-in' building where packaged technical components arrive preassembled for slotting into place. This set of buildings will see many changes during its life-span; an investment has been made in creating an exoskeleton structural frame which supports a complete system for natural and mechanical ventilation, which is enclosed in an envelope system that provides full control of daylight entering the building. External forms and internal spaces are fitted within the exoskeleton; the exoskeleton is designed for the full lifespan of the building, while the mechanical and electrical equipment that is fitted around it is considered to be installed for the shorter term. This is very much an exoskeleton design-driven project. The exoskeleton supports the stairs, the roof glazing and inclined glazed areas at high level. The exoskeleton comprises a mixture of a steel frame and large-scale steel-framed shear walls on both sides of the building, joined by beams and trusses. Stairs are formed in prefabricated metal, delivered to site and installed as part assemblies.

This approach to prefabrication is very much about the near future of the building, where changes are already anticipated as part of the changing work environment. A key part of the structural strategy is that the building frames are designed to adapt to change without requiring structural changes or demolition.

The structural strategy for Option 1 is of a steel frame with columns at each corner, forming a series of floor plates which are stiffened by an

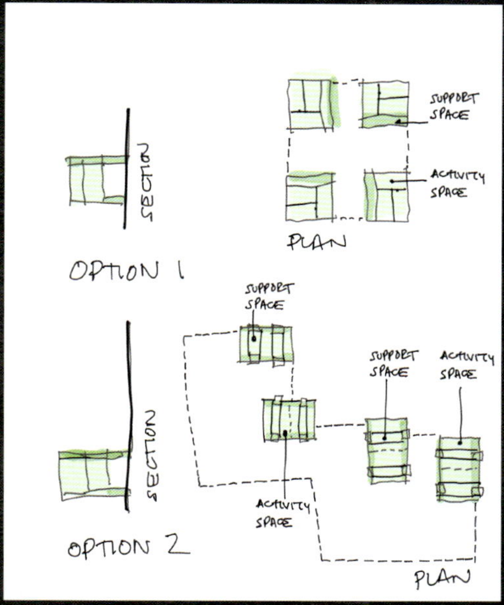

Structural strategy: Zones
Option 1 provides a steel frame with columns at each corner.
Option 2 has the service core extending down to the ground level.

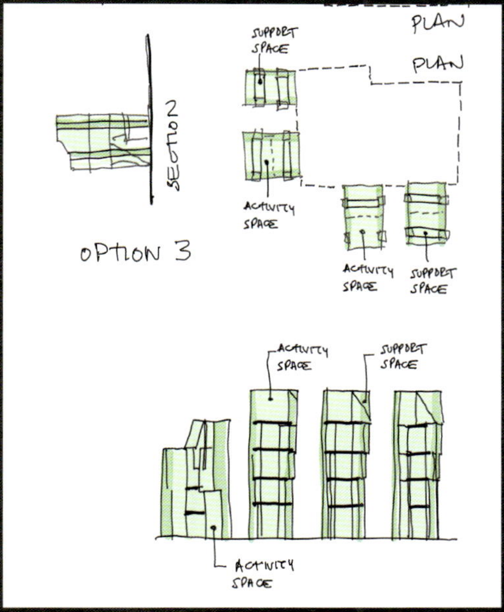

Structural strategy: Zones
Option 3 responds to the requirement for enclosed 'back-pack' services units on the sides of the building.

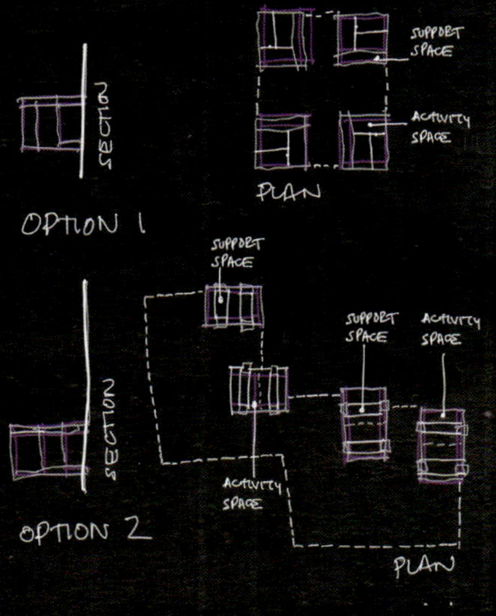

Structural strategy: Framing
Option 1 creates a series of floor plates which are stiff-ened by an addi-tional column in the support spaces.
Option 2 provides an environmental transition zone from outside to inside.

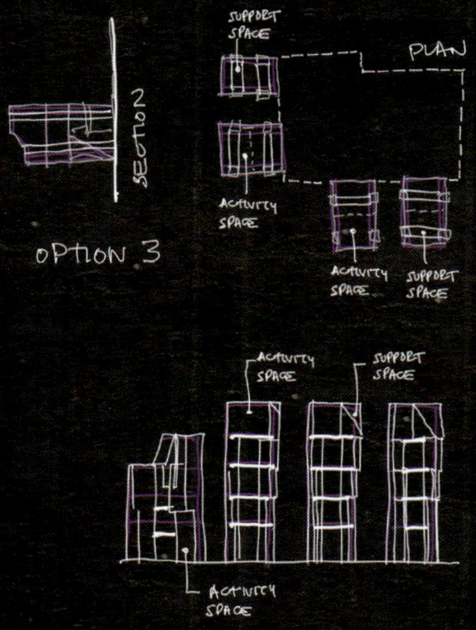

Structural strategy: Zones
Option 3 encloses the columns for the steel exoskel-eton with ser-vices units.

additional column in the support spaces which enclose the circulation, washrooms and storage space.

Option 2 has a similar structural solution, but the service core extends down to the ground level and provides an environmental transition zone from outside to inside.

Option 3 continues the theme of a transition space from outside to inside by having staircases which are semi-external and which are heated or cooled by the recycled air from the spaces above. The structure responds to that requirement by having enclosed 'backpack' services units on the sides of the building which also enclose the columns for the steel exoskeleton. This frame has two corner legs; one side at ground floor, and the service core columns for enclosure on the opposite side at ground floor level. All structures are thermally insulated by their facades.

The exoskeleton is created as a single form of equal size in all four pavilions. The structural walls beneath, continuous with columns beneath that take the structure down to ground level, vary in dimensions in response to the height of the tower. The structure also responds to the modules of interior space that are set within the exoskeleton form. The exoskeleton becomes standardised at this higher level in order for its form to be repeated in all buildings, all of which have a different overall height. This approach allows the structure to remain economical; columns at ground floor level have a different configuration of structural wall behind them in order to create an efficient continuity between them through the full height of the building. The structural frame introduces an effect of the sculpting of its form in order to reduce the quantity of material used, and to create a visual expression of the functions behind that of the spaces within each pavilion. The exoskeleton structure comprises a twin-wall diaphragm configuration in the upper parts and reverts to regular reinforced concrete wall at the lower level. This language of exposed reinforced concrete follows that of neighbouring buildings; a white architectural finish of the pavilions follows that of other buildings around the campus which have some three-dimensional modelling of spaces linked to secondary steel framing.

1A. Isometric view of environmental volumes

1B. Isometric view of environmental volumes

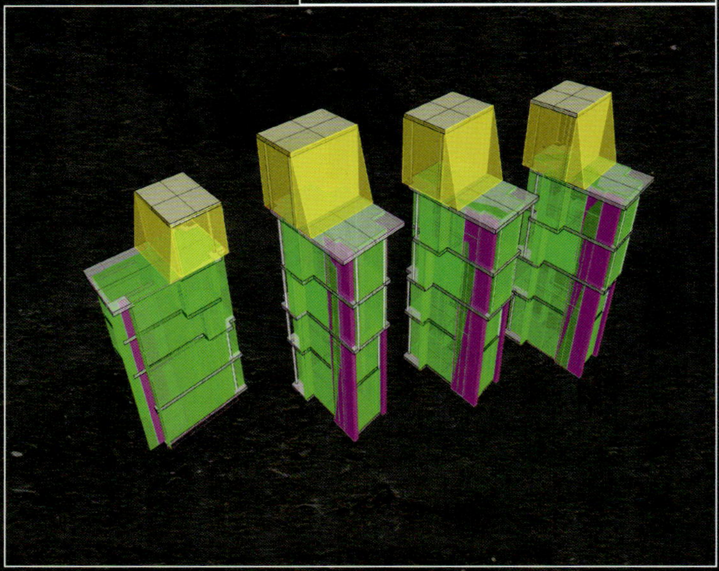

2A. Isometric view of environmental volumes

2B. Isometric view of environmental volum

Environmental strategy

The environmental strategy is one of having the mechanical and electrical equipment components externally mounted in the voids between elements of the structure. The mechanical and electrical equipment is mounted externally in order to minimise the size of the building enclosure. Four primary environmental zones are formed in each tower: the wedge-shaped enclosures contain technical equipment; laboratories form the second zone; meeting rooms and offices form the third zone; partially accessed areas of the ground floor form the fourth zone. The zone accommodating meeting rooms is supported by and accessed from the arm-like elements of structure, forming bridges to the main floor plates. The primary structure also supports the zone enclosing the laboratory spaces. Only the communal areas and staircases have night-time cooling for natural ventilation. There are no natural ventilation spaces other than the communal areas in Pavilions 3 and 4.

Mechanical ventilation, laboratory air-filtering equipment and heat recovery devices are set at high level within each pavilion. Additional 'backpack' service units are fixed to the facades of each pavilion; each unit contains mechanical equipment that is ventilated directly to the outside of the building.

Since the facades require little glazing, the opaque areas of wall are used to accommodate the service routes of the externally fixed mechanical and electrical equipment which extend down the external walls; mainly vertical air ducts and electrical risers.

The environmental strategy for Option 1 is to divide the support space and the activity space; each pavilion has a similar arrangement, but their different orientations create the need for different environmental zones in each; in each building the effects of solar gain will be slightly different as a result of the orientation of the building in relation to both the sun and to shadowing effects created by taller neighbouring buildings.

In Option 2, the same strategy is used for environmental zones; each building is divided into activity space and support space. The two pavilions in each row have similar solar exposure, but one building partly overshadows the other; the environmental conditions in each are simi-

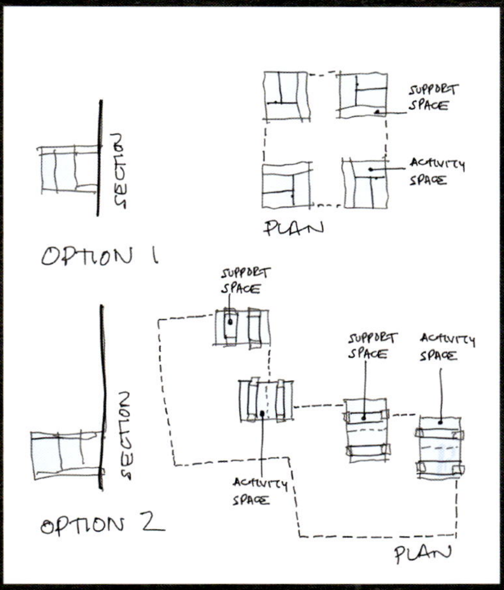

Daylight strategy for direct sunlight and cross ventilation.

Option 1
Daylight and air are admitted to all activity space; daylight is subdued in linear support spaces.

Option 2
Daylight and air are admitted to all activity space; daylight is subdued in support spaces.

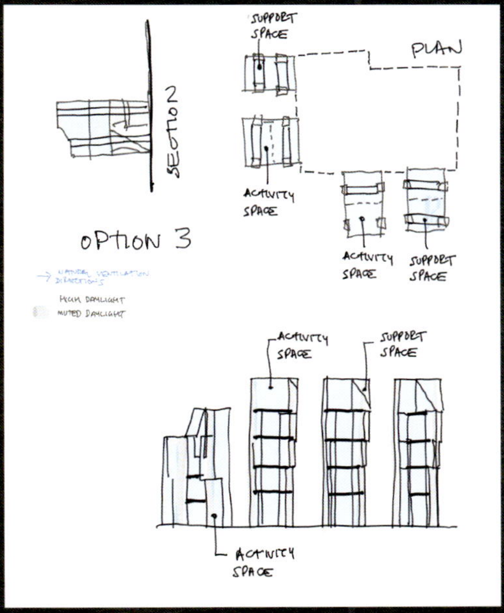

Daylight strategy for direct sunlight and cross ventilation.

Option 3
Daylight and air are admitted to all activity space; daylight is subdued in 'backpack' spaces.

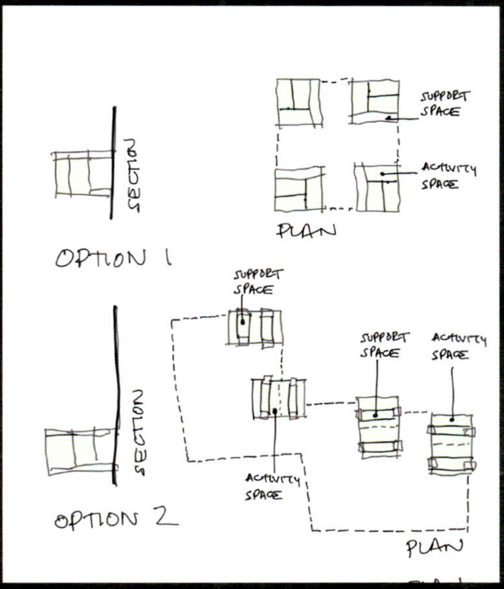

Environmental zones

Option 1
divide the support space and the activity space; each pavilion has a similar arrangement.

Option 2
the same strategy is used for environmental zones; each building is divided into activity space and support space.

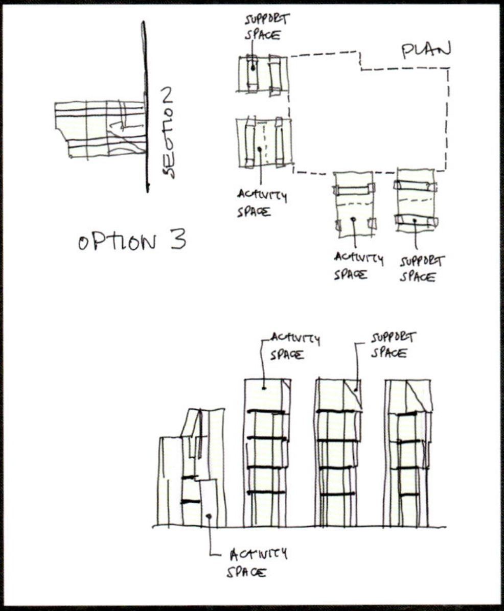

Environmental zones

Option 3
has all four buildings in a similar orientation to Option 2.

lar in each row of buildings. In Option 3 all four buildings have a similar orientation to Option 2; the single activity zone has similar environmental conditions in each building with a modest change in shading effect from that of Option 2. In all three options the service zones and support zones have opaque facades, which provide both thermal insulation and protection from the effects of the sun. As a result, the other three sides of the building in each of the options provide the low levels of daylight required in spaces; achieved by external shading devices fixed to the facades.

From an environmental point of view, each of the three towers in this design has a slightly different geometry in the way the spaces are organised as pavilions. Pavilion 1 has laboratories and associated equipment on the side facing the courtyard; while mechanical and electrical equipment is set on its opposite side. Pavilion 2 has a large external staircase in which services are enclosed; these buttresses accommodate the service spaces within them. Pavilion 3 has additional meeting rooms at the top of the building, with a large glazed space on the side facing the courtyard; primarily for meeting and greeting visitors. The third pavilion has stairs integrated at the lower levels. The large frame seen at the top of each building is used to enclose roof glazing as well as accommodate stairs. The fourth mini-tower has a wedge-shaped profile at its base where the inclined wall at the lower level allows visitors to pass through a ground-level door; gaining access to the lift/elevator and up to the upper reception space.

Since all internal spaces are required to be both sealed and ventilated mechanically, the opportunity for natural ventilation exists only at ground level. The ground floor external areas are shaded from the effects of the sun for most of the year, providing space for staff and employees of the neighbouring building to use the external space; tables and chairs are provided for both informal gatherings and meetings. The stepped arrangement of the buildings creates an individual space in front of each building in order to avoid a continuous colonnade that would be used as a path rather than as a sequence of external shared spaces used to work and to rest.

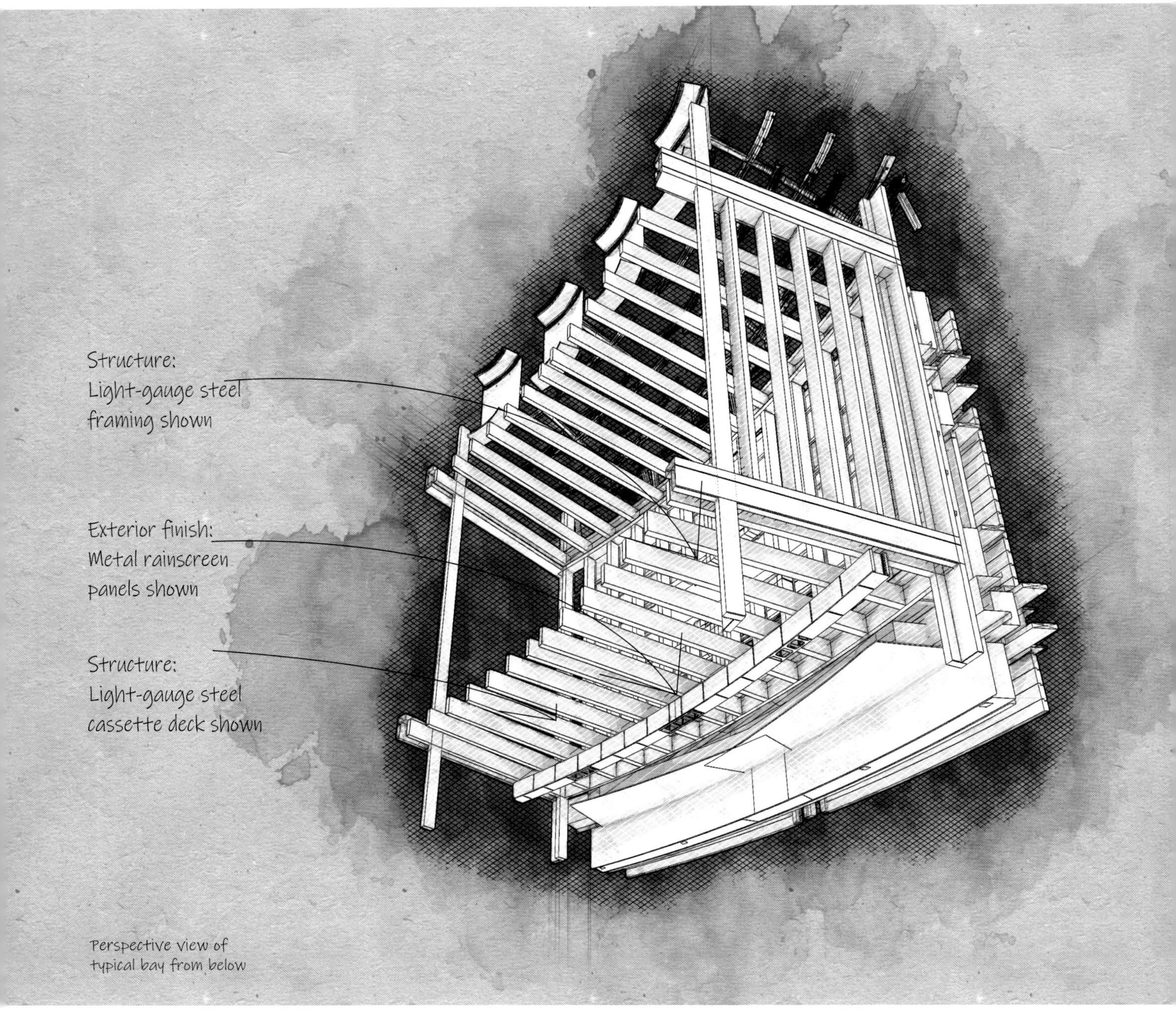

Perspective view of typical bay from below

Typical bay adjacent to facade

The external envelope is formed as an exposed structure with cladding panels; the finishes to the exoskeleton create the general form of the external envelope. Its double-wall, twin-wall configuration has thermal insulation inserted into the void between the two layers of the diaphragm wall.

The two parts of the diaphragm provide the external and internal wall finishes, which are of smooth concrete panels. The joints in the outer surface of the exoskeleton formed by the formwork used to cast the concrete in other parts of the exoskeleton are constructed as steel frames. These frames are clad in metal panels which are not required to be thermally insulated as they extend outside the building. The internal volumes are created as prefabricated modules, assembled in areas which accommodate ducts and associated pipes on the facade. These walls are largely opaque, but are highly modelled, with glazing introduced to provide a visual 'reading' of the interior of the building on the lower floors, orienting the visitor. Inclined areas of glazed roof are provided where they form glazed slots in what is otherwise an opaque roof. The exoskeleton on the vertical faces at the top of the building has glazed slots introduced into the projecting element; little transparency is required in this building. Panels enclosing the inclined roof are of opaque glass with an off-white colour to continue the visual language of the concrete exoskeleton. The structure which is visible on the outside and which holds the prefabricated modules in place is the driving factor behind the envelope design. The prefabricated modules which form typical bays are supported and stabilised by the exoskeleton frame.

A key consideration in the design of the external envelope was to enclose the structural connections between the external concrete base structure and an external steel structure; the internally located modules are clad externally. Modules are enclosed to provide both weather protection and thermal insulation. An outer protective layer of panels which is dark in colour contrasts visually with the concrete structure and is similar to the bronze-coloured panels used on some other buildings in the campus. The cladding is required to connect to the external reinforced

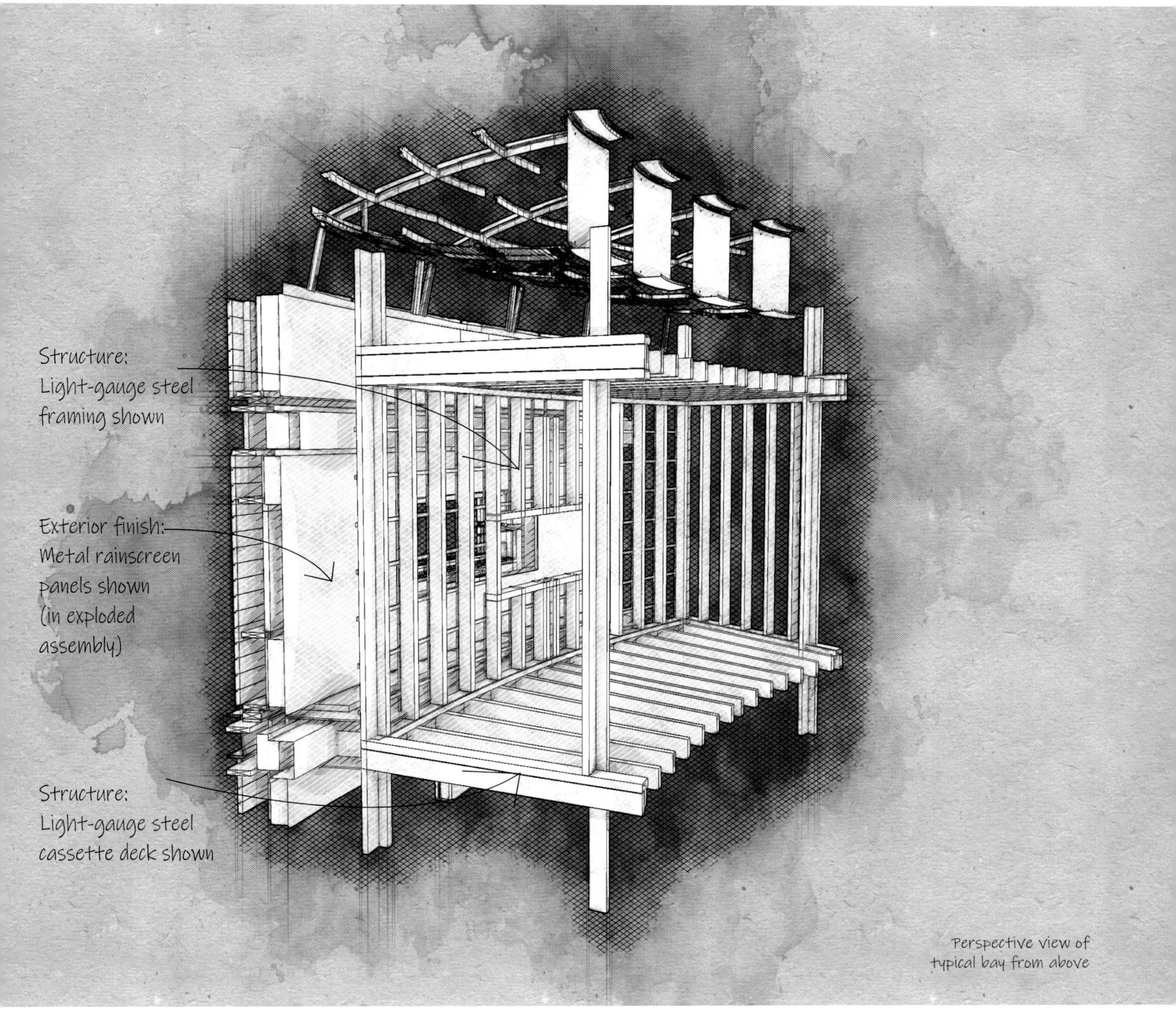

Perspective view of typical bay from above

concrete structure, which is weather resistant in its own right, but the connectors are made from steel and are required to be weather protected and thermally insulated. This configuration is independent of the choice as to whether the connectors are made from stainless steel or mild steel, as both require weather protection for durability. This project uses a mixture of mild steel and stainless steel fixings with provisions for isolators in place to avoid bimetallic corrosion.

All fixings are required to have thermal insulation and be weather protected; consequently, they make use of the same language of panels as those used for the cladding panels of the prefabricated modules. Joints are formed in a similar way at the interface of the glazed panels on the vertical cheek and the exoskeleton at high level. Inclined roofs, which are a mixture of glazing strips and metal panels, use a variation of this principle. Those joints requiring connectors between the exoskeleton and the elements described have a weathertight seal between the two elements, which avoids there being complete separation between exoskeleton and building, which would have made the concept uneconomical. The exoskeleton has recesses set into it for weather protection strips to be inserted and sealed, such that overlaps make use of a method used in traditional building construction, and which has been used successfully in some of the reinforced concrete frame buildings on the same campus, constructed some years ago.

The connectors between the primary external exoskeleton and the internal modules, as well as some of the external steel structure, have thermal breaks between the outside and inside faces. This use of thermal breaks ensures that the level of thermal insulation provided meets performance requirements and also avoids potential interstitial condensation occurring within the construction as it passes from outside the building to inside the construction.

The external envelope is formed in two parts: the exoskeleton, which is not thermally insulated as a result of it being set outside the thermal envelope; and the thermally insulated modules that contain internal spaces, for which support is provided by the exoskeleton. A primary advantage of this approach is of prefabrication in their construction.

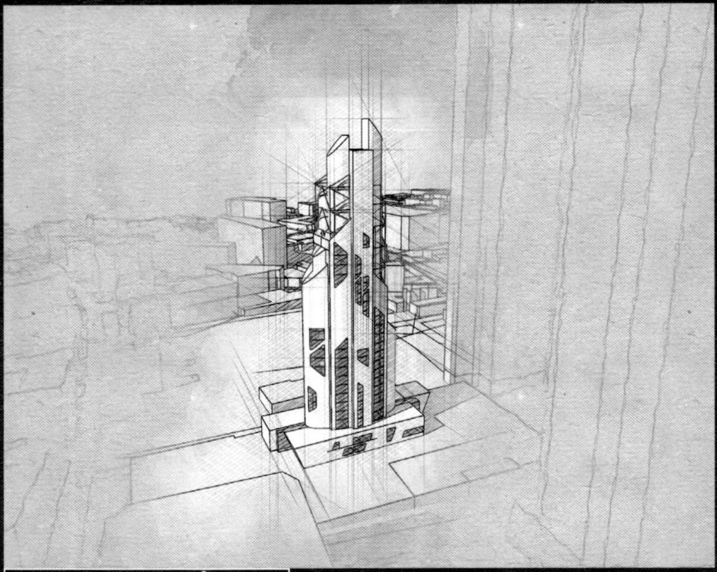

1. Perspective view of exterior

2. Perspective view of exterior

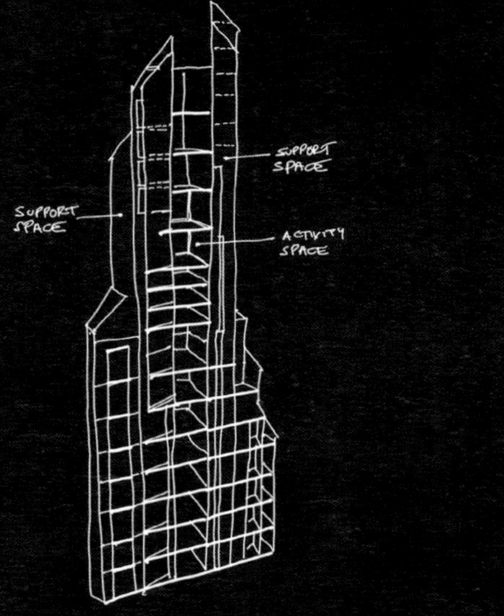

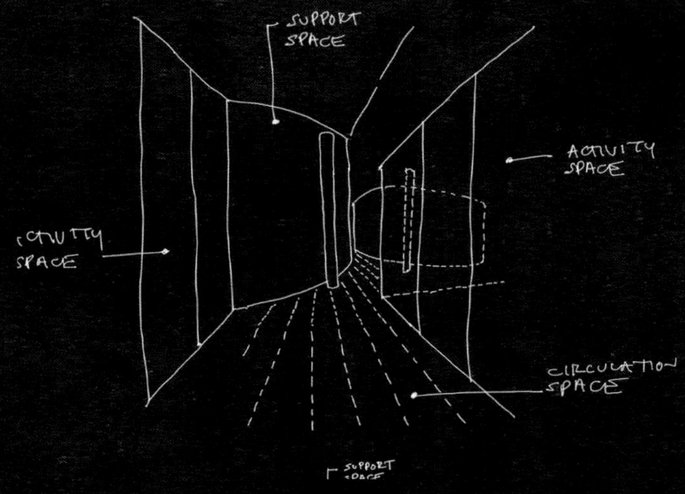

Visual variations of scale in primary steelwork

In order to integrate smaller-scale steelwork to support the facades with the primary structural steelwork, variations of scale are incorporated in all the structural framing by the use of a single structural type across the project. This structural approach is made, not by introducing an additional layer of secondary structure, but by introducing variations in the scale of structural components and assemblies to incorporate facades and secondary structures in a continuity with primary structure.

An alternative view would have been to consider the complete structure as being all 'secondary' structure, with framing that would have been associated with the external envelope providing the necessary points of support for the facades, roofs and internal features such as glazed walls. This results in the structure being 'revealed' to view rather than being 'wrapped' in thermal insulation and an outer protective layer. Both glazing and opaque panels are set on the external facades between the structural components; this results in a structural frame with window openings set within the frame. This is a construction technique that requires accuracy and verification in the dimensional setting-out of structural members and facade members; there is a hierarchy in the differences in construction tolerances of primary steelwork and of aluminium panels fitted to it, for example, that makes possible the close physical connections between glazing and opaque cladding.

This approach of a uniformity of structure used as load-bearing walls and load-bearing cores allows the building to be modular, with the advantage of having a form of framed construction allowing the complete configuration of the building to be modified and optimised during all design stages in order to reduce the amount of external facade, for example. The building can be configured such that there are service cores and support space at the sides of the building instead of within a single volume. In addition, building units can be grouped into a set of four, with some support spaces for a core in the centre, and with two service 'spines' set on the outside of the building for support space. All these configurations are modular and use a 'kit of parts' developed for

. Perspective view of exterior

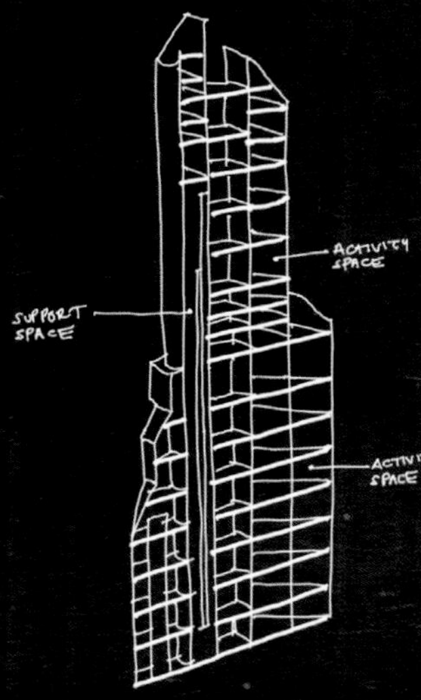

4. Perspective view of exterior

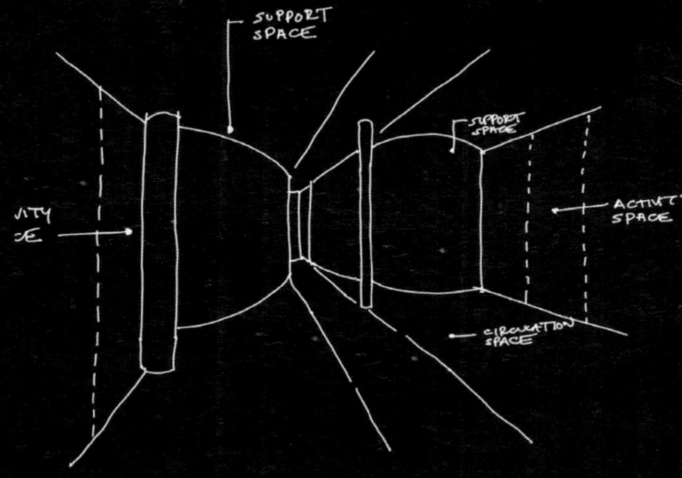

this building design. The use of a smaller-scale structural system, with interchangeable components, allows this structural design approach to be used across a single large-scale project; this approach can be made to be economically viable when applied to a single project only. This mini-tower appears to be sitting on a set of two parallel blocks of building which are set immediately beneath the tower, but which extend outwards from the tower. The podium blocks appear to support the mini-tower visually, allowing the two towers to appear to come to rest on its small podium structure. The mini-tower has the advantage of being able to have most spaces daylit from opposite sides of the tower, which gives rise to its longitudinal form having a long, thin form in plan which brings the benefit of natural cross ventilation; valuable in a residential tower.

There is a requirement for rooms to have clear views out from the building, which is achieved in the living, kitchen and dining spaces as well as in the more enclosed spaces comprising bedrooms and bathrooms. Many of these spaces extend across the width of the building and have some bathrooms next to the facade in order for them to be ventilated naturally and to benefit from fresh air directly from outside. Areas of opaque facade have a mixture of metal panels which are translucent and which are set forward of the partly glazed facades. Areas of opaque facade are clad in metal panels; the aim is to reduce the amount of glazed facade where it brings no benefit environmentally or from the point of view of their utility.

Areas of facade which enclose air-handling ducts are visually expressed through five parallel sets of structures, each expressed as a separate volume, which comprise the overall form of the building. Each volume is offset from its neighbour, resulting in the characteristic appearance of five bundled tubes that form the mini-tower. The service spaces extend above the top of the building, where the structure is expressed visually but is partially concealed from view by facade panels which are modelled to follow the form of the structure behind, resulting in the projection of panels around the structural framing as a result of the thermal insulation that surrounds these elements.

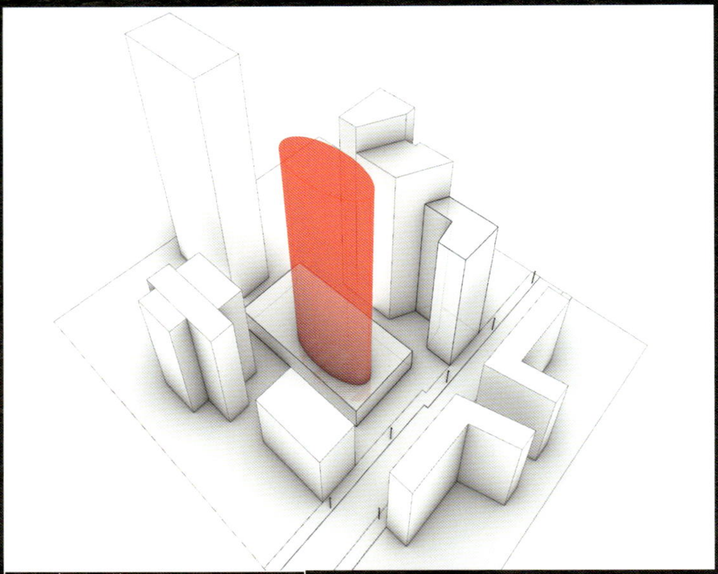

1. Project volume in site context

2. Project volume in site context

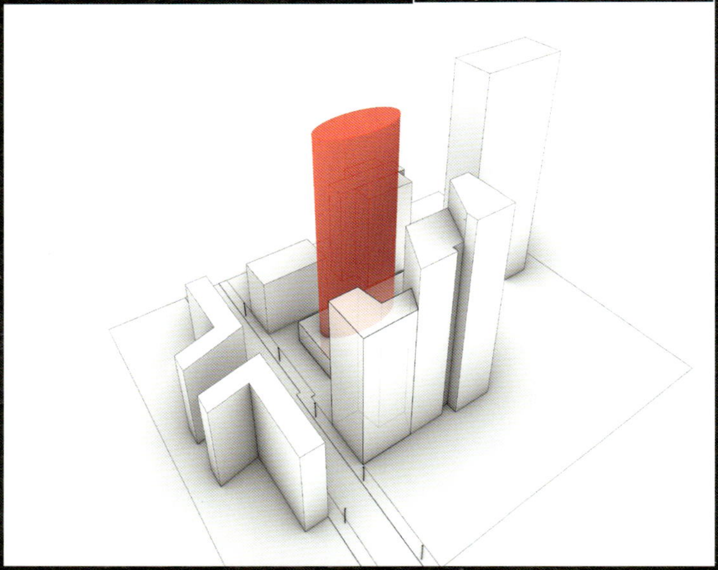

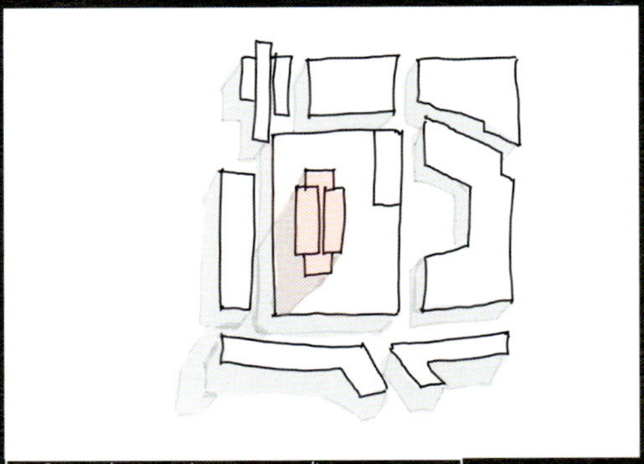

1. Site plan with surrounding context

Option 1 sets out the two apartment blocks that comprise the tower into two blocks with a central circulation space.

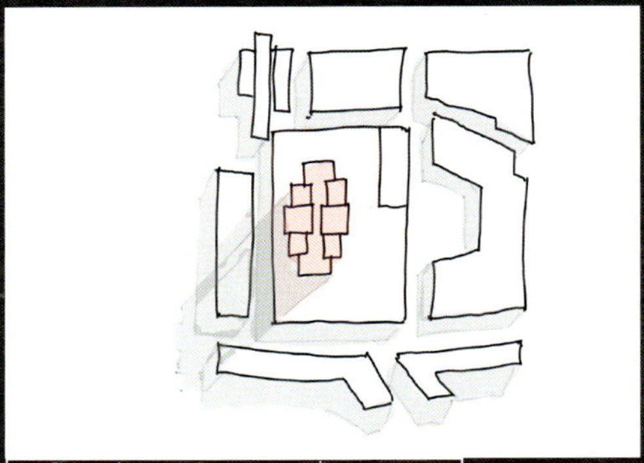

2. Site plan with surrounding context

Option 2 reflects the aims of the brief/program more closely and provides an easier subdivision of space within each apartment.

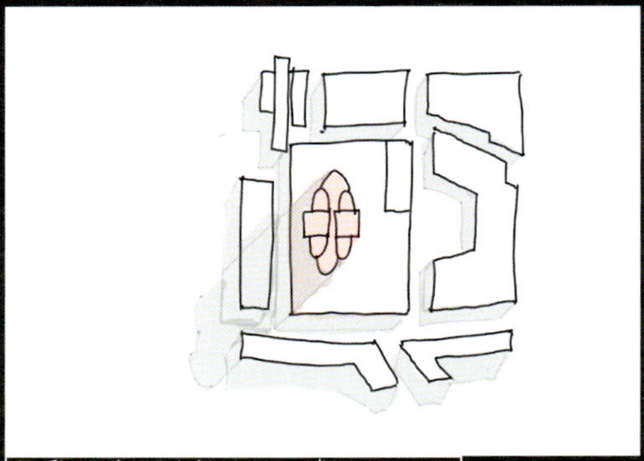

3. Site plan with surrounding context

Option 3 provides differentiated space, with the apartment blocks set out with curved ends for each block.

Site context

This tall apartment building comprises a cluster of five towers, set in a hot, dry, city environment. The cluster of towers is divided into five sets of spaces on each floor plate; spaces are linked together to form an oval-shaped tube that includes lifts/elevators and stairs, together with localised mechanical and electrical equipment installations on each floor. The towers, of different height, are arranged in order to support each other structurally to form a single tower. The shape of each of the five towers is oval shaped in order to create internal spaces with curved edges that facilitate a limited flow of naturally ventilated air through the facade by considering the effects of local wind turbulence around the facades. The cores set between each tower have banks of lifts/elevators that serve sets of five-storey zones that extend over the full height of the building, serving sky gardens at 10-storey intervals. The building is arranged over five-storey groups of floors that are self-contained sets of spaces. Option 1 sets out the two apartment blocks that comprise the tower into two blocks with a central circulation space. The blocks are rectilinear and have a rectilinear podium at ground level. A greater differentiation of apartment size is shown in Option 2, which reflects the aims of the brief/program more closely and provides an easier subdivision of space within each apartment. As per Option 1, a rectilinear podium is shown with rectilinear blocks. Option 3 uses a similar arrangement to Option 2 with differentiated space, but the apartment blocks are set out with curved ends for each block; the podium is also curved. In each option, the footprint of the building on the site is the same.

There are two parts to the site context; that of the podium and that of the tower that rises above it. The preference for a podium and mini-tower is driven by the need to have small-scale spaces in the tower and significant retail activity in the podium. This approach suits small stores in the podium, and well-daylit and well ventilated, but solar protected, apartments above. One approach which was outlined at the beginning of the project was to have a single block rising up to 10 storeys above street level that would have light wells within it and have apartments

. Perspective view of structure

2. Perspective view of structure

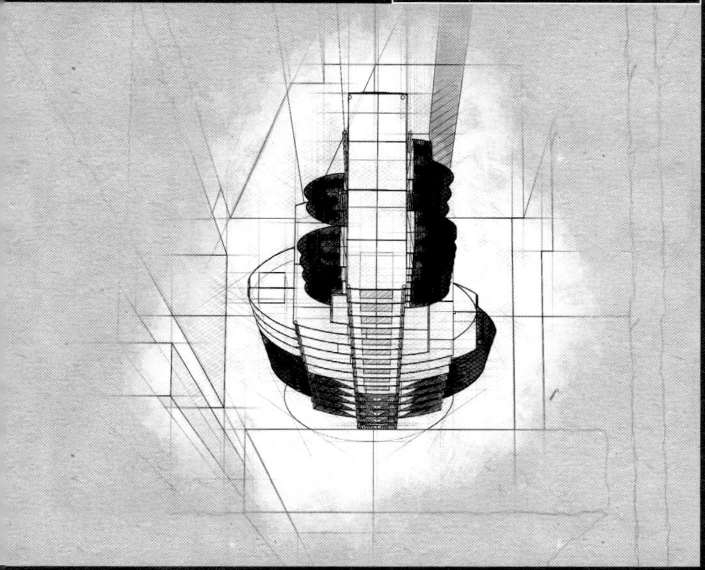

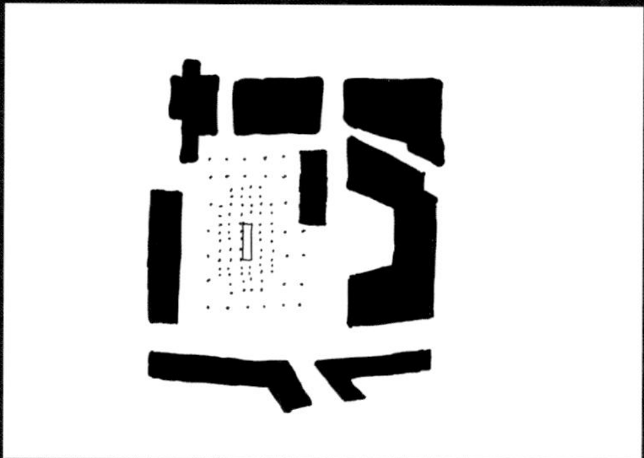

1. Figure-ground diagram showing project structure

Option 1
The structural grid is rectilinear but is distant from neighbouring buildings.

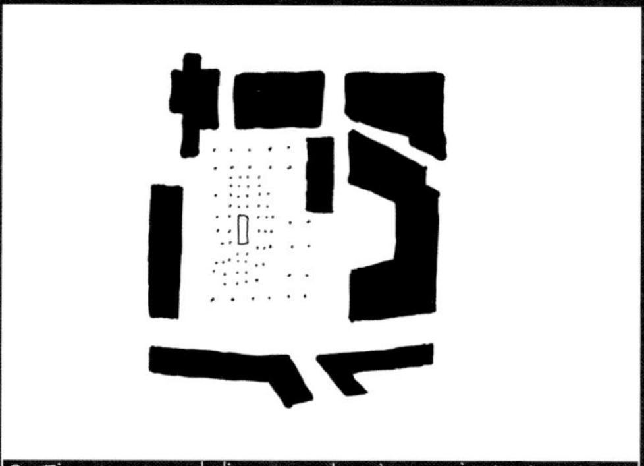

2. Figure-ground diagram showing project structure

Option 2
The structural grid is rectilinear but takes up gentle curves in response to the podium being surrounded by roads.

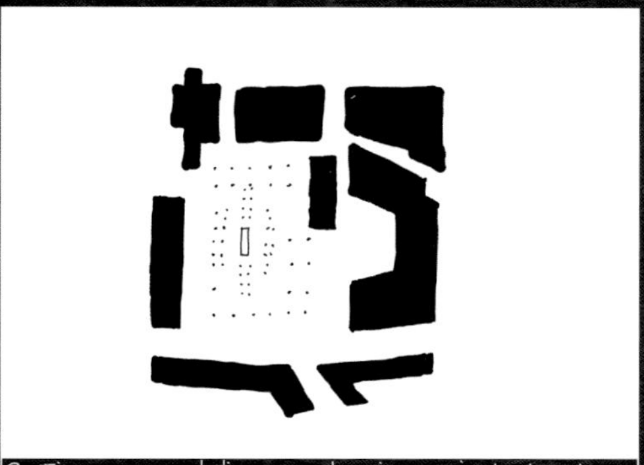

3. Figure-ground diagram showing project structure

Option 3
The structural grid takes up the gentle curves of Option 2 but provides greater transparency in one direction at ground floor level.

with views onto the street as well as into a lightwell. This approach did not take advantage of the opportunity to be able to build higher and allow the retail activity and commercial activity to be established at the podium level. This approach also made it difficult for apartments to be provided with fresh air and be above street noise and its associated activities. The development of the mini-tower concept allowed single apartments to either occupy a complete floor or be subdivided into smaller apartments. Some apartments extend over two floors; the distribution of different apartment sizes results in the characteristic form of bundled volumes attached together, which provides a differentiation of space within the building with opportunities for every room to be set on an external wall, with its advantages of natural ventilation. This approach was taken in the mini-tower configuration instead of a 10-storey building occupying the full extent of the site, with the associated disadvantages of the latter option. The mini-tower also provided opportunities for health club and restaurant facilities for the residents within the tower. A small shop for day-to-day needs is set at the lowest level of the tower, separated from the commercial and retail activities of the podium beneath. The response of the design to the site context is one of setting a podium at street level that makes use of an existing road pattern and has a mini-tower that is similar to the forms of neighbouring tall buildings, but that is more highly articulated and is at the scale of its immediate surroundings. Advantages of the mini-tower are its ability to control the admission of daylight and ventilation in a way that avoids mechanical ventilation for parts of the year, as well as provide adequate daylight that avoids the need for electrical lights to be used during daylight hours. This aim is achieved by using a highly modelled building form. This mini-tower demonstrates how complex arrangements of different apartment sizes and arrangements can be accommodated in a form which is shaped to reduce turbulence around the corners of the building. This approach increases comfort for residents who require some natural ventilation during mid-season and winter. The mini-tower is a typology that is emerging in cities around the world, arranged in groups; providing more amenity for its residents.

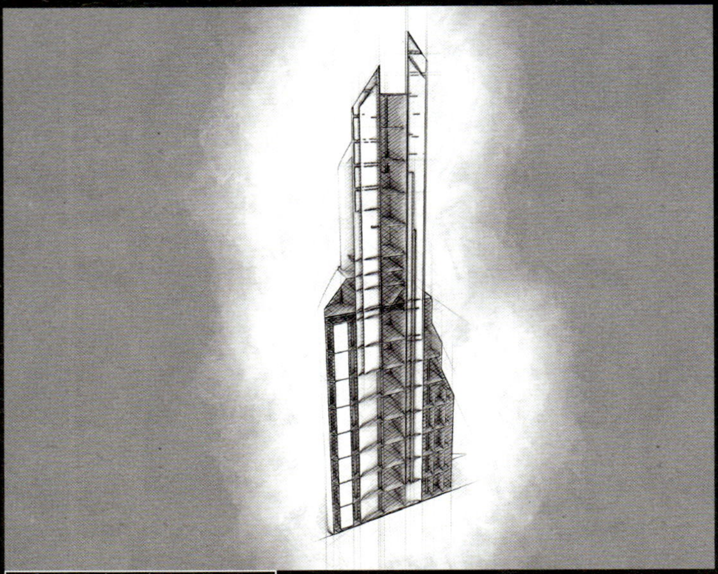

1. Sectional perspective view

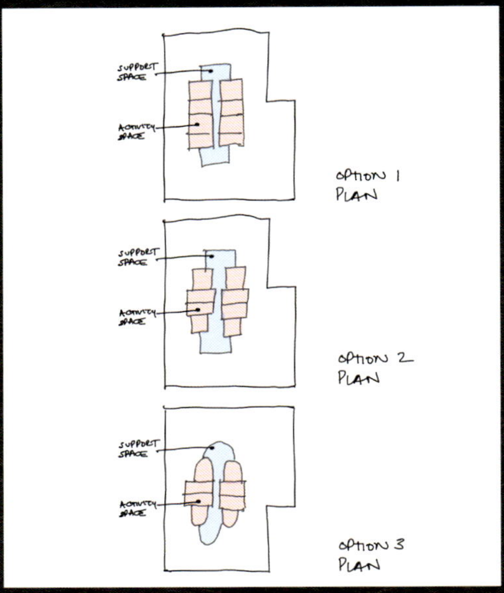

Spatial arrangement Option 1 in plan, is of two clusters of rectilinear apartments which rise to the full height of the building. Option 2 has the circulation space set at the centre between the two blocks.

2. Sectional perspective view

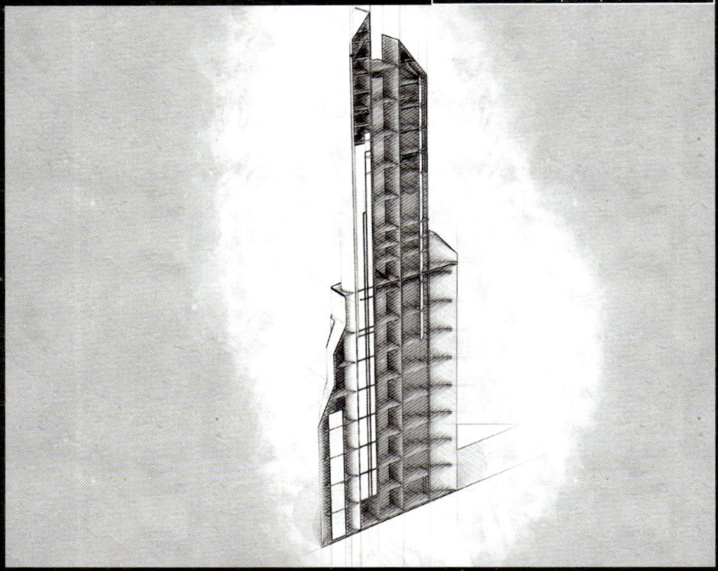

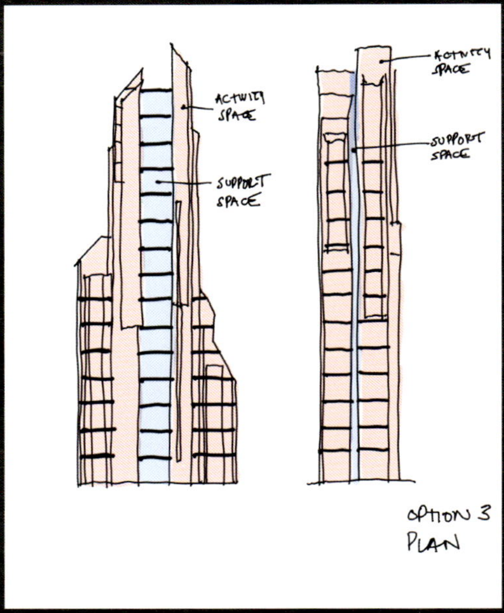

Option 3 adds curved ends to the blocks as a development of Option 2.

Spatial arrangement

This project comprises a podium and a mini-tower. The podium is arranged around a central circulation spine which passes beneath the central line of the tower above. Retail and commercial spaces are arranged on both sides of this spine route, which connects two adjacent streets. The partition walls between the retail spaces can be moved to suit the changing needs of tenants and owners. The management company for the tower occupies space on the top floor of the podium, which is divided into two visually separate halves. At the upper levels, the two podium blocks are linked by bridges; the floors are accessed by an escalator and lifts/elevators. The mini-tower is generally arranged as apartments occupying two symmetrical halves of each floor plate. Apartments can occupy two floors, or a complete floor plate.

The spatial arrangement for Option 1, in plan, is of two clusters of rectilinear apartments which rise to the full height of the building. Access to apartments on all floors is set in the space between the two blocks. The podium rises through three floors; the top floor of the podium steps back to create external terrace spaces for restaurants as the tower rises. One of the two blocks rises higher than the other, and each block steps back in an asymmetrical manner in order to create a visually thin top to the tower by providing more smaller apartments which occupy a complete floor at these upper levels. Option 2, viewed in plan, is a variation on the arrangement of Option 1, with the circulation space set at the centre between the two blocks; a smaller core is positioned in the centre. When seen in section, the floor plates of Option 2 step back as they rise through the building, which is similar to Option 1 but not as steeply arranged as that option. The plan of Option 3 has the curved ends to the blocks as a development of Option 2. As the strands of each cluster rises, the roofs are angled in order to give a geometry to the roof that is sympathetic to that of the glazing and of the steel framing that is visible at the top of the building. The framing at the top of the building encloses the glazing; its primary supporting structure behind the glazing is triangulated for an economy of structure which resonates with the overall visual forms of the mini-tower.

Perspective view of interior

2. Perspective view of interior

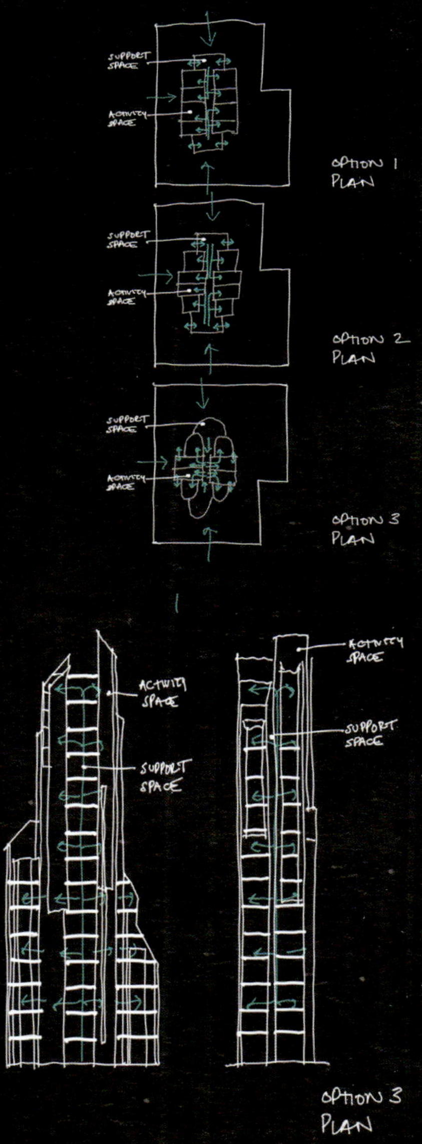

Internal circulation

Option 1 has entrances at both ends of the podium which are accessed through the podium space below.

Option 2 creates more vertical access, less shared access, and more use of bridges for circulation.

Option 3 has two sets of bridges, one at each end of the 'canyon' block between the two apartment blocks.

Internal circulation

Circulation within the podium is provided by the spine route, which is effectively a covered passage with retail units arranged across several floors. As the passage connects neighbouring streets, the use of this space is both for passing through and stopping at the retail and commercial units, or for making a specific visit. Circulation through the minitower is made through the vertical cores which are set between each of the components of the tower. Access is via private lobbies set on both sides of the podium spine. The cores comprise stairs, lifts/elevators and storage space. There is a transfer lobby running across the building on this narrow side, with stairs at each end on all floors. The building also incorporates both direct and reflected daylight, which is reflected from sky gardens down into the internal cores and into the internal circulation spaces in the building. These are typically arranged in three-floor clusters throughout the height of the building. Clusters are stacked vertically in order to bring light into the building through the facades and into the core areas; and in some cases through the perimeter spaces into the central, transversely orientated corridor. The internal circulation for Option 1 has entrances at both ends of the podium which are accessed through the podium space below. The central slot between the two apartment clusters has bridges across each of the three-storey voids, creating villages between groups of floors. In Option 2 that concept is developed further by creating more vertical access, less shared access, and more use of bridges, which provides access directly to apartment doors rather than from a shared corridor. The amount of horizontal circulation is reduced in order to allow each apartment to have its own access area. In Option 3 this principle is taken further, where there are only two sets of bridges, one at each end of the 'canyon' block between the two apartment blocks. These spaces are also closed at three-storey intervals to create the characteristic spaces within each area, side-lit by vertical glass walls, all of which enclose the slot between the blocks at each end of the building. The circulation is further reduced in this layout, as apartments are accessed with their own lobby space.

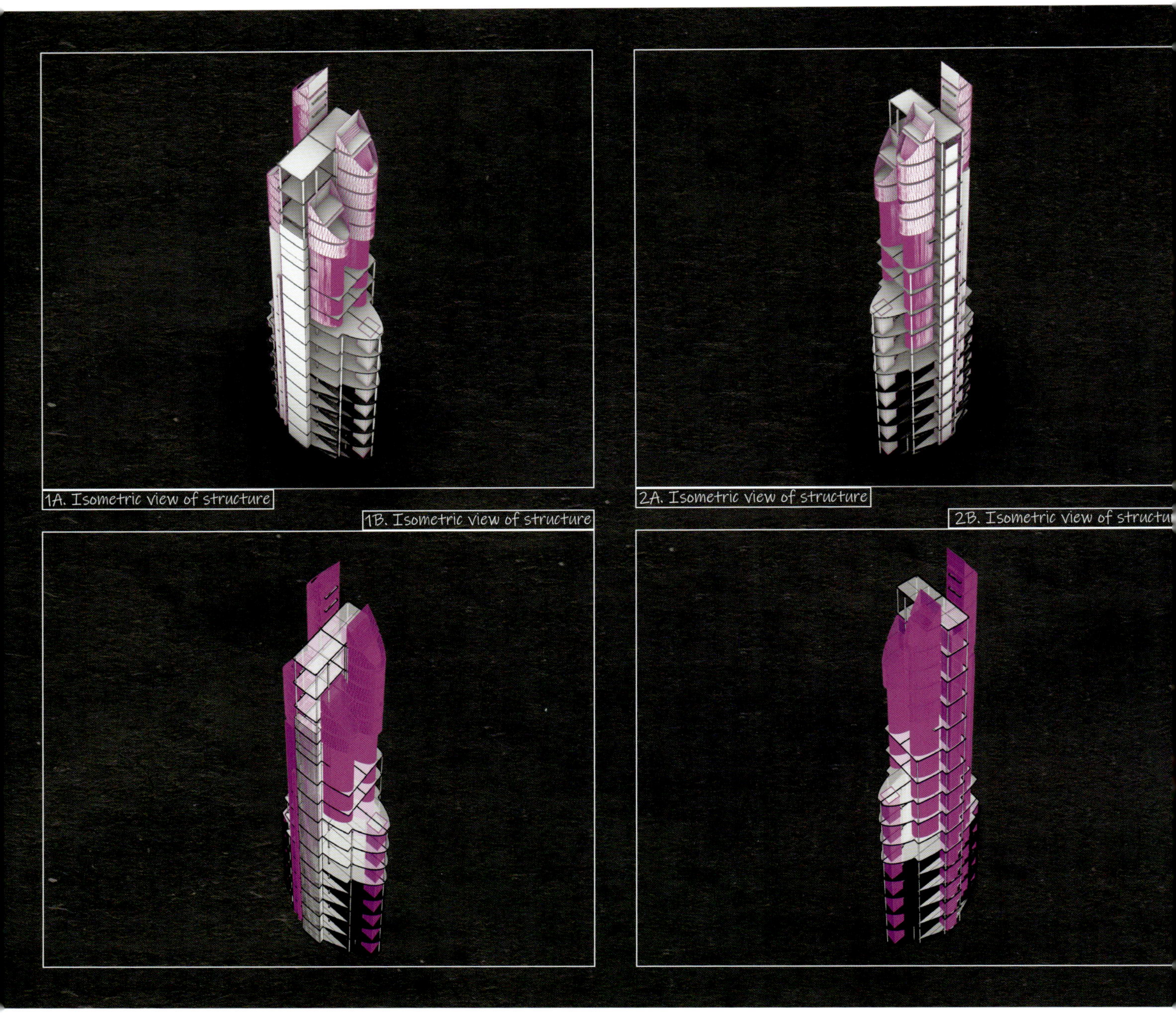

1A. Isometric view of structure

1B. Isometric view of structure

2A. Isometric view of structure

2B. Isometric view of structu

Structural strategy

The structural concept is of five mini-tower volumes which connect as a single structural form. The five mini-towers terminate at different heights throughout the building. Only one of the five towers reach the highest point. The structure is formed as a concrete frame in the lower part of the tower cluster, and a steel frame in the upper part.

The structural strategy for all three options is based on a structural frame that follows the geometry of the building. To create the characteristic form of the building, steel members are used to create the curved forms at floor level. The inclined angle of the structural forms of the sloping roofs at high level are created with curved beams at the edges.

At an early stage of the design, the building was conceived as a 10-storey block, with openings in this block forming lightwells. This solution was considered unsatisfactory as a result of the poor provision of daylight and fresh air that a taller building could achieve. The challenge that was presented with a taller building was to create a mini-tower that uses a similar quantity of facade materials to the 10-storey block solution. This aim was achieved by ensuring that the apartments were supported on the primary structure, with a steel frame and floor plates where the service spaces would not be a part of those areas. As a result, staircases and services were attached to the outside of the building instead of being set within. Secondary structures, as well as staircase structures, were introduced to support the equipment. The approach taken was one of an efficiency in the use of the primary structure to support floor plates and not to support openings in floor plates for staircases and vertical service zones, for example.

Although this added cost to items such as staircases, the value added to those staircases was comparable to the additional cost of creating floor structures with voids cast into them. This approach avoided secondary beams and associated column supports that would have been required. The complete structure aims to 'shrink wrap' the habitable spaces in the apartments, using the facades of the curved forms to enclose stairs and service equipment as prefabricated units. The

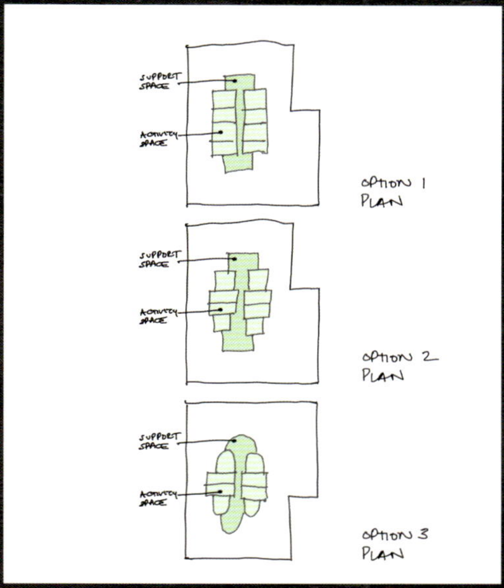

Structural strategy: Zones
Option 1 has structural walls which traverse the narrow width of the building.
Option 2 has a frame that provides structural stability to the concrete cores.

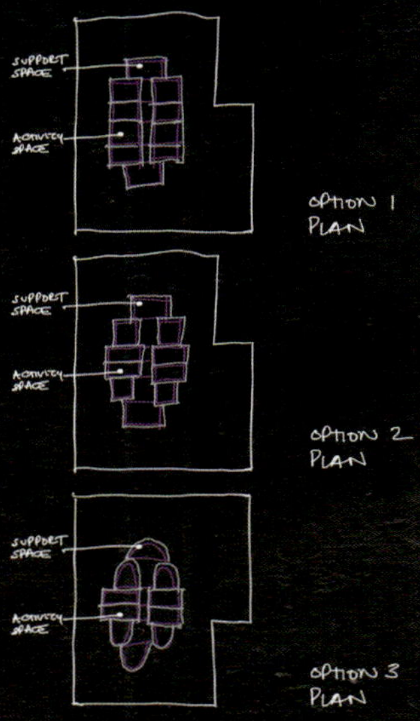

Structural strategy: Framing
Option 1 encloses the core spaces and dividing space into sets of apartments.
Option 2 is largely a steel frame.

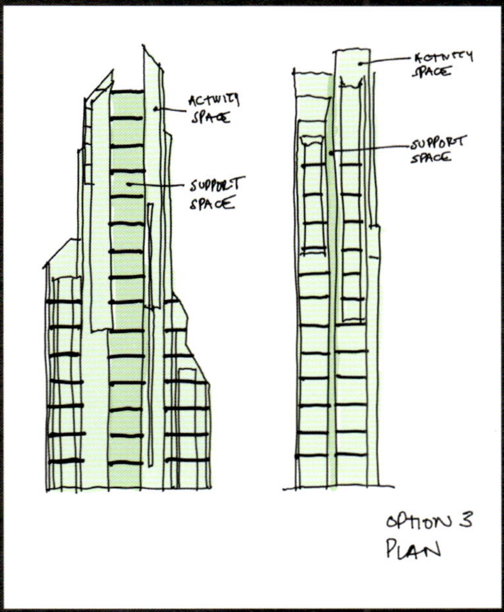

Structural strategy: Zones
Option 3 has structural walls formed as cores and mixes those with columns in the accommodation spaces.

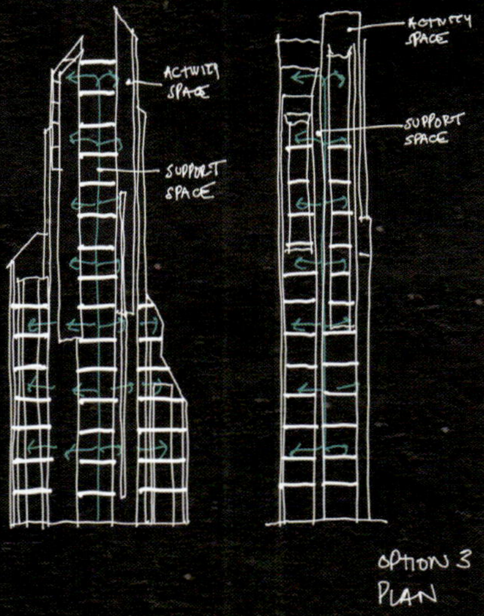

Structural strategy: Zones
Option 3 is a mixed solution of reinforced concrete and steel.

structural members on the edge of the tower which support the facade cladding have additional members accordingly; connections between facades and structure occur not only at floor slab edges. This is a result of the geometry of the cluster of mini-tower forms which are highly modelled in order to visually express the volumes of internal space behind them.

The addition of inclined roofs and recesses that create the overall form resulted in additional vertical and inclined structural members being used to create a braced frame which makes more use of the perimeter of the building than would typically be the case. The external structure of the building is a 'trussed hull' form; a structure which integrates its structural components within the facades to avoid the need for perimeter columns in addition to facade framing.

A structural core carries loads at the centre of the building, allowing the interior between core and facade to be relatively free of structural columns. This approach gave considerable flexibility to the arrangement of the apartments, with their different formats and sizes of apartment. In this configuration, structure is moved to the edge of the building while retaining it within the central cores, allowing these structures to be more efficient as a result of using a design that would have required secondary framing and supports, which in this design become primary structural members. This method provides the opportunity for increasing the service equipment used for passive ventilation as well as providing a different location for technical equipment within the building – not on the roof deck but instead distributed through the height of the building; the external service spaces are positioned on one side of the building and extend through its height.

This approach provides a flexible approach to services and environmental controls in the building, allowing a close interaction between active environmental control and structural efficiency. Services can be accessed and maintained independently of both the apartments and the common areas between apartments. Given the need for regular inspection and maintenance of these installations, the separation of services from apartments provides an amenity to maintenance staff.

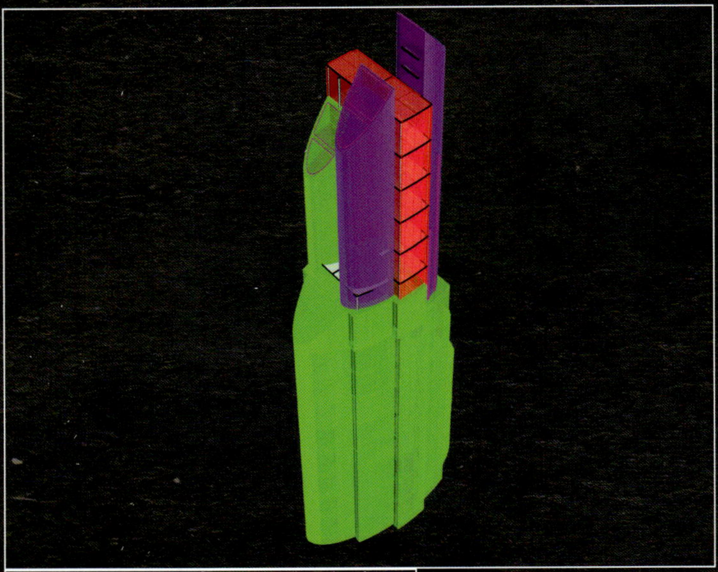

1A. Isometric view of environmental volumes

1B. Isometric view of environmental volumes

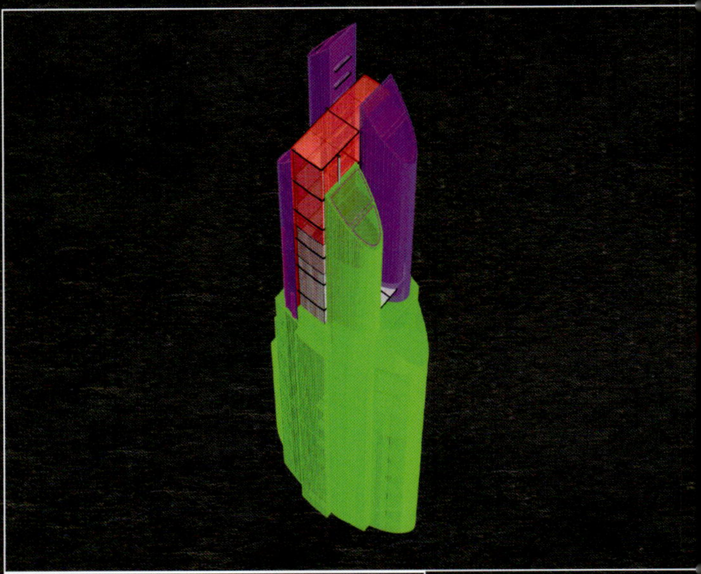

2A. Isometric view of environmental volumes

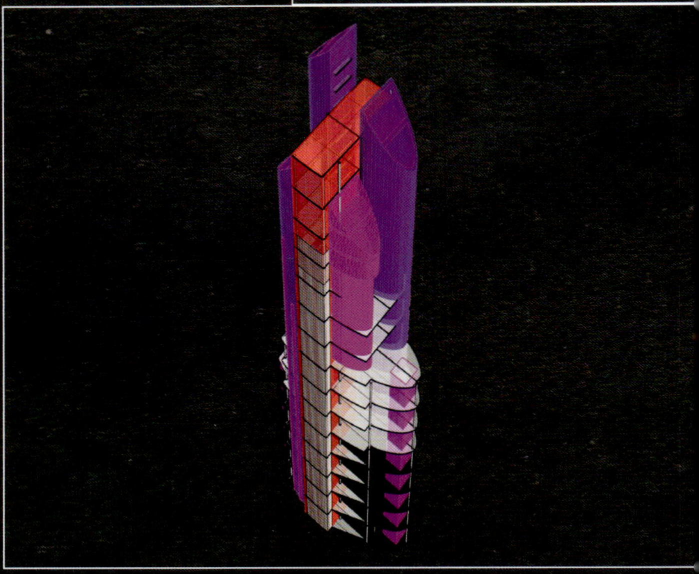

2B. Isometric view of environmental volum

Environmental strategy

The mechanical and electrical equipment design is based on having the services that comprise ducts and servicing equipment mounted on the outside of the building to create a fine visual weave of elements; an arrangement of pipes and ducts that provide services to this building. Pipes and ducts are partially concealed behind perforated metal screens. Within the building, services are integrated within ceiling zones, supplied from external maintenance decks; allowing apartments to be maintained entirely independently of their state of occupation. Efficiency of the floor layouts is maximised by grouping services and vertical circulation; there are no large areas of voids in floor plates or ceilings being used for the distribution of services inside the apartments. Mechanical and electrical equipment is accommodated in self-contained modules within each of the three-storey villages. Chillers are set at roof level, away from points of access and away from where residents would be disturbed by the sound generated by them. As a result, the building offers considerable flexibility for the rearrangement of spaces within each apartment. The interaction between the twin-wall 'maintenance facade' enclosure and its environmental controls is very visible on the outside of the building, allowing the space inside to be as clear as possible.

The distribution of ducted and electrical services throughout the height of the building is achieved by setting air-handling ducts and electrical supplies on the external facades in areas of opaque panel between the external wall and the outer screen within the twin wall configuration. In some cases, the duct casing forms an extension of the opaque panels, which gives rise to some of the angled glazed openings, with air being distributed down ducts which are set into inclined areas of facade. This approach avoids the need to take up valuable space inside the building for ducted air and services, allowing internal volumes to be cross ventilated as single spaces or as linked spaces instead of them being subdivided as a result of the need for air-handling ducts. Staircases are arranged as independent elements, positioned in the centre of the building. The upper floors in the tower, above the levels of apartments,

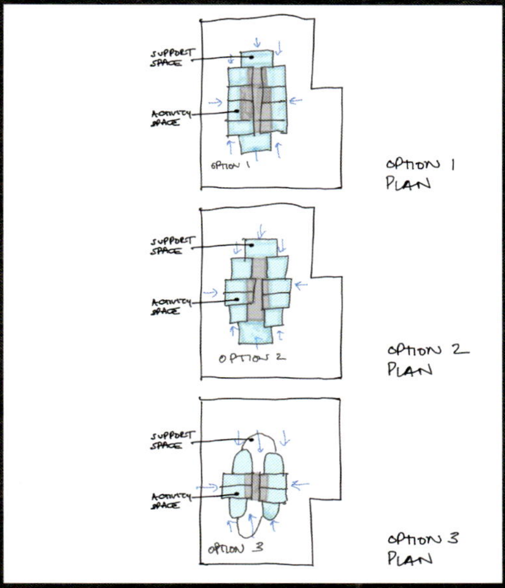

Daylight strategy for direct sunlight and cross ventilation.
Option 1 admits daylight and air to apartments on two sides of each space.
Option 2 admits daylight and air to apartments on three sides of each space.

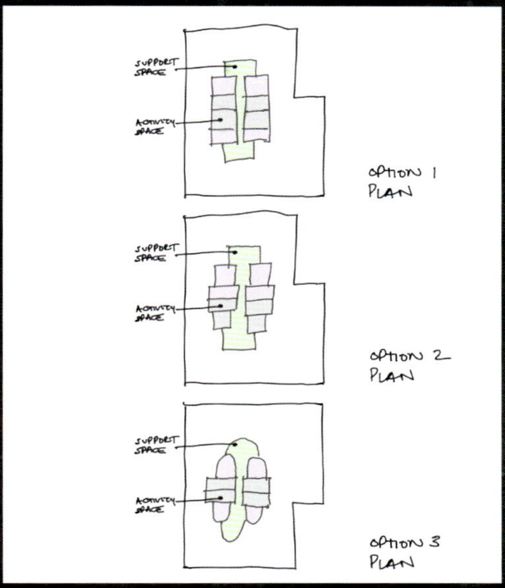

Environmental zones.
Option 1 has two end groups of apartments forming a single environmental zone, as apartments are exposed to similar effects of the sun.
Option 2 has a similar podium arrangement to Option 1, but on the upper floors the grouping of the apartments changes.

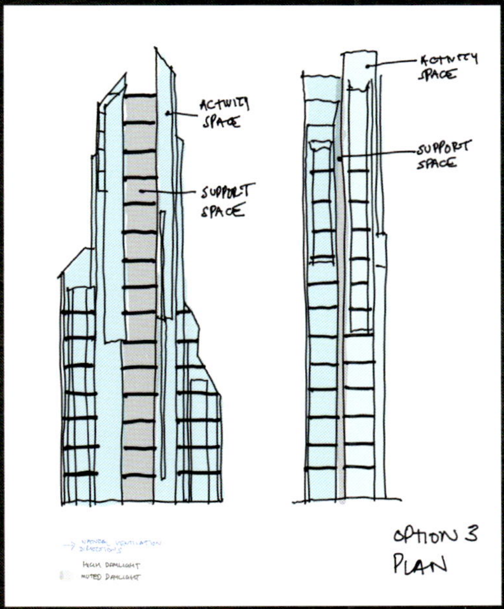

Daylight strategy for direct sunlight and cross ventilation.
Option 3 admits daylight and air to apartments on three sides of each space and provides greater possibilities of natural ventilation.

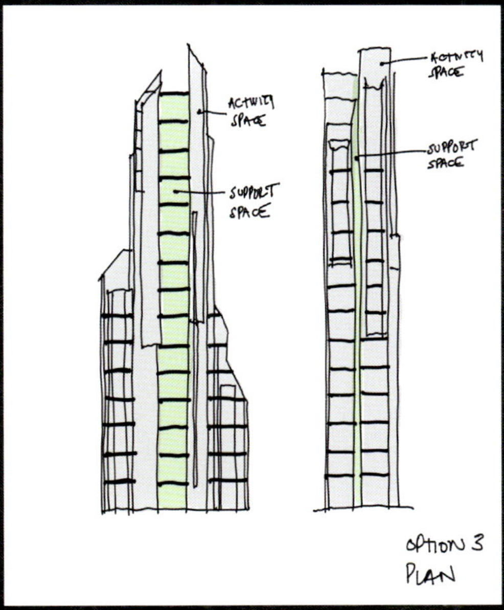

Environmental zones
Option 3 has the curved end strands of the clusters as one environmental zone where strands are exposed to similar effects of the sun.

have framing in place for additional passive cooling equipment which is expected to be installed in the near future. This extra capacity of fans will allow air to be drawn into circulation spaces in order to reduce the cooling load in those areas. The design anticipates the swapping or the exchange of energy-intensive cooling devices for a mixture of passive and active cooling devices. The framing installed on the tower anticipates their arrival.

The environmental strategy for Option 1 is of having the two end groups of apartments to form a single environmental zone, as they are exposed to similar effects of the sun. The two apartment block groups which form party walls with the end blocks form a separate zone; they are more internally oriented and have only one glazed facade, rather than the three glazed facades of the end blocks. There are separate strata; there are separate environmental zones. The third environmental zone, formed by the retail and restaurant spaces of the podium, requires high air changes per hour. Option 2 has a very similar podium arrangement as an environmental zone, but on the upper floors the changing grouping of the apartments is such that all the apartments have more external wall and more views out, and therefore are more susceptible to the effects of the sun. Consequently, there is an environmental zone across both sides and across both towers. The other end of the towers, in plan, form another environmental zone; three environmental zones in total. Option 3 has the curved end strands of the cluster as one environmental zone where strands are all exposed to similar effects of the sun; while the two inner blocks that have only two facades and one long facade, which is glazed, and two return facades, which are also glazed. These blocks form a second environmental zone. In addition, there is the third environmental zone of the podium.

The mini-tower benefits from a lower requirement for electrical lighting during daylight hours, as well as mechanical ventilation to corridor spaces. The contrasting areas of opaque facade and glazed facade allow the internal spaces to be highly thermally insulated where required; this residential tower does not require a highly glazed set of facades.

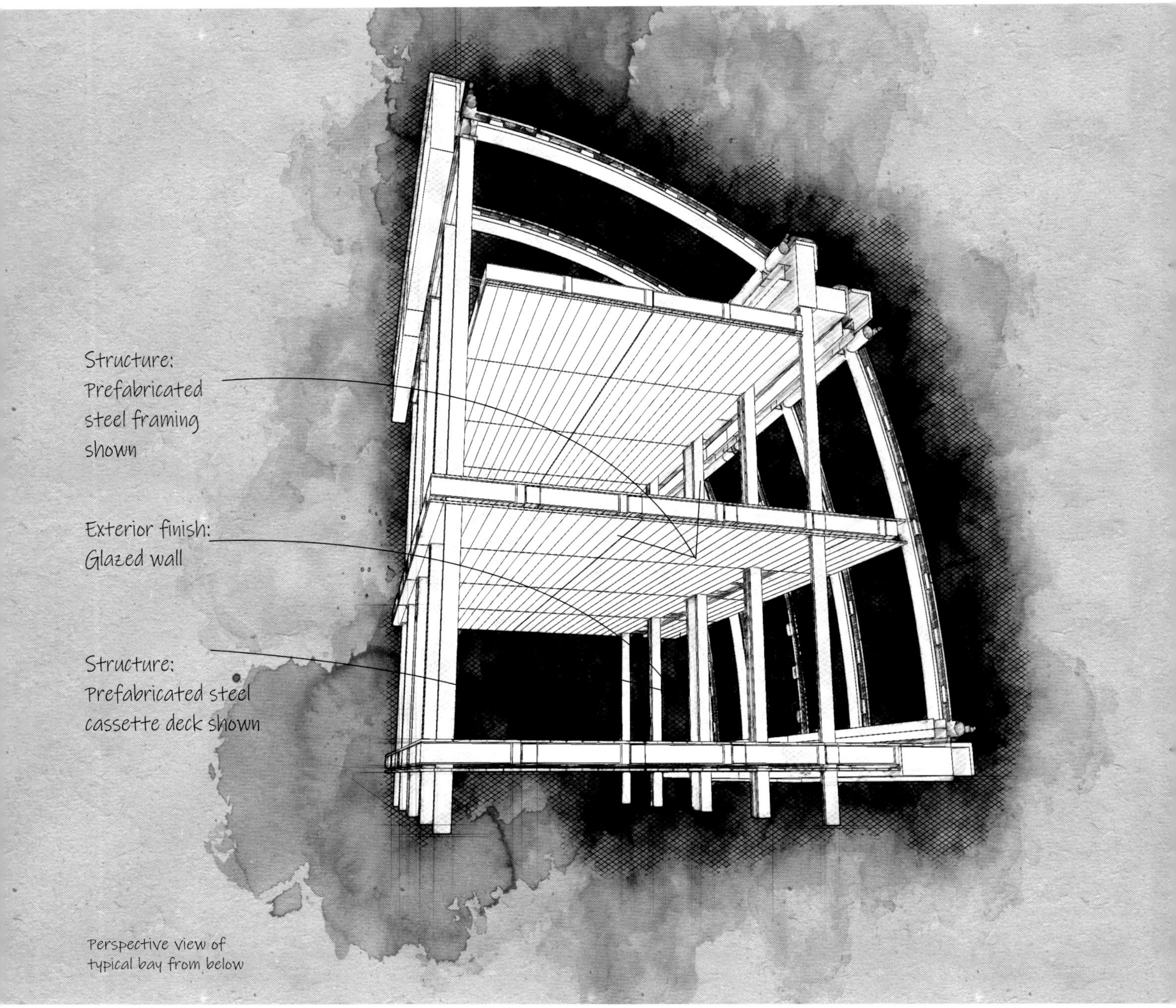

Perspective view of typical bay from below

Typical bay adjacent to facade

The envelope of the mini-tower has a mixture of glazed twin-wall facades, with an outer screen of perforated metal panels. There is more open glazing on the lower part of the tower than on the upper part; a mixture of translucent elements and opaque elements comprise the remaining areas of facade. There is no single opaque area or single glazed area; all areas of facade have some glazing with a mixture of opacified glass in some areas and metal screens set forward of the facade to provide both fixed and vertically sliding solar shading panels.

Different levels of translucency across the building provide the ability to respond to specific requirements of the brief/program in different parts of the tower, without taking away a visual unity that is provided by the translucent screens. At ground level the visual complexity continues with a mixture of retail, restaurants and boutiques. The facade panels on both the mini-tower and the podium are expected to be replaced and recycled during their lifetime. In some cases, panels will be removed, refurbished and reinstalled onto the building. This is achieved through the use of durable materials on the external envelope which have highly modelled concrete-based panels; their presence adds durability to the building. The glasses used in glazed areas of facade will be replaced during their lifetime and the glass will be recycled. The main change in the tower will be the framing to glazed elements which enclose the transparent parts of the of the facades and these have a shorter life than the concrete panels, which are expected to last the lifetime of the building.

The retail podium uses facade systems which are suited to retail and commercial use and which can be replaced easily with components being recycled in most cases. The podium envelope is conceived for ease of the mounting panels independently of adjacent panels; damaged units can be recycled with ease.

The visual language of the facades allows for the visual expression of a rich mix of functions and spaces without the need for a single identity of facade being imposed through the height of the tower. Instead the visual expression is one of near-future change and development; the

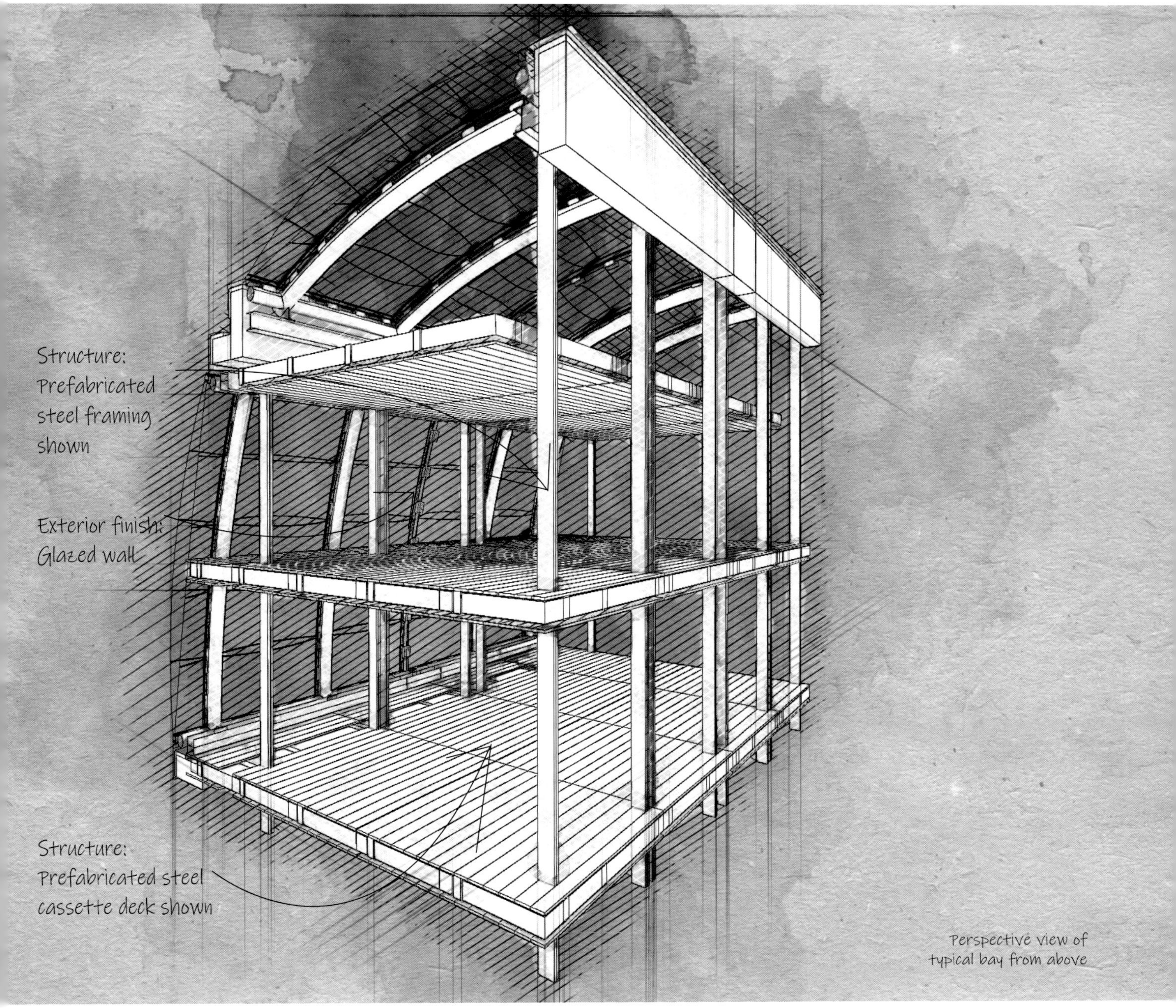

Perspective view of typical bay from above

cluster of towers are in a state of continuous visual development.

The solar shading forms a key part of the strategy for protecting glazed facades from the effects of sunlight. In order to suit those different requirements, some parts of the facades have overlapping panels that are oriented to fulfil the primary purpose of solar shading in areas that require control of sunlight and daylight; allowing other areas to have partial exposure to the sun in order to create shadows within the apartments.

There are typically one or two layers of solar shading of perforated metal panels which follow the curve of the building. They follow the geometry of the building so that the structural form and the architectural forms are revealed. The facade design takes the opportunity to visually express some of the mechanical and electrical equipment areas and screened terrace areas, giving the building its specific identity. Light is also allowed to be admitted deep into the plan by having open slots which form visual 'crevices' within the facade, allowing light to be reflected downwards. This approach allows for double-skinned facades to bring natural ventilation deep into the floor plates rather than to the perimeter only, allowing fresh air to be introduced in the mid-season. As a result, ventilation is doing more than providing night-time ventilation, with air coming in through the facades; drawn across the building and released to the outside at the other side of the apartment. This allows for fresh air to be brought in during the day to all parts of the floor plates.

The facade incorporates the primary structure of the building. Additional lattice framing on this outer grid forms part of the facade. The facade structure varies; as a lattice form that is made denser to meet the needs of supporting stairs and air ducts, for example, but doing so with a secondary structure that is integrated with the primary structure but is visually differentiated from it.

A grid-like set of secondary structures is used externally: sizes of components and assemblies from large scale down to small scale, using a consistent language of components to provide support for key elements of both floor plates and facades.

1. Perspective view of exterior

2. Perspective view of exterior

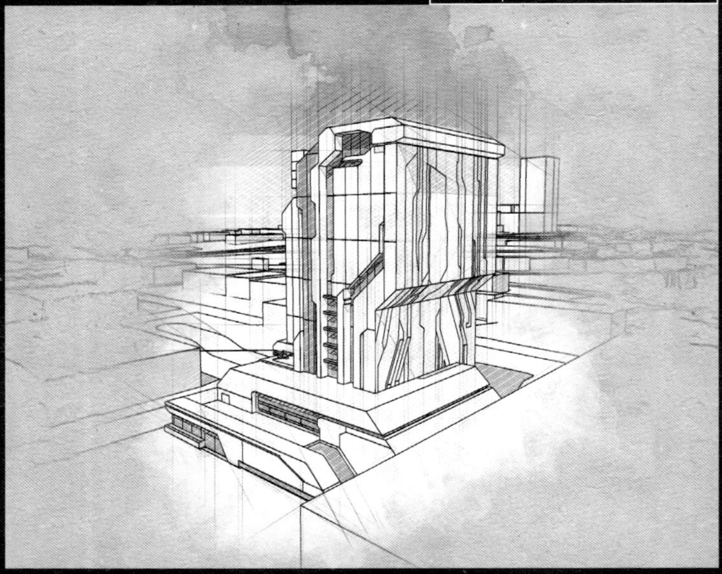

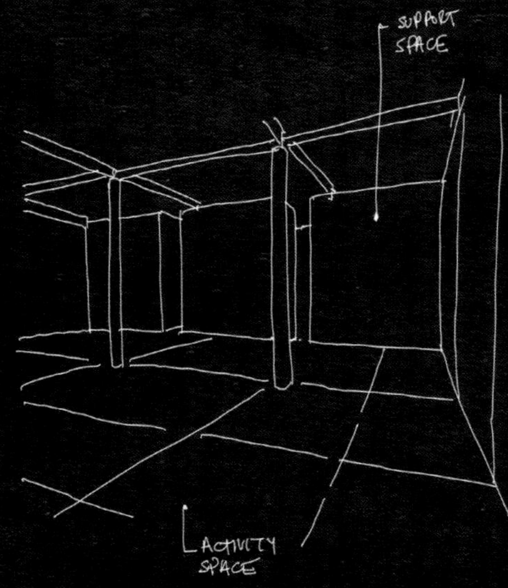

The 3D digital twin and its use in prefabrication

This project involved the development of a set of 3D models that could be used as a 'digital twin' during the lifespan of the building. These 3D models look forward to the near future of building construction; increasingly automated, with digitally controlled fabrication and drone technology to create a highly controlled construction environment. Robotic technologies are to be deployed for heavy lifting and hazardous tasks, reducing the construction time and increasing build quality and safety. Fewer cranes will then be required during the construction process. This development is anticipated within the digital twin with provisions for moving complete parts of the building assembly by machine in the near future.

Digitally controlled and drone technology provides an opportunity for maintenance and cleaning of the building envelope and structure both during and after construction; these will eliminate the use of cradles and remove workers from hazardous working environments. This provides the opportunity for greater integration with other control systems, creating a digital network for the transfer of building information, increasing the efficiency in its operation and maintenance and providing increased precision and safety. The construction methodology of this project shifts towards on-site prefabrication; taking place on site, the fabrication will be undertaken by digitally controlled machines in specific construction labs. The needs of these labs are anticipated in the ordering of information in the digital twin 3D models.

These labs will produce the components as required based on construction sequence data supplied by the project's common data environment. This correlated fabrication sequence will reduce the quantity of material stored on site and also eliminate delays caused by late delivery of components. The need to transport only the raw material reduces the environmental impact relating to transport, as the raw materials can commonly be tightly packed compared to fabricated elements, reducing the transported volume. Data from digital twin 3D models will directly inform the fabrication and assembly of the system, through the use of site-based, computer-controlled technology, improv-

. Perspective view of exterior

4. Perspective view of exterior

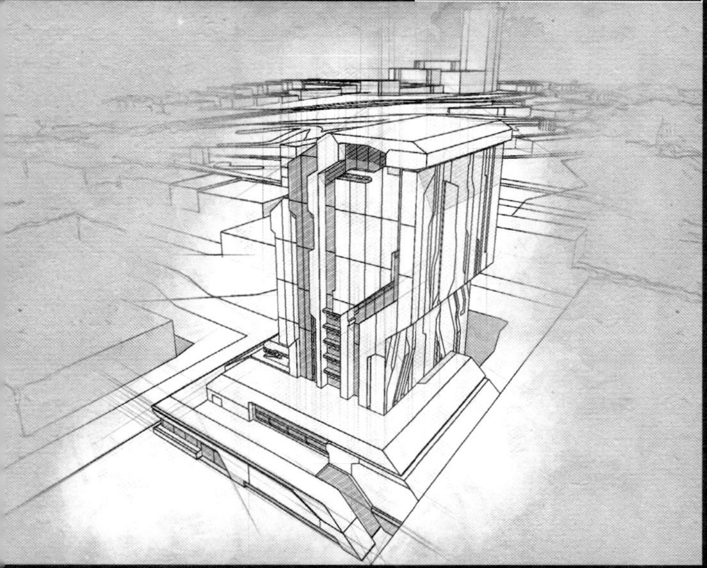

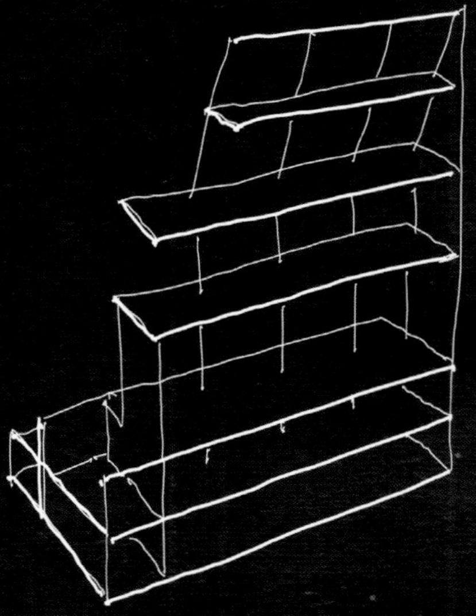

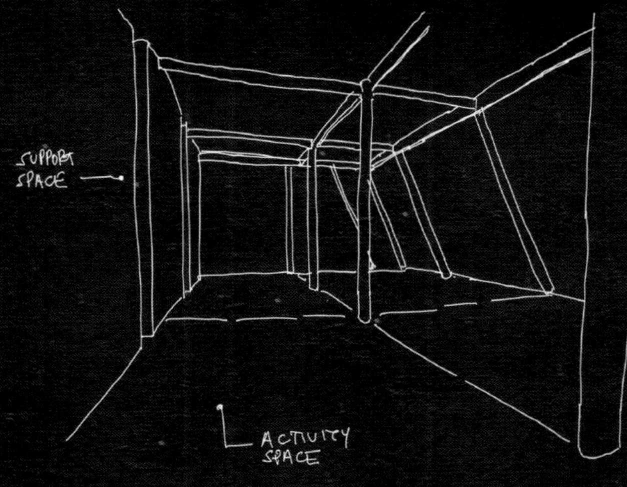

ing system quality and reducing the construction program. Inputting the fabrication directly from the integrated system model reduces the probability of error and increases the precision of the components produced; it also enables automation of the process. This automation combined with the close proximity of the fabrication lab will reduce the manufacture-to-assembly timescale, saving time in the overall project program.

Using a digital twin 3D model for the coordination of fabrication and maintenance allows for greater integration with other control systems. A digital network is created for the transfer of building information increasing efficiency, precision and safety. The increased generation, recording and sharing of data is becoming commonplace, the issue then becomes utilising this data in an effective manner to produce the greatest impact on a project. This can be achieved by integrating all the building data into a digital twin model. The model is then connected through a digital network to the relevant contractors and all the machines working on the building. The project has a 3D digital twin model which can be used as the basis for construction; the model contains all the project components and specifications acting as a living user manual. The digital twin can be paired with Lidar-equipped drones which provide a three-dimensional as-built of the system. This can be compared with the designed model and used to inform updates in the construction progress. The two models can be examined through virtual reality technology, allowing them to be compared against the ongoing construction to highlight issues and reduce delays. This as-built model also provides information on the site conditions, identifying potential issues such as gaps in structure and misplaced material. The identification of these issues allows them to either be resolved or documented, increasing site safety. This process of examining and reflecting on site-based construction activities in real time can result in forming direct links with the factory, where components are fabricated and partially assembled. An immediate benefit of the enhanced communication between factory and building site will be a reduction of the energy used to fabricate and transport materials, components and assemblies.

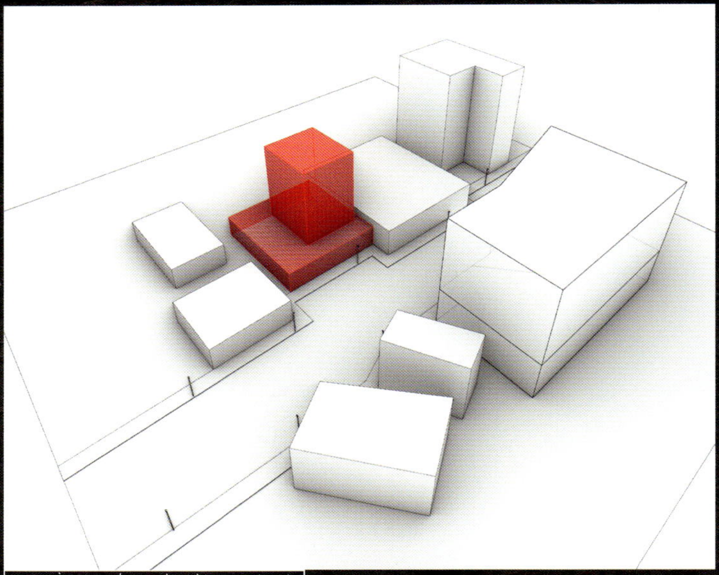

1. Project volume in site context

2. Project volume in site context

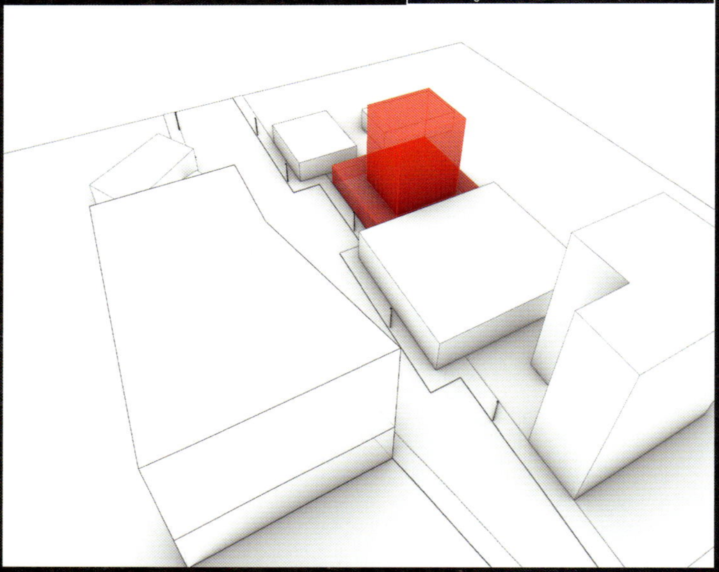

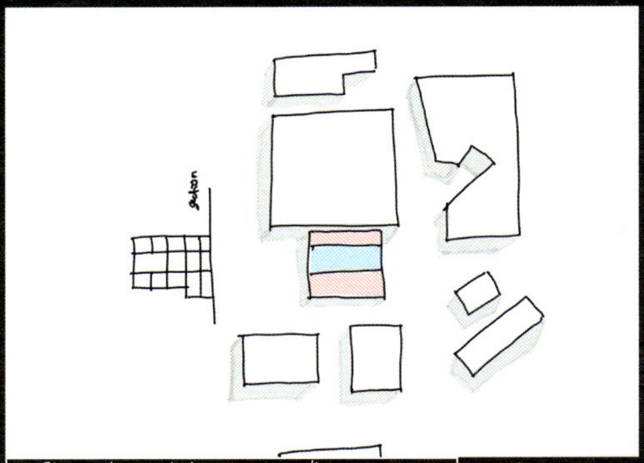

1. Site plan with surrounding context

Option 1
sets the external forms and internal spaces next to one another in a row.

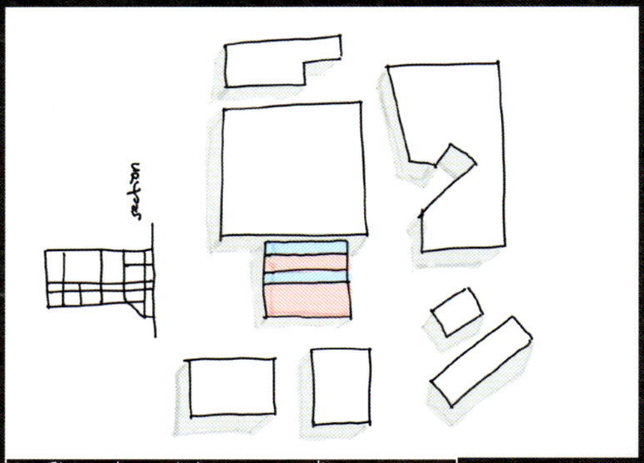

2. Site plan with surrounding context

Option 2
has functions with greater differentiation that are mixed on the first two floors at ground level and higher.

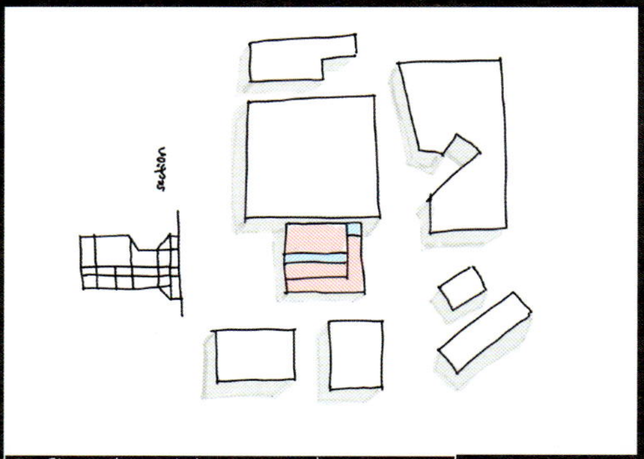

3. Site plan with surrounding context

Option 3
has lower floors accommodate a mix of functions; the upper floors above a two-storey podium are differentiated into separated functions.

Site context

The brief/program is for this design and fabrication facility, aimed at start-up company tenants who require a part-share of fully equipped workshop facilities and design studios. This building is positioned at one end of a site occupied by an unused light-industrial building owned by a public body. The brief/program requires no separate administration space or specific meeting spaces; the administration for the new facility is also managed as a start-up company within the building. The working community in the facility is based on the need for small enterprises to have ready access to a fully functioning facility that offers flexible agreements for tenants. This approach required a paring down of the assembly of the building using self-finish construction for structure and walls both externally and internally.

The project is set in an industrial area of large-sized, low-rise warehouses and production-based buildings. The building is arranged with a geometry that is consistent with the needs of the anticipated near future link with the adjacent large-sized single-storey building. This adjacent storage building is currently occupied for a different use by the same public authority. The new building is required to provide tall workshop spaces that permits the moving of equipment within the space in the near future. Despite its height, the building is suited in appearance to its immediate built environment but stands taller than its neighbours: Circumstances have made a greater density of use on this site necessary; the building's height does not exceed that of the immediate neighbouring building opposite the site.

Option 1 sets the external forms and internal spaces next to one another in a row; three blocks are set next to each other such that the different activities are arranged side by side and set out vertically through the height of each block. Each set of spaces is independent in its functioning; access is provided at each end of each volume at ground level. Option 2 takes that concept and develops it by having functions with greater differentiation that are mixed on the first two floors at ground level and higher; with a simplified arrangement, in plan, on the upper floors. Option 3 combines the benefits of Options 1 and 2 so that the

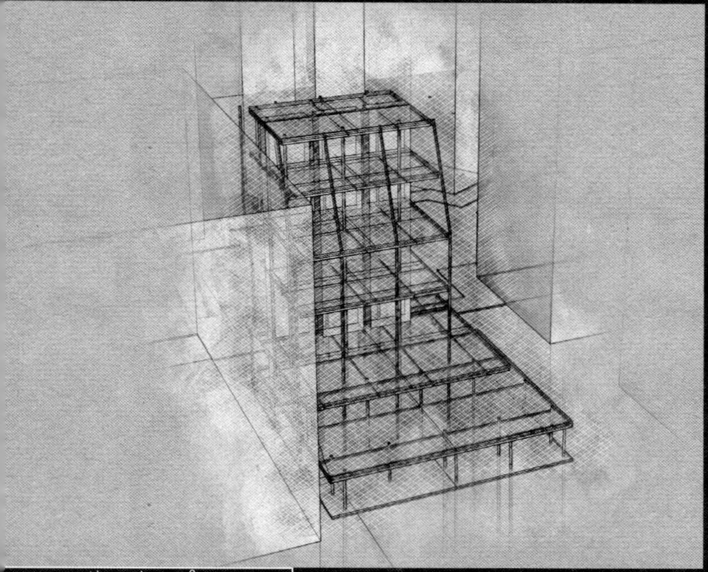

. Perspective view of structure

2. Perspective view of structure

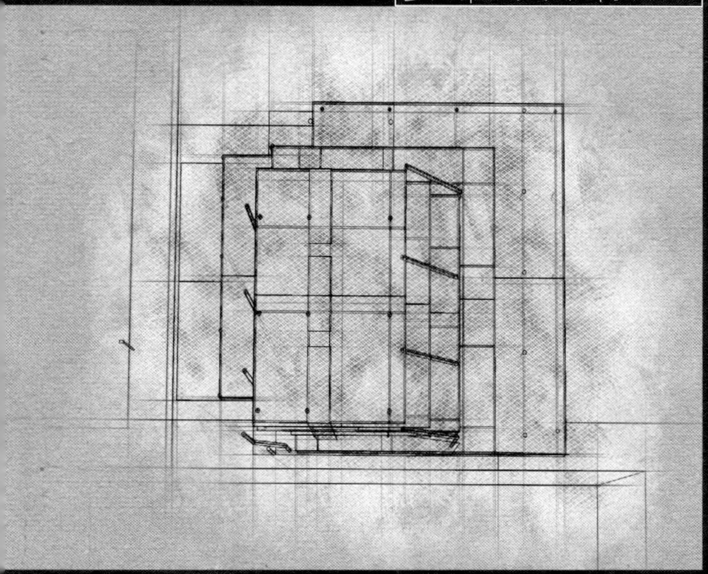

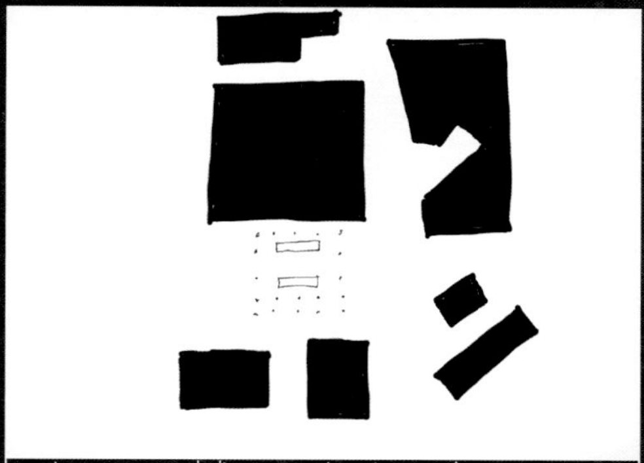

1. Figure-ground diagram showing project structure

Option 1
The structural grid follows the geometry of adjacent buildings, with wide spans between columns.

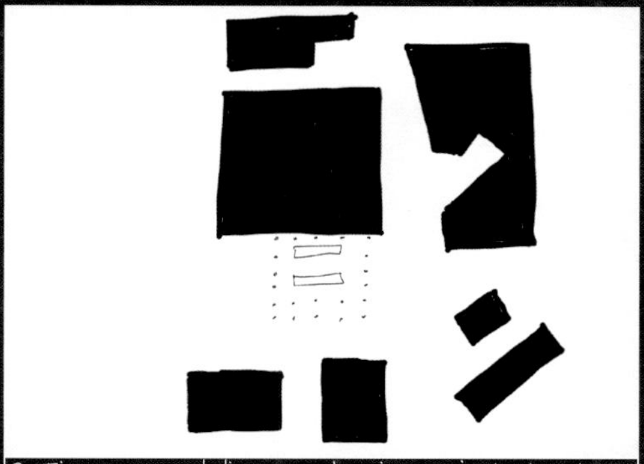

2. Figure-ground diagram showing project structure

Option 2
The structural grid follows the geometry of adjacent buildings, but does not incorporate cores into the geometry.

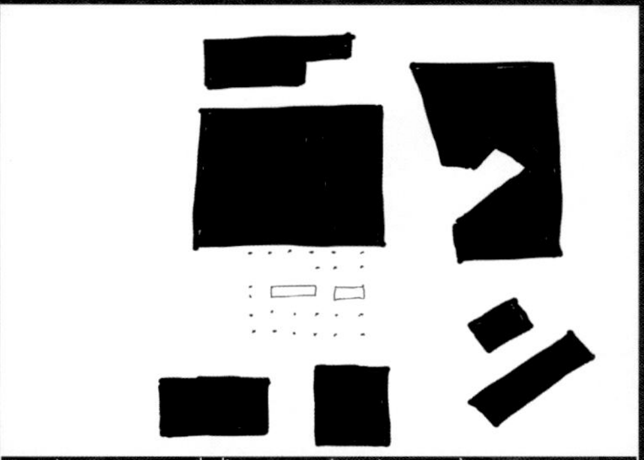

3. Figure-ground diagram showing project structure

Option 3
The structural grid follows the geometry of adjacent buildings, incorporating cores into the geometry.

lower floors accommodate a mix of functions; the upper floors that project outwards above a two-storey podium are now differentiated into separated functions, when viewed in plan, such that the links between different activities in the building are horizontally based, with each zone being stacked vertically.

From an urban design point of view, an essential function of this building is to be a beacon of regeneration and redevelopment in this light industrial area. The owner occupier wanted to increase the density of the use of space by making the building taller; not to allow an increase in the number of people working at the facility, but instead to provide greater facilities for more effective workshop-based activities where design and manufacturing are able to be brought together in a single building. The concept was to have design studios on the lower floors, with workshops on the upper floors. The workshops contain 3D printers and digitally controlled cutting equipment for the development of prototypes, but is not intended to be a facility for the mass production of components or assemblies; that work will be carried out elsewhere.

The owner-occupier sees the future in this part of the city as having a focus on the implementation of the design skills that are abundant in this locality. The building is part of the regeneration of design, innovation and product development in the region rather than manufacturing, which is increasingly automated in factories. The building can be seen as a 'map of functions' which are expressed visually in the short end elevations, where there is an expression of the interior spaces behind. The fabrication halls were required to be larger than the design studios beneath; studios require high levels of daylight while the halls above require little daylight. This arrangement resulted in the characteristic form of the building which allows machinery to be lifted up from the ground, for reasons of safety. As a result, the building form is one which encourages an interaction with the public at the ground floor; a transition space that visitors are invited to see but where manufacturing machinery is kept at a higher level. Workshop safety and access was considered to be important for all users of the building and visitors; the volumes in the building are arranged accordingly.

1. Sectional perspective view

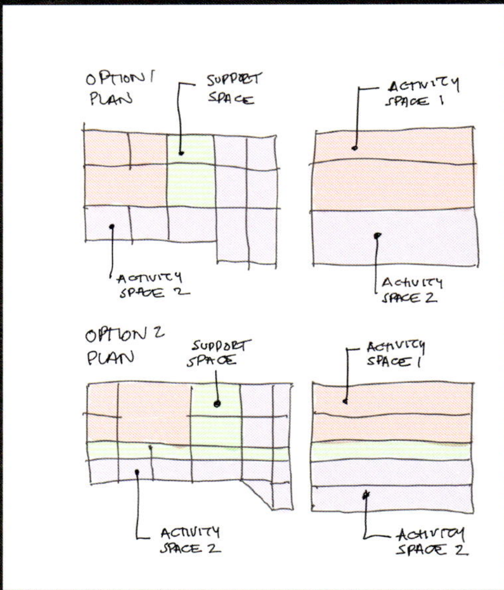

Spatial arrangement

Option 1 has workshop facilities set over four floors in the centre of the building.

Option 2 provides greater differentiation between functions and sets production on the side of the adjacent building.

2. Sectional perspective view

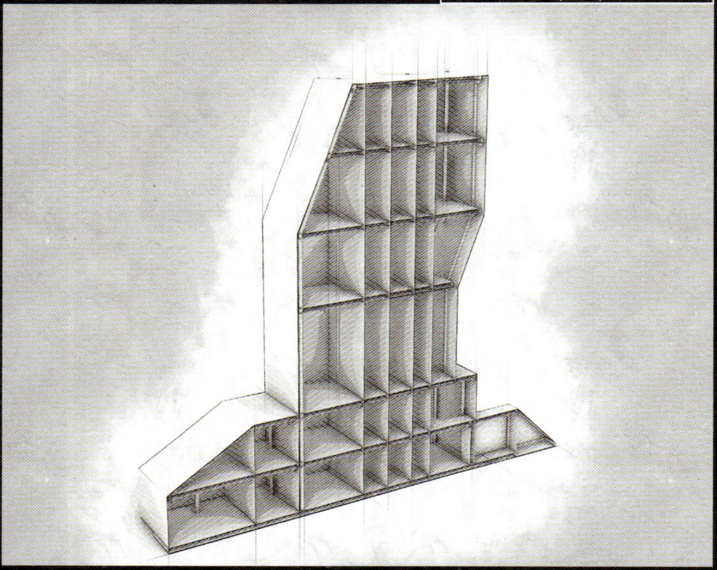

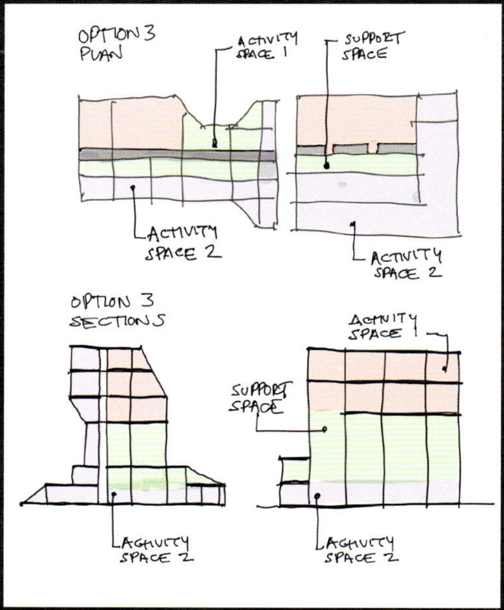

Option 3 places the workshop spaces over several floors.

Spatial arrangement

This building comprises four separate volumes which are set side by side. Access to the building is from one end of each block; stairs and a lift/elevator are set at each end. Above the mid-height of the building the floor plate extends out, supported by 45-degree structural consoles in order to provide a large single space at the top of the building, which cannot be accommodated at ground level due to safety requirements. The ground floor spaces extend across the full width of the site and extend into the tower, which is arranged over five storeys. The podium is set over two storeys and comprises three workshops; the workshop at higher level is accommodated within a double-height space. The two workshops located at ground-floor podium level are arranged side by side. The spatial arrangement of Option 1 is of workshop facilities set over four floors in the centre of the building, with supporting functions positioned on each side so that support spaces, such as stairs, lift/elevator and washrooms, are provided on one side of the building. Design studios and small workshops are arranged on the other side of the building. Option 2 provides greater differentiation between functions and sets production on the side of the building adjacent to the existing single-storey production buildings. These spaces are 'stacked' vertically, one above the other, with their own service support spaces set adjacent to the main spaces. Design studios are positioned on the opposite side of the floor plate, with its support spaces of washrooms and stairs and a lift/elevator set adjacent to it.

Option 3 combines the benefits of the first two options by placing the workshop spaces over several floors; the largest workshop space being on the upper floor, which functions independently of the mixed functions of the two lower floors. Podium-level design studios are set on the opposite wall, such that natural daylight is admitted on one side, with support spaces for both workshops and design studios having shared washrooms, stairs, lift/elevator and storage space. Option 3 is effectively a combination of Options 1 and 2, with mixed functions at the podium level; and those functions extending up through the height of the building.

Perspective view of interior

2. Perspective view of interior

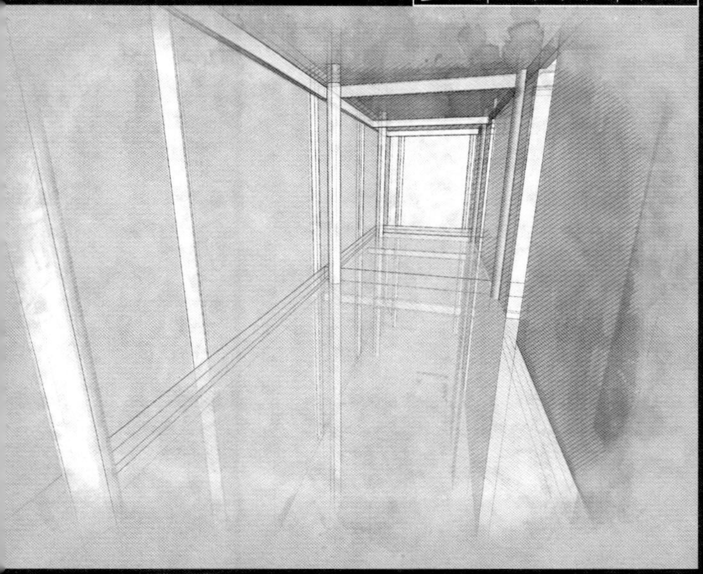

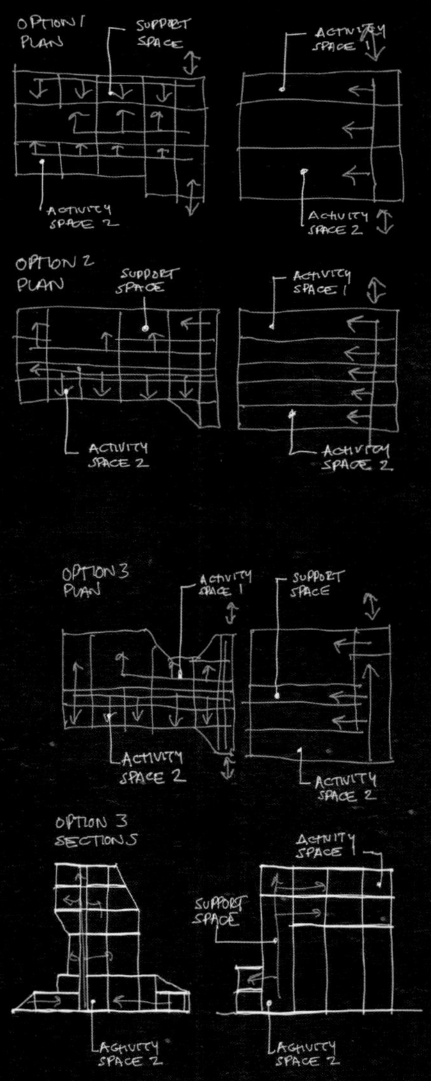

Internal circulation

Option 1 has access into the podium from two sides of the building; from both the outside and from a 'link' into the adjoining single-storey building.

Option 2 uses the spine route across the building on all floors.

Option 3 provides a greater differentiation of circulation routes..

Internal circulation

This building has no large central space as might be imagined from its general form. Instead, the volumes of the building are accessed by following a 'route' of circulation spaces which are set at the ends of the building. The building is arranged as a 'terrace' or 'row' of three buildings which weave together visually at the upper levels in response to the needs of the different sizes of spaces within. The functions within each of the subdivided volumes are expressed visually in the facades on the short sides of the building. Spaces within the building are accessed sequentially and vertically, from floor to floor, rather than horizontally; the circulation routes are arranged as vertical 'ladders' of stairs and lifts/elevators rather than as an arrangement of corridors on each floor level. In other words the main circulation routes around the building are vertical rather than horizontal; each floor plate is formed as a single space.

The internal circulation of Option 1 has access into the podium from two sides of the building; from both the outside and from a 'link' into the adjoining single-storey building. The circulation route between the main entrance and the 'link' forms a spine down one side of the building at ground level, from which access can be made into the building's three zones of workshops, design studios and service space. Internal circulation is similar to Option 2, using the spine route across the building on all floors, but has the advantage of a greater differentiation of space such that there is a direct link to the lower levels. The ends of the spine accommodate vertical circulation; stairs and lifts/elevators provide access to workshops, design studios and service spaces.

In Option 3, a larger space is provided in the middle of the building which closely meets the requirements of the brief/program. Internal circulation is slightly different from Option 2, but there is a progression from Option 1 to Option 3 of providing a greater differentiation of circulation routes; from a straightforward concept of a spine route through the building and then vertical circulation, to spaces branching away vertically from that horizontal spine.

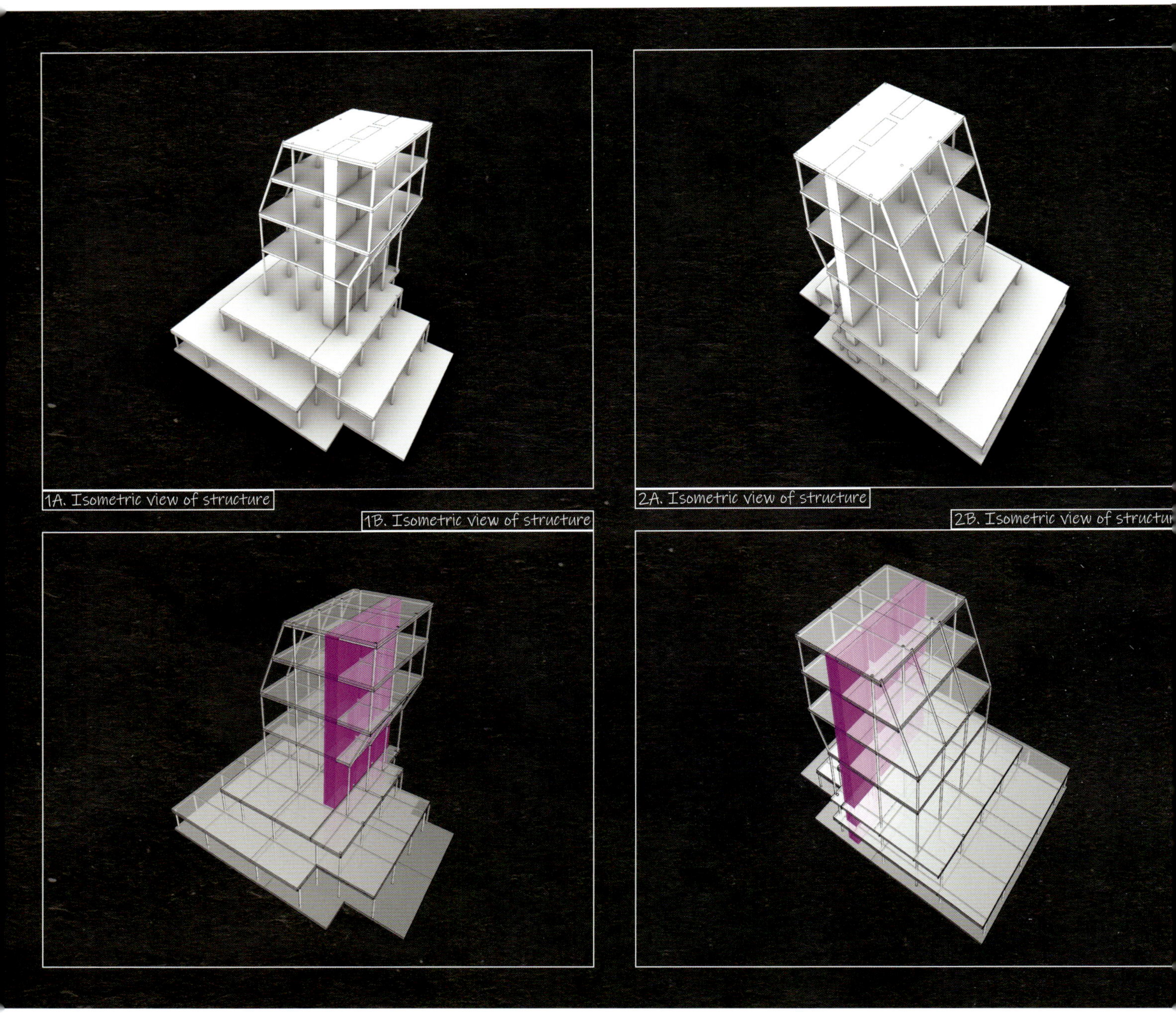

1A. Isometric view of structure

1B. Isometric view of structure

2A. Isometric view of structure

2B. Isometric view of structur

Structural strategy

The structural strategy for all three options is of a reinforced concrete frame with reinforced concrete floors which support the loads of machinery and overhead gantries in the workshops, while providing the required sound insulation and avoiding the transmission of structure-borne vibration through from workshops to design studios.

Option 1 uses a reinforced concrete frame set in a rectilinear arrangement, and has a projecting block at ground level with associated sloping inclined roofs also formed in reinforced concrete that provides sound insulation from inside to outside. Option 2 uses the benefits of reinforced concrete to create large spaces inside and with more sculpted volumes. Option 3 benefits from reinforced concrete, with a cantilevered upper level in the building that allows a workshop to be positioned in a way that is optimised for space distribution and spatial arrangement.

The development of the structural design took place in parallel with the evolving options for the form of the building, as a result of the alternative arrangements being generated for the distribution of spaces within the building and their internal circulation. The structural design is a key driver in this project due to the height of the building and the preference for having manufacturing equipment located at a high level within the structure, resulting from a preference to keep machinery away from areas of general access within the building and away from ground level locations. Due to safety reasons, the design aims to accommodate these workshop spaces at a high level with machinery that is repaired with small components and not large assemblies. Similarly, the items produced are prototypes which are expected to be disassembled and transported by road vehicle. These will be small-scale assemblies of relatively high value; the main structural loads within the workshops are those generated by relatively light equipment used in the workshops. No heavyweight equipment was envisaged in the brief/program; nor is heavy equipment planned to be installed. The three options for the design focussed on different levels of complexity in the building, when viewed in drawn section. Flexibility in the ability to service the spaces

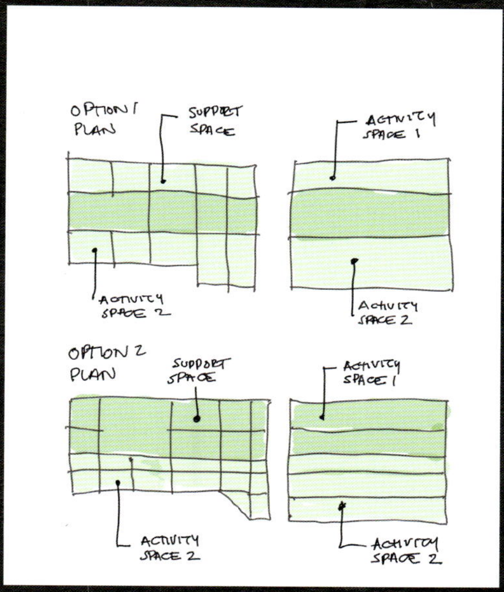

Structural strategy: Zones
Option 1 has structural walls which traverse the narrow width of the building.
Option 2 has concrete cores that provide structural stability to the concrete frame.

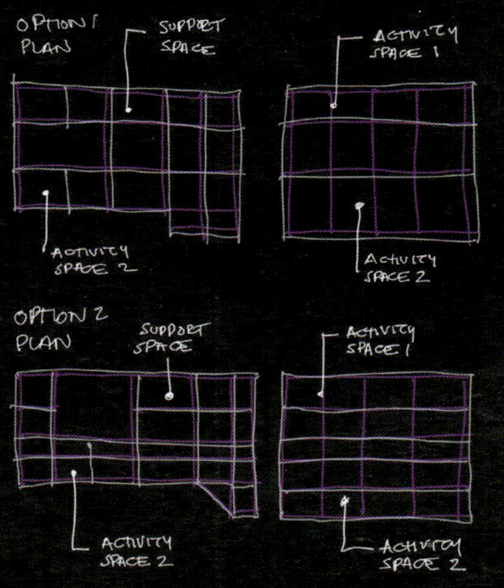

Structural strategy: Framing
Option 1 uses a reinforced concrete frame set in a rectilinear arrangement.
Option 2 uses the benefits of reinforced concrete to create large spaces inside and with more sculpted volumes.

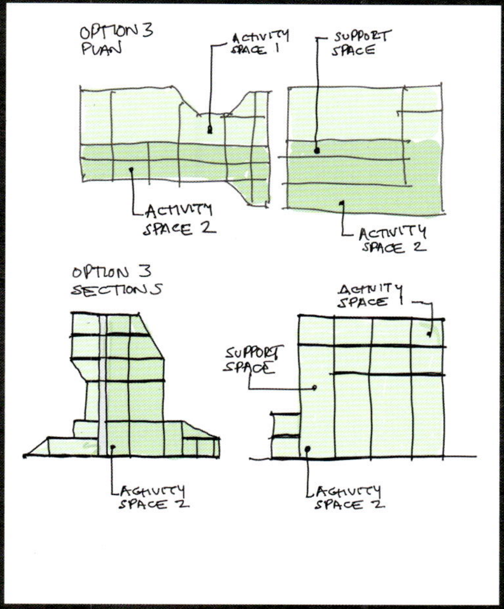

Structural strategy: Zones
Option 3 has structural walls formed as cores and mixes those with columns in the activity spaces.

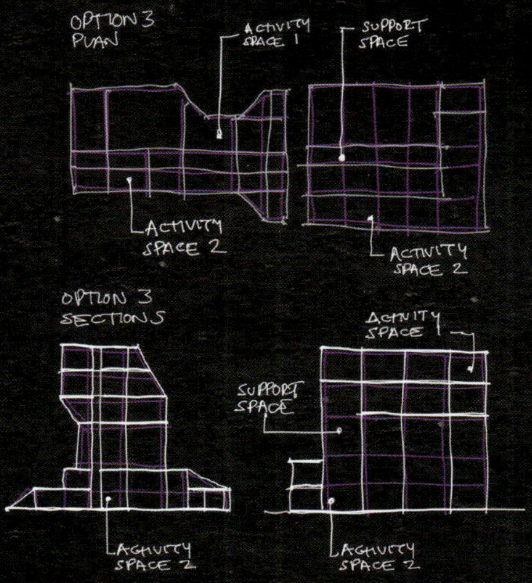

Structural strategy: Zones
Option 3 has a cantilevered upper level in the building.

within the building was considered to be a key design driver, both in terms of the delivery of equipment and the safe environments created for those working and visiting this building.

Option 1 is a rectilinear box-like form which provides little flexibility of use and fewer opportunities for access. The arrangement of spaces sets the design space adjacent to workshop space. Option 2 addresses the need to mix functions on a single floor, but still has design and workshops on the same floor. Option 3 separates functions completely, as is required in the spirit of the brief/program and brings with it the challenge of a cantilevered volume that requires structural support. The projecting volume of the workshops could have been supported on columns beneath, but the omission of columns in that area provided an ease of using the space beneath to both lift and lower materials, components and assemblies in the manner of a traditional warehouse, unimpeded by adjacent columns that would be externally located. At an early stage of the project it was decided that reinforced concrete would be the primary material used for the construction of the building to accommodate the weight and behaviour of machinery. The building was selected at an early point in the project. All three options use reinforced concrete columns, floor slabs and structural walls as part of the overall design. The primary aim of using reinforced concrete is to provide protection from noise and vibration coming from the machinery space within the building and to isolate noise and vibration from adjacent buildings. The choice of reinforced concrete allows the spatial forms to take up the geometries of the internal spaces and create an efficient economical structure and enclosure for the building. Although this building is divided into four typical bay segments, it is effectively four separate and repeated building 'bays' set side by side, to be constructed as a single structure for efficiency and economy. The structure of the podium is a reinforced concrete frame with concrete, roof deck and floor slabs; the upper podium level has a cantilevered reinforced concrete roof, which serves as a canopy to protect those using the ground floor pedestrian routes beneath, linking the building visually to its adjacent storage facility currently under common ownership.

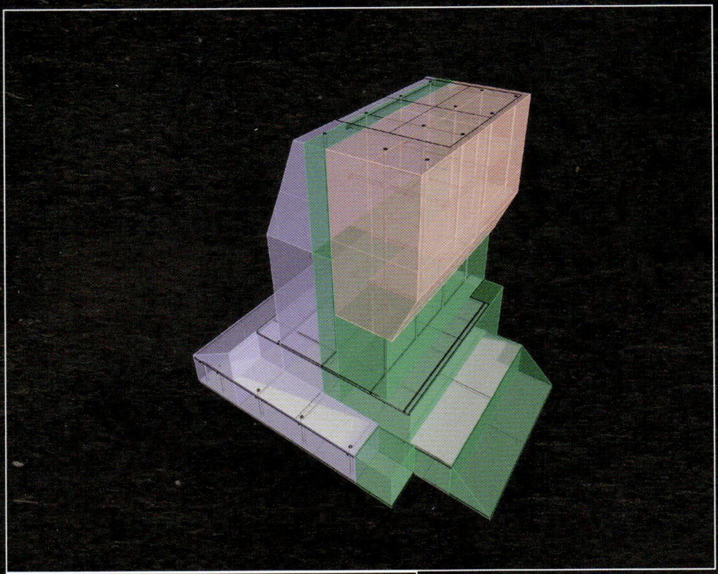

1A. Isometric view of environmental volumes

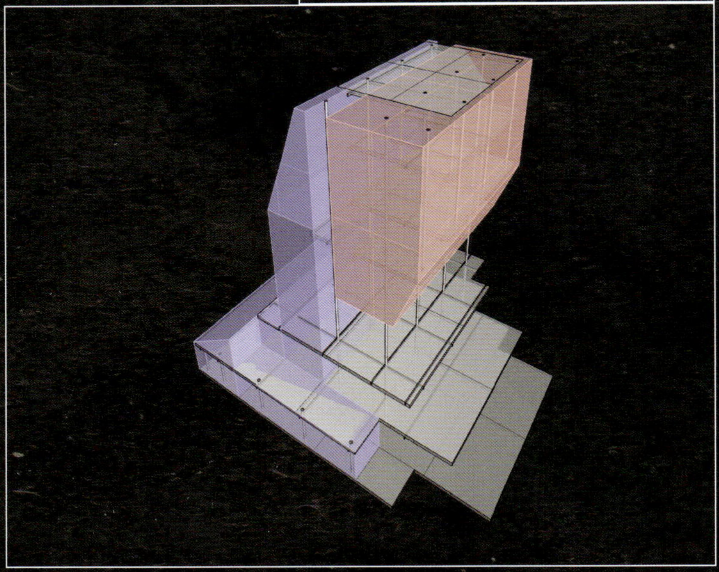

1B. Isometric view of environmental volumes

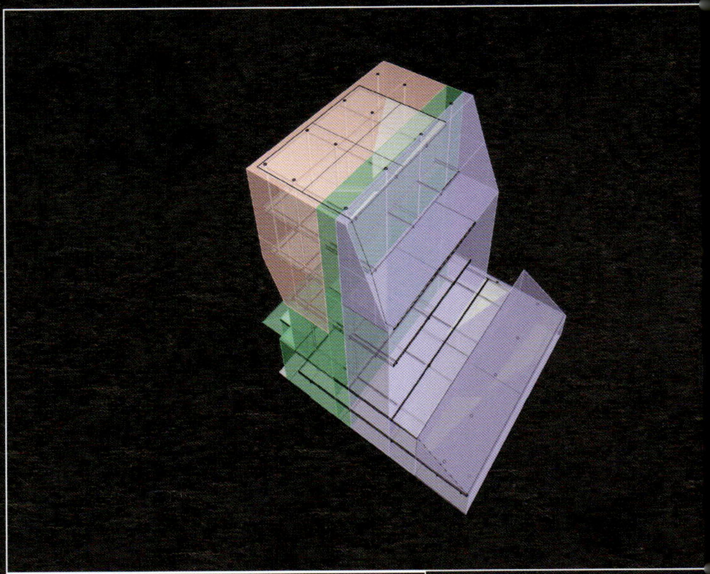

2A. Isometric view of environmental volumes

2B. Isometric view of environmental volume

Environmental strategy

The environmental strategy for Option 1 is to have a single environmental zone for the workshop spaces, which are in constant use, and a set of smaller environmental zones for the design studios, which are used less frequently. Service spaces to the workshops form a third environmental zone; these are largely opaque volumes which are highly thermally insulated, have low air changes, and are ventilated both naturally and mechanically.

In Option 2, environmental zones are created as a result of the vertically oriented arrangement of spaces with their independently functioning 'blocks' of workshop spaces and design studios; the two support spaces each form a single environmental zone. Option 3 simplifies the environmental strategy by creating a single environmental zone for the upper level workshop through its complete height; a second single environmental zone is added for design studios, together with a third environmental zone for support spaces. The environmental design is based around the need for user safety as well as the energy efficiency in operating the building. A decision was made at an early stage to create tall spaces for the workshops in order to provide flexibility in the use of the equipment and to be able to allow for tall assemblies to be fabricated within the workshops. This approach led to the workshops becoming enclosed in order to conserve the heating and cooling that would be required when in use, which excluded the use of access doors and roller shutters. This approach would have been the case if the workshop facilities were set on the ground floor, with the connection always to the outside.

The possibility of doors being open during the day would have made the design far less energy efficient than was envisaged. The decision to put the workshops at a high level ensured that the arrival and departure of assemblies and materials for incoming materials and outgoing parts assemblies would be done in a highly controlled manner using the panels integrated into the projecting workshop spaces. This approach allows the workshops to be well lit and airy without requiring the full extent of the floor plan to be occupied. It was envisaged that few people

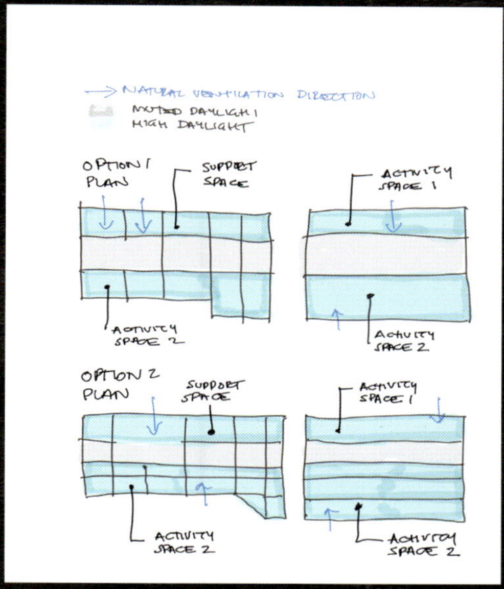

Daylight strategy for direct sunlight and cross ventilation.

Option 1 admits daylight and air from two sides of the building.

Option 2 narrows the central circulation space in order to add more daylit activity space on each floor.

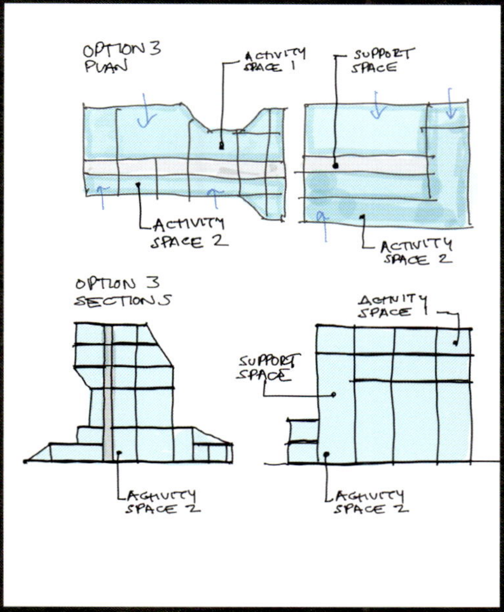

Daylight strategy for direct sunlight and cross ventilation.

Option 3 adds more daylit space at upper levels where the floor plates can be increased in size.

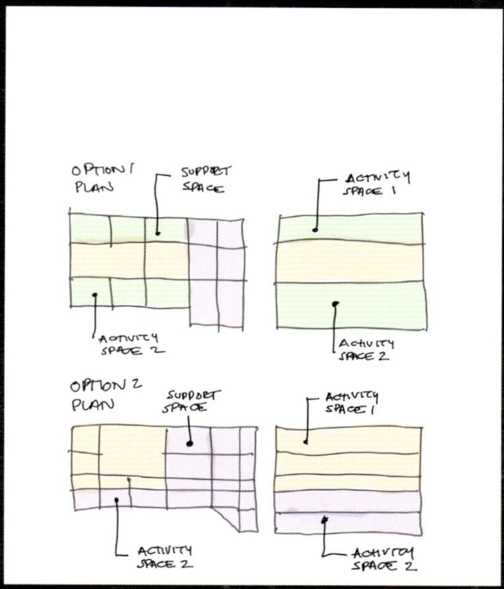

Environmental zones

Option 1 has a single environmental zone for the workshop spaces and a set of smaller environmental zones for the design studios.

Option 2 has environmental zones created as a result of the vertically oriented arrangement of spaces with their independently functioning 'blocks' of space.

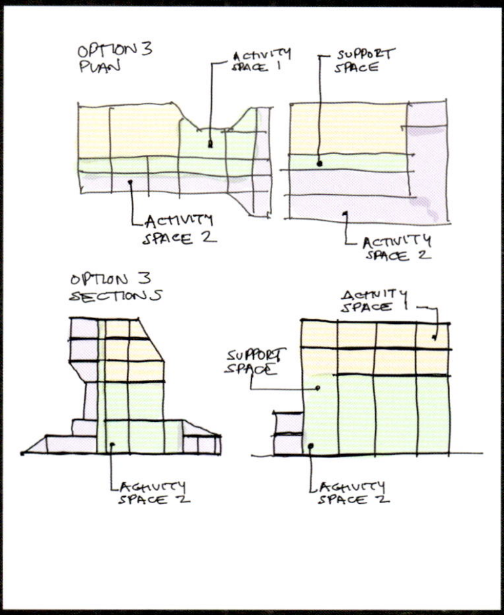

Environmental zones

Option 3 creates environmental zones for the upper level workshop, design studios and support spaces.

will occupy the workshop at any time but lightweight equipment will be used for the fabrication and production of components used in prototype assemblies, together with test versions of assemblies for products that will be manufactured elsewhere. The interaction between the design spaces and the workshop spaces was considered to be important in essence, but they are required to be separated for safety reasons. Internal spaces are cross ventilated for night-time cooling, for day-time thermal comfort and to make a provision for fresh air from outside where possible. In this temperate climate, a key aspect of the design of the environment is safety for users of the building and their visitors, as well as for organised visits for the public.

As a result, the concept for the environmental zones is one of creating small groups of space, each with its own environmental zone such that the workshop spaces can be subdivided and closed off environmentally between spaces. This spatial arrangement will allow specific areas of the workshop immediately above the associated design studios to protect the health and safety of building users and visitors. As a result, the environmental zones are divided per typical bay. A bay corresponds to one of the four building segments containing a design studio, a workshop above, and ground floor and office space. The building is considered as three environmental zones for each of the four segments of the building; a total of 12 linked environmental zones. The advantage of this approach is that the heating energy or cooling energy of the ducted air that circulates around the building can be redirected from space to space as required. This approach is not one of dividing the building into standalone units, but instead aims to share the energy requirements across the extent of the building. Only those parts of the building in occupation are supplied with tempered air when occupied. The use of deep reveals surrounding glazing both in the podium and in the building above it resulted in avoiding the need for additional solar shading. The orientation of the glazing was such as to benefit from the passive shading created by the reveals which allow views in and out of the building, providing a level of transparency in the podium as part of the welcoming strategy of the community.

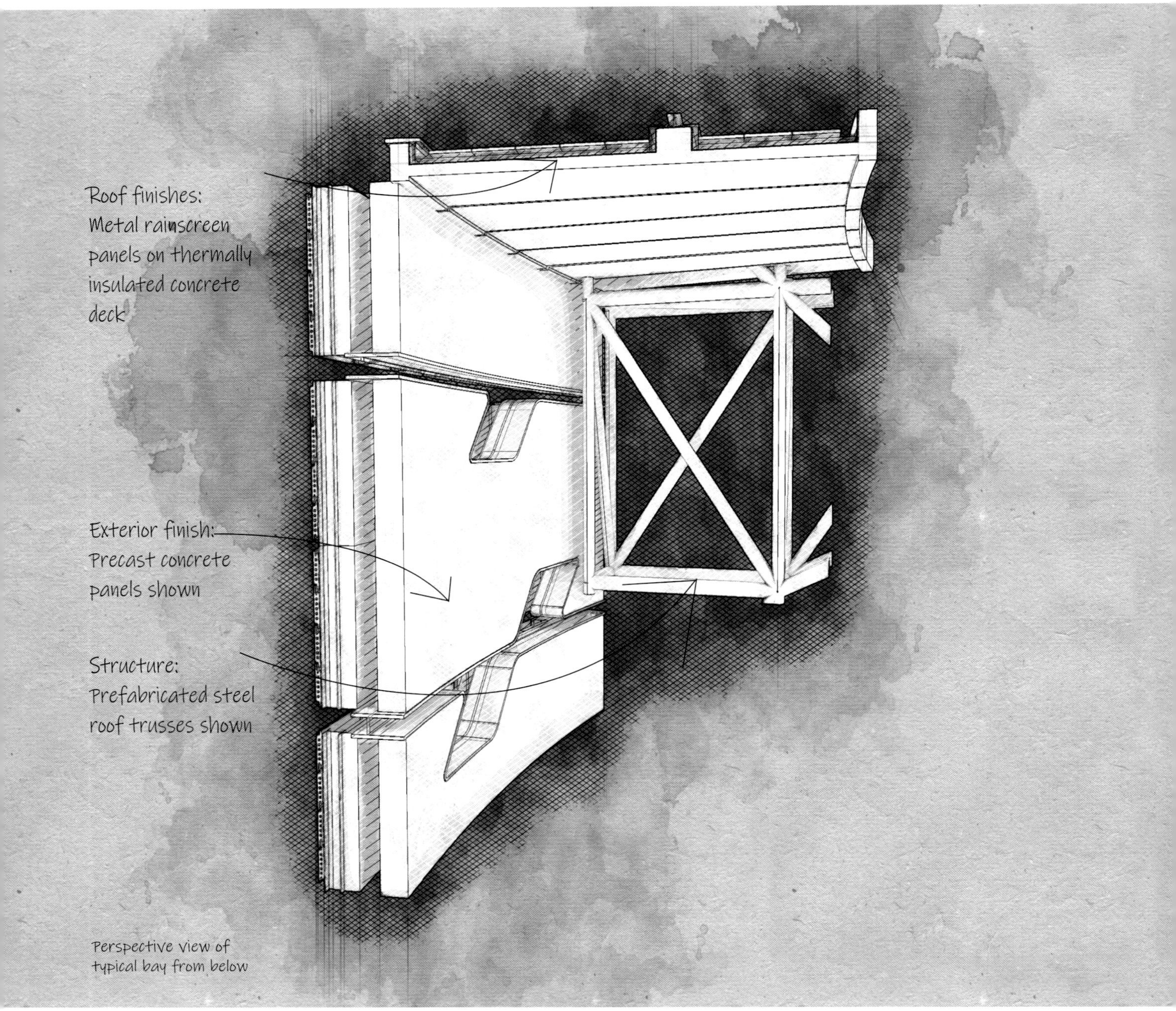

Typical bay adjacent to facade

The envelope of the building provides a set of typical bays that effectively form full 'slices' through the building: 6.0 metres wide and which are joined together, end to end, to create the characteristic section and shape of the building. The typical bays are largely opaque facades punctuated with glazed strips between each bay; admitting daylight to all spaces within the building. The facade treatment of the design spaces is similar to that of the workshop spaces, as both spaces use computers and monitors to interface with the equipment.

The workshop space is an open assembly area with seated laboratory-style areas where small-scale materials are available in stored racks and shelves. Those working in the workshops are generally seated at long benches where they can interact and operate the equipment from computers. The two sets of spaces for manufacturing and design having similar requirements for their occupation; the facade treatment is the same across both functions. Glazing is set in vertical strips between modules which have typical bays arranged as square-shaped glazed panels that provide ease of opening for natural ventilation both at a high level and at a lower level inside each of the spaces. Both spaces are cross ventilated to encourage energy efficiency within the building. Ground level spaces are occupied in a podium of glazed areas and areas of opaque panel. The glazing is formed from clear glass for full transparency; solar shading is provided by the deep horizontal overhangs that surround the glazed openings. The opaque panels are formed as cementitious panels, which are used for their relative ease of replacement as well as for their ability to be modified and changed in response to emerging requirements in the near future. Exposed concrete used at a high level was less practical than the evolving requirements of the podium's facades required.

The underside of the inclined walls to the workshops set at a high level have an external envelope of metal panels which can be lowered into the vertical plane in order to allow materials to be delivered and for goods to be dispatched from the building. All materials and assemblies or manufacture are light in weight and are loaded into vehicles

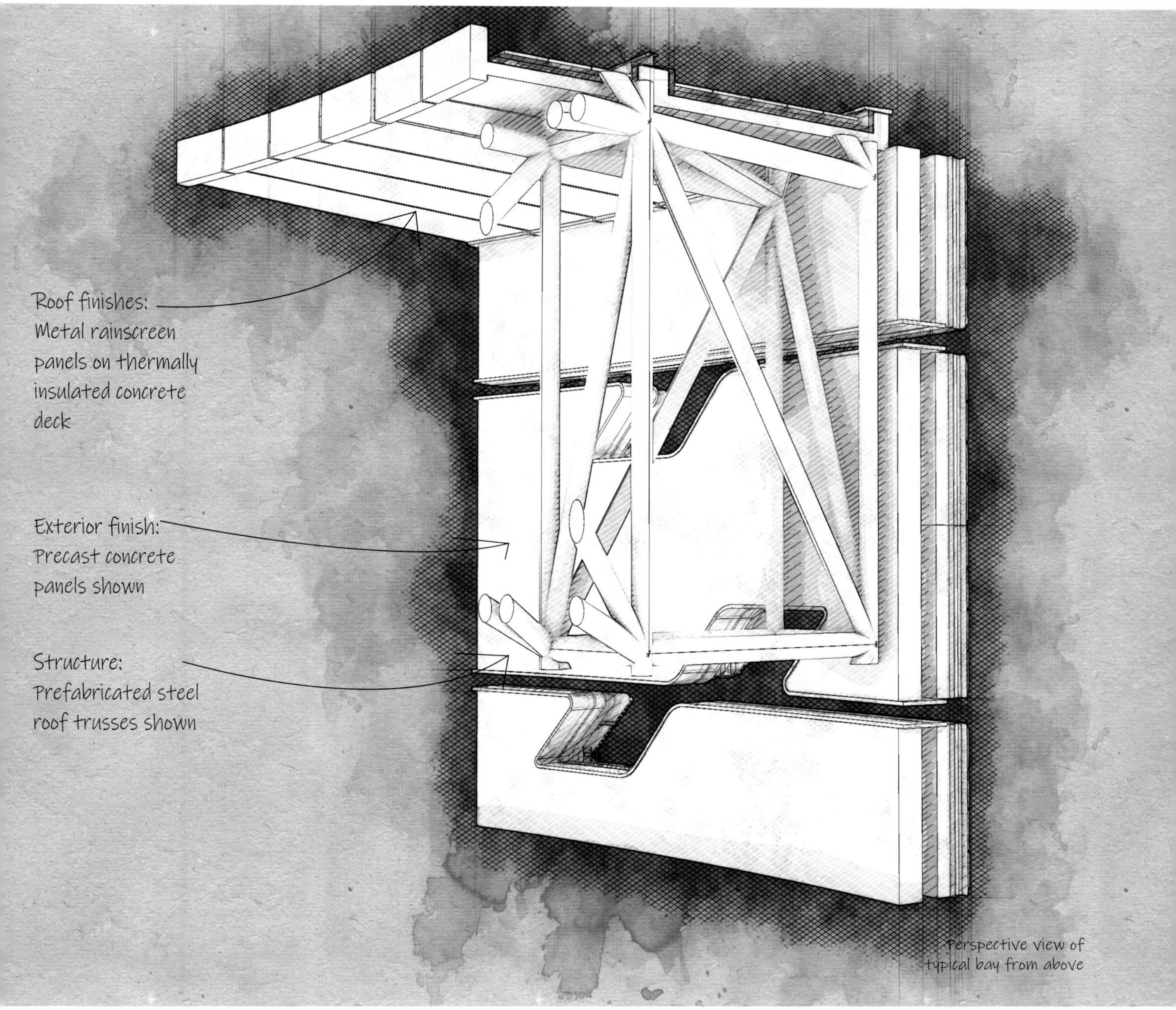

Perspective view of typical bay from above

for transport to their final destination. The expectation of lightweight assemblies of relatively high value and of generally small size appears to be a key part of the future of manufacturing planned for this region. The roofs of the podium and the main spaces above are not accessible as roof terraces; they comprise membrane-covered roofs with maintenance-type paving slabs set onto them in order to protect the membrane. It was decided that roof terraces for access by building users were not a requirement in this design.

The typical bay which forms a set of spaces on different levels is repeated down the length of the building in order to provide an economical solution as a result of the repetition of the construction of the structure and the external envelope. The end elevation of the building extends from ground level to the top of the building, and is a mixture of reinforced concrete columns and structural walls which are combined in places to create the required enclosure for the building. The glazing is silicone-sealed stick curtain walling, with clear glass that provides a visual continuity across the glazed elements of the facade. Glazing has solar shading in areas where the space behind is office or design studio. Part of the workshop has an inclined glazed roof to provide daylight to areas where assemblies are to be viewed. The envelope aims to integrate exposed concrete walls in a diaphragm configuration with stick curtain walling to provide an economical envelope that combines structural components with envelope assemblies. Interior surfaces to the opaque external walls are formed as large panels of in-situ cast reinforced concrete, as are the outer walls.

This approach to the opaque part of the internal walls provides a robust, durable set of working surfaces. Reinforced concrete columns are also exposed. Shadow gaps are introduced to articulate walls and connections between beams, columns and external walls. The strategy allows wall fixings for shelving and equipment to be installed without additional framing. A key part of the design was to ensure that the concrete-like appearance of the external walls is robust, durable but will weather elegantly without the surface staining that has made concrete a less attractive material for buildings in past years.

1. Perspective view of exterior

2. Perspective view of exterior

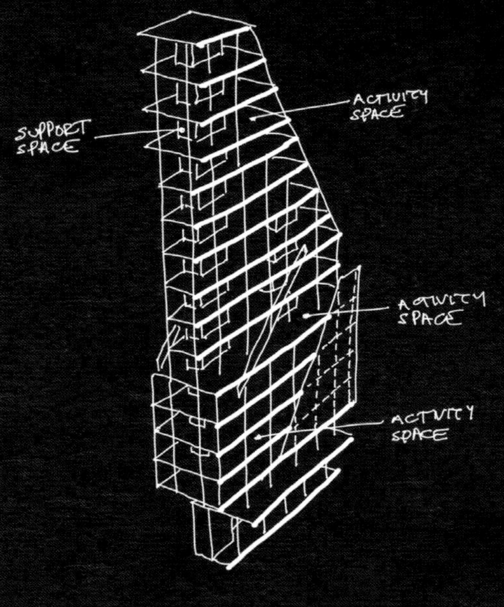

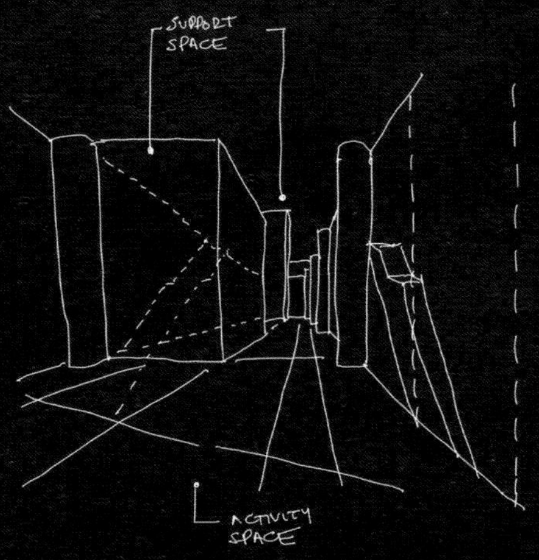

Structural design from a comparison of typologies

A key part of the structural approach to this design was to begin with an arrangement of a simplified grid of columns, beams and floor slabs that would bring a high level of efficiency in the use of energy and of material. Options were prepared for both reinforced concrete and steel. The basis of the structural design was determined by the spatial requirements of the brief/program.

A second iteration of the structural design was based on taking each internal space created by the first iteration, redesigning the structure forming each space or set of volumes, and considering each set of spatial volumes as a separate structure. The two design iterations are then combined and coordinated as three options that use the least amount of material, that take the least amount of time to construct and have the least embodied energy. The three options were compared with the energy and use for each of those solutions, given the anticipated occupation times and frequency of each of those buildings. The resulting structural options were based on these principles. In this project, a first option was considered as a structural arrangement of regularly spaced columns, with beams and floor slabs that follow the shape of the building. In addition, columns were set on the edge of the floor slab, on the structural grid, regardless of the distance to the nearest column that is 'on grid'. There are large-sized 'cutouts' in the slab for the entrance area and non-rectilinear-shaped floor slabs. A second option considers the forms that comprise the building as being three separate structures; a tall block at one end, an entrance area in the centre, and an end block with an inclined vertical wall. These three structures are considered to be positioned next to each other; double rows of columns are set out where each structure meets its adjacent structure.

This approach creates the possibility of introducing additional smaller structures by sub-dividing these three blocks; the block that sits inside the opening formed as a main entrance is an example. Greater differentiation can be created in the overall structure of the building without imposing the need for a continuity of beams and columns in places

. Perspective view of exterior

4. Perspective view of exterior

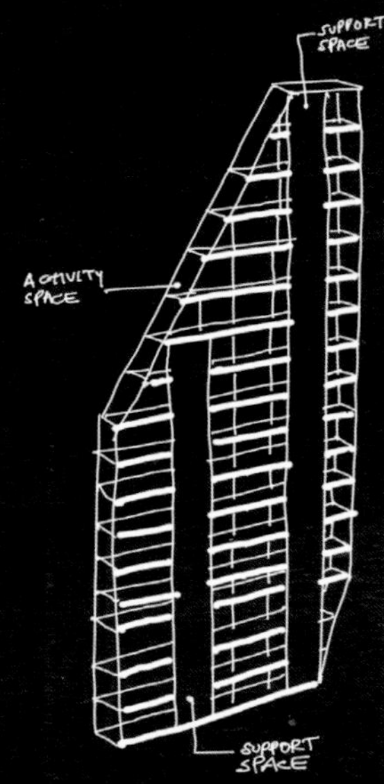

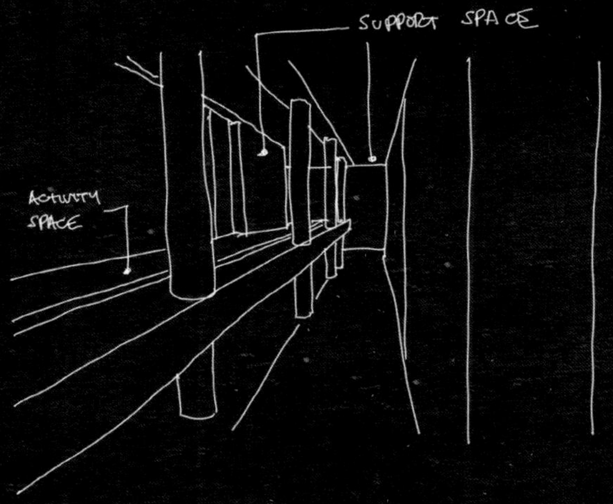

where they are not required, as illustrated by the entrance space. The sub-division of the structure into smaller structures allows the design to respond as closely as possible to both the preferred spatial arrangement and to the vertical circulation of stairs and lifts/elevators inside the building.

In Option 3, the structural form responds closely to the requirements of the corresponding Option 3 for the spatial arrangement. The design is one which mixes an internal concrete-framed structure with an external steel-framed structure. The inner structure has reinforced concrete structural walls at ground floor level, rising through the building vertically. An outer steel frame moves in and out of the building, as seen in vertical section, to create the characteristic profile of projecting circulation spaces. The upper floors of structure have externally fixed steel frames set at an angle of 45 degrees in order to draw ducted air up and down the building without the need for internal ducts and cores. The larger projecting volume extending from Levels 2 to 12 encloses washrooms, staircases and service ducts that provide fresh air within the opaque facades in levels 4 to 12. Air-handling equipment is mounted at the top of the inclined arms that extend from Levels 4 to 12; the associated services distribution is visible on the facades as a result of the projecting bands of horizontal forms. The upper triangular-shaped volume is supported on columns that extend up through the height of the building. Service equipment is accessed by way of an access deck at the level of the base of the triangular form.

The structural design includes reinforced concrete structural walls that form part of the opaque walls seen on the long elevations. This structural design draws different structural types and assemblies together in a single design. The spatial design is optimised for the efficient use of space where ducts – which are extensive in this building – would have occupied a higher proportion of floor area than an equivalent office building; for example as a result of the ventilation requirements for the densely arranged laboratory spaces within this building.

Ventilation from laboratories to the outside air is by way of filters at roof level which undergo continuous safety checks.

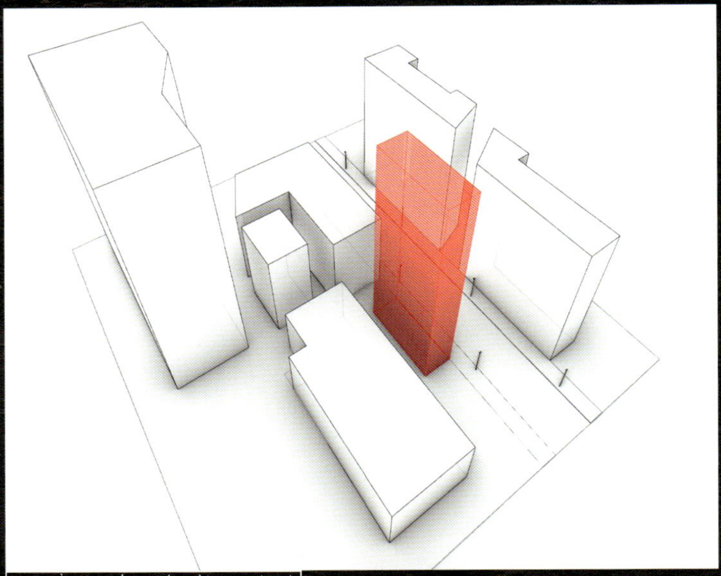

1. Project volume in site context

2. Project volume in site context

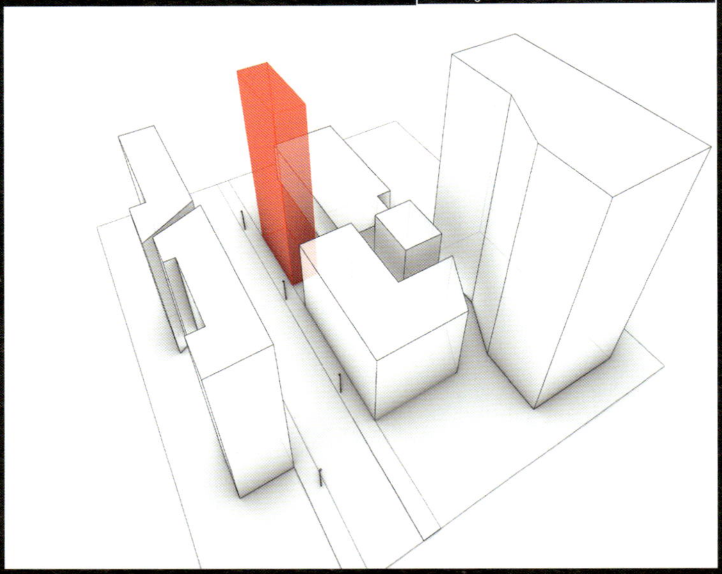

1. Site plan with surrounding context

Option 1 considers the building as two blocks linked in the middle.

2. Site plan with surrounding context

Option 2 creates a more indented form of three blocks in order to respond to the need for ground floor activities.

3. Site plan with surrounding context

Option 3 arranges spaces in horizontal bands, with public functions on the two lower floors.

Site context

The site context for this medical research building is in an urban centre, adjacent to a university campus that is surrounded by office buildings of different height, ranging from an adjacent 10-storey building to rows of two-storey buildings. The footprint of the building fills the extent of the site in order both to provide the accommodation required in the brief/program, and to reinstate the 'missing' urban block which is currently used for ground-level-only car parking.

All three options have the same overall form of the building; each option is differentiated by the arrangement of cores, which are visible at roof level. Option 1 considers the building as two blocks linked in the middle. Option 2 creates a more indented form of three blocks in order to respond to the need for ground floor activities. Option 3 arranges spaces in horizontal bands, with public functions on the two lower floors, and office space for research at the top of the building with views out. The main laboratories are set in the middle part of the building. Option 3 is preferred as it avoids vertical stacking of spaces. This option provides an urban form that fits in with the neighbourhood rather than appearing to be a series of opaque facades that turns back on the urban blocks around it.

The research centre aims to contribute to its urban environment by suggesting a way to integrate buildings which link university research departments and programmes with the commercial environment. This building sits at the interface of a business district and a university campus. The research centre aims to show how research tasks and production tasks can be integrated into a working environment which is efficiently organised into small working spaces.

The design provides views out to adjacent buildings as well as across the building itself to provide an efficient arrangement of spaces. This building has no meeting hall, theatres for presentation or relaxation areas for the academic staff. The spatial arrangement is a 'lean' environment for conducting research in a way that communicates the way work is being conducted across the building, and is visible to those who work there. Specific spaces for socialising within teams did not form a

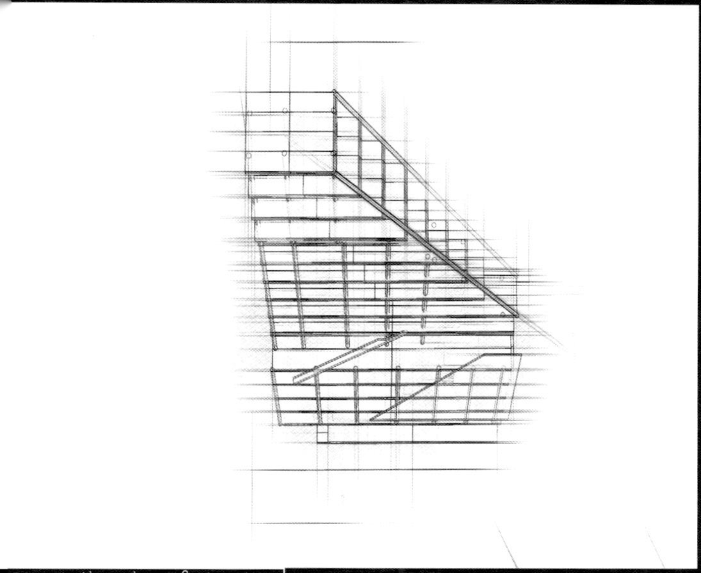

Perspective view of structure

2. Perspective view of structure

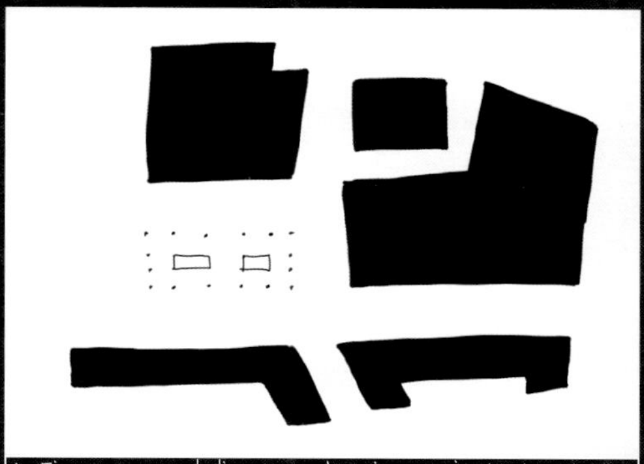

1. Figure-ground diagram showing project structure

Option 1
The structural grid positions cores in the centre of the floor plates, in the long direction.

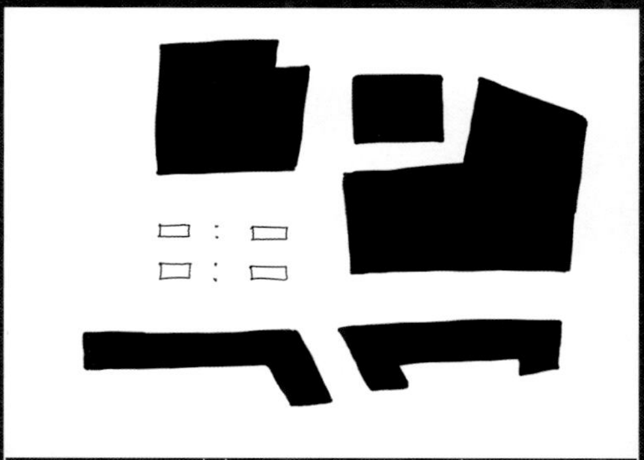

2. Figure-ground diagram showing project structure

Option 2
The structural grid positions cores in the centre of the floor plates, in the short direction.

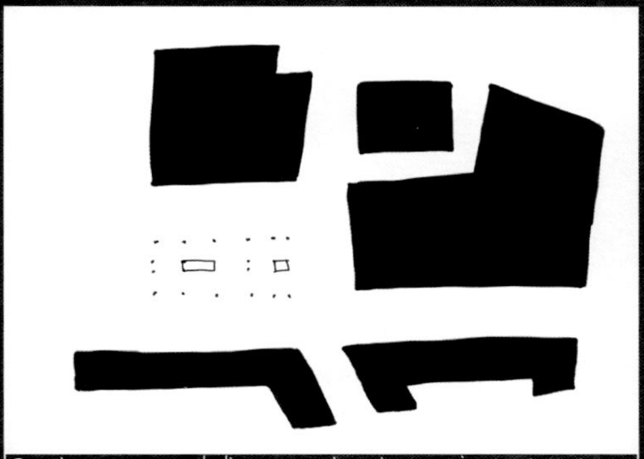

3. Figure-ground diagram showing project structure

Option 3
The structural grid positions cores in the centre of the floor plates, in the long direction.

part of the brief/program as these specific facilities are considered to be already provided by businesses in the neighbouring district. Instead, the building accommodates exhibition space and interaction space in an outreach program to the wider business community.

In terms of massing, the building takes up a shape that is determined largely by the brief/program, which aims to have part of the building looking onto itself, such as the views downward from the inclined glazed wall of the triangular volume at the top of the building. Views are created up and down the circulation spaces, which are set at the edges of the building in an interface zone between inside and outside; between inner and outer structures.

This approach allows the brief/program to be implemented on a long, narrow site, with spatial arrangements that were 'carved' into the volume of space available. During the early stage of the design, spaces were rearranged to arrive at the characteristic form of the current set of spaces. The options for arrangements, structural options and environmental options within that form were pursued in the design of the three options. Spaces for interaction are those in the corridors, which connect different departments on the edge of the building; areas where the volume of the building extends outward. Floor plates are self-contained in a mix of academic and commercially based activities within the building, in that research outputs would eventually be realised as commercially available products.

The building reaches out to the business community at the ground floor level by providing an interaction space which forms a showcase for the activities within the building. Site context is more geared towards reaching out to the mindset of those working in surrounding buildings, as part of a working community, rather than trying to fit the building in quietly to its urban context. The building reaches out visually to those in surrounding buildings and to the wider business community. The ground floor interaction space of an exhibition and showcase becomes the shared space between the local business community and the building, as well as the wider business world, through podcasts delivered from the ground floor space.

1. Sectional perspective view

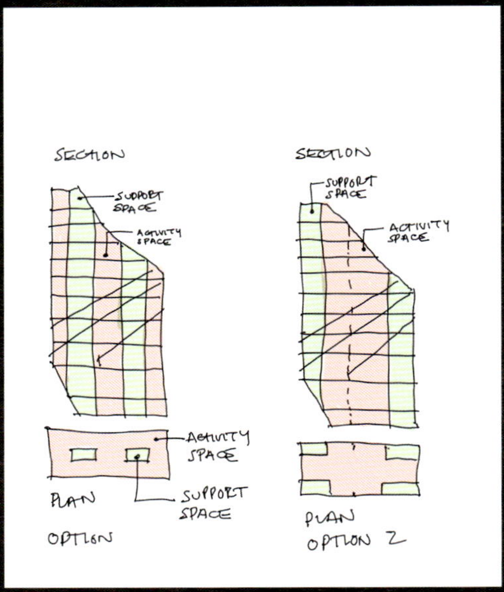

Spatial arrangement

Option 1 has vertically stacked office spaces, public-facing functions and laboratories.

Option 2 has some public functions at ground level; laboratories are positioned on upper levels.

2. Sectional perspective view

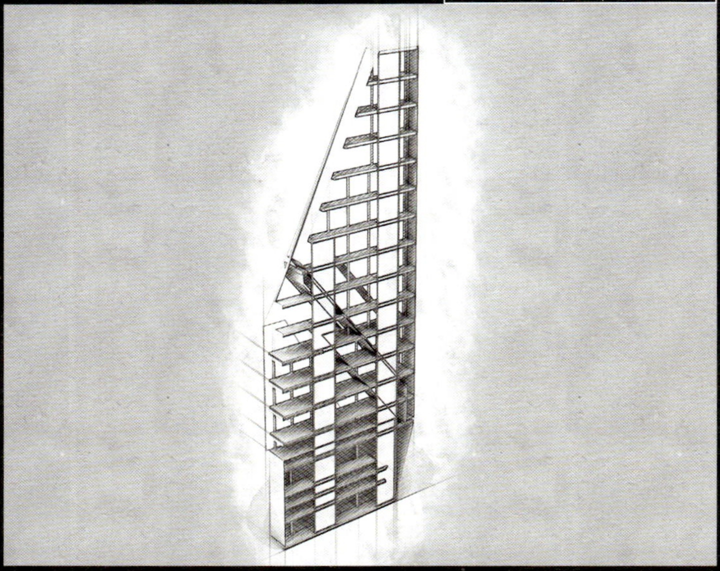

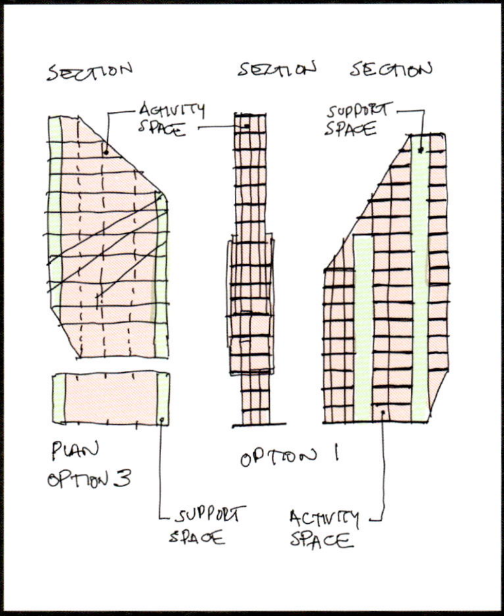

Option 3 creates a more dynamic relationship between the spaces and allows views from the upper offices down onto the shared roof deck on top of the laboratories.

Spatial arrangement

The spatial organisation of the building is derived from the need to accommodate different research activities in areas which are isolated from each other. As a result, the public areas and general administration areas are set on the ground floor and the floor immediately above. Research laboratories are located immediately above those floors in two zones which are divided by an inclined service duct and air-handling ducts which ventilate those spaces. The research offices are in two zones; the top half of the triangular form of the building accommodating offices for one research group, with a corresponding research lab, and associated research offices linked to the work being done in that research lab. The offices have corresponding services zones set on the outside of the building, which follow an inclined route across the facades.

The spatial arrangement of Option 1 is of vertically stacked office spaces, public-facing functions and laboratories; this option is arranged economically but is spatially dynamic, with its stepped arrangement of spaces that generate a sloping profile when viewed in section and in elevation. An exoskeleton of buttresses is set externally and traverses the sequence of spaces within the building. Option 2 addresses that issue by having some public functions at ground level; laboratories are positioned on upper levels, but the constraining form makes them difficult to arrange and to provide a degree of transparency. Option 3 addresses this by creating a more dynamic relationship between the spaces and allows views from the upper offices down onto the shared roof deck on top of the laboratories, by bringing transparency and views into the building where they can be provided.

The spatial organisation of the building is one of having administration and exhibition space on the ground and the floor immediately above. Workgroups are divided by the inclined service core that extends through the building. Above that zone, where the triangular part of the building begins, a second services zone rises at an inclined angle, providing support for two research zones and two sets of supporting office space.

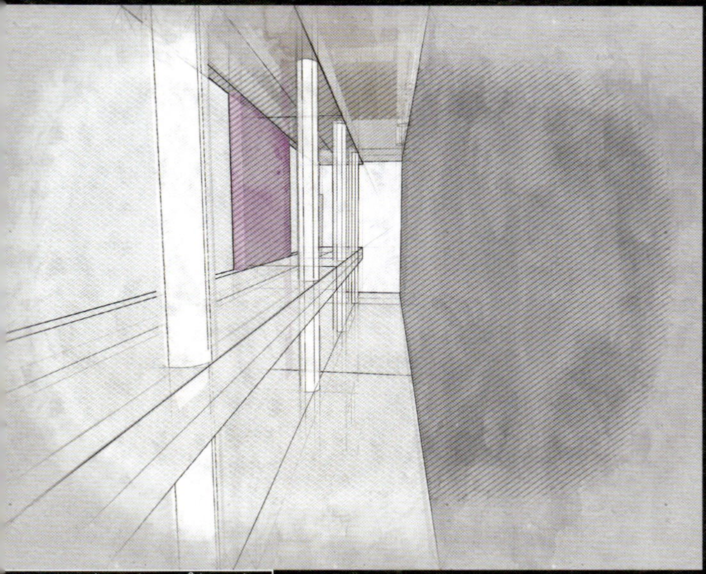

Perspective view of interior

2. Perspective view of interior

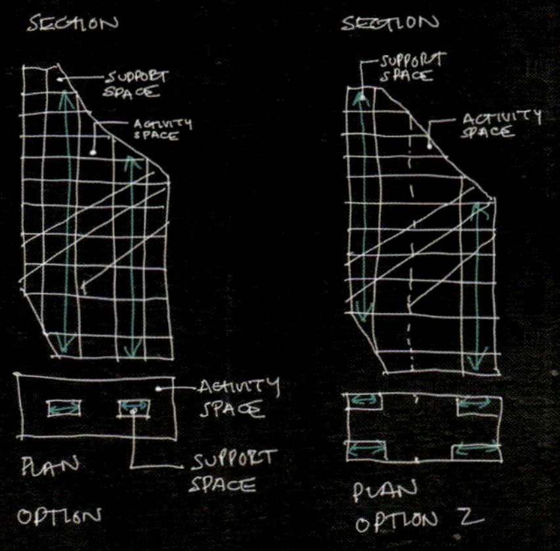

Internal circulation

Option 1 arranges circulation in the centre of floor plates.

Option 2 arranges circulation at the corners of the building.

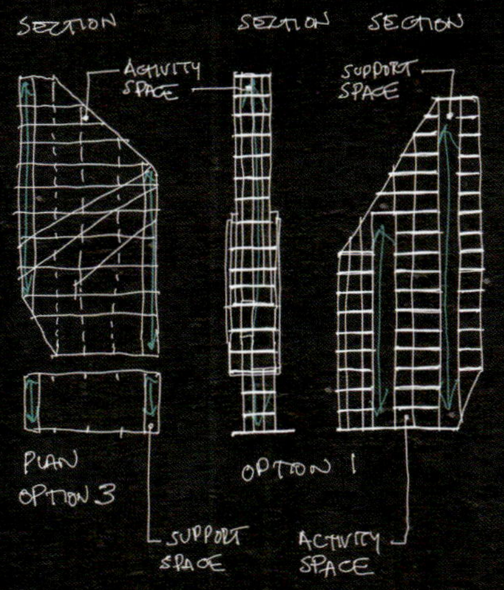

Option 3 arranges circulation at the two narrow ends of the building.

Internal circulation

The internal circulation of the building is organised around the principle of a set of spine cores extending the full height of the building; their presence is revealed in each of the options presented. Lifts/elevators, washrooms, service risers and stairs are set at one end. Meeting rooms are set in spaces between the cores. Spaces in the upper triangular volume of the building accommodate research and development; workspaces are divided into offices for individual research work as well as open plan areas for teamwork. The emphasis on team organisation in this part of the building is of individual researchers and team leaders occupying their own offices. There is a limited number of open spaces, which are mainly in the core in the middle of the building. Elsewhere along the length of the building, a set of individual offices are accessed by a corridor that is adjacent to the void between the core and the edge of the floor slab. The stairs are arranged on each side in order to provide communication between teams on different floors while benefitting from the large-scale open-plan arrangement of low-rise buildings.

The organisation of the low-rise campus-based laboratory has been, conceptually, turned on its side to form a vertically arranged series of spaces.

The internal circulation in the building follows the diagonal bands shown in the illustrations; spaces are accessed either across the length of each floor plate or diagonally to follow the inclined plane of the roofs. The visual link between diagonally arranged spaces, and the visibility between spaces, aims to encourage interaction between research work being carried out in individual office spaces and that which is carried out in laboratories. The sense of one community of a body of people working together with a common set of goals is an essential part of the brief/program.

Where the building projects beyond its general line of columns, a second set of circulation spaces is introduced in order to gain access to the double-sided laboratories which extend only to the centre of the building. Access to the laboratories is on both sides of the building in order to create a sense of community within each work group.

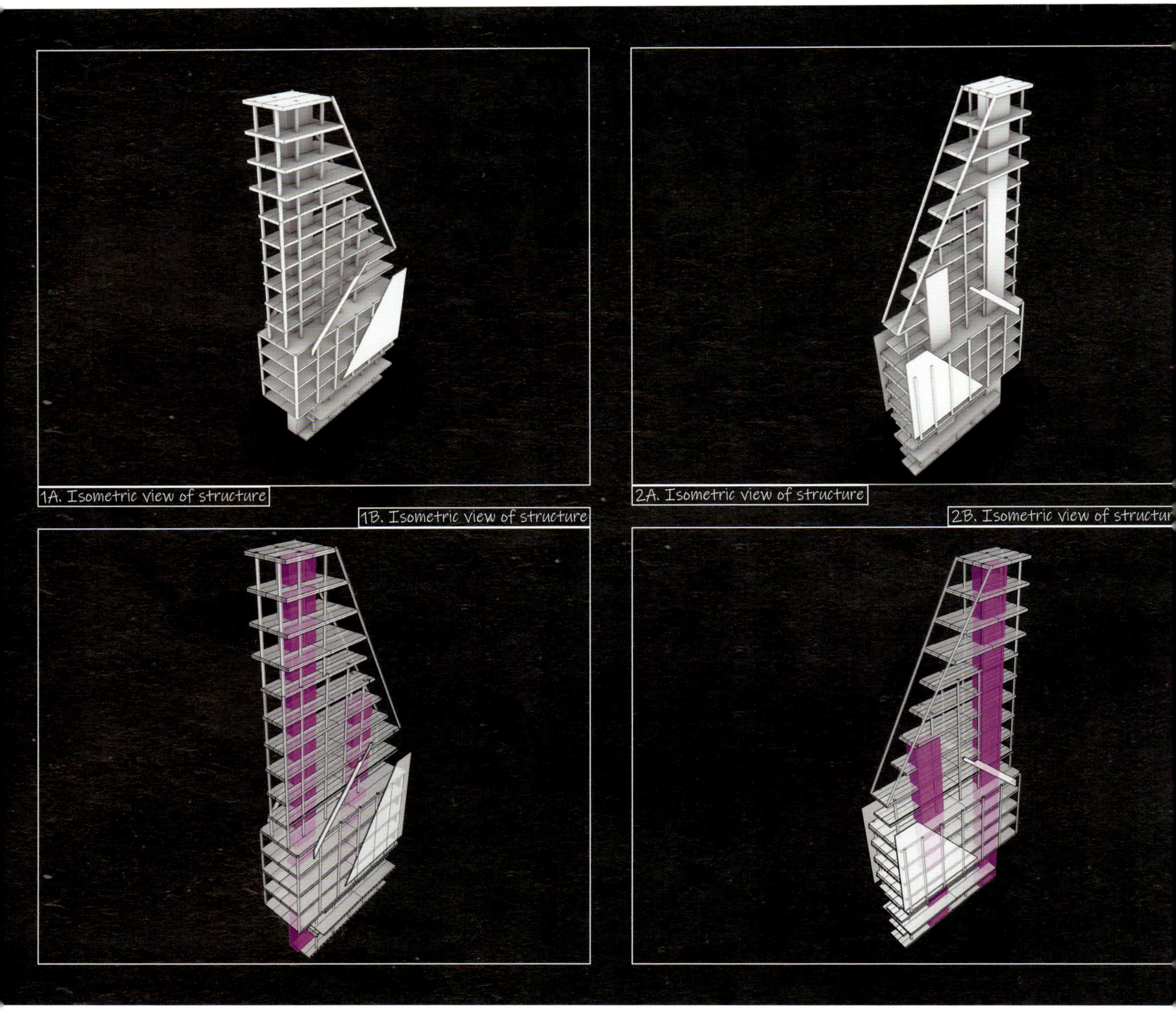

Structural strategy

Structurally, the building comprises a set of container-like assemblies; a set of ready-assembled part-wall spaces for offices or laboratories. The structure is formed from a prefabricated series of pods which are linked together. Modules are up to 12.0 metres long and 2.5 metres wide such that they can be delivered to site by road and be installed in place to form the external form and the interior spaces. The structure comprises a mixed frame of reinforced concrete and steel, into which the prefabricated container-like assemblies are inserted, including cladding panels.

The structure is partly an exoskeleton comprising an externally set structure which provides a large-scale frame into which a set of site-assembled smaller-scale frames are installed. The outer frame is primarily a mixture of steel and concrete, while the inner frame of floors and columns is formed in steel. This differentiation of structure between an outer primary exoskeleton and an internal smaller-scale structure allows greater interaction between primary structure and facades in the smaller-scale steel structure, as well as the ability to thermally insulate and clad the exposed primary structure.

The structural design comprises an inner structure and an outer structure set at the edges of the building, adjacent to the facades. Columns are set within the floor plates to support the mid-span of the floor plates between the two outer edges where they meet the facades. The inner and outer structure is formed as a double row of columns, set 3.0 metres apart within the depth of the plan and spaced 6.0 metres apart when viewed along the length of the facade.

This arrangement provides a circulation zone for the areas where this double structure projects out on the lower floors. The structure returns to the inner part of the building; the columns in some places continuing vertically down to the ground. Columns are inclined in areas of circulation in order to to connect the projecting structure back to the second row of the structural grid. The structure is braced by a set of cores which are set within a central service spine; forming a zone that accommodates office spaces and support spaces within the laborato-

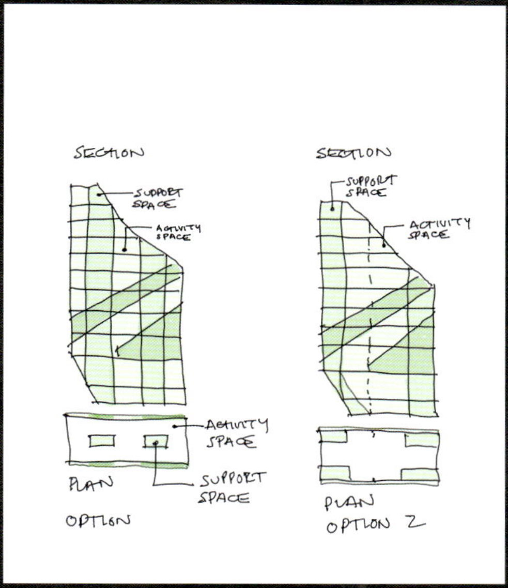

Structural strategy: Zones

Option 1 arranges cores in the centre of floor plates.

Option 2 arranges cores at the corners of the building.

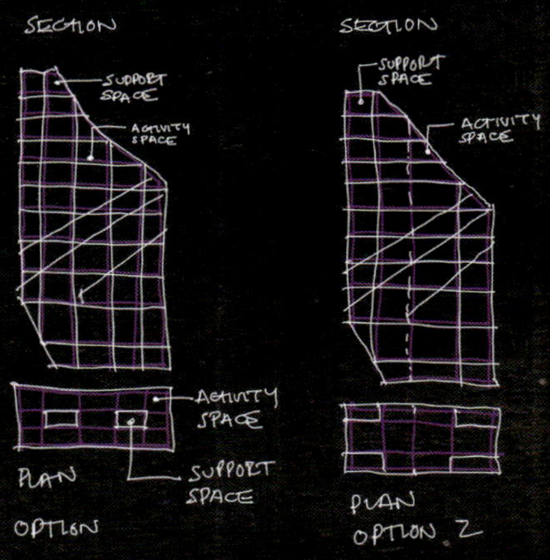

Structural strategy: Framing

Option 1 encloses the core spaces and dividing space into sets of activity spaces.

Option 2 is largely a steel frame.

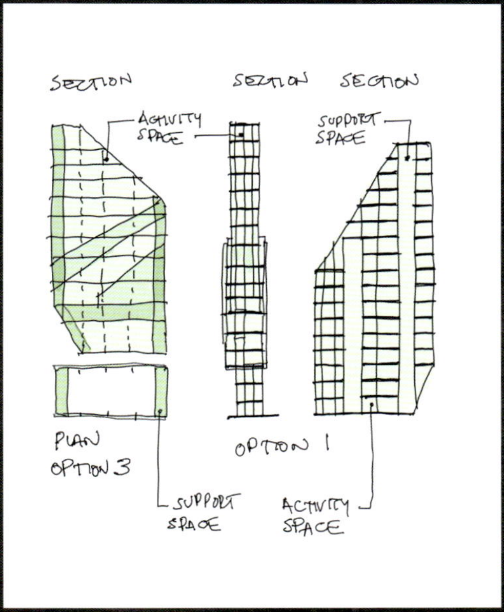

Structural strategy: Zones Option 3 arranges cores at the narrow ends of the building.

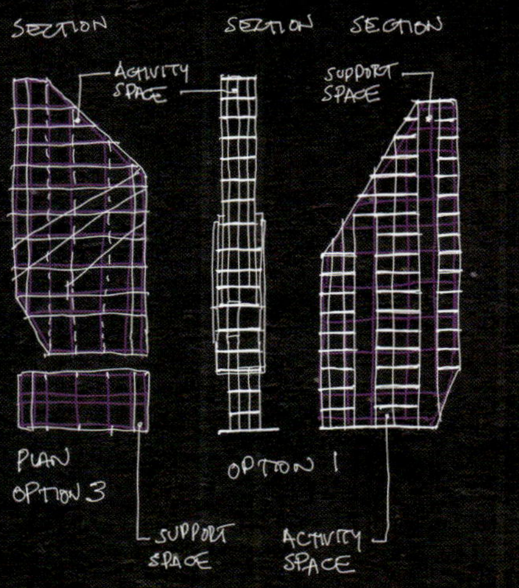

Structural strategy: Zones Option 3 is a mixed solution of reinforced concrete and steel.

ries themselves. A laboratory is set on each side of the spine, which accommodates mostly preparation rooms or investigation rooms that require an inner enclosed space. This space can be shared across the laboratories on both sides of the spine, ventilated to filters as part of their own ventilation system. The laboratories on both sides are ventilated with ducts which are inclined; the structure is organised with columns to enable this spatial organisation to be formed. Columns set on both sides of the spine define the zone of the spine; providing the area for staircases at the ends of the building. The ducts and air-handling items are supported on secondary steelwork which is attached to the outer row of perimeter columns. This building has a denser arrangement of columns than in comparable buildings of this type. The spaces required by the client are not a 'loose fit' set of moveable partitions but instead form a fixed set of 'modules' of laboratory space that can be joined to form a double module; extending from one side of the building to the other. Alternatively, adjacent modules can be joined to form a spatial group. The structural design is driven by the need to accommodate these modules of space, not by an optimisation of cubicles within open floor plates.

The subdivided nature of the building provided opportunities for a smaller structural grid to be used than that associated with office buildings. This approach resulted in smaller structural members than might be expected, and the opportunity to integrate small-scale inclined members to support services that are attached to the outside of the building; a structurally integrated, small-scale approach to a complex form.

It was decided not to make any of the walls that divide laboratories into structural walls in order to provide the ability to combine laboratory modules. In addition, the use of floor plates and columns allows complete areas of laboratory to be refitted to suit the evolving needs of the organisation. This approach was one of embracing change in the near future and discouraging a sense of 'permanence' within teams and within groups of which it forms a part, in collaboration with commercial organisations. A future for spatial design in the partnership of public and private organisations is set out in this design.

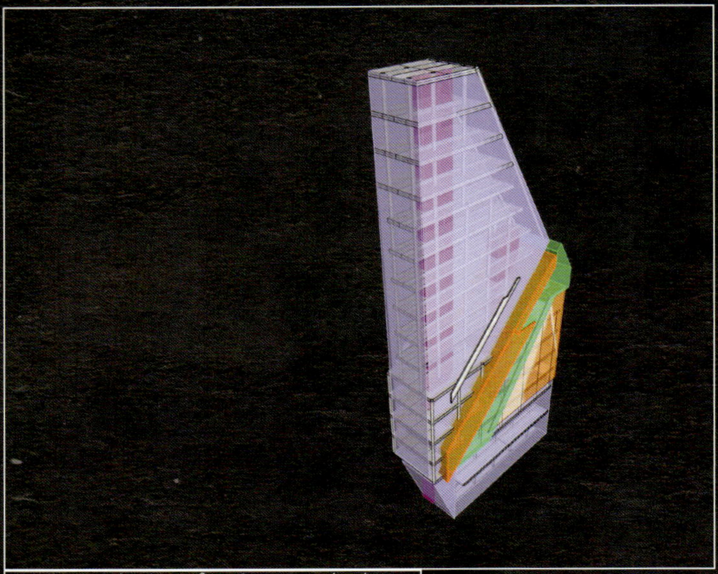

1A. Isometric view of environmental volumes

1B. Isometric view of environmental volumes

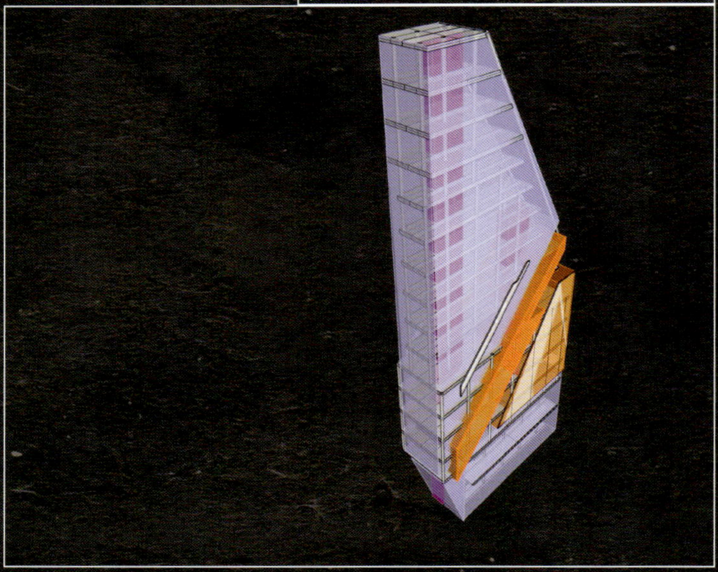

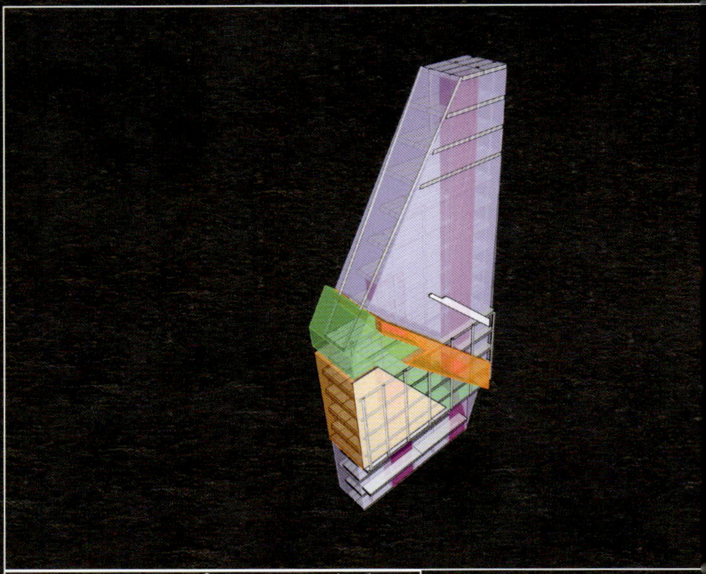

2A. Isometric view of environmental volumes

2B. Isometric view of environmental volume

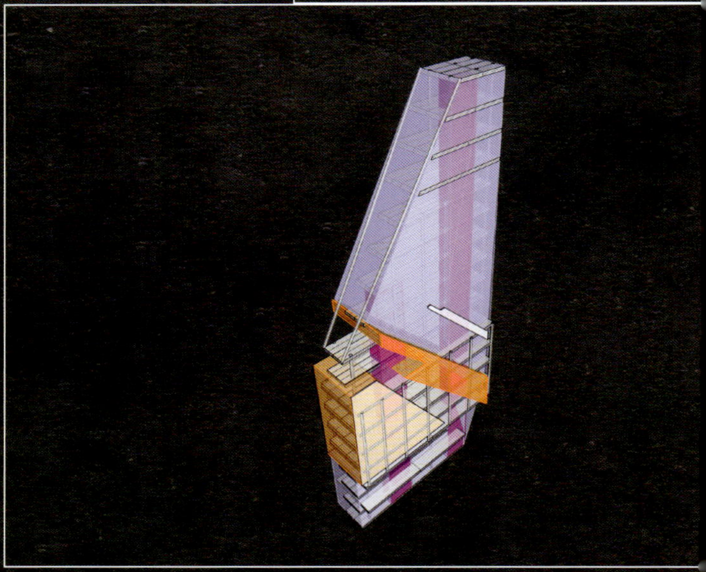

Environmental strategy

In addition to the criteria for establishing environmental zones set out earlier in the description of this project, different levels of daylight form part of the grouping of environmental zones in this building; a set of parallelogram-shaped spaces are grouped as zones. Circulation spaces are considered to be interstitial spaces which are supplied with fresh air from the main spaces of the offices; laboratories are ventilated separately.

The key component in the environmental design of this project is the use of daylight reflectors to draw light down deep into the sides of the building, which is required to be generally top lit rather than side lit in the centre of the building as a result of external wall space being required for working laboratory use, rather than for admitting daylight and creating a visual connection with the outside. The geometry of the building is based on the need for deep daylight penetration provided by openings in the spaces and associated reflectors. The building is highly thermally insulated in response to the climate, which is very cold in winter and mild in summer. The external form of the building has openings in it which extend all the way through the building, in order to provide top light into spaces below. Daylight also penetrates the building vertically by way of glazed slots set on both the narrow 'end' facades and the inclined areas of glazing on the long 'side' facades.

The environmental zones are formed from a patchwork of spaces across the building, which are associated with common levels of daylight, thermal insulation in the external envelope and ventilation requirements. A gap of 1.0 metres is set between the outer exoskeleton structure and the outer face of the inner facade; floor plates extend beyond the facade to be supported by the exoskeleton.

The mechanical and electrical installation makes use of prefabricated assemblies; mechanical ventilation units are local to each floor, and local to the zone in which they are accommodated. Equipment is accommodated in the 1.0 metre external zone between the outer facade, aligned with the exoskeleton, and the inner facade. Air is drawn down through the facade zone through individual ducts before it is intro-

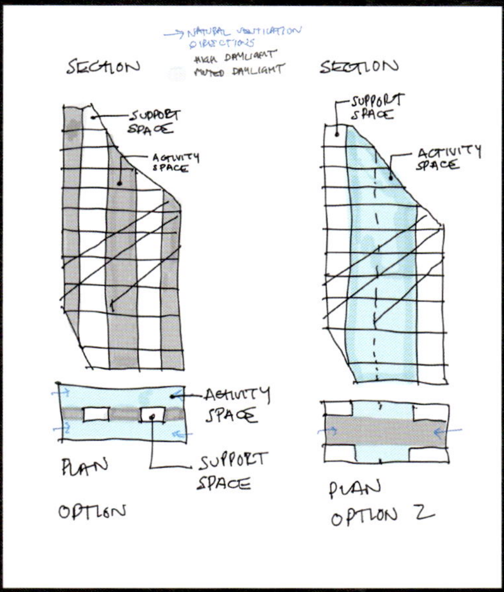

Daylight strategy for direct sunlight and cross ventilation.

Option 1 admits daylight and air on two sides of the building.

Option 2 admits daylight and air on four sides of the building.

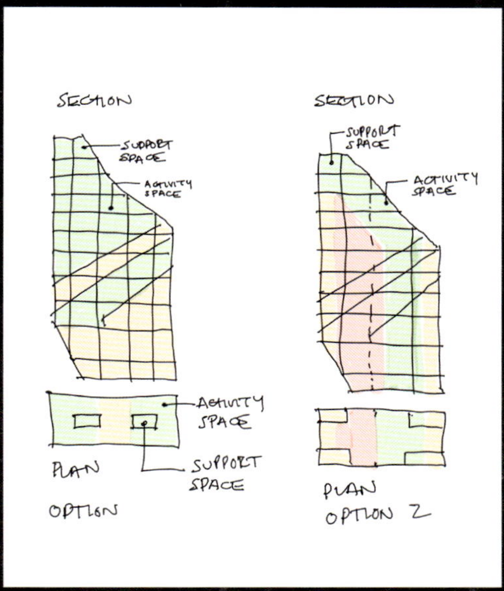

Environmental zones

Option 1 is one of three vertically stacked environmental zones for public spaces, offices and laboratories.

Option 2 provides a different arrangement of 'support' spaces, or service spaces.

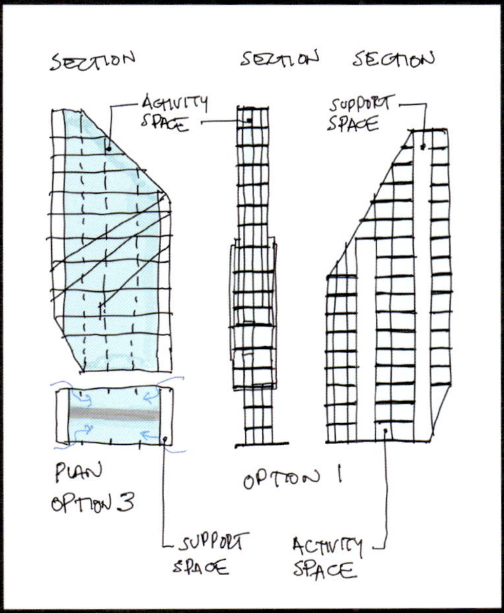

Daylight strategy for direct sunlight and cross ventilation.

Option 3 admits daylight and air on two sides of the building and admits air via the support spaces at the narrow ends of the building.

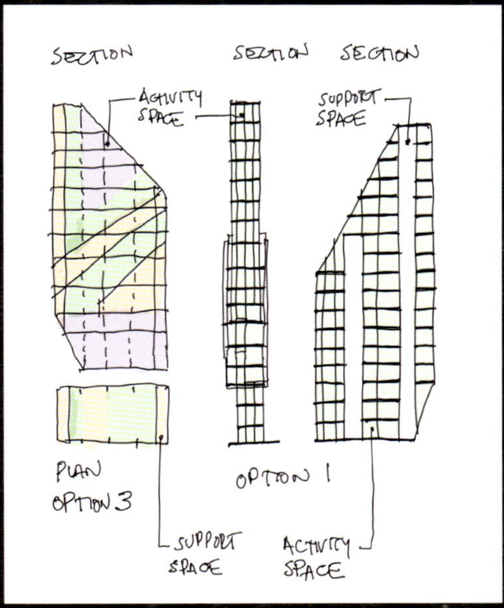

Environmental zones

Option 3 has external spaces around the edges of the building at upper levels; forming a set of 'inhabited' roof terraces.

duced into each floor plate; there are no vertical service risers within the building. Natural air for ventilation is drawn through the facades into internal spaces; the heat or cooling energy already in the space is recovered and recycled.

The environmental strategy for Option 1 is one of three vertically stacked environmental zones for public spaces, offices and laboratories. Option 2 has a similar arrangement, following a different arrangement of 'support' spaces, or service spaces, but where the office spaces are brought into the zone occupied by the laboratories in order to provide a sequence of spaces that is connected more easily by circulation routes.

Option 3 develops this theme of a sequence of spaces further, with more external spaces around the edges of the building at upper levels for building users to mix in an outside environment as a set of 'inhabited' roof terraces, which is a characteristic of this urban environment. The environmental design is based around zoning each of these types of space. The office space, situated in the volume within the tower above the middle podium, is the area that requires the greatest amount of tempered air in this climate, which is hot in summer and temperate in the winter with high humidity during summer months. The environmental strategy is one of providing tempered air to office spaces, then distributing them to the secondary spaces in the entrance; the breakout spaces and meeting spaces which occur in the middle podium. A zone of open office space which is mainly glazed is located on the upper floors.

Laboratories are serviced separately with their own ventilation system, both for the occupied space and for the ventilation of specific chambers within those working spaces.

Space was allocated for winter gardens, to be implemented in the circulation spaces that project outward from the building. This facility will be created while the building is in use if occupants wish to add this set of spaces in the near future. The structural design ensured that this item could be added at a later date while still forming an integral part of the design.

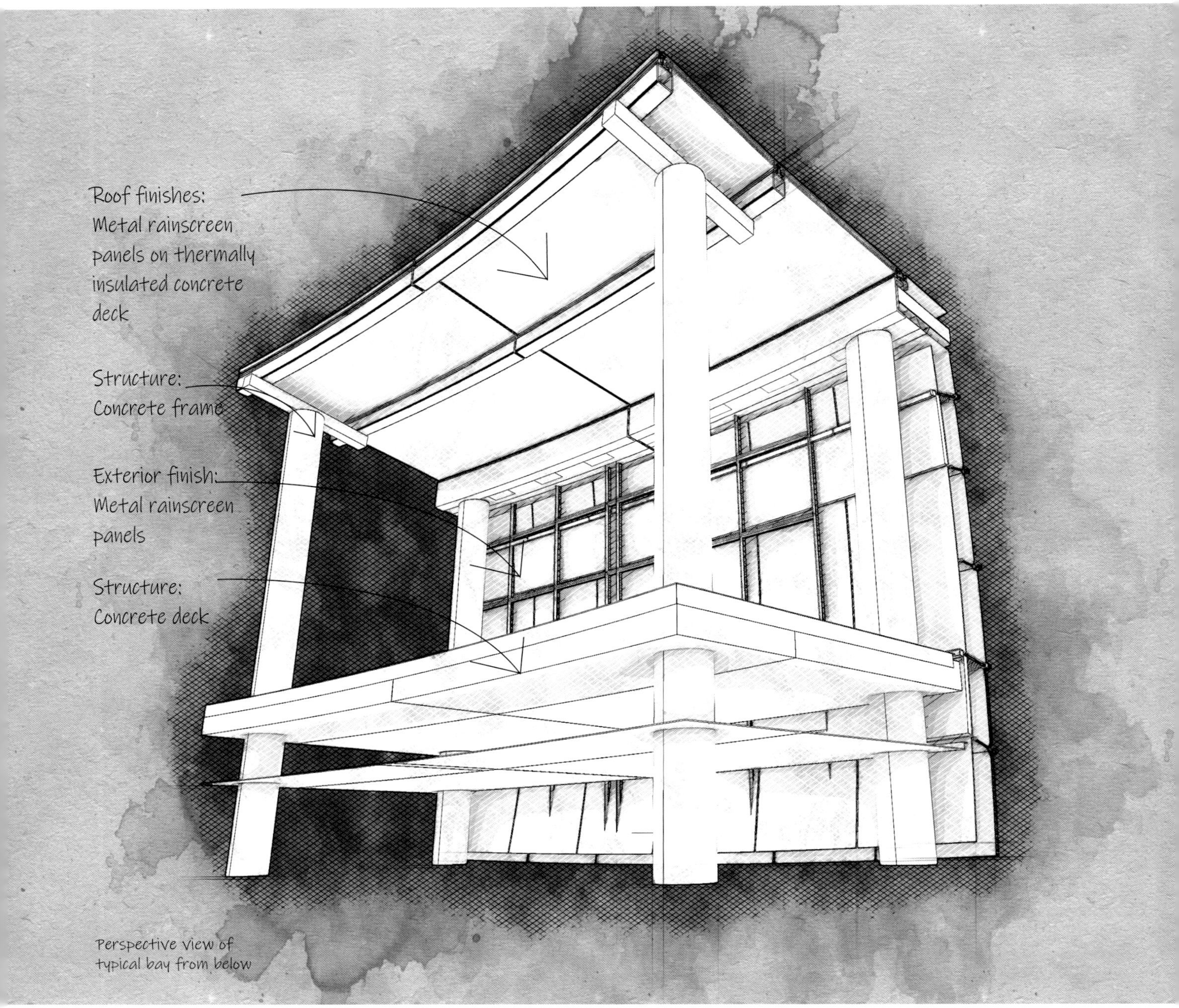

Typical bay adjacent to facade

The differentiated nature of the external envelope results from the overall forms of the building making use of different envelope systems around the building to differentiate the forms as much as possible. The first system has an inner concrete wall and an outer closed wall, divided by a layer of thermal insulation. The second type is of an opaque wall with metal screens in front, and the third type is reinforced concrete exposed over parts of the triangulated structure at the top of the building. Part of the structure in reinforced concrete is exposed, but most is covered in metal cladding panels which enclose air-handling ducts for the mechanical ventilation in the building. Some spaces have exposed glazing such as that of the two thin ends of the building. These glazed facades have horizontal metal panels covering the floor zones. The edges of the structure are picked out in insulated metal cladding panels to provide the required thermal insulation. Parts of the upper triangular form have walls formed in exposed reinforced concrete, which form part of the primary strucutre rather than being a part of the non-loadbearing facades. These walls are in a diaphragm wall configuration. These different configurations are revealed clearly on the end elevation, appearing as a visual 'slice' in both inclined and vertical walls; the facade construction is revealed in those thin end elevations.

In the projecting parts of the building, access to the laboratories is from a corridor on each side of the building, with the possibility of combining the laboratory spaces with a link between the two halves of the floor plan. The corridors have controlled lighting as a result of the tasks that are required to be carried out; most laboratory spaces are lit naturally with borrowed light from the corridor spaces. Daylight can be controlled as required for the research tasks on the two floors which are set between the lower and upper parts of the triangular block. An interaction space and meeting spaces are set up for use by all occupants of the building during the day. The upper floors have the circulation set on the end vertical wall, providing access on each floor to suites of offices which are glazed with views down into the interaction space at the mid-height of the building.

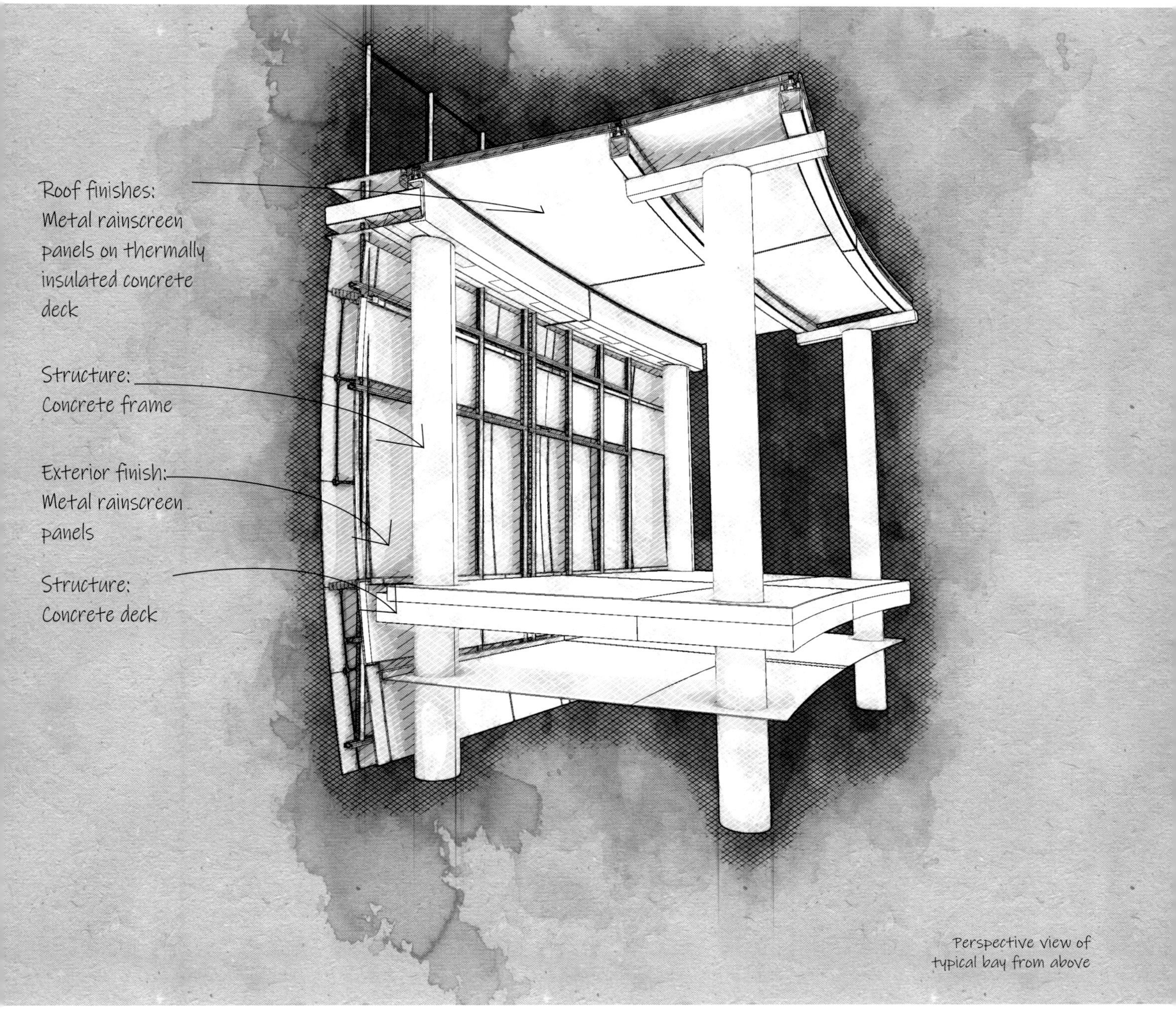

Perspective view of
typical bay from above

The facades have infill panels which enclose the triangulated frame of the support that form the outside of the volumes, throughout the height of the building. These are a mixture of glazed and opacified glass. The daylight requirements in different parts of the building and associated functions was the driver behind the environmental strategy, which is to take each of the office zones and treat them as separate environmentally controlled spaces. The primary energy user is the first projecting volume above the triangular entrance. Chillers are positioned at roof level. Air-handling units are distributed throughout the form of the building and are clad in metal panels.

A key visual aspect of the design is the externally set structure and air-handling ducts which are fixed to these primary members. The design for the structure and environmental design were closely integrated in order to release floor area for the tasks to be carried out in the building. The resulting arrangement forms a visual 'map' of not only the external structure but also the movement of air around the building. The envelope also accommodates the projecting areas of the facades at lower level which accommodate circulation areas on each floor. This element uses the same facade system as elsewhere; emphasising its form in the end elevations instead of changing the language of the external envelope in this location on each side of the building. The projecting volumes of the building have sets of integrated light fittings and mechanical ventilation air diffusers set within the corridor on each floor. This integrated set of service provision forms a part of the external envelope, comprising a blade-like fixing which projects inwards from the external envelope from which it is supported. The light fittings face upwards rather than downward in order to use reflected light from the soffit of the floor slab above. This integrated system aims to reduce the energy consumption of electrical lighting during hours of darkness, avoiding the sharp points of light associated with light fittings adjacent to the external envelope. The integrated fittings are skeletal in form in order to support lights and diffusers with the minimum amount of material while making the assembly easily accessible and easy to maintain; each assembly folds downward for access.

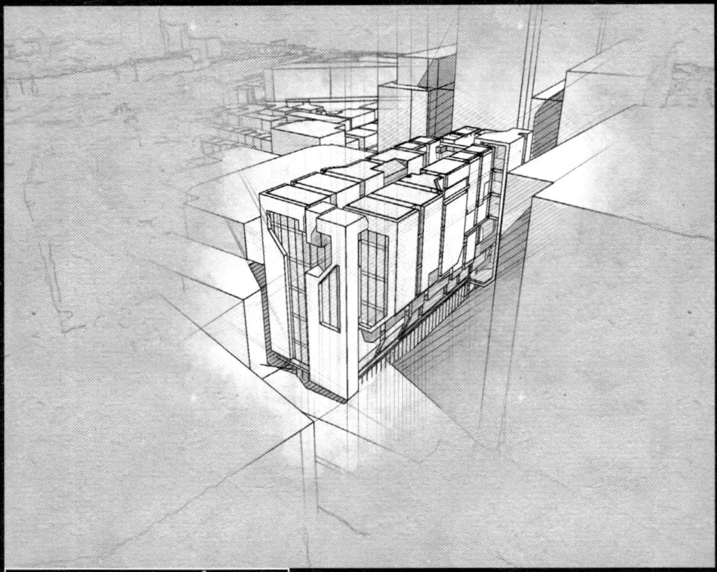

1. Perspective view of exterior

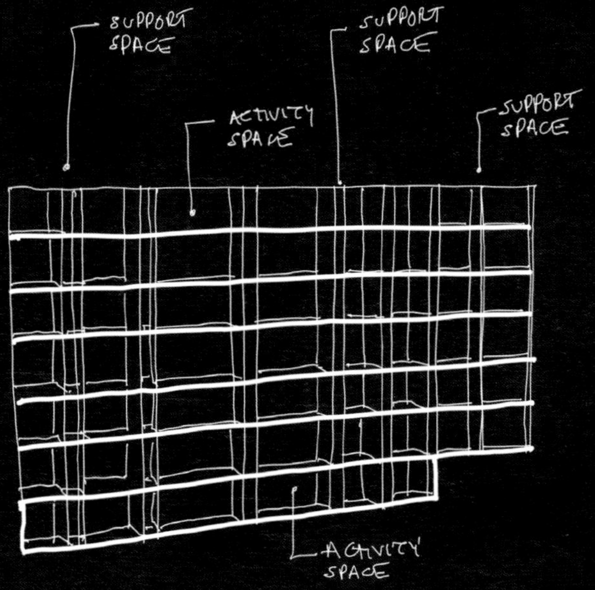

2. Perspective view of exterior

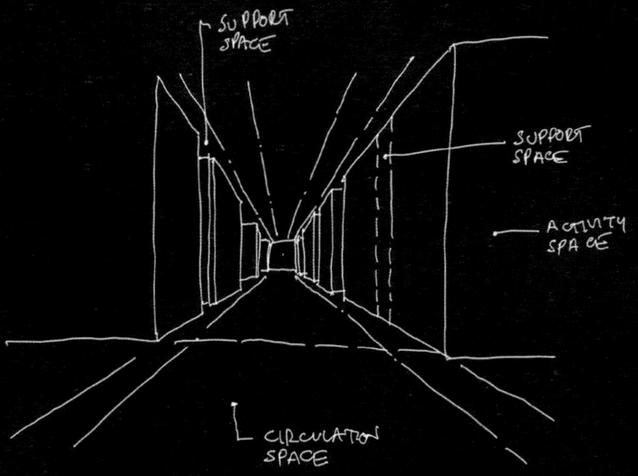

Relationship between facade and structure

A key issue in the design of this school building was the integration of facade and primary structure to reduce and optimise the depth of the facade zone in order to allow the required floor space to be created.

A key aspect of the engineering design of the facades in this project is the interdependence between the assembly of the facades and the structure which supports that envelope. Typically, facades are designed to be supported by the primary structure, which includes floor slabs, at specific points only. In other respects, facades are independent of the supporting structure; they remain self-contained assemblies. Facade systems are usually designed for fabrication in workshops and factories rather than for fabrication on site; they are installed on site, either as complete panel assemblies, or as assembly-ready components for installation. This approach meets requirements across projects for a high-quality construction which is both economic and fast to install on site. To this end, panels are used to form a continuous outer layer, rather than a continuous substrate or surface such as reinforced concrete, that would require only site-based assembly techniques. The size and shape of the outer facade panels is determined by several factors: repeatability, families of shape, size, visual requirements and, if required, ease of installing a second facade layer of solar shading.

Individual panels are often fixed in each corner of four-sided panels, which allows the movements of each panel to be predicted accurately by calculation. Each panel fixing is then secured to the supporting primary structure; fixings form points of contact with the structure, creating a 'cloud' of fixing points. As a result, two design approaches could be taken: the structure could be designed from the arrangement of the facade panels – minimising secondary support – or alternatively the structure could be designed without providing specific points of support for the facade panels.

The specific approach taken for the design of this project was to examine the overall geometry of the external envelope and to search for structural types which would realise that form. This step is required since the facade design is based on the economic use of a single

. Perspective view of exterior

4. Perspective view of exterior

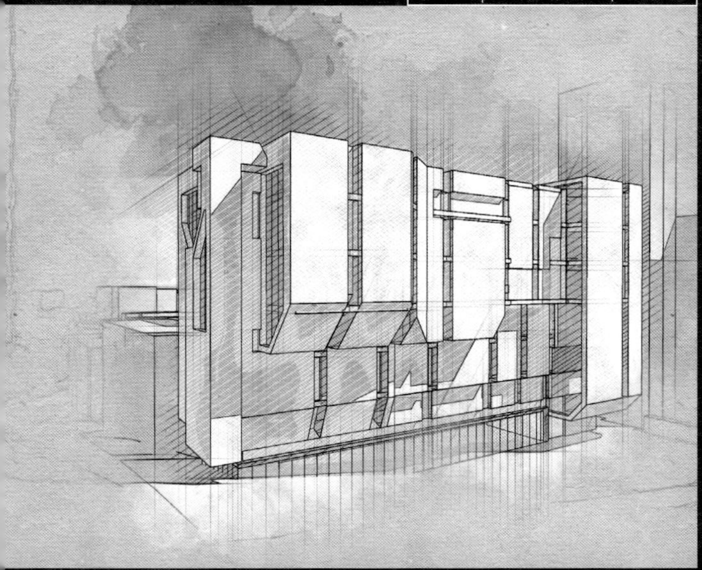

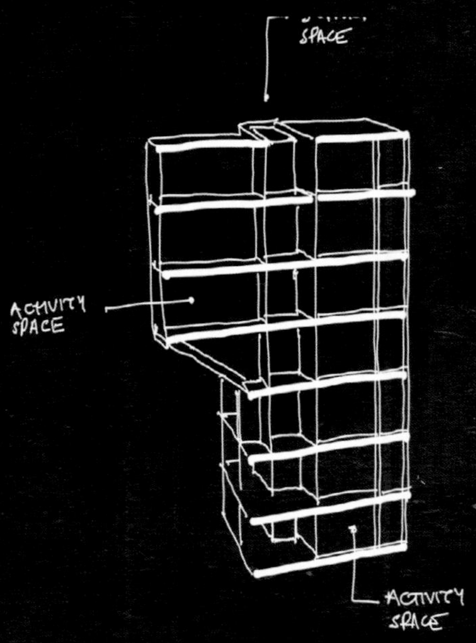

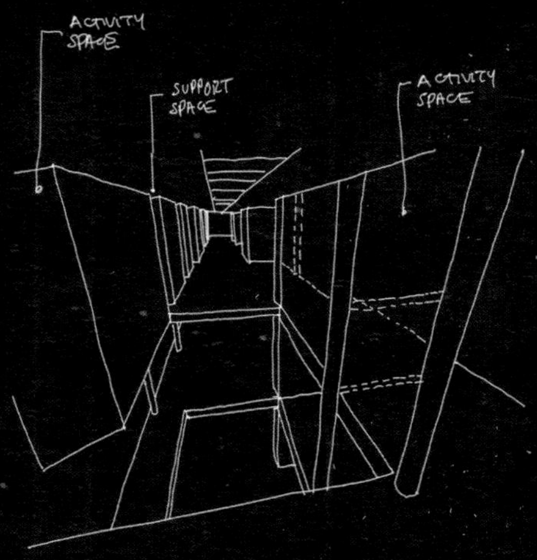

facade system across the complete project – for reasons of economy and speed of construction – resulting in the need for a single structural system to support the facade system. This convergence of facade system and supporting structure as a single combined type is an essential part of this project.

The engineering design of the structure is essentially based around establishing a coherent and economical relationship between primary structure and secondary structure. The secondary structure is used to support the facades and the primary structure is used to support floor slabs and services equipment. The facades divide into two types: one is the opaque walls, the other is largely glazed walls. The opaque walls face out onto the street. The glazed walls are largely slots which are visible on the ends of the building and have narrower slots between projecting classroom blocks. The overall impression of the facades is one of a sculpted surface which is not transparent in order to provide the teaching spaces required within the building, which are to be daylit but have sufficient wall space for teaching and presentation in different orientations and different formats. The opaque facades have a cementitious or concrete-like finish; the glazed walls have a clear glass appearance in order to introduce daylight in a way which is projected along the walls. Daylight is washed along these surfaces to create controlled reflections rather than admit constant amounts of daylight through horizontal bands, as is more customary.

As a result, the structure needed to be adapted to suit the opaque walls which require support for the panels at each corner. The glazed facades were supported at floor slabs on the edge of those slabs in the classic manner of curtain wall. The rooflight, which extends the length of the central open space, is essentially a glazed rooflight system which functions in a similar way to the curtain wall. From a structural point of view, the structure supporting areas of opaque wall have column centres at 3.0 metres rather than the regular 6.0 metres used elsewhere. Inside the building, this extra steel avoided the need for additional secondary steelwork to support the facades and ensured that the facade zones were optimised.

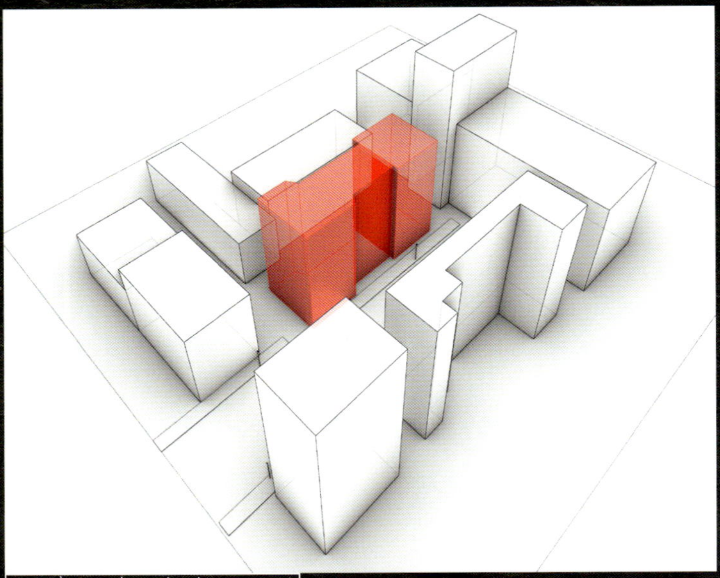

1. Project volume in site context

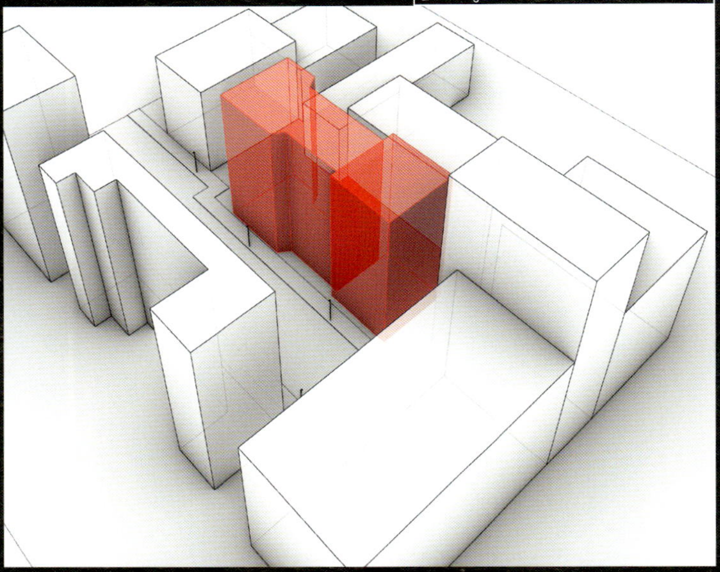

2. Project volume in site context

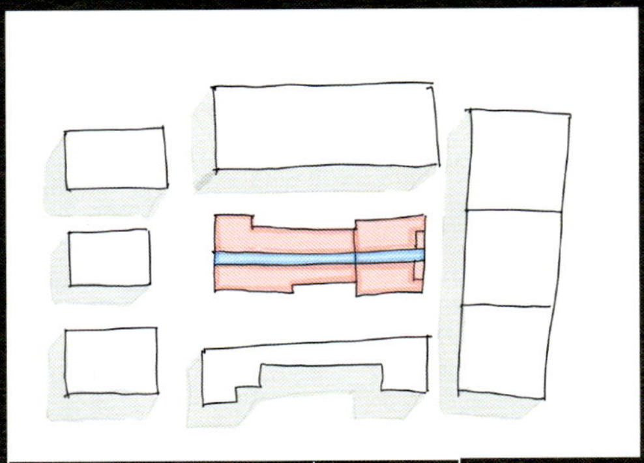

1. Site plan with surrounding context

Option 1 is of an urban area; a mixture of office and residential spaces where the site occupies a vacated block.

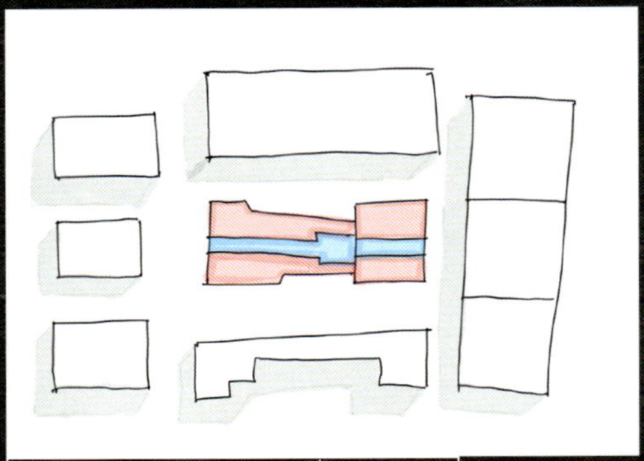

2. Site plan with surrounding context

Option 2 subdivides each block into further interlocking spatial components.

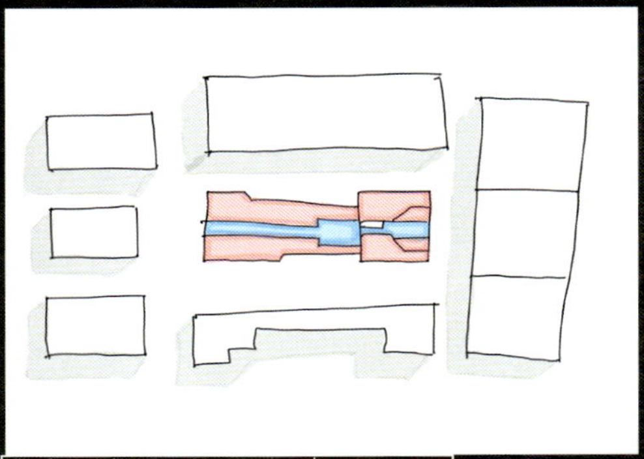

3. Site plan with surrounding context

Option 3 creates greater differentiation in the larger square block at one end of the building.

Site context

The design context is an urban location, adjacent to a large residential area but set close to the edge of a commercial area of the city. The school has been on the site for over 50 years, and was required to be rebuilt: The building had not been in use for many years, as the first facility had been divided among three other schools in the area. As a result of the need to create a new specialist school, a new large-scale school building was required. Its location led to the approach that the school should reach out to the community and should engage with neighbouring businesses, which are office-based enterprises located around the site.

The existing building was no longer able to be refurbished as it did not allow the required spatial organisation to be accommodated within its existing form. A new building was required to inspire the students and the teaching staff which included members of the community. This aim of inspiring those using the building, those occupying the building and those visiting the building generated a building design that had a form and function which was more like an urban university campus than a regular high school. The forms conceived were of a larger scale than those expected in regular schools, with the relationships between the external spaces resulting in an ambiguity in the use of those spaces. The building design responds to these requirements by forming a series of framed openings and glazed slots, which allow views up and down the building from the windows. The community and spaces within the building can be taken in visually from the classrooms as well as from the internal circulation spaces. The framed openings create semi-external spaces which are seen by passers by, as these forms reach out to invite the local community into parts of the building for evening classes. The structure responds accordingly by providing a structural frame that integrates support of the facades and support of floor slabs while remaining economical. This is achieved by integrating staircases into the structure rather than them being separate elements.

This school is arranged on a small site, requiring it to be constructed over several floors. The arrangement of forms, at the urban scale, is

Perspective view of structure

2. Perspective view of structure

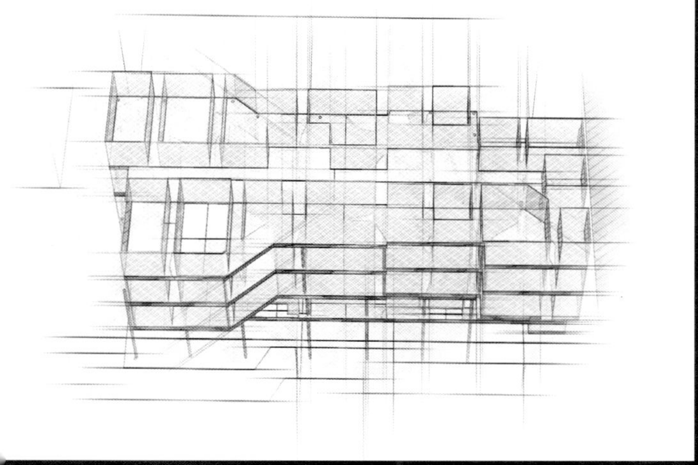

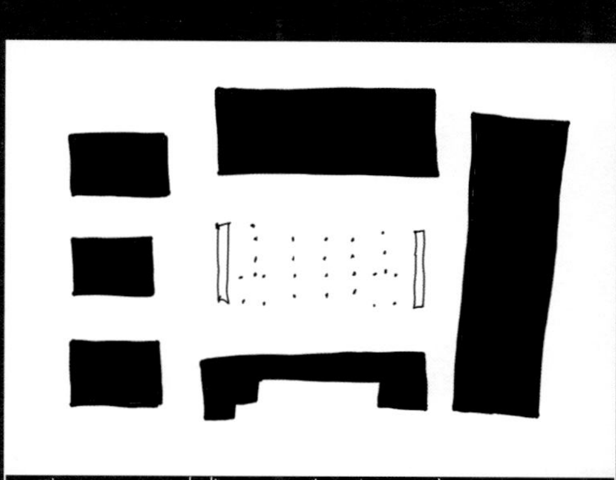

1. Figure-ground diagram showing project structure

Option 1
The structural grid is set parallel with neighbouring buildings on three sides. Cores block views into and out of the building.

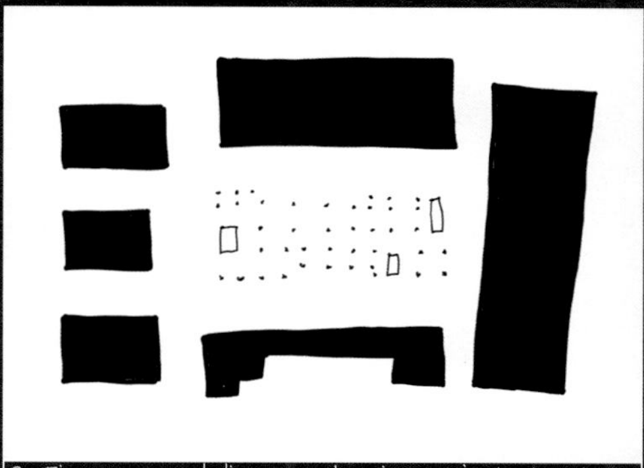

2. Figure-ground diagram showing project structure

Option 2
The structural grid is set parallel with more neighbouring buildings on three sides. Cores are more dispersed.

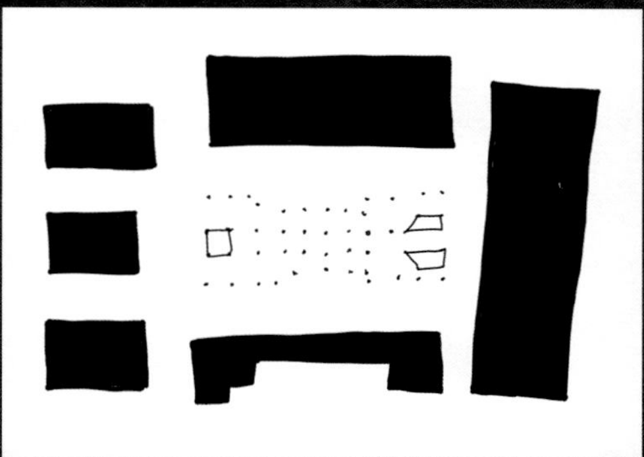

3. Figure-ground diagram showing project structure

Option 3
The structural grid is set parallel with more neighbouring buildings but responds to the functional needs of internal spaces.

of two rectilinear blocks set side by side, which are separated by a continuous open internal space extending the full length of the building. At ground floor level the two blocks are set on horizontal service cores; rectilinear blocks which extend the full length of the building to facilitate entry and exit from the school.

The site context for Option 1 is of an urban area; a mixture of office and residential spaces where the site occupies a vacated block. All three options use the full extent of the site; all options are highly articulated forms that outwardly project the character of the space behind. Option 1 has a continuous strip of roof glazing down the centre of the building; dividing the overall form into two halves, with a full-height atrium space extending the full height of the building; bridges link the two half-blocks that comprise the overall form of the building. Each half-volume is a self-contained block accommodating teaching spaces of different size and different height from floor to ceiling. The shape of the building reflects the sum of the various functions within; creating an interlocking set of differentiated forms that are at the scale of building forms in the neighbourhood, but which are distinct in their language of construction; a language of prefabrication. Option 2 develops the themes of Option 1; creating a greater hierarchy of spaces than Option 1 by subdividing each block into further interlocking spatial components. This option also develops the open central space through the centre of the building to create an internal courtyard. The long strip of roof glazing of Option 1 no longer forms a full-height atrium space down the centre of the building, but instead is formed as a set of galleries on each floor. Option 3 takes this concept further; creating greater differentiation in the larger square block at one end of the building, and articulating the volumes attached to the facades as contemporary oriel windows that provide views up and down the street.

In addition, the oriel windows provide oblique views between the different spaces as seen from the perspective of inhabiting the facade – rather than forming a flat facade, which affords no views along the length of the building – through the sequence of spaces created by continuous views through the teaching spaces.

1. Sectional perspective view

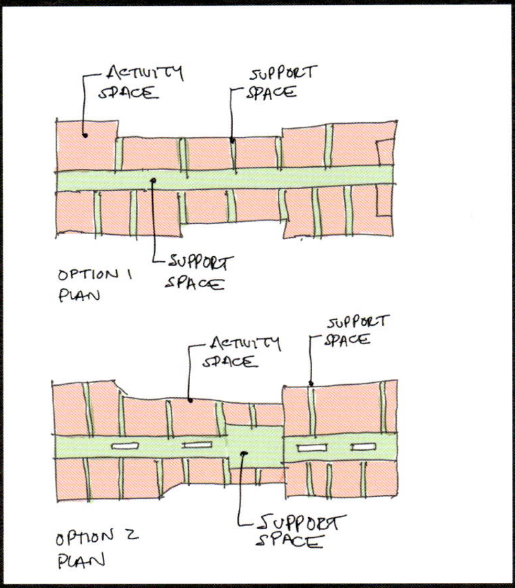

Spatial arrangement

Option 1 has a central circulation space extending through the full height of the building.

Option 2 removes the full-height void of Option 1 and instead introduces light from floor to floor with open stairs; the central void is filled with a circulation 'spine' on each floor.

2. Sectional perspective view

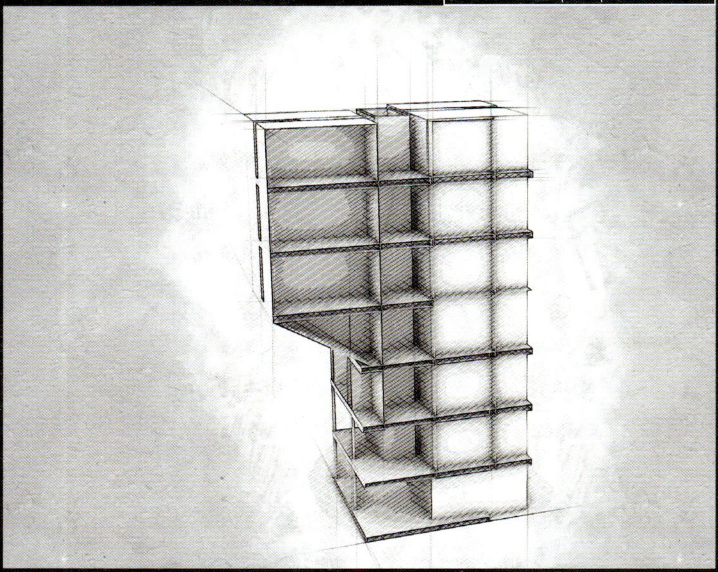

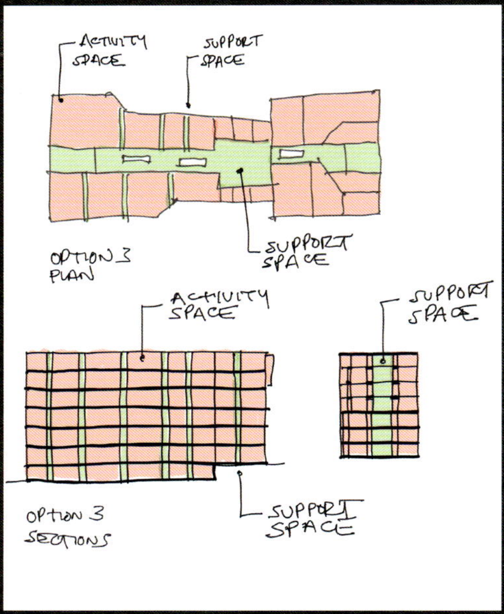

Option 3 creates a further differentiation within each of the spaces of Option 2.

Spatial arrangement

The spatial arrangement is one of teaching spaces being set on both sides of a central space, which in Option 1 extends full height through the building. Options 2 and 3 remove this full-height void and instead introduce light from floor to floor with open stairs; the central void is filled with a circulation 'spine' on each floor. Option 2 has stairs set out along the length of this circulation spine: an open, top-lit space is created in the central building as a courtyard, which can be walked around or, in Option 3, walked across; providing gaps in the central corridor space that create views up and down the stairs arranged along the circulation spines of Options 2 and 3. Option 3 creates a further differentiation within each of the spaces of Option 2; spaces with a music practice room set internally with a surrounding preparation and waiting space. Staff rooms are set at one end of the building, and music practice rooms and assembly rooms at the other end of the building. No external areas for outdoor sport sector activities are provided; these are carried out off site. The general arrangement of the spaces in the building is one of a central spine for circulation through the building, with classrooms arranged as blocks on both sides of the central space. The different sizes of classroom resulted in the strategy of having a straightforward linear spine within the building, which would provide rectilinear circulation space with clear views up and down the space in the centre of the building. This approach resulted in classrooms of different width or length, some projecting out beyond the line of the external walls, others creating a stepped facade. Alternatively, a stepped internal set of circulation spaces could be arranged, with flat, rectilinear facades. The requirements of the brief/program led to the need for both a stepped outer facade and a partly stepped set of spine walls inside the building; the internal circulation spaces could not become straight corridors due to the need for floor space. The width of classrooms could not be changed in order to create a rectilinear box with straight walls internally and a rectilinear internal space. The three options considered the spatial organisation as required in the brief/program, creating three options for spatial arrangement.

Perspective view of interior

2. Perspective view of interior

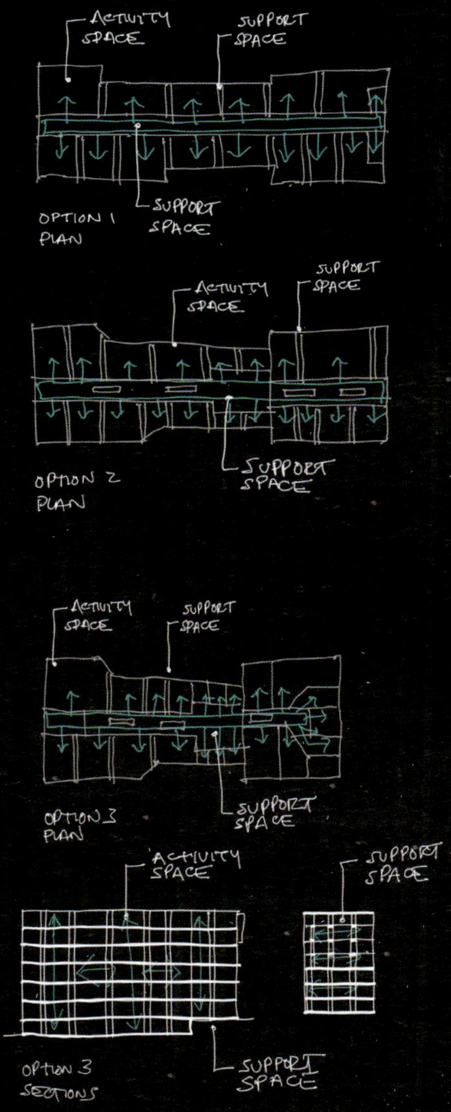

Internal circulation

Option 1 has a central space with bridges spanning across it.

Option 2 arranges the access and vertical circulation at the ends of the building.

Option 3 provides stairs along the corridors which extend down the centre of the building.

Internal circulation

Related to the disposition of teaching rooms and support facilities is the need to accommodate vertical circulation in the building for both stairs and lifts/elevators. At an early stage of the design, it was asked if staircases could be accommodated in the zones between classrooms. However, these spaces were seen as zones that could accommodate storage space and service space for classrooms.

As a result, the staircases were set along the length of the building, in the centre of the spine, to allow easy passage from floor to floor and to allow students to circulate on both sides of the staircases in opposite directions; for ease of flow from floor to floor and from room to room. The three options then focused on internal circulation within the core, and how that space could be enhanced beyond being a corridor. The three options focussed on space for musical and theatrical performances. These spaces are created for less formal events, including exhibitions. The main circulation routes are in the two staircases at the ends of the building as required in the brief/program, with a large assembly hall at one end and associated washrooms and storage spaces being set into the opaque outer walls on the end facades.

The internal circulation of Option 1 is of a central space with bridges spanning across it; providing access directly into classrooms on both sides by way of stairs positioned at the ends of the building. Option 2 continues this theme, but introduces a central courtyard with stairs and a lift/elevator in order to provide primary circulation in the centre of the building.

Option 2 keeps the access and vertical circulation at the ends of the building. In Option 3, stairs are provided along the corridors which extend down the centre of the building; openings in floor slabs, formed by the stairs, provide a more even distribution of movement around the building. The stairs in the courtyard remain in place in Option 3.

A key aspect of this design is to not use staircases as an element to visually animate the elevations. Placing stairs within the building provides the opportunity for teaching spaces to occupy these areas.

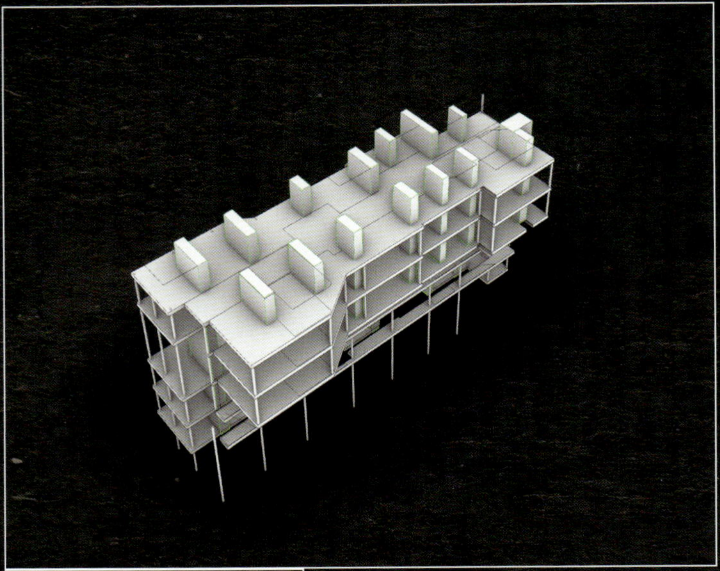

1A. Isometric view of structure

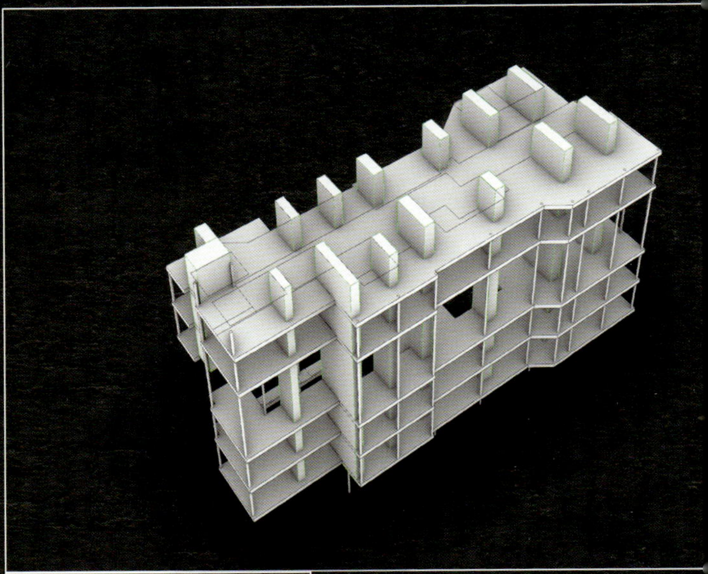

2A. Isometric view of structure

1B. Isometric view of structure

2B. Isometric view of structur

Structural strategy

The structural design for each of the three options is based on the common structural solution of a central corridor space which forms a circulation spine, with openings cut into it for staircases. Floor plates on each side of the corridor are formed as teaching rooms. The plan is not symmetrical, either in its length or across its width. The two sides of the two long elevations of the building have completely different profiles as a result of the classrooms being of a very different size on each side of the building. The general arrangement of the structure is common to the three options in that the arrangement of floor plates and columns is the same for all options. Columns are set at the perimeter of the floor plates in the corners of teaching spaces as a result of a highly indented plan. The indents are cut into the floor plate, serving as slots for glazing panels. The structure is formed with a double set of columns that provides one set of columns for each classroom. Teaching spaces are arranged continuously across the building, with rooms extending outwards from the building differently on each side of the building. Columns are arranged in the corners of each of the classrooms in order to create spaces for the service zones and teaching support spaces, set between each classroom. In the centre of the building, staircases are set into openings which are cut out of the floor slab area that creates the circulation spine which extends the length of the building. No columns are required to support the edges of this opening as the floor slabs and beams in this area span the width of the circulation spine. The column arrangement is consistent through the height of the building as a result of the indents and facade being continuously vertical through its height, such that classrooms are formed as a multiple of the typical floor, modified where needed. This approach is consistent with the requirements of the brief/program. In all three options, columns are arranged on the edges of the slab and on the deep edges of the corridor in order to create a set of columns on a relatively small-scale structural grid that allows the spans to be shorter than might be expected for a building of this size. This approach allows services to pass down through the building and out into the central spine space

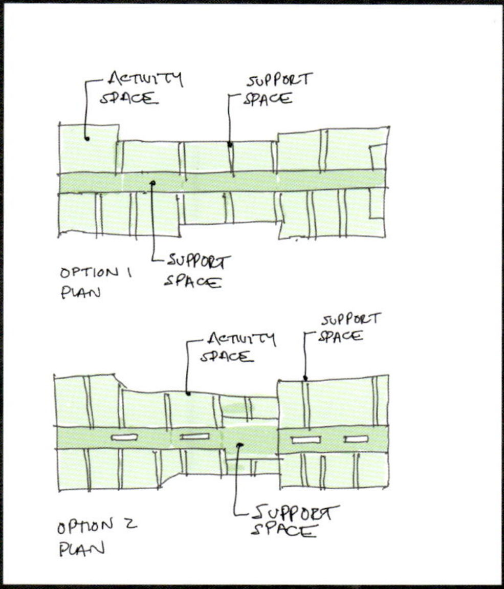

Structural strategy: Zones
Option 1 has structural walls which traverse the width of the building.
Option 2 has a frame that provides structural stability to the concrete cross walls.

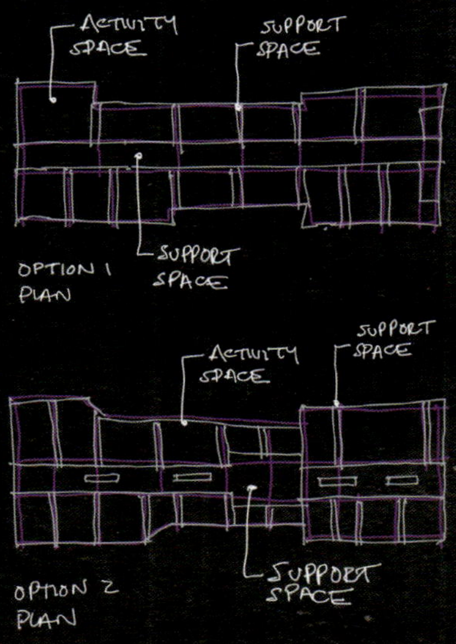

Structural strategy: Framing
Option 1 encloses the core spaces and dividing space into sets of activity spaces.
Option 2 is largely a concrete frame.

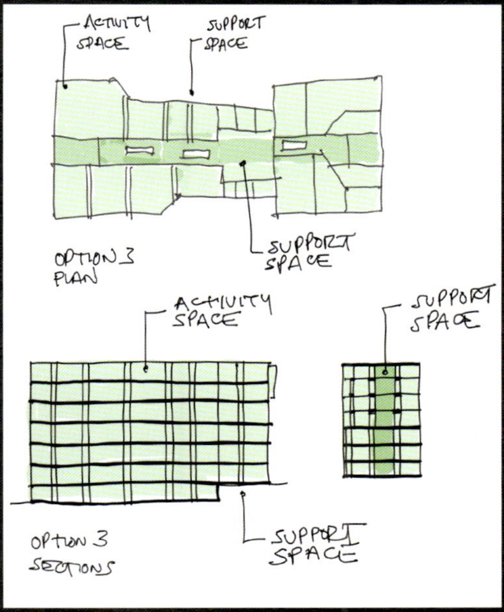

Structural strategy: Zones
Option 3 has structural walls formed as cross walls and mixes those with columns in the activity spaces.

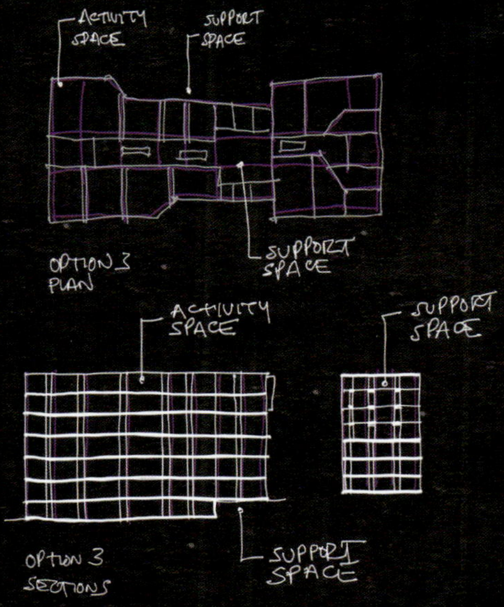

Structural strategy: Zones
Option 3 is a mixed solution of reinforced concrete and steel.

without having to pass through deep beams; this solution allows the structure and services to occupy different zones, which would make this solution less cost effective. It is anticipated that some spaces for teaching will be opened up as shared space for performance and for music teaching; space which would be open to the central corridor. The relatively small structural grid has columns at points of intersection with internal walls; these longitudinal spine walls with column stepping in and out of the continuous grid line were required to form either wider or deeper spaces in the central spine area. Enclosures for lifts/elevators and washrooms are provided at the ends of the building; forming walls which contribute to the structural design of the building. The strategy of using the opaque walls of the washrooms and service spaces is used as part of the structure, but most walls that divide classrooms are non-structural in order to provide the required flexibility in the long-term use of the building. The staircase enclosures were originally going to form part of the structure, but this was not considered to be within the spirit of the functionality of the building and the internal circulation. As a result, elements of structure are located at the ends of the building and between some of the classrooms where changes of use are not expected to take place; modifications are expected in the arrangement of adjacent spaces. Classrooms on the upper levels cantilever out beyond the regular line of the external walls; these walls are supported on inclined columns which carry loads back to the vertical line of structure immediately behind. Floor slabs create large-scale oriel windows in order to provide views along the facades. Vertical columns were not provided beneath these projecting bays in order to create a visual expression of a large-scale window rather than being seen as a projecting element which would have required columns beneath.

The structural strategy for all options is one of a mixed reinforced concrete and steel frame. The design creates two separate blocks linked by bridges, with a rooflight spanning across the void and closing the building at roof level. A reinforced concrete structure is used for the lower floors, and a steel frame is provided for the upper floors. A concrete frame is used where resistance to sound transmission is required.

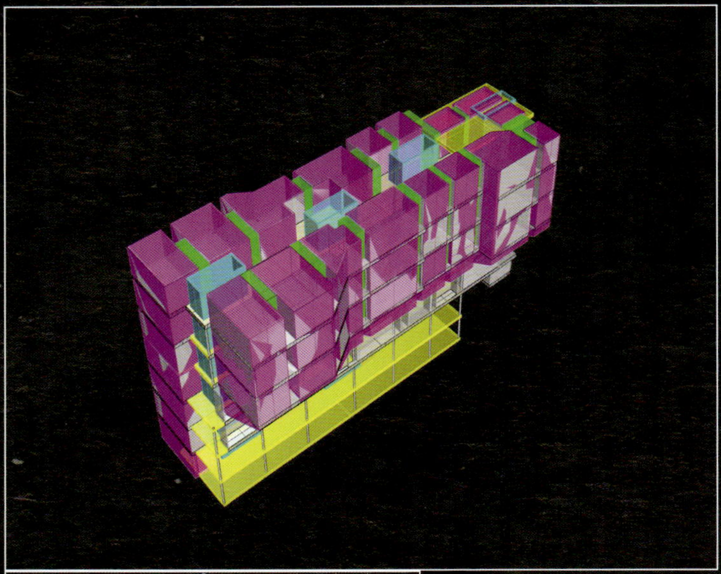

1A. Isometric view of environmental volumes

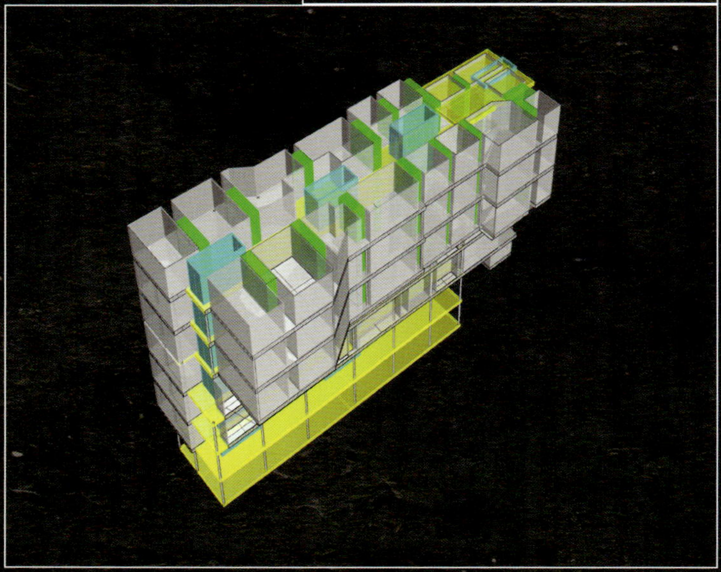

1B. Isometric view of environmental volumes

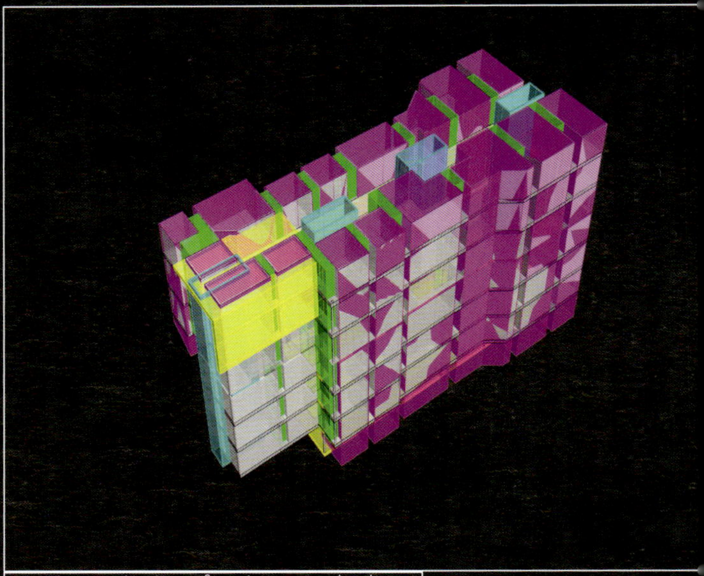

2A. Isometric view of environmental volumes

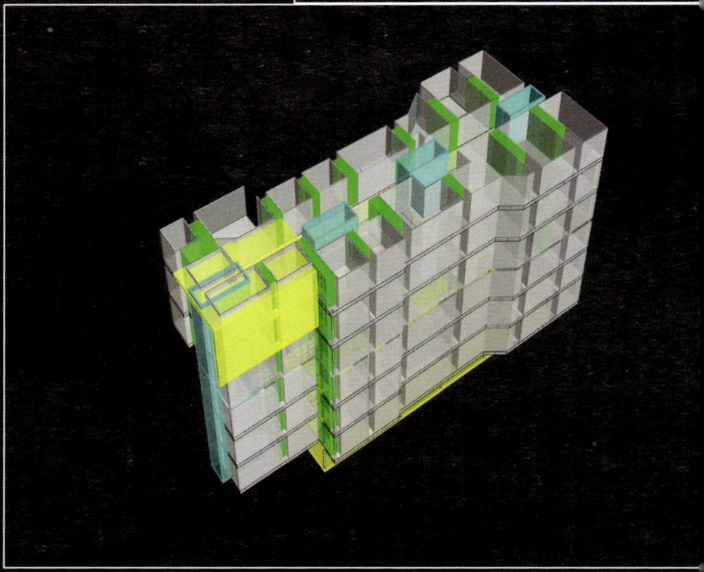

2B. Isometric view of environmental volume

Environmental strategy

The environmental strategy is based on creating environmental zones on opposite sides of the building, along its length, to respond to the different heating and cooling loads resulting from its orientation in relation to the path of the sun. Exposure to solar gain is reduced as a result of the windows being set within the sides of indents within the floor plan, rather than by setting glazing on the outer facades.

The options for environmental strategy are based on a response to each of the different spatial arrangements that form the basis of each option. The overall environmental strategy for this project is based on the introduction of natural ventilation across the width of the building and to make use of this provision for night-time ventilation. Cross ventilation of teaching spaces can be used in the mid-season and summer, which is approximately half the academic year. Rooms are ventilated from vents in the facade elements of glazing. Windows can also be opened in order to admit fresh air, which is then directed and controlled by fans. In the classrooms, teaching rooms do not have suspended ceilings in order to benefit from the height in each room as much as possible. Fans and associated ducting draw fresh air from teaching spaces, across the central space and into the opposite classroom. Air is also directed along the length of the spine areas into ancillary spaces. The strategy for ventilation encourages cross ventilation from one long classroom to an adjacent teaching space, passing across the spine area in air ducts.

The areas for fan-assisted ventilation across the spine were needed to provide a mixed mode of partial mechanical ventilation and partial natural ventilation to suit the general arrangement of this building, avoiding the need for mechanical ventilation as much as possible. A key consideration in the environmental design was to provide sufficient air changes per hour to ensure the freshness of air in the teaching spaces, which is vital for rooms occupied by a relatively high number of people for an extended period.

Another key aspect of the environmental design was the acoustic separation between teaching spaces and between spine, circulation

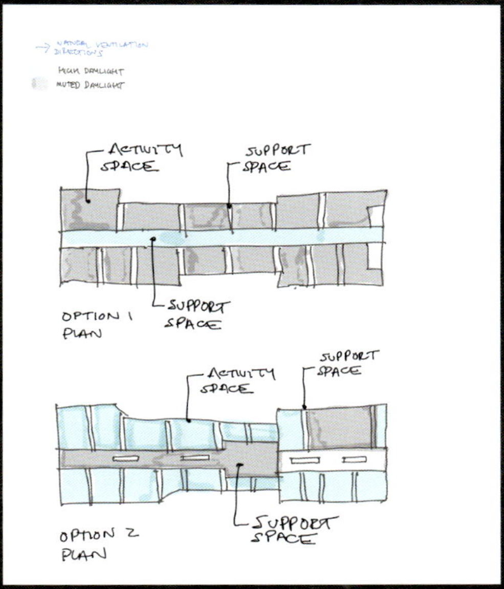

Daylight strategy for direct sunlight and cross ventilation.

Option 1 admits daylight and air through the central irculation space.

Option 2 admits daylight and air through the activity space.

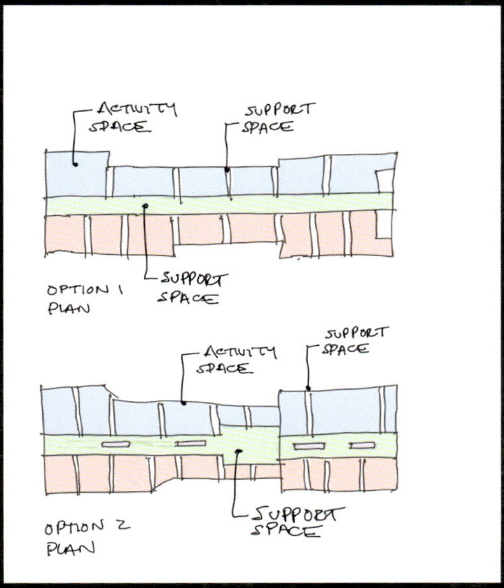

Environmental zones

Option 1 responds to the orientation of the building on the site.

Option 2 introduces secondary spaces in the centre with stairs that create a continuous open zone in the centre of the building.

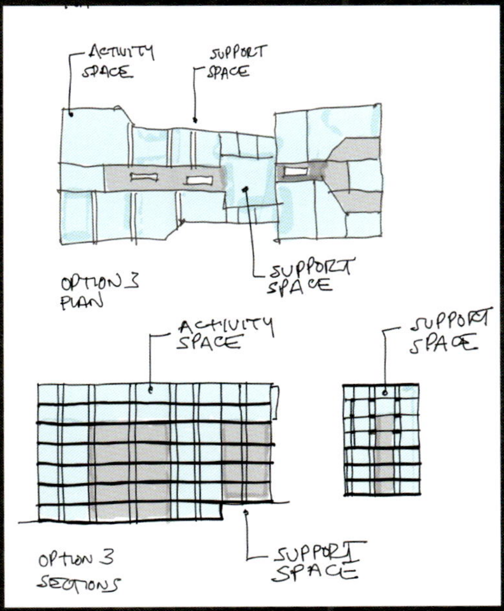

Daylight strategy for direct sunlight and cross ventilation.

Option 3 admits daylight and air through the activity space and through circulation space.

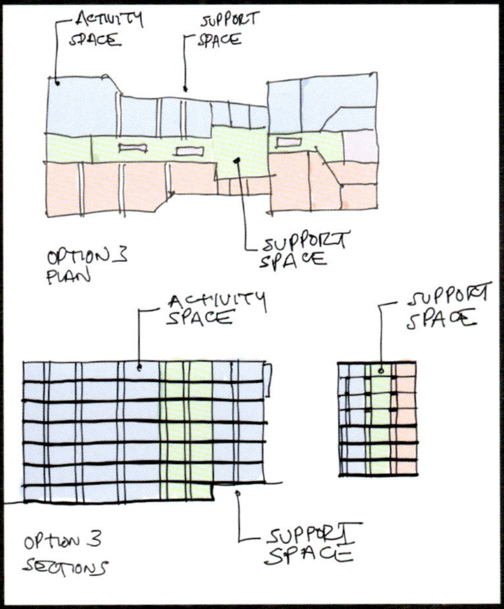

Environmental zones

Option 3 has more enclosed spaces which are smaller; these form a fourth environmental zone.

space and teaching spaces. This separation is required since many of the rooms have audio-visual projectors in which some lessons are delivered through classes presented remotely; an increasing teaching trend in this school. In addition, the need to create a quiet space for the teacher to be heard clearly is essential, as is the need for low levels of outside noise in music practice spaces, which are located throughout the building. The main music practice spaces are set at the ends of the building; teaching spaces for the dramatic arts requires more acoustic insulation than in general teaching spaces.

Mechanical ventilation is provided in the science laboratories, which are set on the highest floor of the school. Glazed rooflights, set at the top of the spine circulation space, have external shading in order to reduce the effects of direct sunlight but also to encourage a certain amount of light and shadow to be incident on floor surfaces in the space during the day. The effects of thermal transmission from solar gain are reduced by the use of solar control glass, with reduced light transmission over regular glass. A high level of light transmission was not required inside the central circulation space; instead a contrast of light and shadow created a greater sense of being in a covered courtyard rather than being in a long, glazed corridor. Additional solar baffles are set on the underside of the rooflight, inside the circulation space, providing additional daylight control.

The environmental strategy for Option 1 responds to the orientation of the building on the site; one side of the building has sunlight incident upon the facades during the school day, while the facade on the opposite side is more protected. As a result, the two halves of the building have different environmental conditions and are tempered differently as a result. Each half of the building is served by the mechanical and electrical equipment, but each zone has different air changes and different levels of daylight required to be admitted. Option 2 introduces secondary spaces in the centre with stairs that create a continuous open zone in the centre of the building, similar to that of Option 1. Option 3 has more enclosed spaces which are smaller; these form a fourth environmental zone.

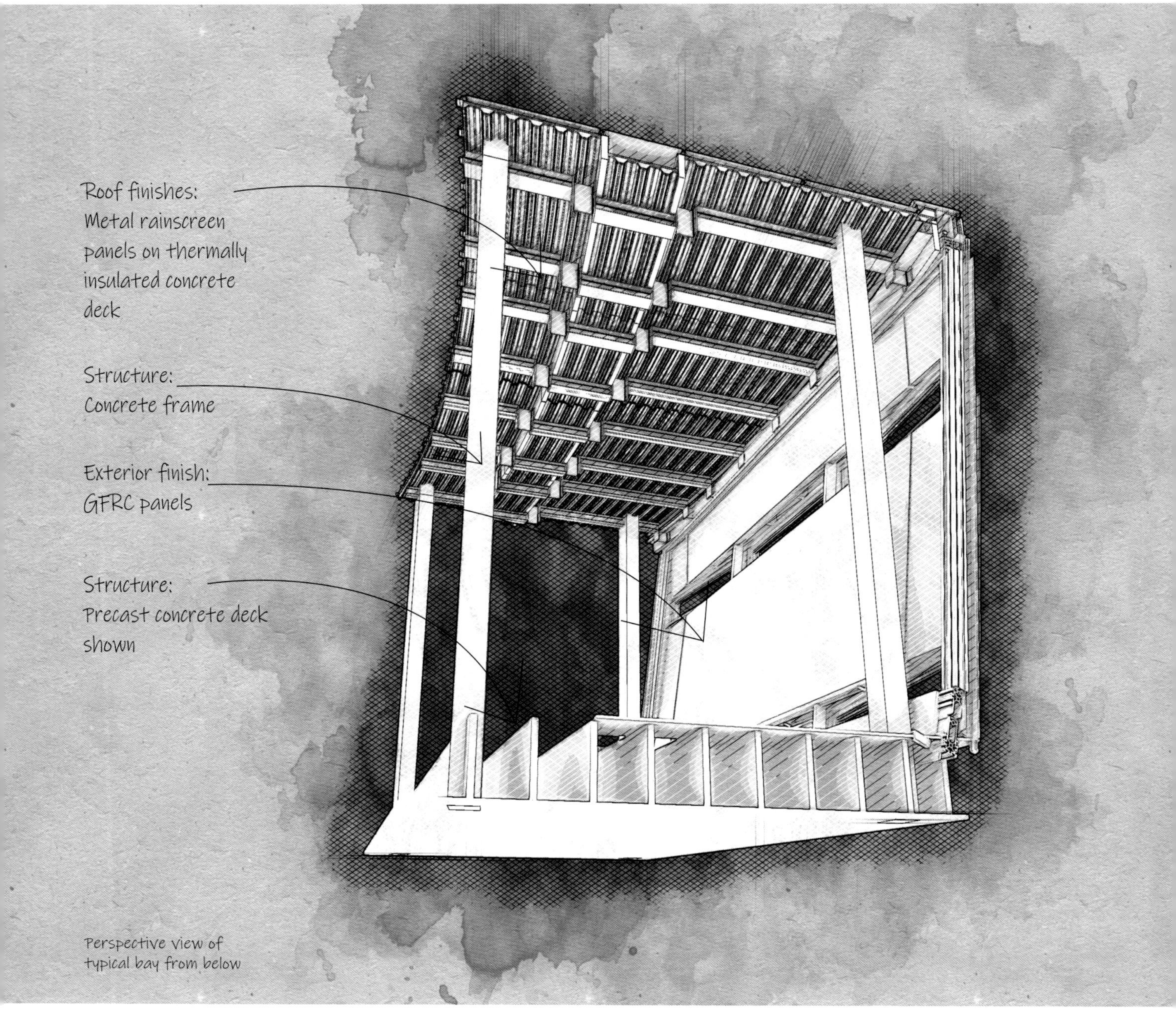

Perspective view of typical bay from below

Typical bay adjacent to facade

Most of the building is enclosed with opaque panels, creating external forms and internal spaces which are separated by glazed vertical strips; a glazed strip of rooflights is positioned in the centre of the building, extending along its full length. The external envelope makes use of large, opaque panels which enclose ducts for the supply and extract of air as well as electrical supply; all located inside the double-skin wall. In some areas, the facade has opacified glazing on the outside; elsewhere clear glazing is used. The upper parts of the building have glazed slots in order to provide directed natural light into spaces.

A facade maintenance unit is mounted at roof level, moving on rails around the perimeter of the building. The unit is rotated when in use; arms are extended and a cradle descends for access for cleaning and maintenance of the facades.

The external envelope forms a key part of the typical bay in this building; the facades having a largely opaque aspect when viewed in elevation, but having a glazed aspect when viewed at oblique angles from the surrounding streets. Glazed parts of the facades admit light in slots set into the sides of the projecting external walls. Within the building, borrowed light from the central spine provides daylight at the corridor end of each teaching space. This borrowed light from the central circulation zone makes use of a light shelf to direct light across the ceiling of each classroom such that rooms are illuminated from both ends of these spaces. The typical bay comprises a classroom with an indented glazed side and a service space for each classroom; adjacent classrooms use the other half of the space created as a buffer zone between classrooms. The typical bay is based on a repeated rhythm of classroom / service space / classroom. The typical bay is also defined by columns set at an edge condition or a facade; an arrangement which is repeated across the width of the building. Ceilings have exposed soffits that reduce the need for additional internal finishes. Higher floor-to-ceiling heights are provided such that each classroom enjoys a more general sense of space than is typical in classrooms. of this type. Music practice rooms and rooms for the teaching of the dramatic arts have

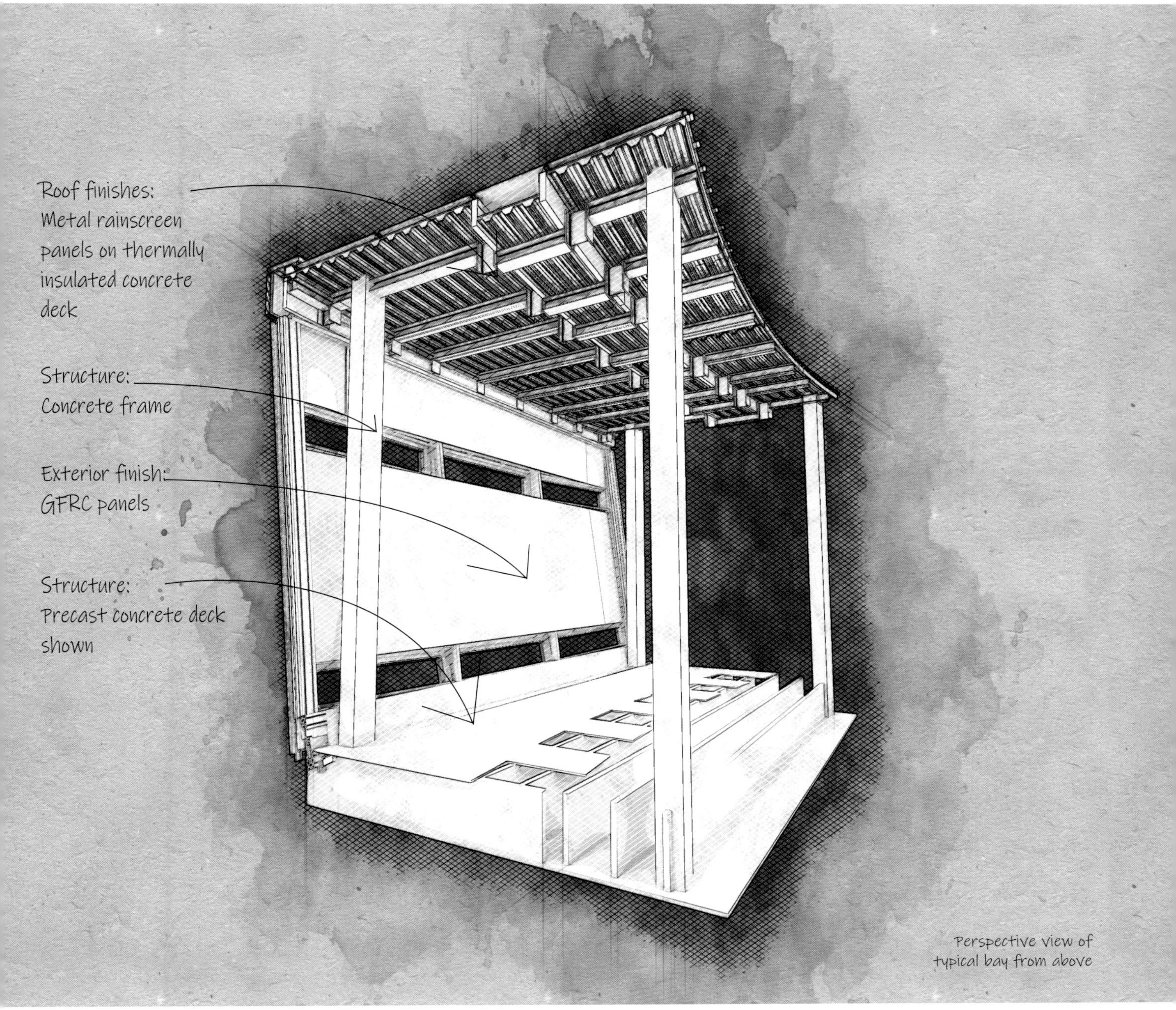

Perspective view of typical bay from above

acoustic insulation panels added to the ceilings to suit the expected use of that space for teaching and performance. The facades comprise exposed reinforced concrete walls externally, which are formed in glass fibre reinforced cementitious panels. The panels vary in both shape and relief in order to create a sense of the entire facade being a low-relief design reminiscent of a sculpture, rather than being flat facades that create only a single rhythm. The slots between classrooms punctuate the low relief visually; the design of which is mapped onto the undulating surface of each long facade. The facades of the two end walls of the building have smooth GFRC (glass fibre reinforced concrete) panels in regular rectilinear arrangements that contrast with the low-relief panels along the long elevations. These panels on the short-end elevations provide a visual frame into which glazed panels are arranged. The ground floor of all elevations comprises bands of glazing that extend the length of both long sides of the building. Some glazed panels are opaque and others vary in transparency to suit the functional requirements of the ground floor spaces of this school. When transparent glazing is introduced into the long facades, characterised by the low relief cementitious panels, the glazing is arranged as long, thin slots with windows offset from each other, and inclined from the vertical plane in order to follow the lines of the low-relief modelling of the facades. Small areas of glazing are used to create points of visual emphasis across the facades. Windows are needed in spaces for music practice, dramatic performance and administration offices. These windows are positioned away from the external face of the facades in order to provide specific areas of natural daylight within the building, while providing points of electrical light outwards during evening hours of darkness; seen by those occupying surrounding buildings as pockets of light contrasting with the strips of electrical light seen in the glazed slots between areas of low-relief facade. Thin GFRC cladding panels are used to enclose the opaque facades in preference to structural reinforced concrete because of the need to keep the facades as light in weight as possible in order for the oriel window/projecting bay to be economical as well as 'inhabitable'.

1. Perspective view of exterior

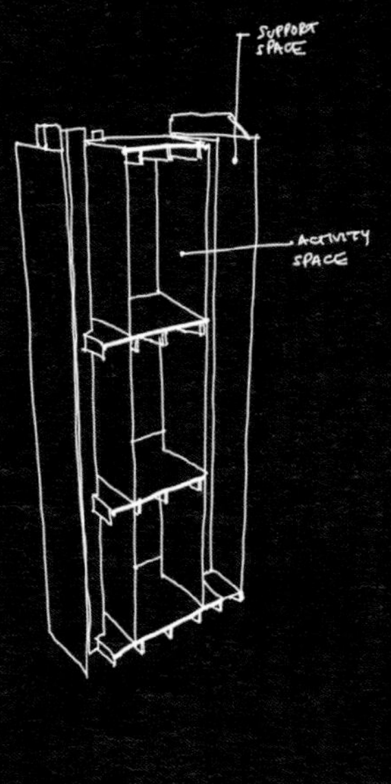

2. Perspective view of exterior

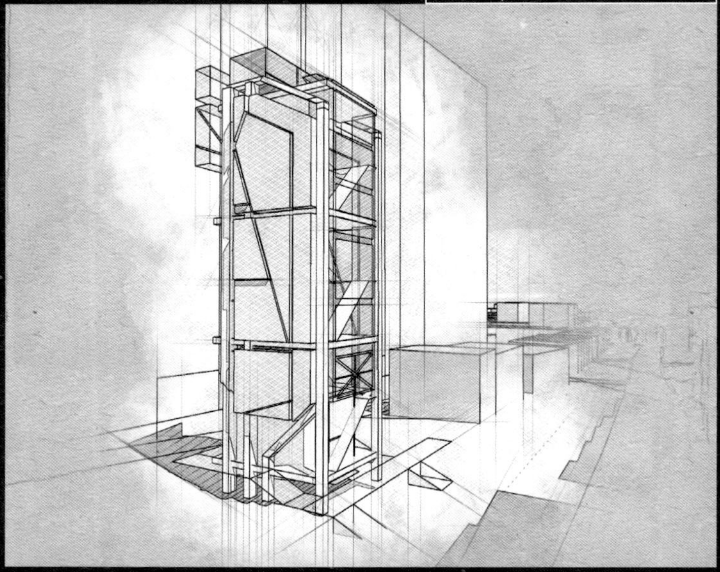

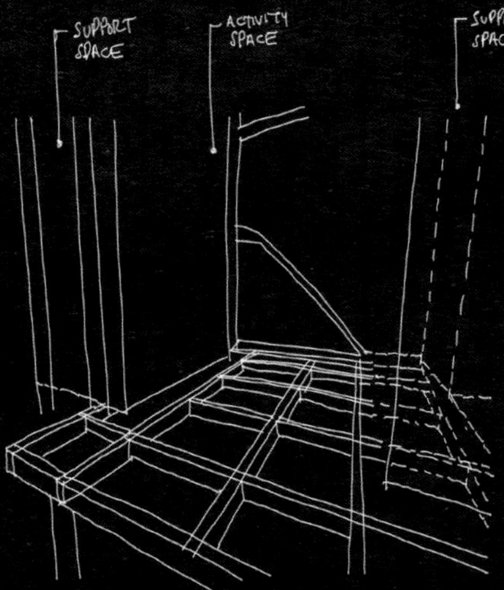

Prefabrication and reduction of embodied energy

This project is one of a temporary exhibition building which moves from city to city as part of a marketing strategy for a technology-based company. A key consideration is the energy used in transporting the building in disassembled form, when moving from city to city. A key requirement of the brief/program is to provide a building which reflects the levels of technology being promoted by the company, and also a building that performs highly in being able to deal with a range of climatic conditions. As a result, a fabric-based enclosure or an inflatable enclosure were not considered to meet those requirements. The design is one of a 'kit of parts' of external weathertight forms and internal serviced spaces which can be assembled and disassembled with ease and at speed. The kit of parts comprises components which are prefabricated, which fit onto a set of road vehicles and can be transported by road and by sea in containers.

A key theme in the building's environmental design is in making a comparison between the energy required to fabricate and construct the building, and the energy required to move the temporary exhibition building from place to place; from city square to city square. As a result of the energy expended in road transport, it was decided to use materials that provide components which are light in weight, durable and which can be replaced quickly if damaged during transit. It was agreed that having a set of 'back-up' components ready in storage was not a practical way to replace components that are damaged during cycles of assembly, disassembly and transportation. Instead, a set of workshops is made available for the repair and replacement of components, which are taken to the immediate location of the tower.

In addition to seeking the expenditure of a minimum of energy used in the day-to-day functioning of the exhibition spaces, the environmental strategy is one of expending the least amount of energy in the fabrication, assembly, disassembly and transportation of the complete tower experience, including interior displays and digital displays fixed to the outside of the structure. The tower is designed as a steel frame with aluminium-based secondary assemblies of stairs, hoist and display

3. Perspective view of exterior

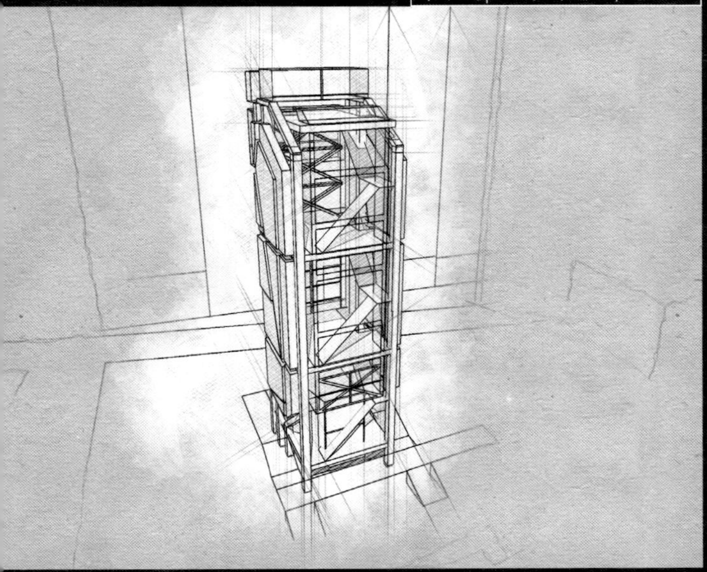

4. Perspective view of exterior

furniture. Floor decks are a mixture of hot rolled steel sections and pressed steel sections which are light in weight when compared with 'permanent' construction, and which are mounted in place as prefabricated cassettes. Washrooms and ancillary spaces are accommodated in separate prefabricated modules set at ground floor level.

At an early stage of the design, a comparison was made between the use of heavier-weight materials and those materials which are lighter in weight. The heavier-weight materials selected were determined, generally speaking, to have less embodied energy than the solution that used materials which were lighter in weight. However, it became apparent that for the purposes of assembly, disassembly, loading and unloading to and from road vehicles, the reduction of weight and the need for durability were primary considerations over the weight of the assemblies themselves. The lighter assemblies that were used, of light gauge steel framing, were found to require less energy to assemble and disassemble in their transportation to successive locations as a result of the reduced weight in each of the road vehicles. Comparisons made between different combinations of lightweight construction showed that the choices of lightweight construction provided similar levels of embodied energy, such that the overall weight of the assembly became a guiding principle not only for road transportation, but also for the overall weight of the building whose movement would be required, both assembled and disassembled.

It was found that, had a precast concrete hollow floor system with steel beams been a practical solution (which was not the case, due its self-weight) then the embodied energy would have been around 30% less than that of light gauge steel floor cassettes and steel beams; not taking transportation from location to location into consideration. Levels of embodied energy could have been further reduced if recycled steel or multi-cycle steel were used. The choice of bolt-fixed glazing as the glazing system had the advantage of having a level of embodied energy roughly 1/3 that of a regular aluminium-framed curtain wall. All values did not take into account the recyclability of materials, which vary considerably.

1. Project volume in site context

2. Project volume in site context

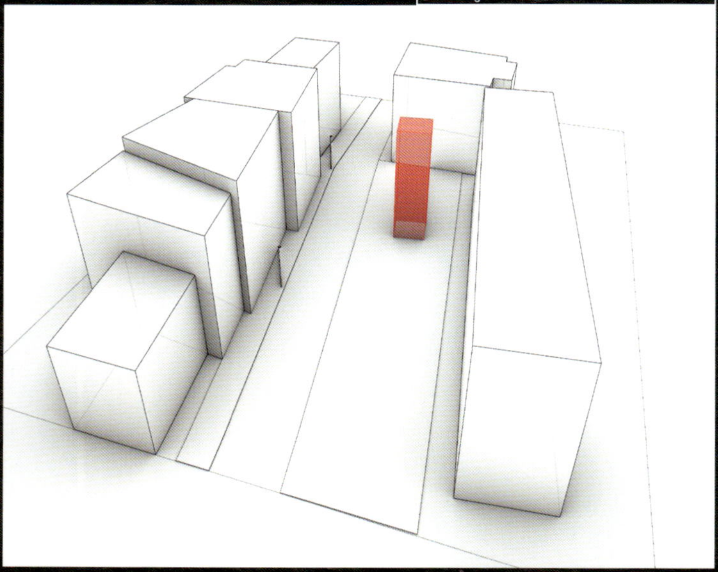

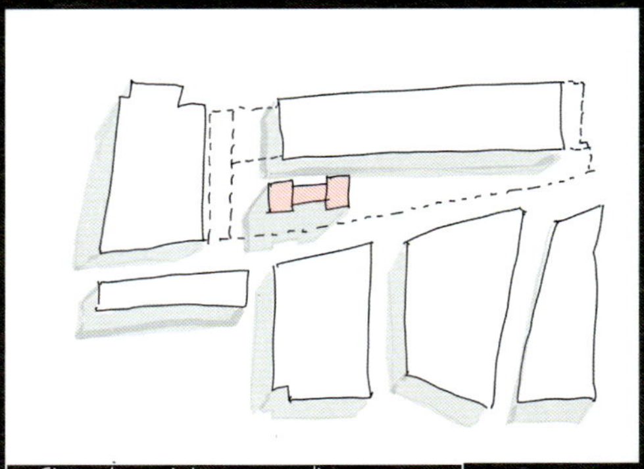

1. Site plan with surrounding context

Option 1 separates out the functions within the tower on a floor-by-floor basis.

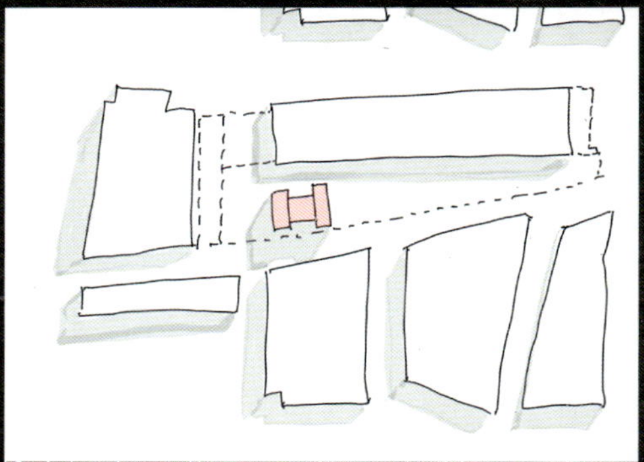

2. Site plan with surrounding context

Option 2 forms a single space, set in the middle of the arrangement, with two 'support cores' of staircases and digital displays to be fixed to the tower.

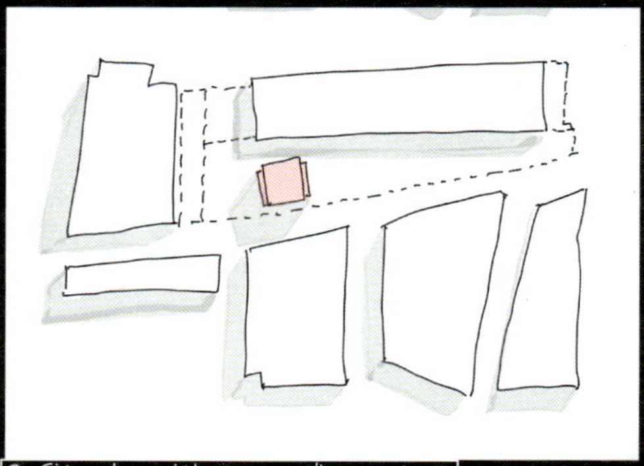

3. Site plan with surrounding context

Option 3 supports cores which are fixed like 'backpacks' to the side of the building in order to give the tower greater presence.

Site context

This temporary building is conceived as an urban 'performance in the round' in that visitors have smaller-scale interactions with the exhibition around the main tower; the urban context is of an experience of digital information being presented from screens set up on all sides of the tower, like a theatre performance 'set in the round'. The urban space surrounding the tower becomes part of its temporary integration into the cityscape. The tower interacts with its immediate urban environment while it is present on each site. The tower does not attempt to fit-in, complement or complete the existing arrangement of urban forms; it is not a piece in an ever-changing urban puzzle. Instead, the building is part of a travelling show that transforms the use of its surrounding public space for a few weeks at a time; then it moves on to a new urban space.

This small-scale tower for travelling exhibitions can be assembled, disassembled and transported by road between exhibition locations of city-centre public spaces. The purpose of the building is to present new technologies to the public through a series of four vertically stacked exhibition spaces and interaction spaces, where new products can be experienced first hand. In addition, there is a set of more intimate spaces for each of the new technologies on display. This approach contrasts with the large single-space pavilion used in previous iterations of the travelling temporary exhibition building. In this travelling tower, the four vertically stacked spaces can be rearranged both vertically and horizontally to respond to the technologies being exhibited. The tower can be arranged in a different order vertically, in response to different priorities in communicating the exhibited products from place to place; alternatively, the tower modules can be assembled horizontally to form a set of ground-level pavilion experiences, linked in different combinations of space.

All three options for this exhibition tower are based on being located on sites where it can be seen as a 'temporary' experience; an event, a moveable set of structures which are self contained and self supporting but which reach out to both the passing public and to visitors. Each

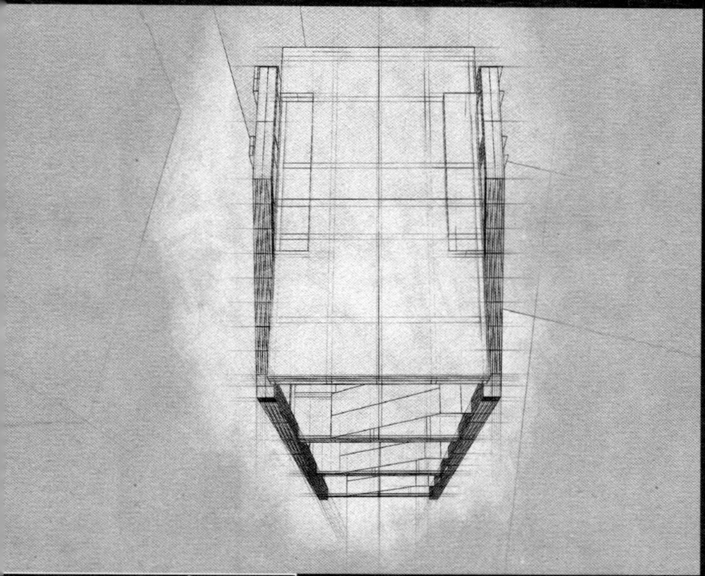

Perspective view of structure

2. Perspective view of structure

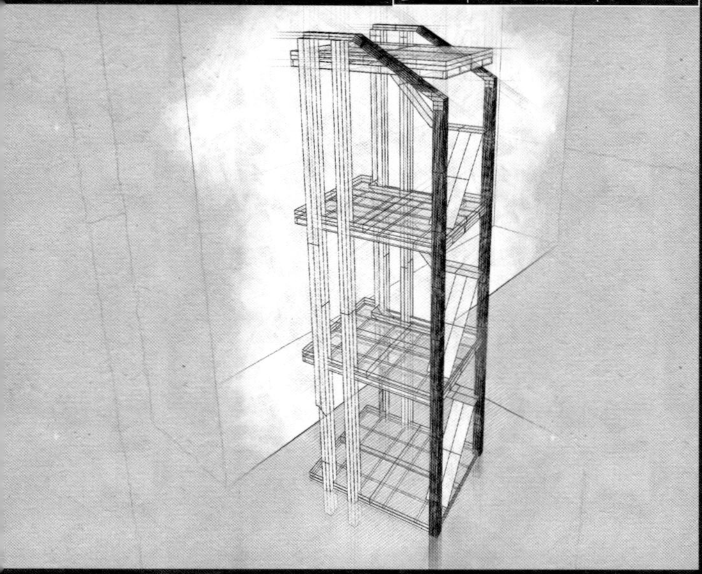

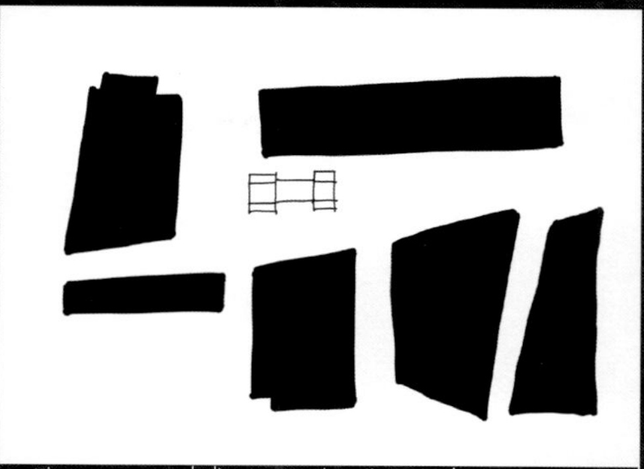

1. Figure-ground diagram showing project structure

Option 1
The structural frame of the tower is distributed in three parts; two mini-towers linked by a set of rectilinear spaces.

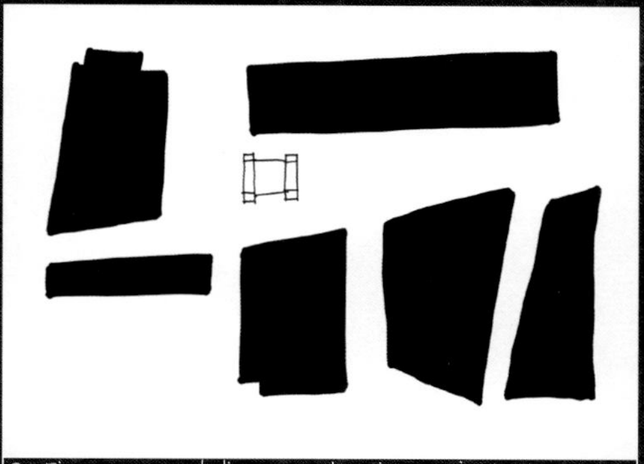

2. Figure-ground diagram showing project structure

Option 2
The structural frame of the tower is distributed in three parts; two cores linked to a set of rectilinear, vertically arranged spaces.

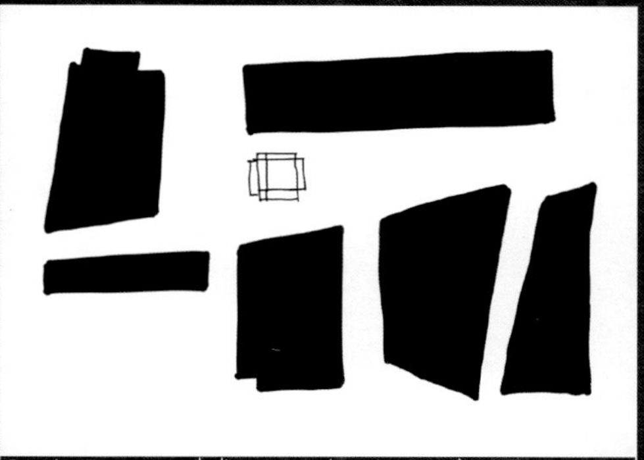

3. Figure-ground diagram showing project structure

Option 3
The structural frame of the tower forms a single tower object; restrained at street level.

site requires space around it for visitors to gather, and to be able to enjoy the difference in the urban space made by its temporary presence for up to a four-week period. Movement around the outside public space is expected to be a similar experience to the traditional use of public squares; open spaces which are framed by buildings and which are easy to access as part of an inner-city infrastructure.

Option 1 separates out the functions within the tower on a floor-by-floor basis; creating an indented form which can be assembled to form a set of three distinct spaces. Option 2 forms a single space, set in the middle of the arrangement, with two 'support cores' of staircases and digital displays to be fixed to the sides of the tower. In Option 3 the support cores are fixed like 'backpacks' to the side of the building in order to give the tower greater presence in each of its urban environments; the tower seen as an object with four open spaces which are superimposed to create a distinctive form. The support cores provide access to the top to create views of the activity taking place in the surrounding urban space. The tower in Option 1 is set at the centre of the public space which it inhabits in its movement from place to place. In Option 2 the tower is aligned with one of the edges of the notional 'site'; aligned with both one edge of the site and with one side of an adjacent building, such that the tower creates a temporary urban space in front of it. Option 3 aligns the tower with the road and one edge of a building in order to create a framed urban space on either side of the building, such that any installation of this tower creates a clear set of urban spaces that surround it as a result of its presence. Option 3 is a compact built form that can respond to a wider range of urban contexts by creating 'framed' outdoor spaces where they currently do not exist; being in place only for the duration of the exhibition in that part of the city.

It is envisaged that more pavilions will be fabricated for display together in a single group. A set of four pavilions can be used to create temporary corners for a rectangular-shaped city space, or can be located in nearby locations to set up a route from one pavilion to another; each pavilion having a different set of products on display.

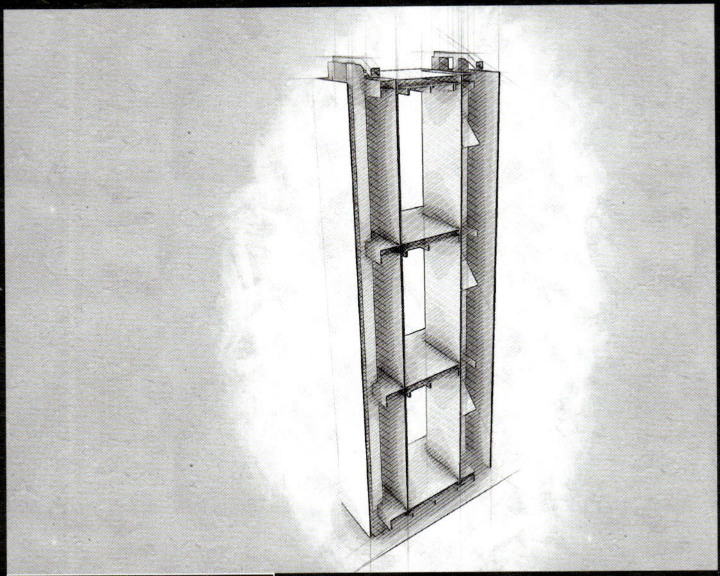

1. Sectional perspective view

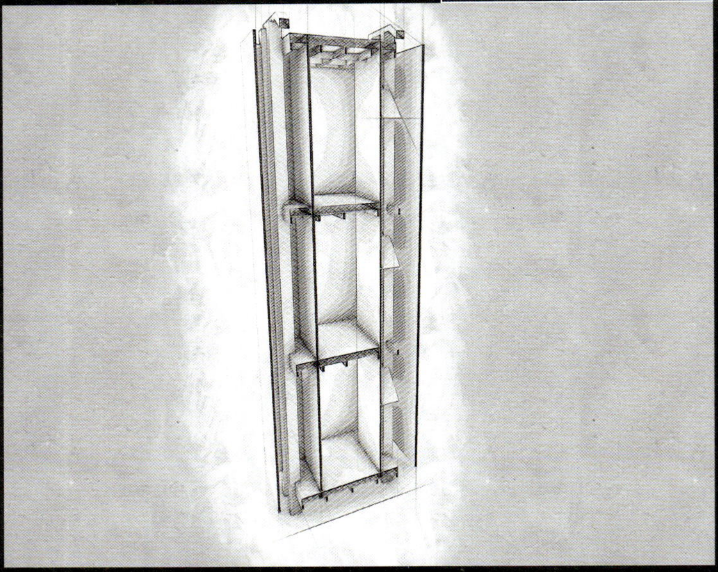

2. Sectional perspective view

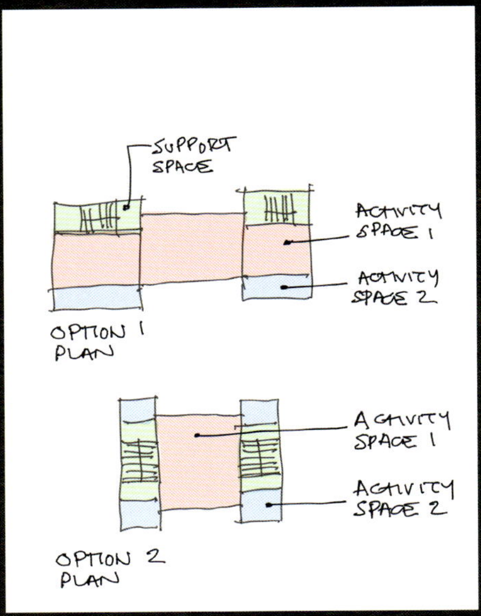

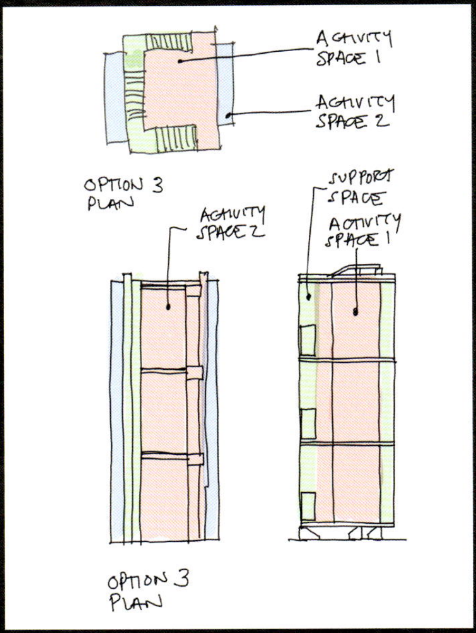

Spatial arrangement Option 1 creates a rectilinear space with service spaces and stairs in the corners of the floor plate, and clear space in the centre of the arrangement.Option 2 has support spaces to the main 'activity' space which are grouped as cores.

Option 3 creates framed views out from the tower at corner slots only, allowing more of the wall space within each stacked exhibition module to be used as opaque screens.

Spatial arrangement

This mini-tower consists of three floors of a relatively small floor plate where the stairs are mounted on its two opposite sides: The side used for access is clad in opacified glass; the other side used for an exit has metal screens that provide clear views out.

This is a mobile exhibition building with exhibitions on the lower floors in an enclosed 'black box' environment; an exhibition space is located in each of the three stacked spaces that comprise the tower. A small terrace is provided at roof level, accessed by hoist from the ground floor level. The hoist is a simplified lift/elevator device that is suited to rapid assembly and disassembly. A further lookout space, a belvedere, is positioned at the top level of the tower.

The towers can be rearranged; for example, modules can be set side by side as two vertical arrangements of two modules each, or as a group of four individual pavilions that form the corners of an amphitheatre-like space for performance-based events that form part of the exhibition. With an additional lighting tower and sound towers, the modules can be arranged for product shows. This is very much a mobile building for the communication of products with an urban-based public; assemblies can be assembled and disassembled from road trailers that arrive in a given sequence for installation or removal from the site. As a result, assemblies are fabricated in materials which are as light as possible.

The spatial organisation for Option 1 is to create a rectilinear space with service spaces and stairs in the corners of the floor plate, with clear space in the centre of the arrangement; creating a single rectangular space. The single space can be subdivided into three smaller spaces; creating a set of three square-shaped spaces as an alternative. In Option 2 the support spaces to the main 'activity' space are grouped as cores, with stair and hoist set on both sides of a central space whose floor plate is equal to that of Option 1. This arrangement of activity space creates views out of the tower on two sides only; focussing attention more on the exhibits. Option 3 creates framed views out from the tower at corner slots only, allowing more of the wall space within each stacked exhibition module to be focussed on the visitor

Perspective view of interior

2. Perspective view of interior

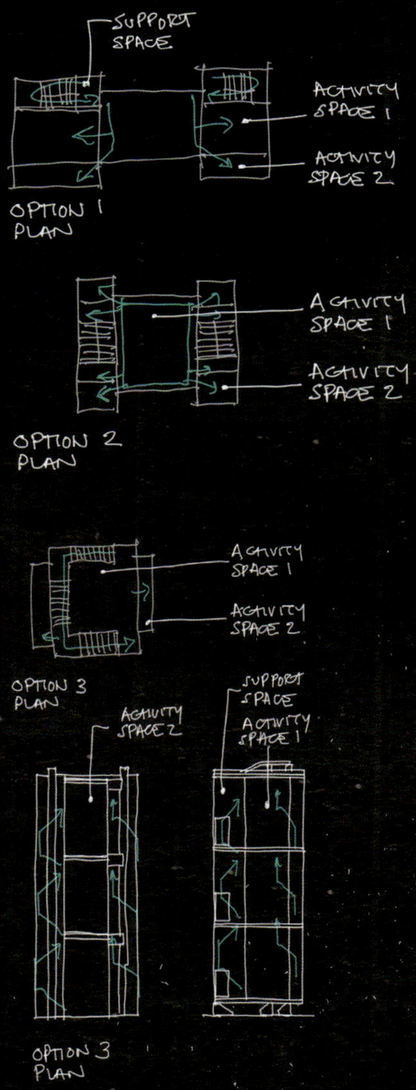

Internal circulation Option 1 and Option 2 have two sets of stairs; one set for rising up through the building and the other for descending to the temporary external spaces that are created alongside the tower.

Option 3 provides the ability for a single staircase to be shared, and for the experience of both rising up through the building to explore spaces at each floor level as well as of moving freely between those spaces.

experience; a secondary part of the spatial experience is to be able to climb stairs or hoist in order to experience both the tower, how it is fitted together, how the technology-driven tower fits in with the surrounding urban landscape and how that experience is reflected in the products and services on display.

Internal circulation

Circulation within the pavilion can be made either by stair or by mechanical hoist. The hoist can be hand-operated through a set of gears which allow the machine to avoid the need for electrical power.

Stairs are set between the two console brackets that form the basis of structural support and stability for the tower; providing an entry on one side of the tower, and an exit from upper levels on the opposite side. Entrance and exit, as well as general circulation, follow the route of a circuit going up the tower in one direction and down the other side. The internal circulation for all three options is based on being able to use the building in a way which is interactive with its urban environment; a temporary urban environment is created as a result of the short-term presence of the tower and the events immediately surrounding it. Access to the exhibition spaces is provided by a stair and lift/elevator. The internal circulation experience is one of anticipation of arrival at the space beyond, rather than as a passage that links internal spaces. The three options for internal circulation are based on enhancing the experience of moving around the four forms of space that are juxtaposed, either vertically, horizontally or as a mixed volumetric arrangement from space to space.

The internal circulation for Options 1 and 2 has two sets of stairs; one set for rising up through the building and the other for descending to the temporary external spaces that are created alongside the tower. Option 3 provides the ability for a single staircase to be shared, and for the experience of both rising up through the building to explore spaces at each floor level as well as of moving freely between those spaces such that the arrangement of stairs and hoist forms a spatial experience of its own.

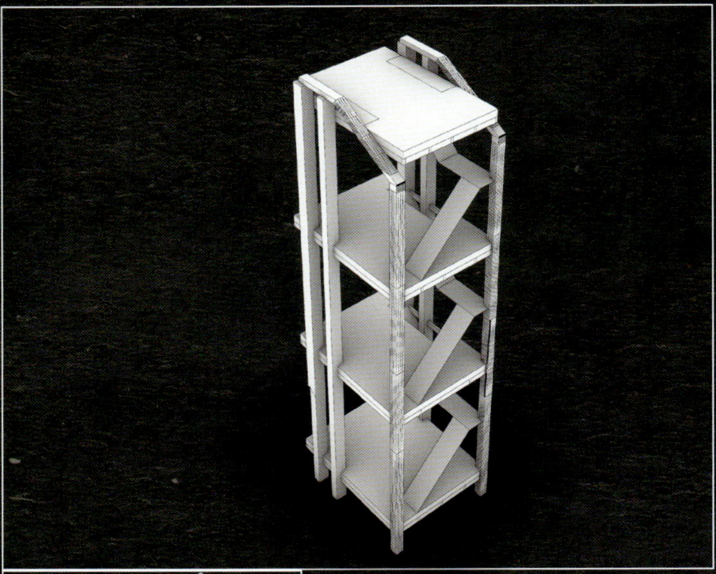

1A. Isometric view of structure

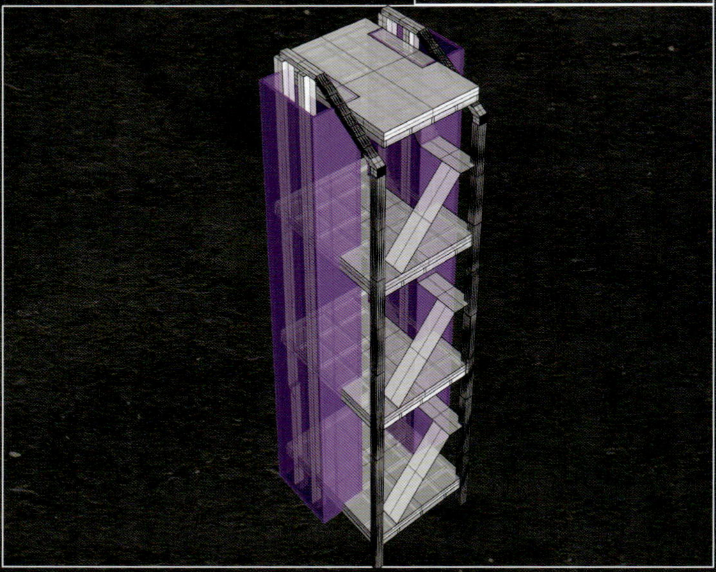

1B. Isometric view of structure

2A. Isometric view of structure

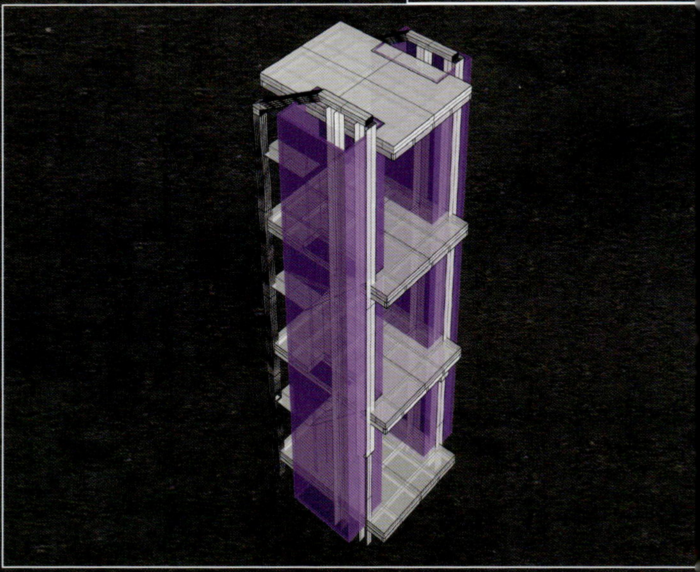

2B. Isometric view of structu

Structural strategy

This building is a small-scale, steel-framed structure which stabilises part-prefabricated modules which are set within the frame. Modules are site-assembled and comprise a set of four external forms with integrated walls and floors. Each module is light in weight and is constructed from cassettes of wall and floor panels using pressed metal sections with a sheathing board on either side. The modules are designed to be self-supporting when assembled; stacked one above the other such that loads are taken at points around the perimeter of the modules down to floor level, where they are supported on pads to suit the conditions of the existing public space in each location where the building is to be assembled. The prefabricated modules are secured by an external steel frame which is restrained at ground level by consoles and outriggers that provide stability. The same structural components are used to support both stairs and the exhibition-space modules. Each module of exhibition space has 'pop-up' display case units which form a key part of the rapid fit-out of each space; fixing points at floor level and ceiling level are introduced into the supporting cassettes for floor and wall as part of the structure.The sequence of assembly of the primary structure in each urban location is of installing modules in a stacked arrangement: Modules are assembled in-situ, like building a timber framed house; they are not pre-assembled as a complete unit, then hoisted into place. The latter arrangement would require heavy lifting equipment which takes away the benefit of enabling the 'kit of parts' to be assembled as individual components and sub-assemblies which are light in weight. The assembled modules are stabilised internally by the floor panels and diaphragm infill facade panels. The complete tower assembly is stabilised by the external steel frame; fabricated from hot rolled square hollow box sections. To summarise: The structural concept is one of an externally fixed primary steel frame which provides stability to the set of superimposed, stacked structural modules. Modules arrive on site by road trailer, and they are offloaded and connected together directly in place. A large space for the laying out of components, prior to their assembly in situ, is not required. The structural

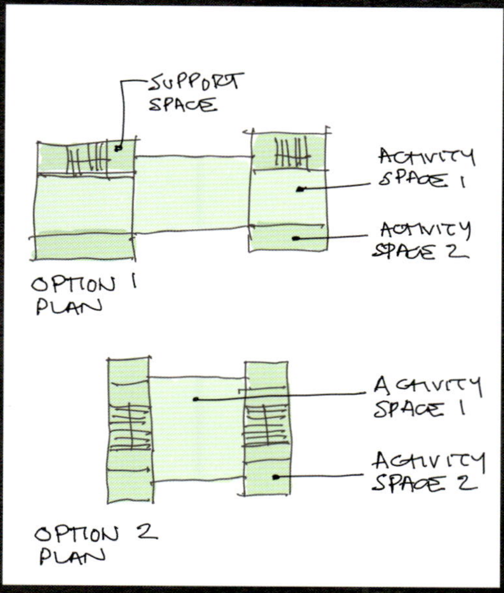

Structural strategy: Zones

Option 1 makes use of a steel frame which is divided into three parts.

Option 2 has service cores grouped together with a central space.

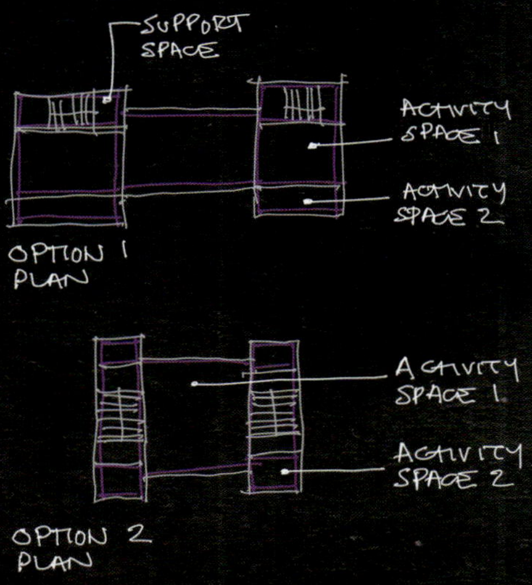

Structural strategy: Framing

Option 1 has one steel frame for each section of the tower, on a floor-by-floor basis.

Option 2 has an alignment of beams and columns, and a prefabricated set of frames.

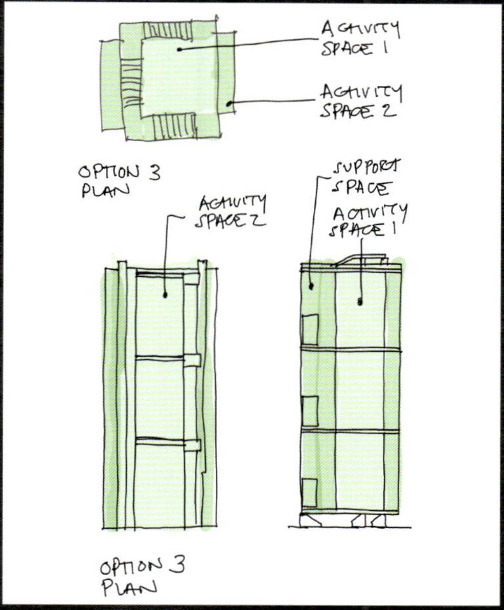

Structural strategy: Zones Option 3 creates a set of 'nested' modules that are fixed together on each floor, and which stack vertically.

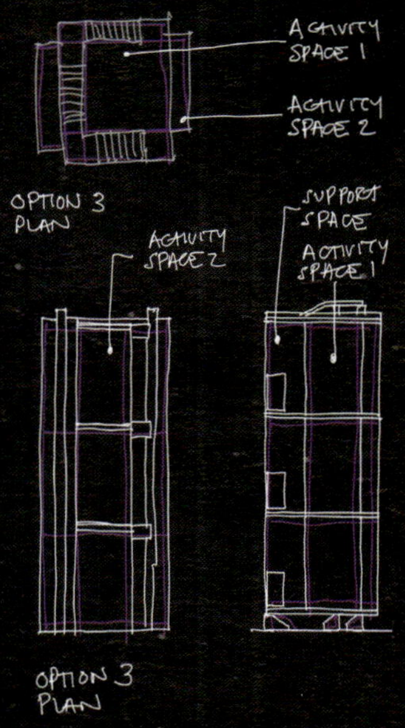

Structural strategy: Zones Option 3 has a double layer of structure corresponding to the inner layer of the stairs and the outer layer of the stack of stairs, creating a rigid frame through the height of the tower.

strategy for all three buildings is of the building as a series of modules that can be assembled and disassembled on any site, and are made from a steel frame. The steel frames are installed as modules which can be transported by road, then lifted into place by crane and bolted together; similar to a construction crane being assembled on a building site. Option 1 makes use of a steel frame which is divided into three parts, one for each section of the tower, on a floor-by-floor basis; two parts for the end volumes and one part for the central space. The difference in form between the end towers and the central space makes it necessary to have more steelwork than would be the case if they were all combined. An associated need to reduce the steelwork is answered in Option 2, where service cores are grouped together with a central space, such that a straightforward alignment of beams and columns, and a prefabricated set of frames, can be deployed. Option 3 simplifies the structure to create a set of 'nested' modules that are fixed together on each floor, and which stack vertically. This arrangement has a double layer of structure corresponding to the inner layer of the stairs and the outer layer of the stack of stairs, creating a rigid frame through the height of the tower. The mechanical and electrical equipment 'backpacks' are attached to the sides of the building. They are prefabricated steel frames which are clad and thermally insulated, and are attached to that frame.

The tower supports a set of digital screens which are used for the projection of images during hours of darkness. Associated lighting equipment is installed at roof level to support events which take place next to the tower at ground level. The structural concept for the pavilion building is one of a demountable set of floor decks and columns which provide a lightweight deck and frame solution to the structure of this building. The frames are cross braced on the opaque wall sides and by the lift assembly on the rear of the pavilion. The cross bracing is concealed behind the stretched fabric frames. The vertical supports are concealed behind cladding, where they are enclosed in order to avoid a visual clash between the glazed side which presents products on display inside the pavilion.

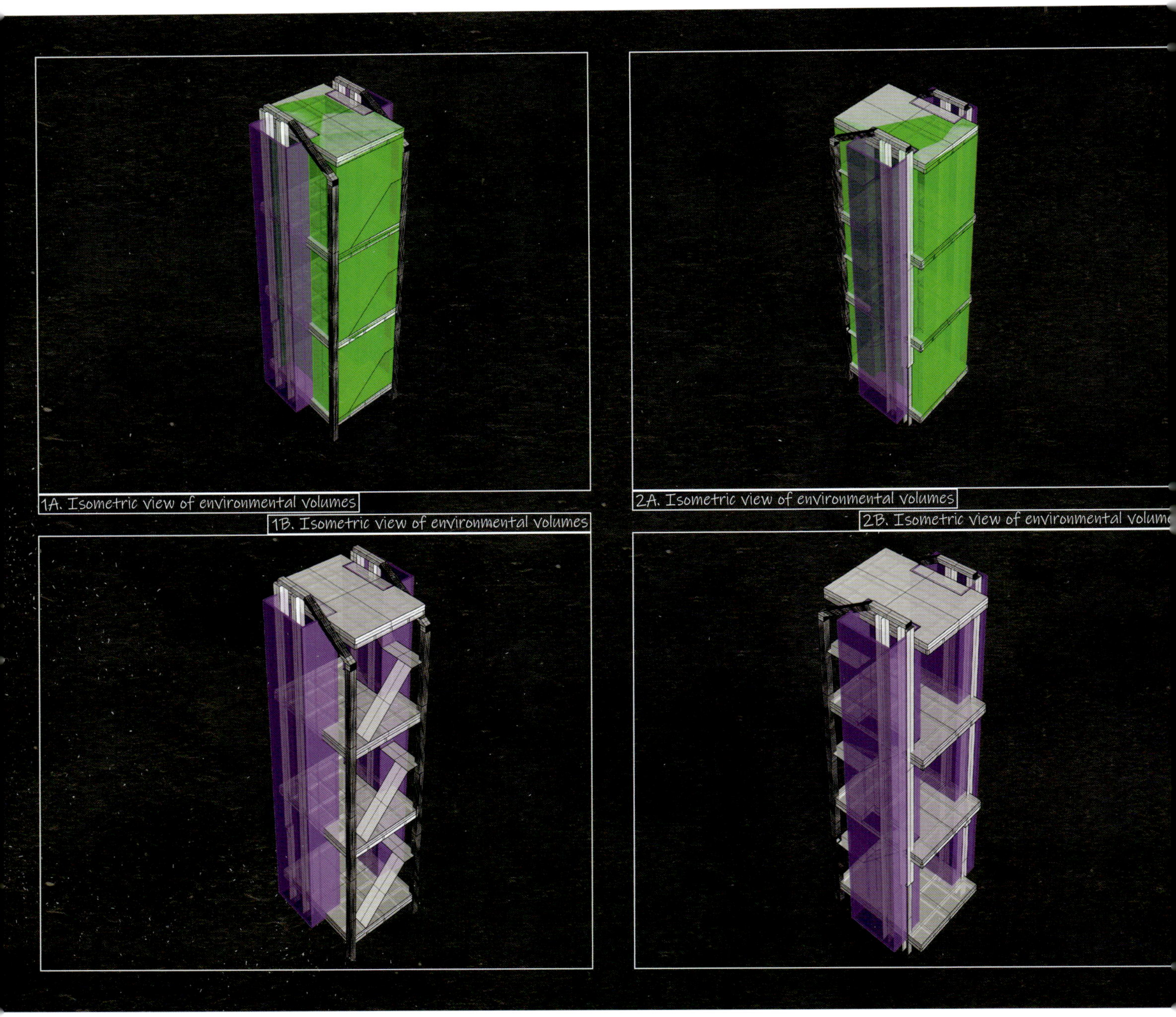

Environmental zones

Environmentally, the project is conceived as a daylit 'light box', illuminated from the outside rather than only providing light from the inside to the outside during hours of darkness. Daylight is directed by way of reflectors that draw natural light through the height of the tower that allows opaque walls to be washed in daylight and shadow. An essential aim is to show how spaces can glow in ordinary daylight, without the need for electrical lighting during hours of natural light. In addition, this tower is very much a low-energy building where spaces are naturally ventilated during the day and maintained close to the external temperature; the comfort experience of being in the tower is similar to that of being in the temporary urban spaces created at street level. As a result, a modest set of mechanical ventilation equipment is provided to temper spaces; this installation is fixed to the facades.

The building is set out as three environmental zones; a lower-level zone with external access into the inside where there is a semi-controlled environmental space; a mid-height zone which has independent control of temperature and humidity for each floor, and a high-level zone at roof level comprising an external terrace and a semi-environmentally controlled space immediately behind.

The environmental strategy for the three options is to create spaces which can be open and naturally ventilated at certain times, and be closed and mechanically ventilated, for specific exhibitions and events, at other times. This mixed mode of use may switch from one to the other during the course of a mid-season day, or might be fixed over the course of a single city 'visit'.

The pavilion building is heated from equipment mounted at roof level; warmed air is directed down ducts to two upper floors. Only a small amount of heating is required for these small spaces. Warmed air is also directed into the external facades; into the two layers of fire-resistant polymer-based sheet that form opaque cladding panels that provide a thermal buffer during cold weather. No cooling is provided during summer months as the external air temperature, in this generally temperate zone, was deemed sufficiently low to create a passive

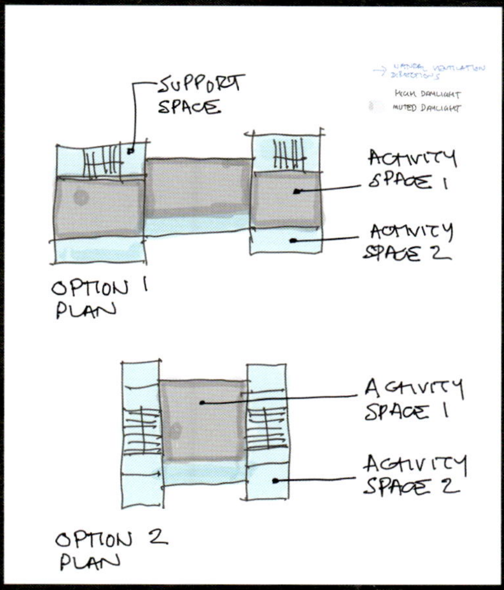

Daylight strategy for direct sunlight and cross ventilation.
Option 1 admits daylight and air on the two long sides of the tower.
Option 2 admits daylight and air into the circulation areas of the tower and within the main display area.

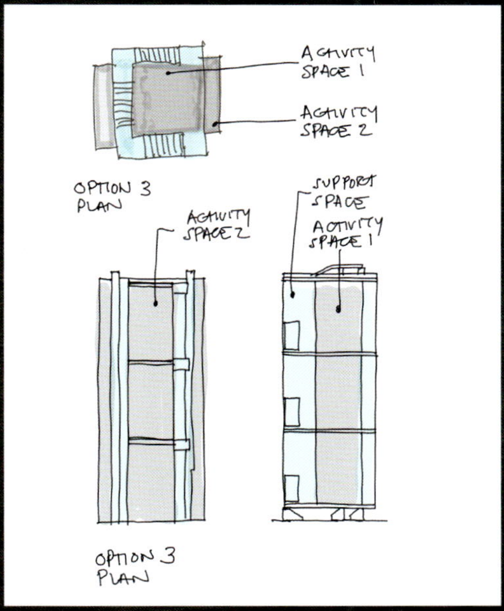

Daylight strategy for direct sunlight and cross ventilation.
Option 3 admits daylight and air into the circulation areas of the tower.

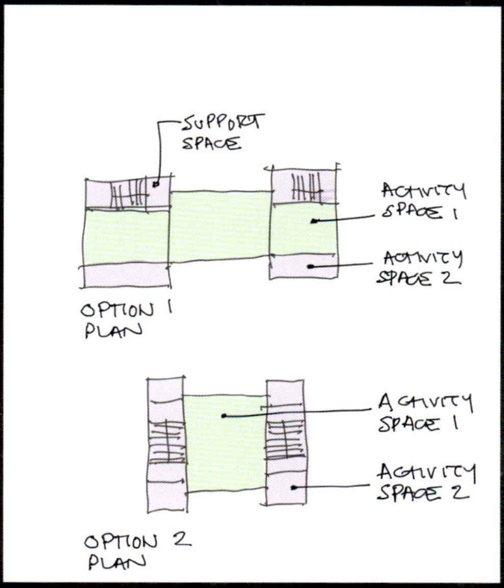

Environmental zones
Option 1 allows the stairs and hoist to form a single environmental zone, which can be both heated and naturally ventilated.
Option 2 brings the environmental zones together into two zones rather than four for the support spaces.

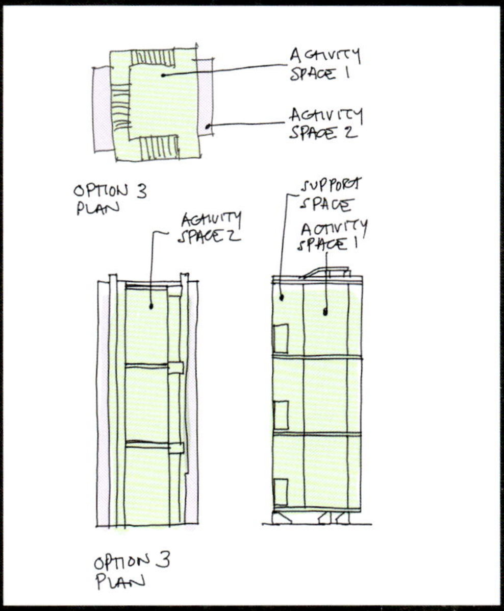

Environmental zones
Option 3 distributes environmental control to each of the four prefabricated modules.

cooling and refreshing effect for visitors inside the building. As a result, the energy required to warm the building during colder months is low; the ground floor entrance area benefits from the ducted warm air generated at roof level. Electricity for the heating and electrical lighting for the pavilion is provided from a generator truck positioned remotely from the pavilion. Cables which pass from the pavilion to the truck are protected with temporary deck-like covers which form part of the way-finding for visitors.

The emphasis in this design is to use natural ventilation as much as possible, with only small amounts of additional heating required from ducted air and air curtains in front of the glass facade. The building can perform in completely passive mode during daylight hours, when the building is naturally ventilated and heated. This suits mid-season and summer months. The design uses battery packs for lighting which are attached to the sides of the building behind the fabric frames. It is envisaged that additional elements will be added to the external envelope, for which a provision for fixings has been made.

In Option 1 the stairs and hoist form a single environmental zone, which can be both heated and naturally ventilated. Service spaces are ventilated separately from activity spaces but have less environmental control than that provided in the main exhibition spaces, which are required to meet a range of temperatures and levels of relative humidity for the different locations where the exhibition facility will be in use. Option 2 brings the environmental zones together into two zones rather than four for the support spaces, making the zones more efficient; reducing the effects of external temperature and humidity as a result of the introduction of opaque walls which are thermally insulated. Option 3 distributes environmental control to each of the four prefabricated modules, together with their mechanical and electrical equipment which is attached to the sides of the building. Vertical circulation is now positioned within the main space such that exhibition spaces, stairs and hoist are arranged within a single environmental zone; support spaces are accommodated within modules attached to the side of the tower to further reduce the energy consumption for the building in use.

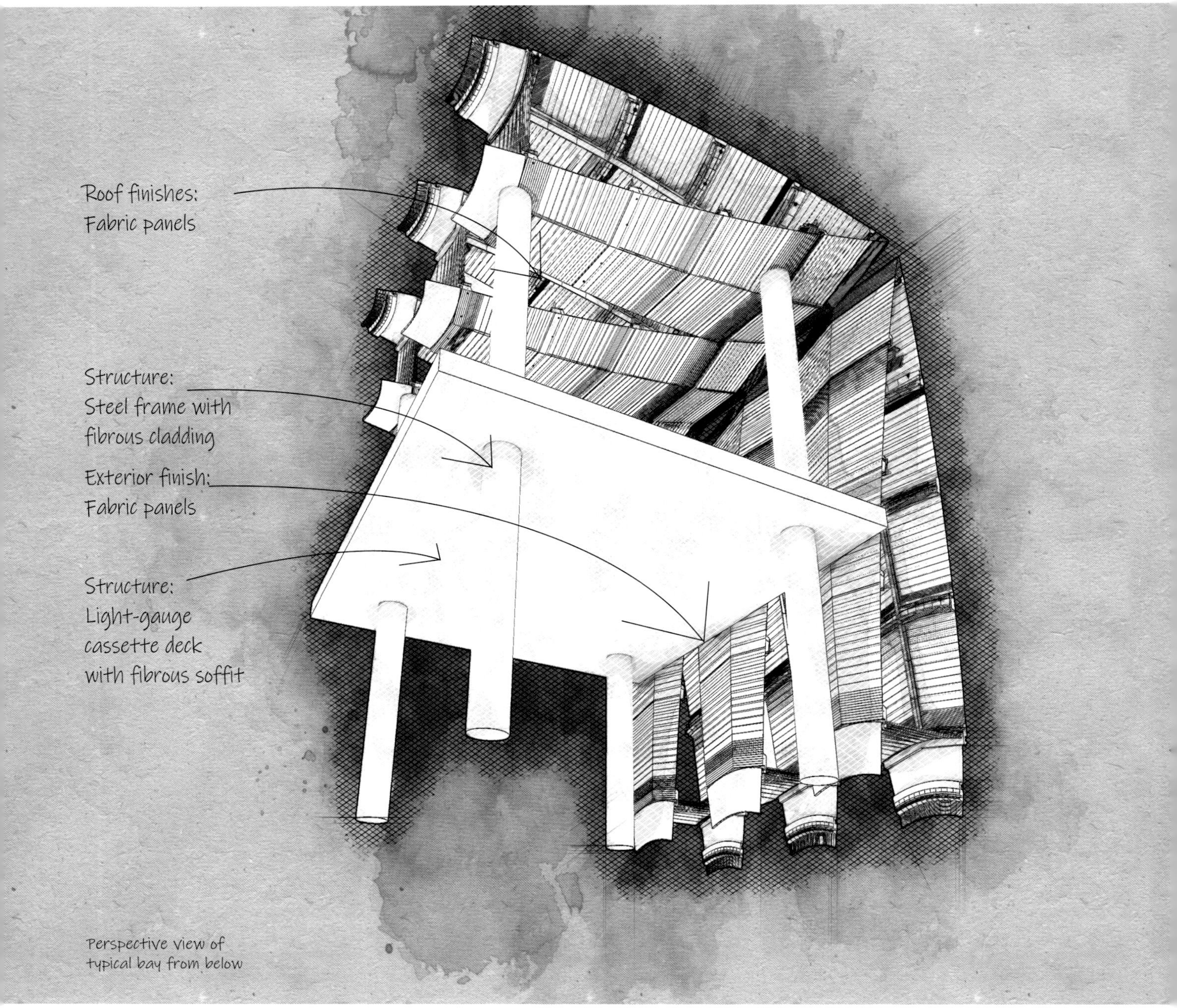

Perspective view of
typical bay from below

Typical bay adjacent to facade

An essential consideration in the design of the external envelope was to ensure that the panels and frame supporting the tower were demountable in a way that is easy to pack and transport by road. The tower components are moved from city to city; the structure and facades are designed to be light in weight. This requirement of weight allows the design to create a certain 'festival' atmosphere in its deployment and when in use, rather than the pavilion using the language of a permanent building.

The outer panels have stretched fabric frames which are light in weight and which are used to enclose the side walls of the pavilion. The front facade is glazed but does not make use of materials that are light in weight, such as air pillows or polymer-based sheet materials like polycarbonate, as the primary purpose of the pavilion is to exhibit products inside – partly as shopfronts and partly as exhibition spaces. The front elevation is a set of display cases formed in glass with single-glazed sheets of clear glass. These panels are fixed back to frames with a point-fixed glazing system which provides them with physical robustness when the building is in use. This glazing system provides a transparency and a sense of visual lightness of weight across an arrangement of glazing which extends across several floors. The glazing can be assembled and installed on the pavilion when the structure and deck are in place.

The building is assembled one floor at a time; each floor can be considered to be a typical bay, with floors repeated in their construction from floor to floor. Staircases are arranged in prefabricated half-flights; the stairs and envelope assemblies are installed when the frame is already assembled in place. The floors that are installed with the staircases provide platforms from which to fix the fabric-based side panels. Panels behind the glazed shop fronts, on the opposite facade, are also formed from fabric-based panels. The panels on the three opaque sides of the pavilion are thermally insulated, with an air curtain introduced in front of the single glazing, inside the pavilion, in order to overcome the downdraft of cold air; use of double-glazed units was considered to be too

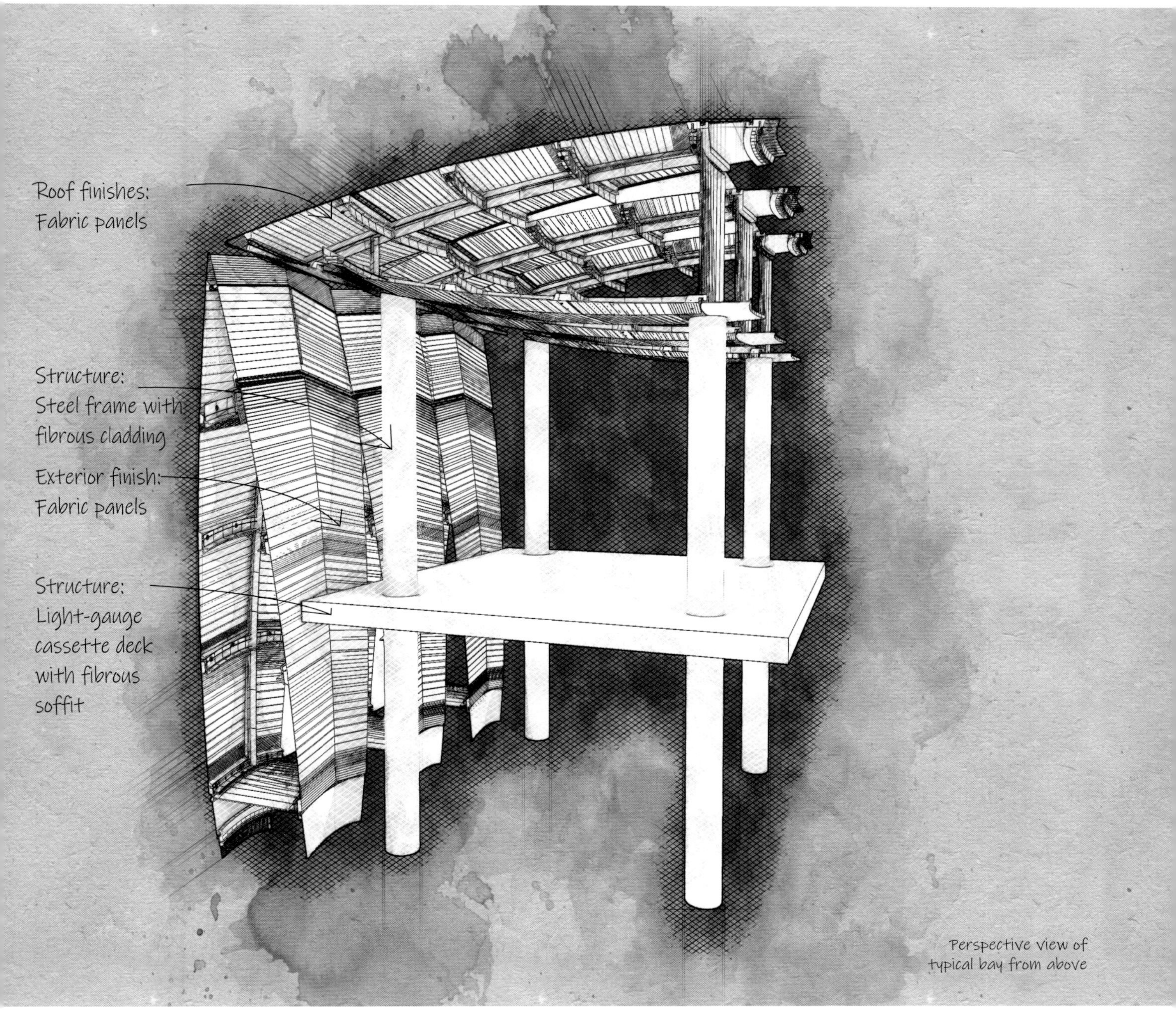

Perspective view of typical bay from above

heavy for regular transportation and would have reduced the effectiveness of the 'shopfront' concept of the building. Polycarbonate panels were considered but the lack of flatness of the material was problematic, and other materials such as acrylic were not considered to be sufficiently durable.

The top of the building has services equipment installed, which is self-contained with its own enclosures, though visually the equipment forms part of the language of the external envelope. The opaque panels on the sides of the pavilion are thermally insulated in order to provide an enclosure which is airtight, watertight and can be heated efficiently. This building is used for only certain hours of the day; its main times of use is in the evenings, when electrical lighting is used to illuminate the building and the facades.

The opaque facade panels appear to glow under the electrical light in the evenings as a result of them being created in two layers of fire-resistant fabric. The gap between the two skins is used to draw warm air from the spaces within the pavilion to create a thermal buffer between inside and outside. The top of the building has a fabric canopy enclosure to create a sheltered environment which is not thermally insulated. The aim for all spaces is to create as much of a sense of a 'pavilion' experience as possible; letting visitors know that there they are not in a regular building. This approach aims to bring a sense of occasion to their visit and to allow the outside spaces, both around the building and at roof terrace level, to be enjoyed as protected outside spaces regardless of the time of year.

The translucent area opposite the glazed front facade is enclosed in a mixture of stretched fabric frames and polycarbonate sheet. This arrangement is also used at the sides of the building in the slots between the opaque stretched fabric frames and the primary structure. This use of polycarbonate is used to create links between different opaque areas and enhance the sense of the building's visual lightness.

The adaptability of the external envelope to near-future changes is a key aspect of the design of the external envelope, with its ability to have new elements of enclosure added to suit evolving needs.

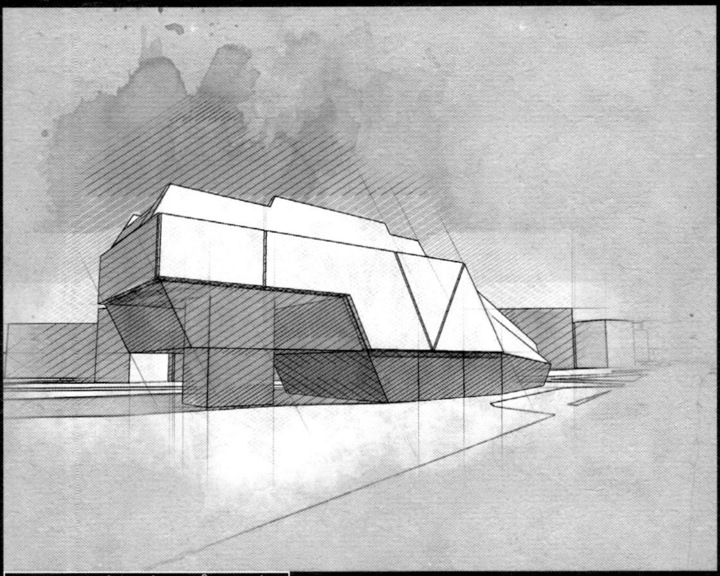
1. Perspective view of exterior

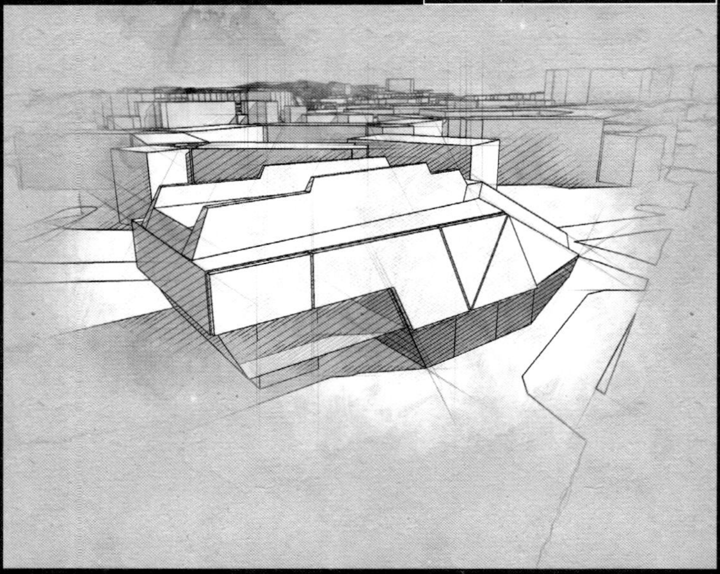
2. Perspective view of exterior

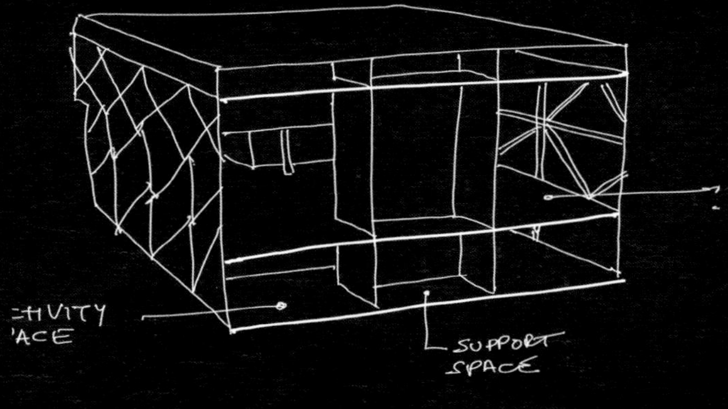

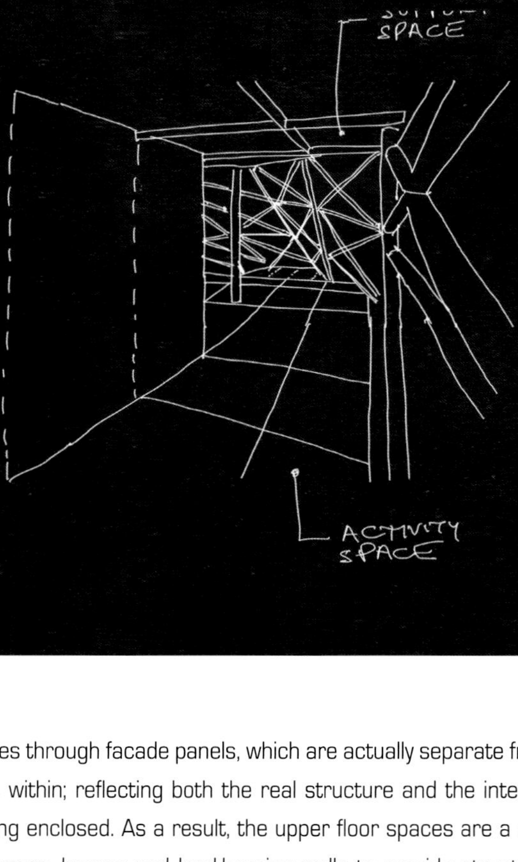

Relationship of interior space to structural system

The structural approach taken for this project started by asking a question: 'Do we need to have the raised volumes, which comprise most of the spaces in this building, supported on columns or can the upper floors be supported another way? The solution is to consider the upper floors as an independent structural element that is supported from beneath on both columns and load-bearing walls.

A propped cantilever was considered in order to visually articulate the structure beyond a regular frame of slabs, beams and columns. The approach taken is one of creating distinct structural forms for each of the volumes that comprise the spatial arrangement of the building, allowing a much greater expression of the functionality and organisation of the spaces within the building. A differentiated structure also allows a visual 'reading' of the internal spaces when viewing the external form of the building from surrounding streets. This approach allows the structure to be expressed visually by showing how the internal spaces are created from its structural elements rather than signalling those spaces through facade panels, which are actually separate from the spaces within; reflecting both the real structure and the internal spaces being enclosed. As a result, the upper floor spaces are a mixture of columns, beams and load-bearing walls to provide structural stability without needing to introduce an internal service core of stairs and washrooms that would have protruded into the spaces of the arts building. This approach has been applied to all three options that were considered; the structural approach is sufficiently flexible to provide an authentic visual expression to the volumes within the building for Options 1, 2 and 3.

The building is required to have an exposed reinforced concrete appearance externally, in order to continue the trend of neighbouring buildings that form part of this university campus. A decision was taken to express that concrete-like appearance on the inside face of the facades. As a result, the building comprises a reinforced concrete structure of a mixture of twin-wall construction in concrete and reinforced concrete frame which is clad externally with GFRC (glass-fibre

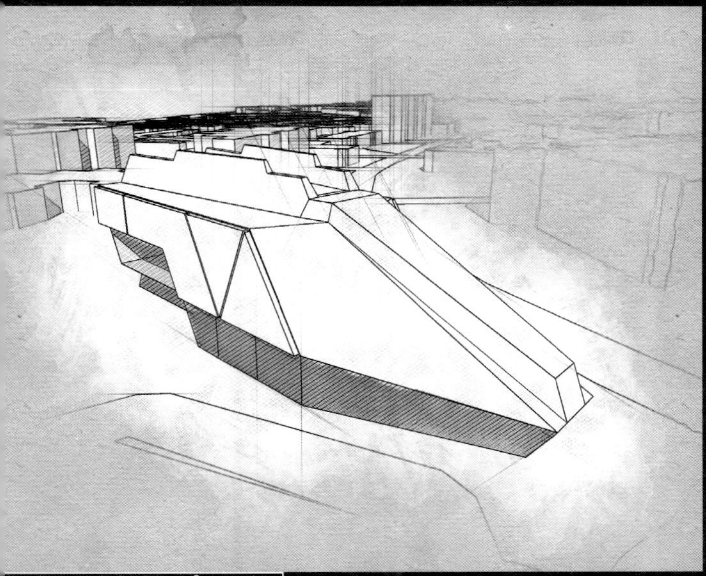

. Perspective view of exterior

4. Perspective view of exterior

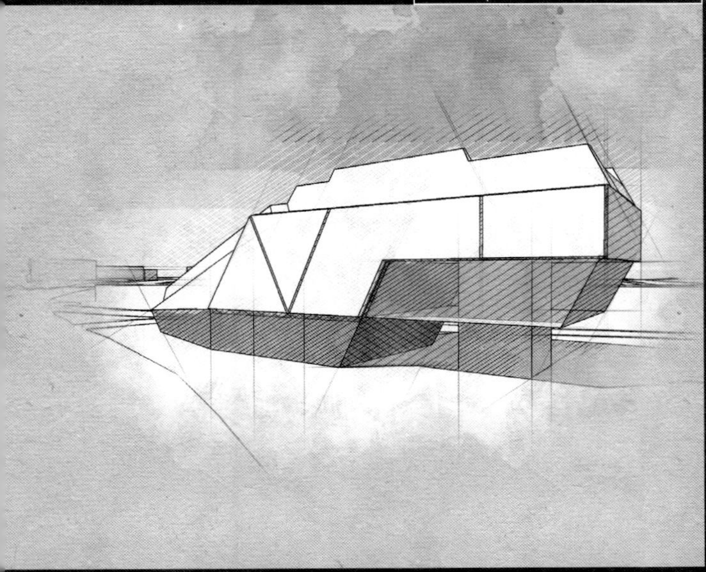

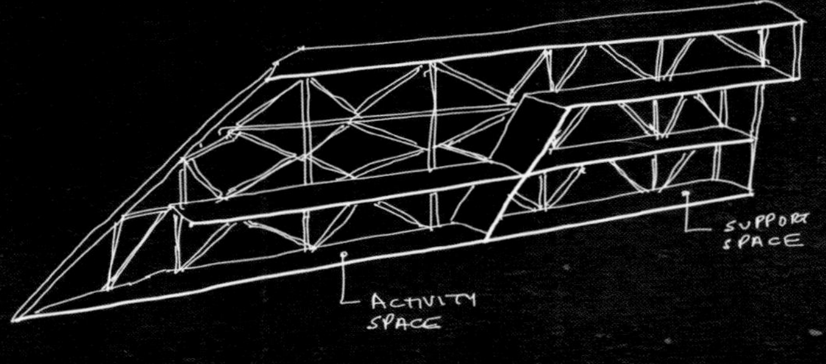

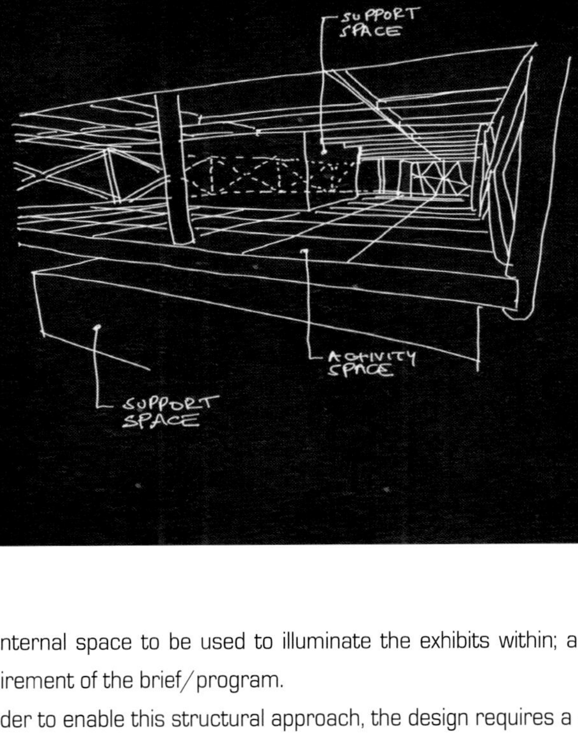

reinforced concrete) panels. The opaque twin walls comprise an inner and outer concrete wall 'skin' with thermal insulation set between them. The two concrete walls are joined by ties to create a diaphragm wall configuration.

This approach avoids the rainscreen facade arrangement that would typically be used for the outer face of a concrete wall. A concrete twin wall, or 'sandwich' wall, provides part of an entirely concrete building seen from the both outside and inside the building. Smaller volumes of the building make it uneconomical to make use of twin-wall sandwich walls; a single thickness concrete wall is clad externally with thermal insulation and GFRC panels.

This approach 'reveals' the structure entirely on the inside of the building; the concrete wall enclosing these smaller spaces is structurally continuous with its surrounding elements of load-bearing walls, beams, columns and floor slabs. From the point of view of the building user and gallery visitor, the continuous, exposed concrete surfaces allow the slow-changing effects of daylight and shadow across the surfaces of the internal space to be used to illuminate the exhibits within; a key requirement of the brief/program.

In order to enable this structural approach, the design requires a specific fixing bracket at the interface of the primary structure and the cladding system; the cladding system is centred around the design of the bracket that is used to fix the cladding panels to the supporting structure. A bracket was designed specifically for the project in order for the solution of structure and cladding to be economical. A key part of this design is to separate the weather seal, located at the joint between the cassette panels to which the outer cladding panel is fixed, from the bracket that supports both the cassette and the outer cladding panel. This bracket would typically penetrate through the seal between the thermally insulated waterproof cassette panel in order to support the weight of the GFRC outer cladding panel.

The facade panels have a low-relief design set into the depth of the GFRC panels that emphasises the dynamic nature of the spaces within, without the need to use an increased structure to create this effect.

1. Project volume in site context

2. Project volume in site context

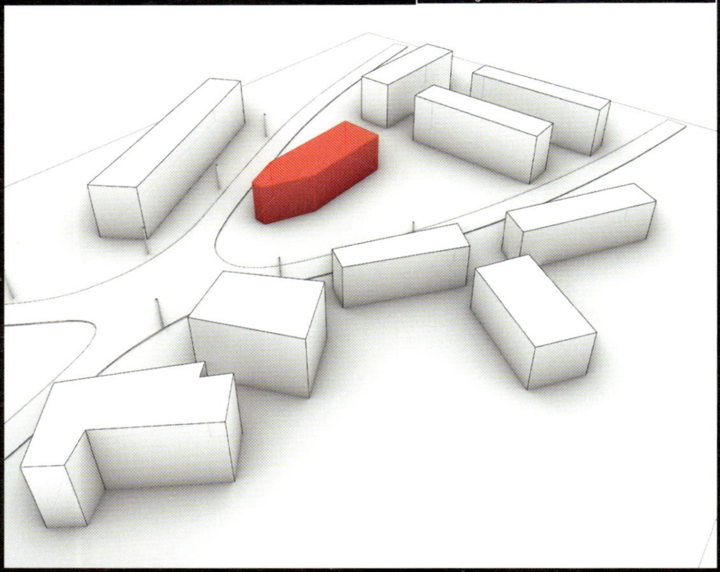

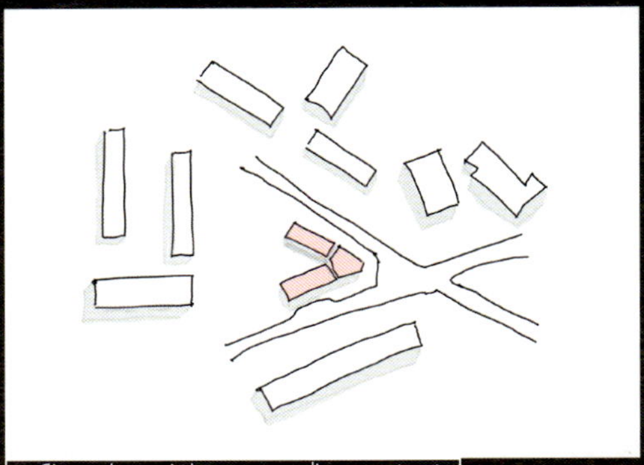

1. Site plan with surrounding context

Option 1 separates the constituent spatial components of the art gallery into an entrance space with two wings.

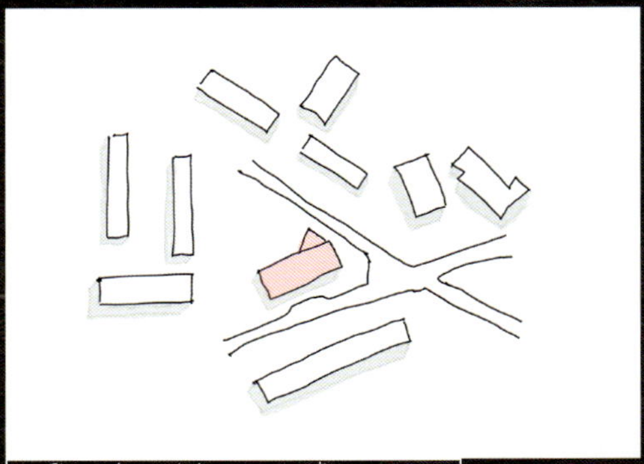

2. Site plan with surrounding context

Option 2 aligns itself with one of the roads at the site boundary to create a 'straightforward' box-like form.

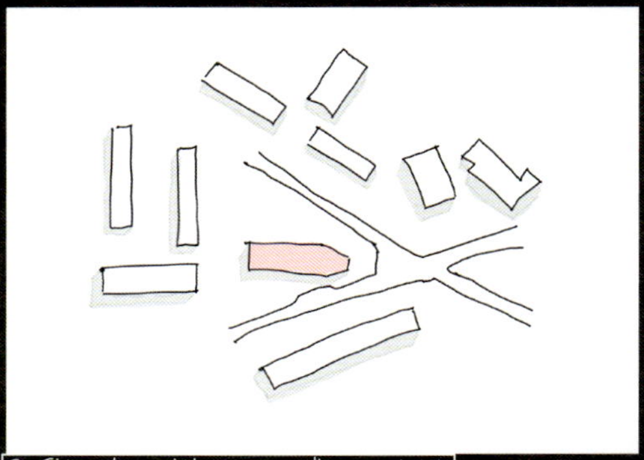

3. Site plan with surrounding context

Option 3 creates a step in the building by aligning the volume with the adjacent buildings; creating a point of inflection at the front of the building.

Site context

This project is for an art exhibition building in a university campus that is used as a gallery to exhibit both paintings and large-scale sculpture. The building comprises a central spine core with side-lit exhibition spaces on the lower floor, and top-lit spaces on the upper floor. This arrangement is one of a sequence of art exhibition spaces, but with the addition of the projecting upper storey which cantilevers out into space providing a canopy for solar shading, as well as the daylighting and electrical lighting of large-scale sculpture pieces which are set externally beneath the canopy formed by the cantilevered floor above.

The building is located at the intersection of two routes that traverse the university campus; the triangular-shaped site is owned by the client. An essential challenge is to give the art gallery a visual significance in its location at a road intersection, but without it appearing to be a building of purely utilitarian purpose, which is the case with the previous use of the site. The relationship to the site is based around how this building is positioned; at the intersection of four streets which are not in a rectilinear arrangement as a result of the roads passing through and around the edge of this university campus. The gallery could have positioned itself as a rectilinear form that is aligned with one or other of the streets; a set of box-like volumes arranged as a set of linked rooms. Instead, the building was considered as a form that addresses its location at the intersection of streets.

The design creates an integrated set of opaque volumes, taking up non-rectilinear forms arranged at angles that respond to the streets that enclose the site. This approach results in a large-scale single volume that has a set of internal spaces which are very different from one another. An essential requirement of the design was that it would be both compact and capable of adapting to changing exhibitions, but there should be a sense of the 'unexpected'; a sense of drama when arriving at the building.

The design provides an unusual spatial experience within a relatively small building. Large, airy spaces were not in the spirit of the brief/program for this compact facility for the exhibition and display of art-

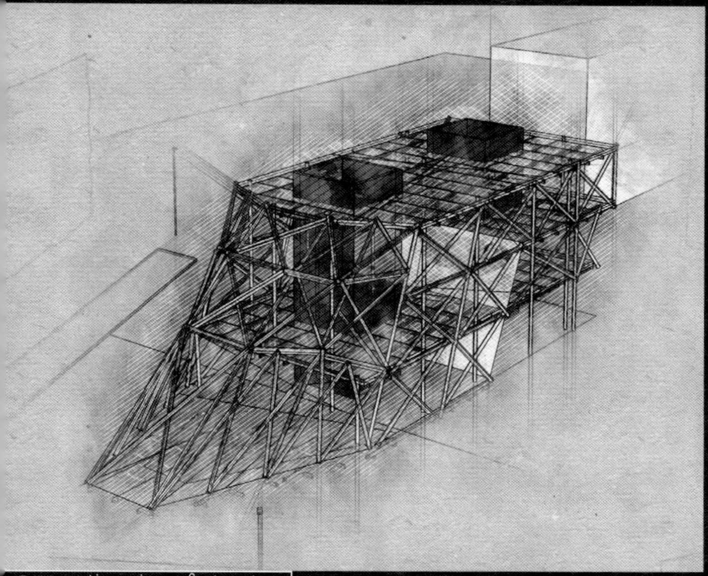

. Perspective view of structure

2. Perspective view of structure

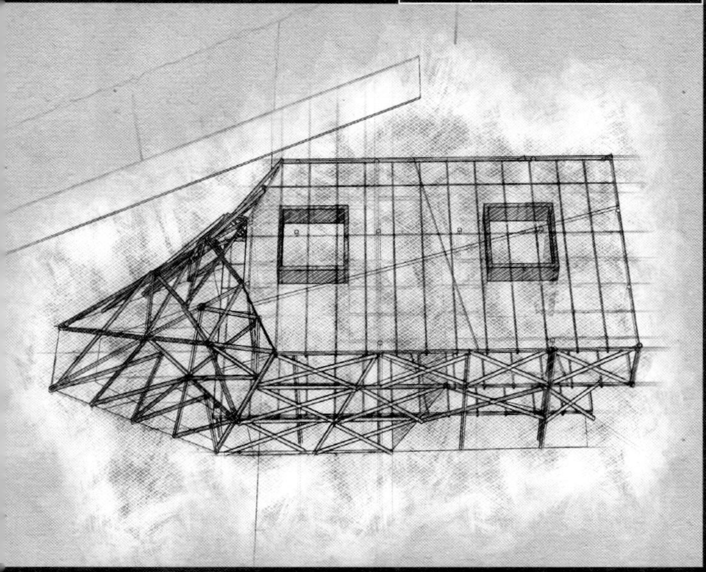

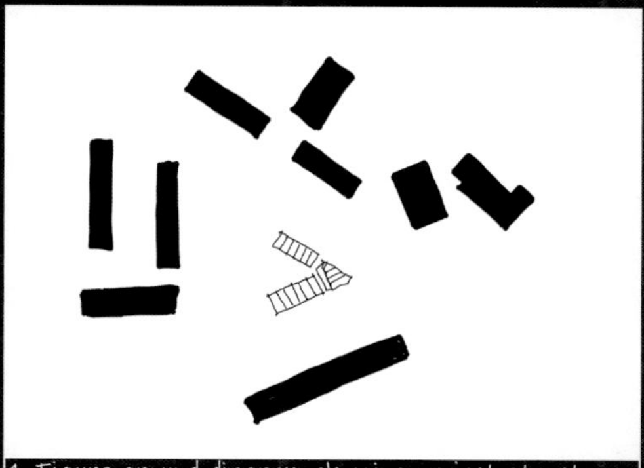

1. Figure-ground diagram showing project structure

Option 1
The structural grid visually reinforces the box-like forms, as the site context is one of distant neighbouring buildings.

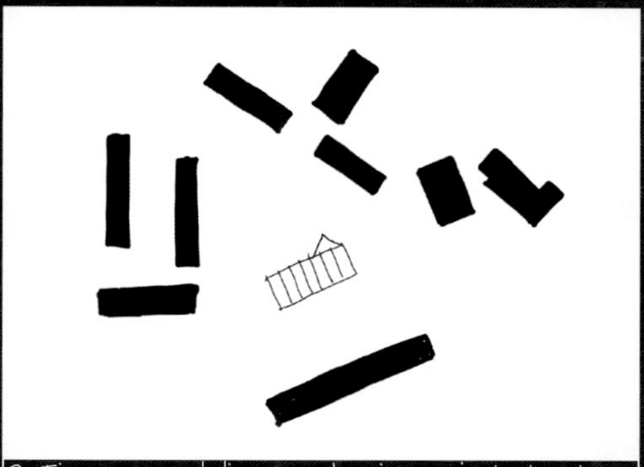

2. Figure-ground diagram showing project structure

Option 2
The structural grid visually reinforces the main rectilinear form; allowing the triangular space to be expressed visually as a separate form.

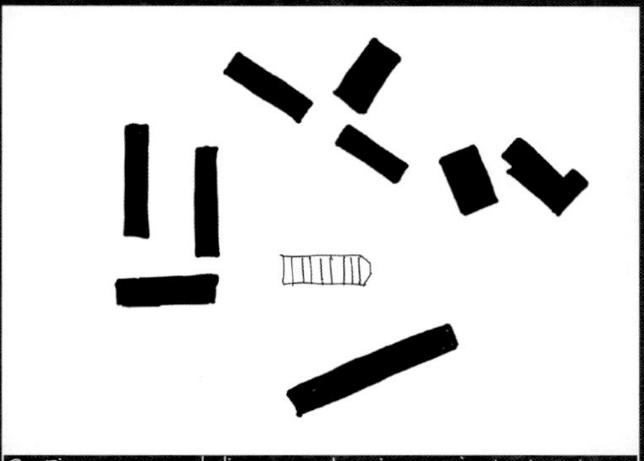

3. Figure-ground diagram showing project structure

Option 3
The structural grid visually reinforces the forms which are rectilinear in plan but are triangular-based in section.

works. Instead, a set of unexpected spaces were created which would contrast with the artworks themselves. The gallery owner did not want a set of regular rectilinear box-like rooms forming tall spaces with white interior walls and a timber-effect floor finish. The lightness of colour was less important than the range of spaces that would take the visitor into an unexpected environment, one which would allow them to see the artworks in a different way. The design is one of entering through a relatively small space at the entrance, then being lifted upward into a set of spaces at a higher level that provided views down to ground floor spaces. The visitor is drawn up into this building in order to experience unexpected views both outwards and into the building.

Although the art gallery sits within its immediate context of a road intersection, the scale of its massing is similar to that of its neighbours, both in height, depth and width. The surrounding buildings occupy three to four storeys, as does this gallery in terms of overall height. This building is a two-storey building but the floor-to-floor heights are generous as a result of the need to exhibit installations and sculptures that require such volume. Although visually bolder than its neighbours, the gallery demonstrates an intention to create dramatic space within a compact form.

In Option 1, the constituent spatial components of the art gallery are separated into an entrance space with two wings that provide the service core spaces of the two exhibition spaces of the brief/program. Option 2 aligns itself with one of the roads at the site boundary to create a 'straightforward' box-like form. This option sets out space on the other side of the building, providing both support space and office space, leaving a clear gallery volume as the service core of the building. Option 3 creates a step in the building by aligning the volume with the adjacent buildings; creating a point of inflection at the front of the building where it aligns itself with the roads at their point of intersection. This arrangement creates views, for visitors arriving, of alignment with each of the two roads that the building forms now address. At the other end of the gallery, the building accommodates spaces for sculpture, and allows for a mixture of top-lit space and side-lit internal space.

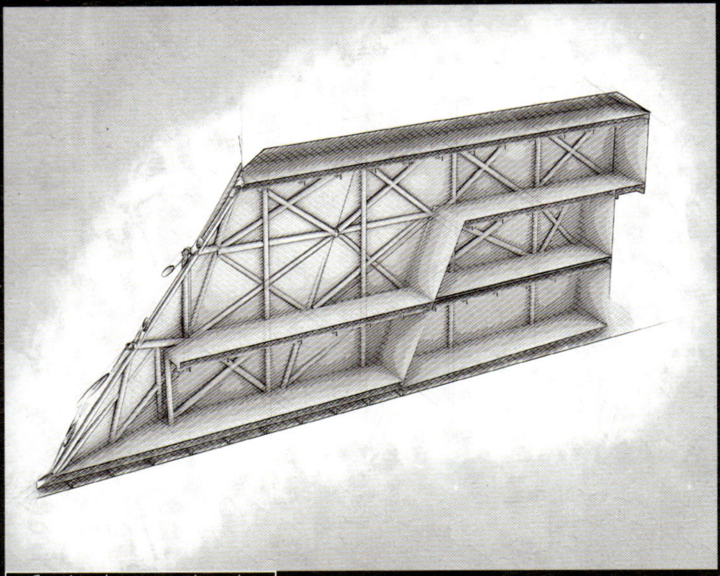

1. Sectional perspective view

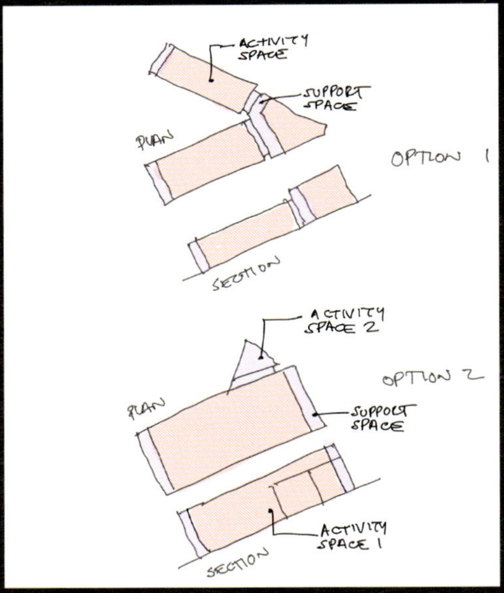

2. Sectional perspective view

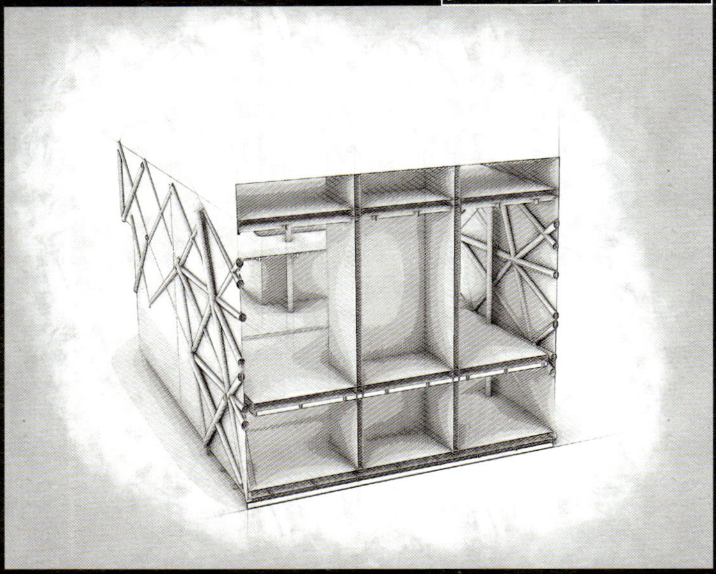

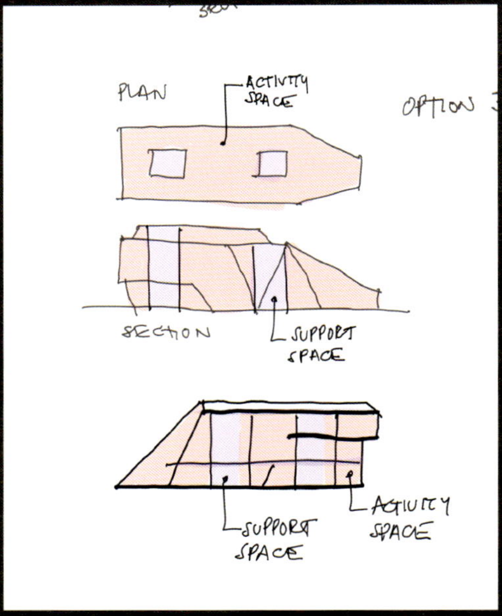

Spatial arrangement

Option 1 provides entry to the building at the corner through a temporary exhibition space.

Option 2 sets an entrance space between a small triangular-shaped gallery and one of the large galleries, on both sides of the building.

Option 3 is arranged over two floors that are staggered in order to create a raised two-storey block at the back that has a top-lit space.

Spatial arrangement

The spatial concept of this building is to introduce a sense of the unexpected when entering this relatively small building, which has an unusual external appearance suggesting that something unusual can be expected within the gallery. The spatial arrangement of Option 1 is to enter the building at the corner through a temporary exhibition space before reaching the reception area. Cloakrooms and the reception desk are positioned immediately behind that exhibition space, within a more intimate-scale space that provides formal welcoming to a visit to the gallery. This reception space provides access directly to gallery spaces set immediately behind. Option 2 sets an entrance space between a small triangular-shaped gallery and one of the large galleries, on both sides of the building. The passage through the entrance space, set on one side of the low-height triangular gallery area, makes the adjacent gallery spaces appear to be even larger upon arrival; the spatial surprise is matched by the visual effect of the internal walls, which are side lit by controlled daylight entering the spaces. Option 3 takes the visual effect generated by the entrance space, taking the contrast one step further; bringing visitors in through the 'pointed' entrance at one end of the building. This configuration draws visitors around a central core which conceals a larger space behind; steps lead up to the gallery on the level above. At the far end of this upper floor gallery, the cantilevered portion of floor provides views down into the external space immediately below. Option 3 is arranged over two floors that are staggered in order to create a raised two-storey block at the back that has a top-lit space. This is double height at the front and cantilevers or projects out of the building. The exhibition spaces have north lights at high level; forming a fully glazed box at the first-floor level. The ground level is open to the outside and has sculptures set externally.

Internal circulation

Visitors arrive through the 'nose' of the building at the small gallery; taking the stairs or a lift/elevator to the upper floor beyond the reception area for ticketing, cloakrooms and washrooms; an adjacent café is

Perspective view of interior

2. Perspective view of interior

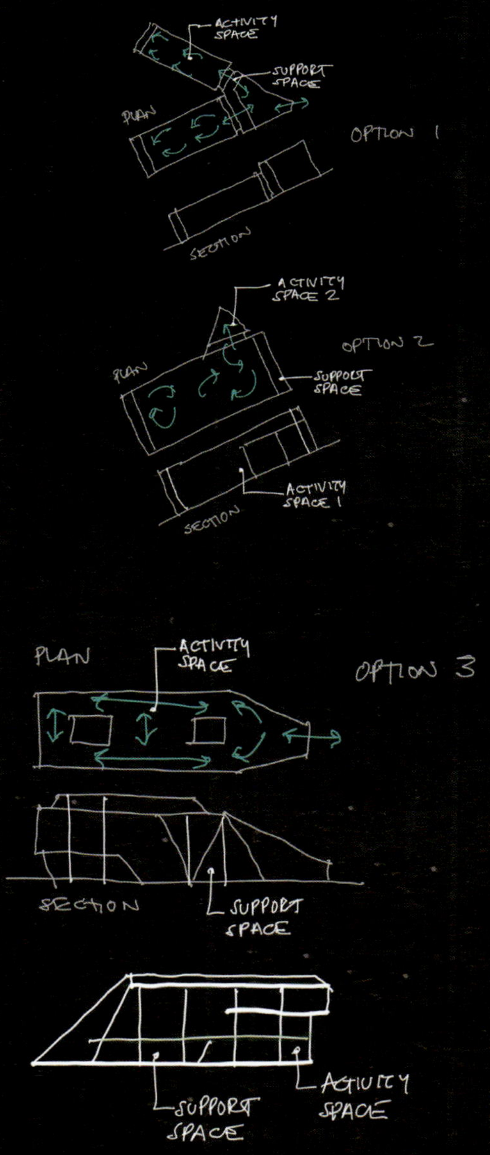

Internal circulation

Option 1 has circulation passing through the volume at the front to spaces behind, where movement becomes circular rather than following straight lines.

Option 2 creates a circuitous movement where the visitor is continually changing direction in order to explore this large gallery space.

Option 3 uses the entrance as a lobby space which opens directly into the gallery when arriving.

located on the ground floor. Visitors move up to the first floor, directly into one of two large exhibition spaces, where they can see through to other spaces ahead of them; arranged in steps, as viewed in section. Looking up, visitors can see the two floors of exhibition spaces above and in front of them as they pass around the sides of the building; two further spaces are revealed at this point. Volumes are linked with an opening in the middle, such that both spaces are revealed at the same time. These two exhibition spaces terminate with a 'drop-down' space which is hung beneath the two spaces at the other end of the building. This drop-down space has a partially glazed floor, set away from the exhibition spaces but still within the circulation route. This partially glazed floor is formed as 'slots' that provide views down to the sculpture court below. Visitors then pass through to the end of the building; descending a stair into the outside sculpture space; re-entering the building at the 'nose' end having followed the display of sculptures in the garden surrounding the gallery. The cantilevered end of the building comprises a set of smaller spaces, nested together, with small openings and angled slots in walls that reveal a set of even smaller spaces behind. The internal circulation for Option 1 is one of passing through the exhibition volume at the front to spaces behind, where the movement becomes circular rather than following straight lines. This strategy is suggested by the internal arrangements of furniture and hanging space; a series of curved screens creates movement around the two larger galleries at the back, set beyond the entrance gallery. Option 2 uses a similar approach of a triangular entrance gallery, after which visitors enter the gallery spaces; screens are arranged in curved forms to create a circuitous movement where the visitor is continually changing direction in order to explore new works in this large gallery space. Option 3 uses the entrance as a lobby space which opens directly into the gallery when arriving. There is no barrier between the visitor and the far end wall of the galleries. The view through the gallery is created by cores which are set in the centre; providing both structural support and creating screens beyond which the full extent of the gallery spaces is revealed.

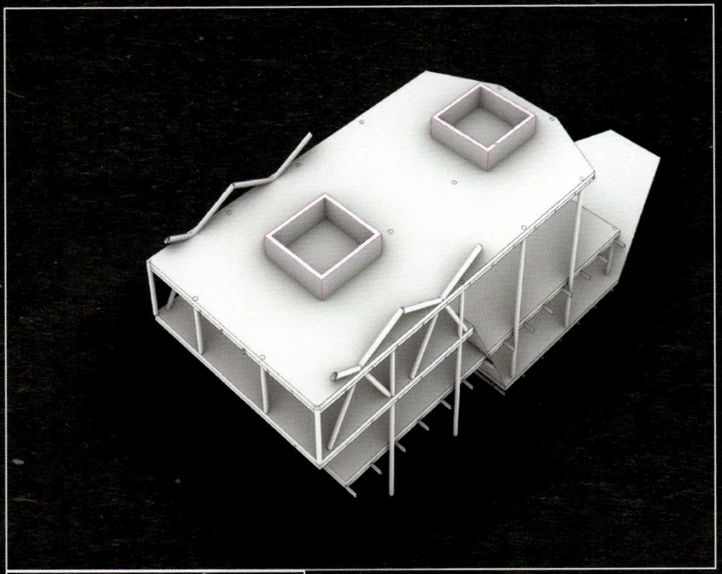

1A. Isometric view of structure

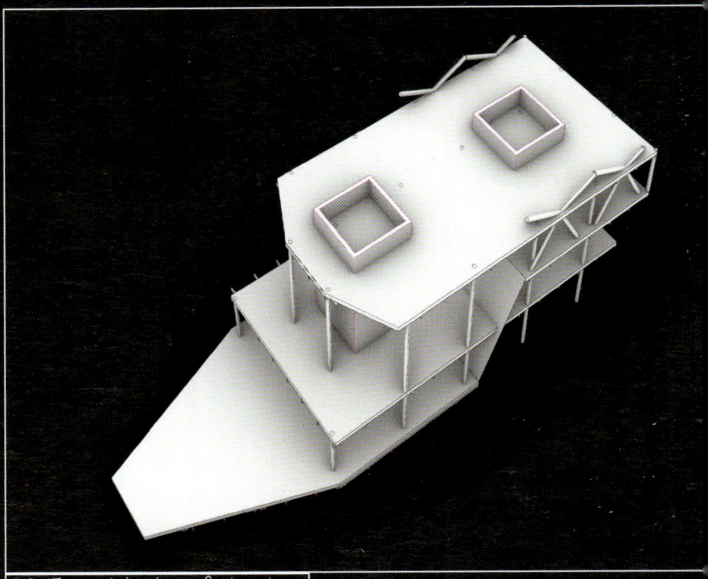

2A. Isometric view of structure

1B. Isometric view of structure

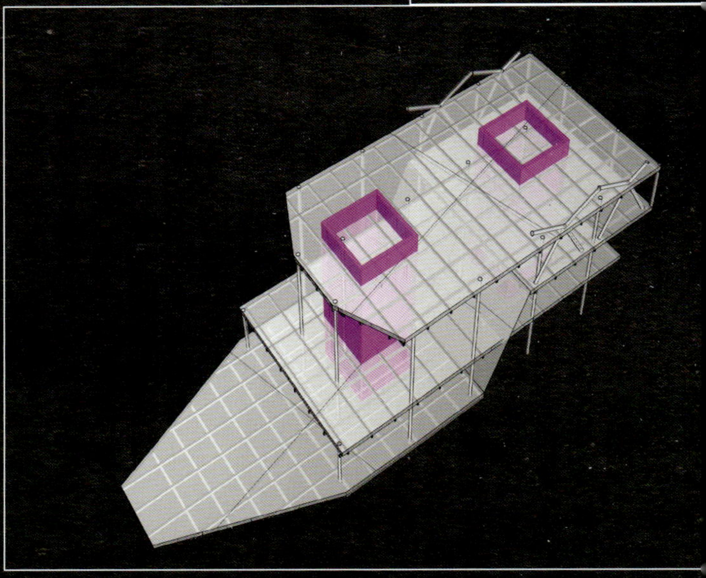

2B. Isometric view of structur

Structural strategy

The general forms of the building create a set of dynamic spaces connected by a route up through the building. The structural design explores the possibilities of creating an envelope that expresses this range of spaces. The solution of a rectilinear box-like structural form, with portions cut out for the external sculpture space, are used as a starting point. The limitations of this approach are that the experience of circulating around such a building, as well as viewing it from the outside, provide no visual expression of the spaces within. The introduction of inclined end walls, the tapering inner wall of the upper gallery and the conical form at the entrance gallery would not have been expressed in the form of the building; a set of interiors would have created a very different spatial form within that would result in a reduced volume of space inside the building had that approach been taken. To resolve this a second interior structure would have been needed to create the forms of the interior spaces, in addition to a box-like form that would have enclosed the outer walls and roof. Instead, the design creates a visual idea of what might be expected within the building upon entering the lobby; the spatial experience being intriguing but inviting. Visitors are encouraged to immerse themselves in the artworks displayed; disengaging themselves from the immediate context of the building's location and context. The resulting structural design is one of using a steel frame which supports floor slabs onto which installations and sculptures form temporary exhibitions. The high loads imposed by some of the artworks that are expected to be exhibited led to a reinforced concrete floor slab design with beams and columns, such that the interior structure is made from reinforced concrete while the outer shell of the building is constructed as a series of steel frames which are clad in opaque and glazed panels, translucent glazed panels and GFRC panels. This approach brought a sense of concrete forms within the building as well being experienced externally. The roof contains most of the environmental controls both for ventilation and light and solar control. The rooftop arrangement of louvres for solar protection arranged in rows is the most obvious signal that this is an art gallery. The control of top

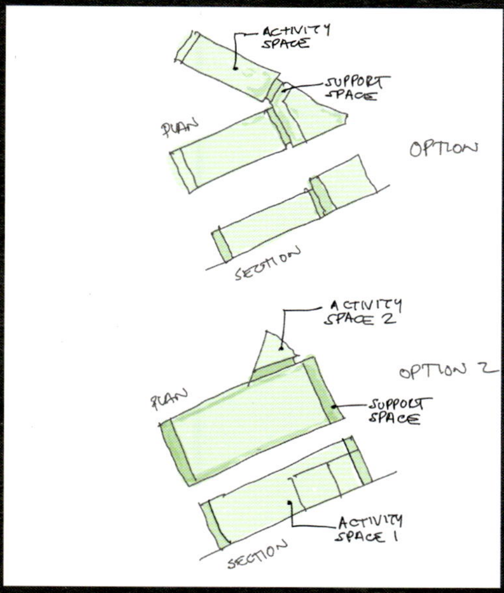

Structural strategy: Zones

Option 1 has three separate structures, which are linked with a set of smaller structures.

Option 2 uses curved members to create the form of the inclined roof of the triangular entrance space.

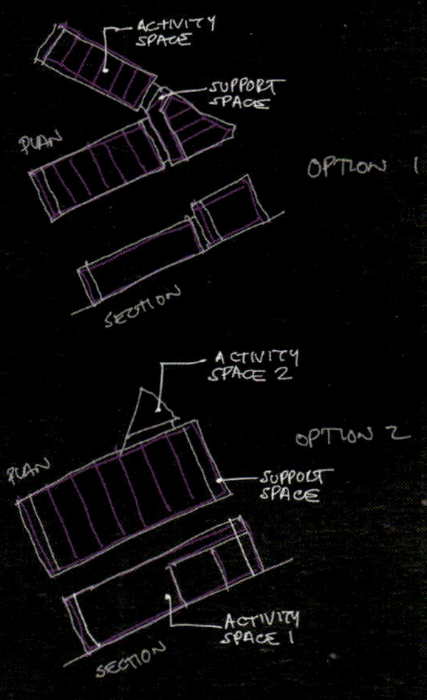

Structural strategy: Framing

Option 1 uses no secondary steelwork.

Option 2 integrates the structure with the geometry of the primary steel members and concrete structure at ground level.

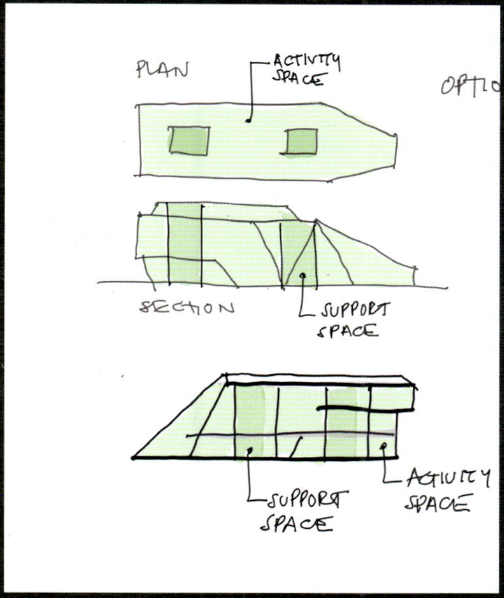

Structural strategy: Zones Option 3 integrates primary and secondary structures to create the complex form.

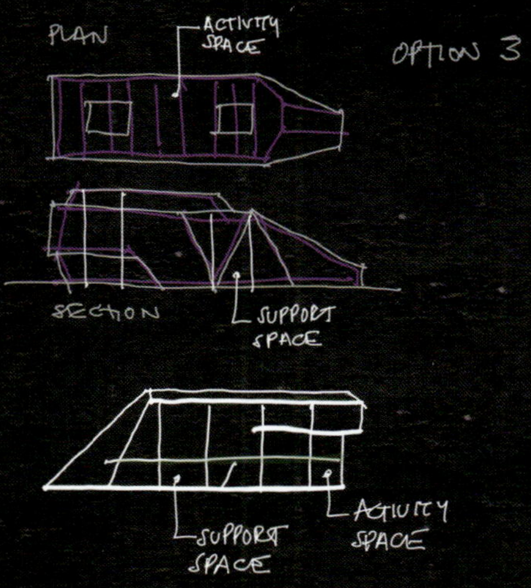

Structural strategy: Zones Option 3 forms curves in the structure at the front of the building without the need for curved structural members.

light is a key part of this art gallery. The translucent and opaque glass walls are less characteristic of this building type, but wall spaces are provided for the exhibition of paintings.

Installations and sculptures often require a mixture of natural light and electrical light. Screens are installed behind the large-scale opaque glass areas that are fixed back to the primary structure; the screens are fixed back to back to the lines of primary structure. As a result, the framing supporting the panels has a role of supporting tall internal screens which can be removed to suit the exhibition. As a result, the complete form of the building is determined by its primary structure; only the secondary structure supports the rooftop solar shading devices.

The structural concept is one of a lattice frame in steel; a reinforced concrete structure is used to form a concrete shaft that props the cantilevered structure at one end of the building, and is used to form a set of consoles at the other end of the building. The steel lattice frame spans between the two sets of concrete consoles; one set for the cantilever at one end and one for the side walls at the front where the concrete is revealed on the outside of the building. The steel frame is formed from straight members only; no curving is required to create the characteristic forms.

The structural strategy for Option 1 is to have three separate structures, which are linked with a set of smaller structures that form links between the three external forms of the gallery. Each external form is a steel frame which varies in geometry to create the form of the building. Option 2 uses curved members to create the form of the inclined roof of the triangular entrance space. In Options 1 and 2 there is a mixture of large-scale steel framing and small-scale framing combined into a single structure to create the form; no secondary steelwork is used, as the structure is integrated with the geometry of the primary steel members and concrete structure at ground level. Option 3 takes a similar approach of integrating primary and secondary structures to create the complex form; curves in structure at the front of the building are created without the need for curved structural members.

1A. Isometric view of environmental volumes

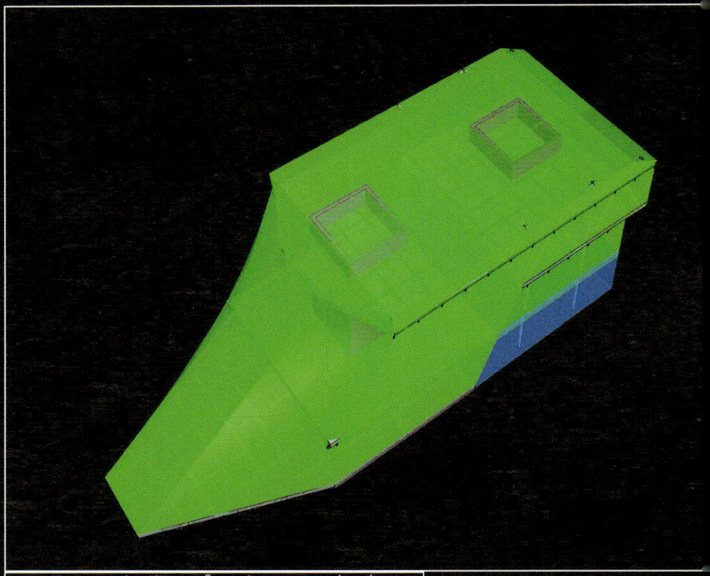

2A. Isometric view of environmental volumes

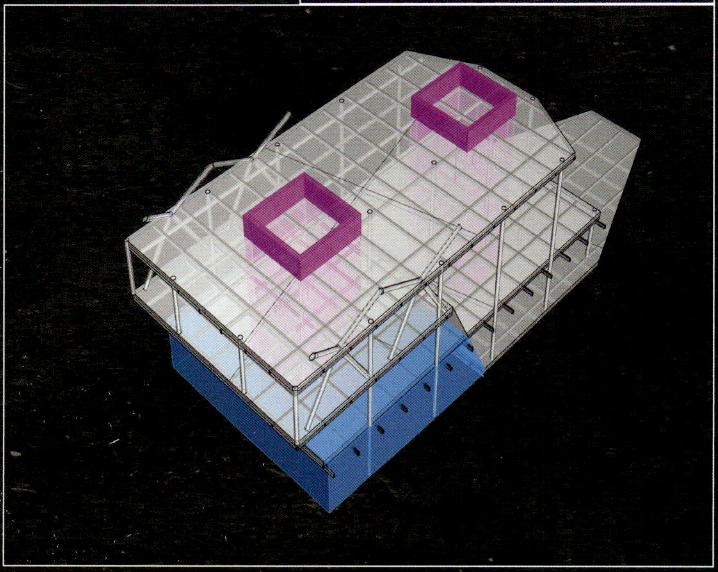

1B. Isometric view of environmental volumes

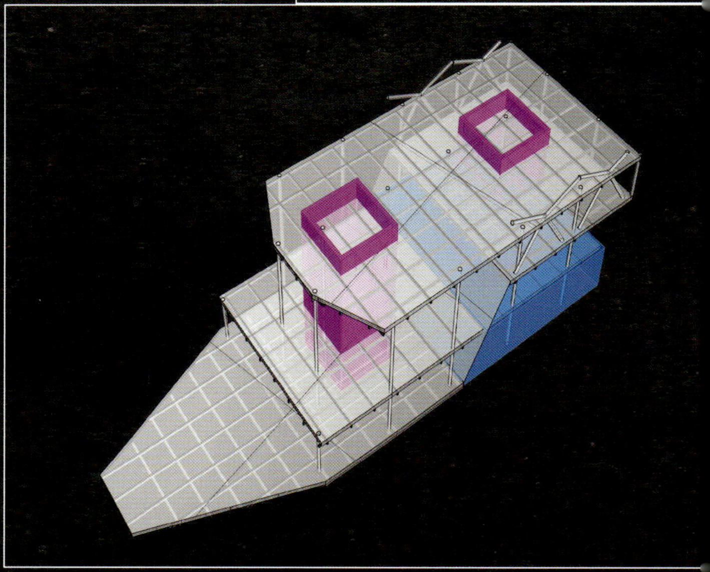

2B. Isometric view of environmental volume

Environmental strategy

The environmental strategy sets all mechanical and electrical equipment and services installations on the roof. At roof level, two wing-shaped ducts distribute air, with air-handling units and chillers set between them. This is a compact arrangement, like the approach of a 'backpack' module, but it is mounted at a higher level. There is a wedge-shaped external form and internal space at the cantilevered end of the building. The upper floor is where the exhibition starts; most of the exhibitions require darkened space, or 'black boxes', which are top lit through the roof. The top-light is expressed by joint lines in panels which provide light that washes down the walls set between spaces; taller spaces are set over two additional floors towards the rear of the building.

The environmental strategy for Option 1 is closely linked to the spatial arrangement in that the reception area, with its administration and café, forms a secondary environmental zone that creates a buffer zone between each of the three exhibition spaces. The exhibition space at the entrance is exposed to solar gain; the space is protected accordingly, with overhanging eaves and solar-protective glass, avoiding the need for external shading that would detract from the spatial experience of viewing the artworks. Option 2 has an environmental zone for the triangular space of the entrance exhibition area, with three adjacent environmental buffer spaces. The top of the glazing provides an entry point for solar gain, but that area is protected by the overhanging eaves. Option 3 uses the external space which is protected from the effects of the sun for most of the day; this treatment with darker glass allows the space inside to become a single environmental zone, with the service cores that provide vertical circulation being enclosed in the secondary environmental zone. No environmental buffers are needed in this solution.

The environmental strategy is based on the control of daylight into the building and the efficient use of mechanical ventilation. Natural daylight is admitted at the top of spaces at a high level, and is controlled with baffles and louvres that protect the artwork within from direct daylight.

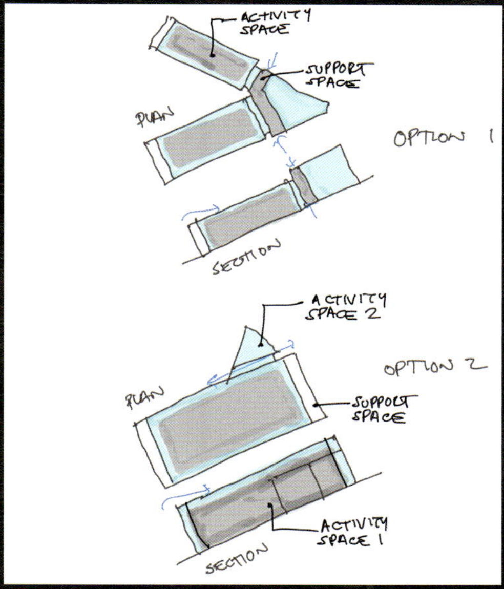

Daylight strategy for direct sunlight and cross ventilation.
Option 1 admits daylight and air through the entrance gallery and at the perimeter of internal spaces.
Option 2 admits daylight and air through the entrance gallery and at the perimeter of the main gallery spaces.

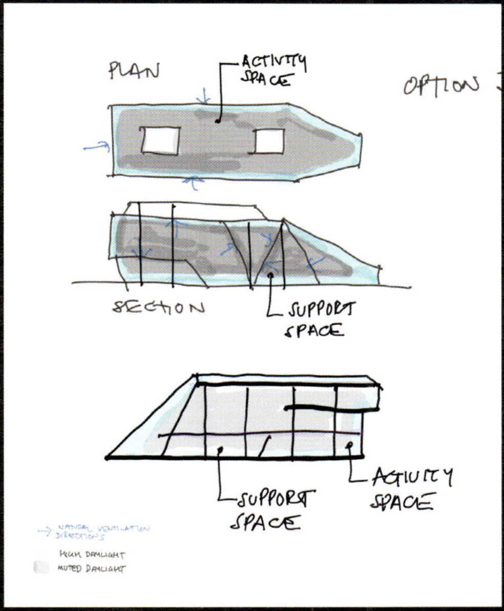

Daylight strategy for direct sunlight and cross ventilation.
Option 3 admits daylight and air at the perimeter of the gallery spaces.

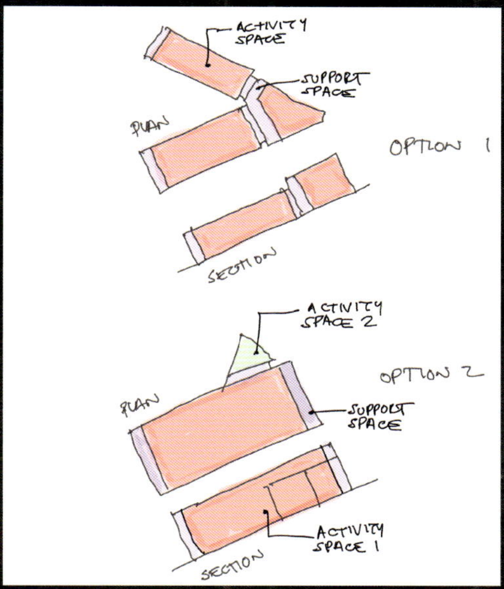

Environmental zones
Option 1 creates a buffer zone between each of the three exhibition spaces.
Option 2 has an environmental zone for the triangular space of the entrance exhibition area, with three adjacent environmental buffer spaces.

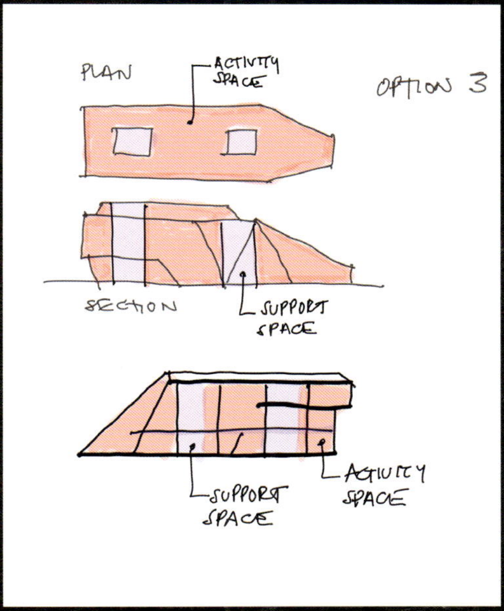

Environmental zones
Option 3 allows the space inside to become a single environmental zone, with the service cores that provide vertical circulation being enclosed in the secondary environmental zone.

The electrical lighting installation in the galleries imitates this source of light by providing light fittings that follow the direction of daylight. The environmental controls for mechanical ventilation are assisted by having a high thermal mass in the construction, which is provided by the inner reinforced concrete frame and GFRC (glass fibre reinforced concrete) external wall finishes that reduce the effects of the external air temperature, which can vary considerably throughout the seasons. This thermal mass is assisted by night-time cooling in the summer months and by thermal storage during colder months. The building has no natural ventilation requirements for exhibiting the artworks, which require close control of both temperature and humidity in most cases. The rooftop zone introduces both daylight and electrical light to interior spaces. The roof area also admits direct mechanical ventilation provided by ducted air, to ensure that the environmental controls are set in one location rather than being dispersed through the building. The environmental design makes use of the rooftop zone in order not to distract visually from the need to create flexible spaces in the volumes below that could be changed or modified in order to suit the specific requirements of a new exhibition. This is achieved without the gallery spaces reverting to rectilinear box-like interior spaces.

The high degree of modelling of the facades allows light from outside to enter from below interior spaces as well as from the rooflights above. Daylight and electrical light enter the building through slots set at both floor level and at ceiling level. These slots can be closed off to suit exhibition requirements. Baffles are deployed below the glazed floor, outside the building, in order to prevent daylight from entering exhibition spaces if required. The triangulated forms of the facade panels assist by allowing narrow bands of daylight, or electrical light, to enter the building. These slots can be closed off inside the building; the glazed slots at floor level are flush with the walking surface; they can be closed only from the outside. The electrical lighting is installed in the areas of the glass slots, such that during hours of darkness the effect is imitated with electrical lighting; the installation comprises sets of point sources of light as well as continuous bands of light to suit exhibition requirements.

Typical bay adjacent to facade

Slots in the facades admit both natural light and electrical light during daytime, and added electrical lighting during hours of darkness. Both forms of light enter internal spaces through glazed slots set in floors and inclined joints in the external wall which are recessed from the outer surface of the facades. The external envelope accommodates a set of slots and grooves that admit light by creating lines of illumination internally; washing internal surfaces with natural light, following the principle of slots formed in the rooflights above that provide top-light into gallery spaces. All panels have horizontal flaps which can be opened outwards to admit light at the mid-height of spaces as narrow bands of illumination. When all external flaps, which are operated manually, are opened, the complete structural frame of the external walls provides the sense of a structure revealed as lines of light. The glazed slots follow lines of structure in both inclined and horizontal directions that pick out the structure with strips of natural light. Electrical light sources imitate the effect of daylight for use during hours of darkness, or alternatively when required during daytime for a specific application. In this situation the outer flaps are closed; the electrical lighting providing the primary source of illumination within the gallery space. These inclined and horizontal bands of light can be used either with, or independently of, the rooftop installation of light baffles. At roof level, the envelope admits light into the interior spaces, picked out as diagonal lines across the roof that follow lines of structure, creating lighting effects which follow the arrangement of the preferred use of space inside the galleries at the upper level. Ground floor spaces have borrowed light admitted through strips set at the edges of floor plates. Flaps can be opened or closed to allow light across more than one floor. Hinged facade panels can be deployed outwards to admit greater amounts of daylight and can be deployed at a 10-degree angle to create more dramatic lighting effects inside the building. The design is very much a deployable 'black box' that admits both natural light and artificial light. The building is not a 'light box' that emits light or a box that has homogenous light entering from above; it is a black box where flaps in the walls and on

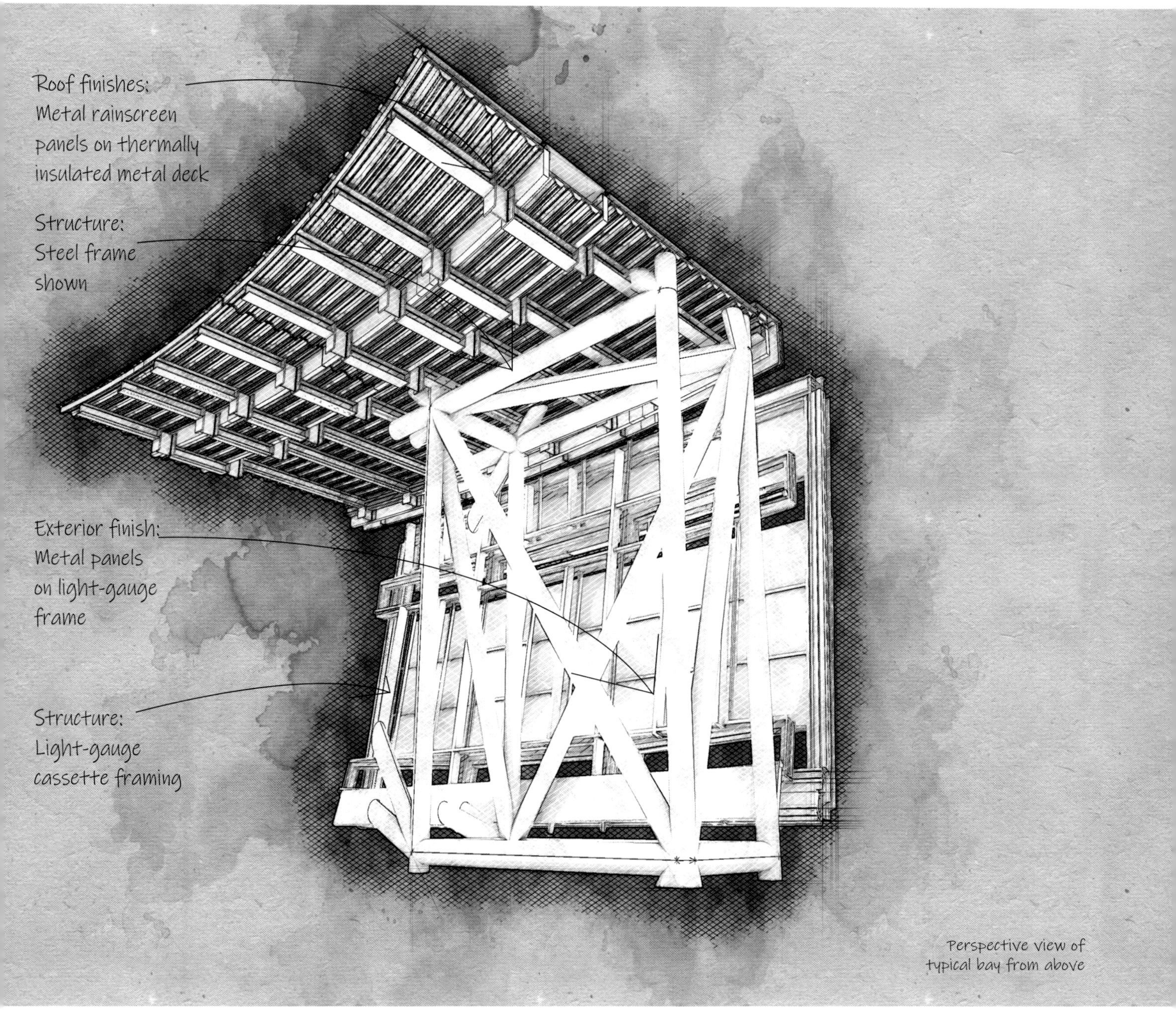

Perspective view of typical bay from above

the roof can control daylight as well as exclude the effects of the sun. The glazed slots and external covers are independent of the ventilation system, as artworks are required to be exhibited in a controlled environment; both from the control of light and a control of temperature and humidity; all three requirements will vary between exhibitions. Air for ventilation passes through the inner part of the external wall in ducts that are directed to diffusers at floor level; the air is extracted at ceiling level. Air at floor level is supplied just below the horizontal bands of glazing that introduce daylight when the external flaps are raised. This installation of ducted air gives rise to the characteristic curved small window-like shapes set at floor level and at ceiling level within the building. When all panels are deployed, the external envelope takes on a different appearance; one of a set of planes that appear to project the building outwards, suggesting that the built form of the art gallery has no simple definition of facades and roof, but is instead a series of planes directing light to the artworks within. The facades reflect the structural design; GFRC (glass-fibre reinforced concrete) panels are used to enclose elements of concrete structure; metal panels are used to clad the steel frame, forming the characteristic triangular shapes of the facades. Roof panels are also enclosed in metal; providing weathertight enclosures for the mechanical and electrical equipment and associated air ducts. No air-handling units or chillers are visible from outside the building; they are concealed beneath the metal enclosures. As a result, no additional secondary structure is required to support facade panels. Glazed openings are enclosed with full-height frameless glazing to provide a continuity of glass, making the elevations as flat in appearance as possible in order to emphasise the external forms of the building. The service zone is accommodated within the 1.0 metre thickness of the facade zone, providing deep reveals in both the walls and the roof. This language of 'depth' of facade allows light slots to be integrated into the depth of the GFRC facades; thin slots and grooves are created as a result of the possibility of modelling the depth of the facades, visually tracking the movement of light into the building and of air for ventilation around the building.

Newtecnic have produced a number of volumes on contemporary building technology which provide reference material for engineering and architecture students and professionals around the world. The Modern Construction Series is published by Birkhäuser Verlag GmbH. In addition, the Facades Technical Review, from RIBA Publications was published in spring 2007.

The text and drawings are by Andrew Watts. The designs and renders are by Andrew Watts and Howard Tee. The book is designed by Yasmin Watts.

David Marold is Acquisitions Editor at Birkhäuser Verlag GmbH in Vienna. He has driven this book from a set of basic layouts to a completed book. He has a passion for books and their design, ranging from their wider content to the quality of print paper, to the overall reader experience.

Andrew Watts an architect and engineer and the author of the Modern Construction series of textbooks. He specialises in the architectural and engineering design of large-scale buildings of complex form. Andrew is a founding director of Newtecnic. He is a Fellow of three UK engineering institutions in recognition of his engineering-based designs; the Institution of Civil Engineers; the Institution of Engineering Designers; the Institution of Engineering and Technology. Andrew holds charterships in the UK for architecture and engineering from the Royal Institute of British Architects and the Institution of Engineering Designers. In the US, he holds an ASCE membership.

Yasmin Watts is a sculptor, architect and creative director of Newtecnic, leading the exploration of links between sculpture, architecture and public art. She is a founding director of the company. Yasmin is focused on creating sculptures in public spaces in the built environment; the fusion of architecture and sculpture has been a key driver in the work behind all the projects shown in this book. She is a registered architect in the UK and is a Fellow of the Royal Society of Arts. Yasmin is continuing her research with the Royal College of Art.

The author would like thank Bettina Algieri for the content and production editing of this book. The author would also like to thank Alun Brown for the task of proofreading the text.

Newtecnic is a world-leading firm of engineering designers and architects led by Andrew Watts, Howard Tee and Yasmin Watts. The Modern Construction series forms an essential part of their research work which informs the design of projects. The Newtecnic approach, based on research from first principles, allows the company to achieve architectural and engineering design solutions for the complete building, from concept to production information, including engineering and architectural design for structure, facades and the internal environment. With 3D BIM, Newtecnic designs models that become highly evolved during the design process. Components of these BIM models, with accompanying text and images to illustrate design processes, form the basis of the Modern Construction Series. The engineering and architectural design skills of Newtecnic allow the firm to communicate, through the Modern Construction series, the key issues of 21st century building engineering and design. Newtecnic are holders of *The Queen's Award for Enterprise* and is ranked in the top three of *New Civil Engineer 100 Companies of the Year 2019*.

Glossary of terms devised for this book

The following expressions used in this book are defined below as working definitions only, and are defined for the purposes of this book only. These terms have been used by the author on projects shown in this book at the early stages of projects among architectural and engineering team designed at Newtecnic. They are not industry-standard terms.

'Activity' space and 'support' space

A key aim of this project is to use the separation between 'activity space', which comprises the main spaces within the building, and 'support' space, which accommodates the technical and vertical circulation spaces. This approach creates zones within a building; typically for 'activity' spaces, or useable spaces, and spaces for the movement of people around a floor plate. 'Support' spaces, or service spaces, are considered here to accommodate stairs, lifts/elevators, washrooms, mechanical and electrical equipment rooms and associated storage space. The spatial arrangements of activity space, support space or circulation space should not reduce their mutual effectiveness.

A **'backpack'** is a set of environmental control equipment, mechanical and electrical installations and associated service equipment which is fixed to the sides of the building, supplied from the roof level. This arrangement is used in preference to locating these installations within dedicated floors of a building or on its roof.

A **'building within a building'** is an architectural concept of having two layers of enclosure which provide a deep facade zone. This zone can be used to accommodate service equipment or internal circulation for example, and provides the opportunity to free internal space from these support functions while also providing opportunities for natural ventilation, controlled daylight and solar shading.

A **'differentiated form'** is one where there is a close interaction between the geometry of the internal space and the external facade or envelope which encloses it. Differentiated forms are those which are differentiated visually on the outside as a result of each volume whose geometry follows closely that of the internal space. An **'integrated form'** is one where interior spaces are independent of the external envelope, such that internal spaces are arranged within a single external envelope. An integrated form combines internal spaces in a way which is not reflected in the visual perception of the volumes from outside the building.

An **'environmental zone'** comprises a space or a group of spaces where a single set of environmental controls are provided. The facade zone, which is considered to be part of the external envelope, can also be considered as an environmental zone or a buffer zone.

An 'environmental buffer zone' is an area or volume within a building which is positioned between the inner spaces and the outside of the building, or alternatively between environmental zones, in order to create an area which is semi-external or semi-internal and may be partially ventilated and partially tempered. A buffer zone reduces the need to provide full air conditioning for example, or full mechanical ventilation or full heating for cooling in an area adjacent to a facade This approach can support the idea of a 'building within a building' from the environmental design point of view.

A **'floor plate'** is the usable area of the floor slabs or floor construction, which excludes the openings created by staircases and services and typically constitutes an 'activity zone'.

A **'kit of parts'** is a set of possibly prefabricated components that may be individual parts of the building as assemblies or as individual components. In the projects shown in this book, building assemblies across structure, envelope and internal environment are coordinated as part of an interlinked set of assemblies.

An **'outdoor room'** is an external space which is defined by the floor and the walls surrounding it. The walls are usually external walls rather than internal walls to form something resembling a courtyard, but which may not be recognised as a specific space other than to provide a continuity from internal spaces to external space; but without a ceiling or roof.

A **'shrink wrap'** describes the tendency for the external volumes of a building to follow closely the internal forms and shapes of the internal spaces, such that they appear to visually shrink-wrap the volumes inside the building, instead of adopting a rectilinear form that is independent of those spaces.

Solar gain is defined here as being the heat energy absorbed, transferred or transmitted into the building. The projects, as described here, control this effect through the use of external shading devices combined with a specific glass type. The types of glass used may be different in each of the individual forms that comprise a single project.

Thermal comfort is defined here as being a range of temperatures and associated relative humidity which is comfortable, as experienced by occupants in going about their normal tasks or routines in the specific internal environment for which thermal comfort is provided. Thermal comfort will be considered to be different in conditions in a gym from those of a desk-based office space, for example.

A **'visual map of spatial functions'** is a visualisation of the spatial arrangement within a building, as experienced visually either from outside the building or within the interior spaces. A set of facades, or alternatively views across open spaces within a building, may provide a visual diagram of where the 'main' spaces are in a building and where the circulation spaces are positioned, such that the building can be understood intuitively by visitors arriving at the building without the need for specific guidance, allowing them to visually take in the overall functions undertaken within the building.

The following bibliography is suggested for further reading, based on the design methods used for the projects in this book.

Construction technologies in architecture
Modern Construction Handbook, 5th edition
Birkhäuser, 2019

Modern Construction Envelopes, 3rd edition
Birkhäuser, 2019

Modern Construction Case Studies, 2nd edition
Birkhäuser, 2019

Structural design in architecture
Structural Engineering for Architects: A Handbook
Silver, P., McLean, W., Evans, P.
Laurence King, 2014

Structural Design for Architecture
Macdonald, A.
RIBA Publishing, 2021

Salvadori's Structure in Architecture: The Building of Buildings
Heller, R., Oakley, D.
Pearson, 2016

Structure as Architecture
Charleson, A.
Routledge, 2015

The Structural Basis of Architecture
Sandaker, B., Eggen, A., Cruvellier, M.
Routledge, 2019

Structural Design for Architects
Nash, A.
Routledge, 2017

Author
Andrew Watts
London, England

Acquisitions Editor: David Marold, Birkhäuser Verlag, Vienna, Austria
Content & Production Editor: Bettina R. Algieri, Birkhäuser Verlag, Vienna, Austria

Layout: Yasmin Watts, London, England
Cover design: Floyd Schulze
Proofreading: Alun Brown, Vienna, Austria
Printing and binding: Holzhausen, die Buchmarke der Gerin Druck GmbH, Wolkersdorf, Austria

Library of Congress Control Number: 2021936963

Bibliographic information published by the German National Library
The German National Library lists this publication in the Deutsche Nationalbibliografie; detailed bibliographic data are available on the Internet at http://dnb.dnb.de.

ISBN 978-3-0356-2141-9 Hardcover edition
ISBN 978-3-0356-2142-6 Softcover edition
e-ISBN (PDF) 978-3-0356-2146-4

P.O. Box 44, 4009 Basel, Switzerland
Part of Walter de Gruyter GmbH, Berlin/Boston

987654321 www.birkhauser.com